Resonance Energy Transfer

Resonance Energy Transfer

David L. Andrews

University of East Anglia, UK

and

Andrey A. Demidov

Northeastern University, Boston, USA

JOHN WILEY & SONS

Chichester · New York · Weinhein · Brisbane · Singapore · Toronto

Other Wiley Editorial Offices

John Wiley & Sons, Inc., 605 Third Avenue,
New York, NY 10158-0012, USA

WILEY-VCH Verlag GmbH, Pappelallee 3,
D-69469 Weinheim, Germany

Jacaranda Wiley Ltd, 33 Park Road, Milton,
Queensland 4064, Australia

John Wiley & Sons (Asia) Pte Ltd, Clementi Loop #02-01,
Jin Xing Distripark, Singapore 129809

John Wiley & Sons (Canada) Ltd, 22 Worcester Road,
Rexdale, Ontario M9W 1L1, Canada

Library of Congress Cataloging-in-Publication Data

Andrews, David L.
    Resonance energy transfer / David L. Andrews and Andrey A. Demidov.
        p.   cm.
    Includes bibliographical references and index.
    ISBN 0-471-98732-8 (alk. paper)
    1. Molecular dynamics. 2. Energy transfer. I. Demidov, Andrey A. II. Title.
    QD461.A54 1999
    539'.6–dc21   98-38684
                                                             CIP

British Library Cataloguing in Publication Data

A catalogue record for this book is available from the British Library

ISBN 0 471 98732 8

Typeset in 10/12pt Times by Pure Tech India Ltd, Pondicherry
http://www.puretech.com
Printed and bound in Great Britain by Biddles Limited, Guildford and King's lynn
This book is printed on acid-free paper responsibly manufactured from sustainable forestry,
in which at least two trees are planted for each one used for paper production

Contents

2   Unified theory of radiative and radiationless energy transfer

3   Dynamics of radiative transport

List of Contributors

David L. Andrews
School of Chemical Sciences
University of East Anglia
Norwich NR4 7TJ
UK
d.l.andrews@uea.ac.uk

Mário N. Berberan-Santos
Centro de Química-Física Molecular
Instituto Superior Técnico
1049-001 Lisboa
Portugal
berberan@ist.utl.pt

Daniel R. Buck
Ames Laboratory – USDOE and Department of Chemistry
Iowa State University
Ames
IA 50011
USA
buckd@ameslab.gov

Andrey A. Demidov
Physics Department
Northeastern University
110 Forsyth, DANA 110
Boston
MA 02115
USA
ademidov@lynx.dac.neu.edu

Sandrasegaram Gnanakaran
Department of Chemistry
University of Pennsylvania
Philadelphia
PA 19104-6323
USA
gnana@sas.upenn.edu

Gilad Haran
Department of Chemistry
University of Pennsylvania
Philadelphia
PA 19104-6323
USA
haran@sas.upenn.edu

Robin M. Hochstrasser
Department of Chemistry
University of Pennsylvania
Philadelphia
PA 19104-6323
USA
hochstra@sas.upenn.edu

Gediminas Juzeliūnas
Institute of Theoretical Physics and Astronomy
A. Goštauto 12
Vilnius 2600
Lithuania
juz@physik.uni-ulm.de

Ranjit Kumble
Department of Chemistry
University of Pennsylvania
Philadelphia
PA 19104-6323
USA
kumble@sas.upenn.edn

Vladas Liuolia
Institute of Physics
A. Goštauto 12
Vilnius 2600

Lithuania
vladasl@ktl.mii.lt

José M. G. Martinho
Centro de Química-Física Molecular
Instituto Superior Técnico
1049-001 Lisboa
Portugal
jgmartinho@ist.utl.pt

Pierre D. J. Moens
Formerly Institute for Biomedical Research and Muscle Research Unit
Department of Anatomy & Histology
The University of Sydney
Sydney 2006
Australia

Eduardo J. Nunes Pereira
Centro de Química-Física Molecular
Instituto Superior Técnico
1049-001 Lisboa
Portugal
epereira@fisica.uminho.pt

Cristobal G. dos Remedios
Institute for Biomedical Research and Muscle Research Unit
Department of Anatomy & Histology
The University of Sydney
Sydney 2006
Australia
crisdos@anatomy.usyd.edu.au

Sergei Savikhin
Ames Laboratory – USDOE and Department of Chemistry
Iowa State University
Ames
IA 50011
USA
sergei@ameslab.gov

Gregory D. Scholes
Department of Chemistry

University of California
Berkeley
CA 94720-1460
USA
gscholes@atto.cchem.berkeley.edu

Oscar J. G. Somsen
Department of Physics and Astronomy and Department of Biology
Vrije Universiteit
De Boelelaan 1087
NL – 1081 HV, Amsterdam
The Netherlands
oscar@bio.vu.nl

Walter S. Struve
Ames Laboratory – USDOE and Department of Chemistry
Iowa State University
Ames
IA 50011
USA
wstruve@ameslab.gov

Gedimines Trinkunas
Institute of Physics
A. Goštauto 12
Vilnius 2600
Lithuania
trinkun@ktl.mii.lt

Leonas Valkunas
Institute of Physics
A. Goštauto 12
Vilnius 2600
Lithuania
valkunas@ktl.mii.lt

B. Wieb van der Meer
Western Kentucky University
Bowling Green
KY 42101
USA
Wieb.Vandermeer@wku.edu

and

Vanderbilt University
Nashville
TN 37072
USA

Rienk van Grondelle
Department of Physics and Astronomy
Vrije Universiteit
De Boelelaan 1081
NL 1081–HV, Amsterdam
The Netherlands
rienk@nat.vu.nl

Preface

It has been our pleasure to work on this book, whose compilation has occupied us over the last few years. One of the driving ideas behind its inception was recognising the need for an amenable modern text offering comprehensive coverage of the subject; a task which would clearly require input from experts covering a wide range of disciplines. It was our delight to find one after another of the key figures in this field willing to contribute to this project. Without their keen involvement, this project would not have got off the ground.

Having tried to properly balance the detailed account of theory with experimental results; it is out hope that the book will be a useful source for all those wishing to become familiar with the subject of resonance energy transfer, particularly in biological systems, both to beginners but also to the specialist seeking a wider perspective. Readers can benefit from the overview given by Graham Fleming in his Foreword, which ably identifies the thread running from start to finish throught all chapters dealing with the various aspects of the subject matter.

In conclusion we would especially like to thank Wiley staff for their support and attention to detail at every stage.

David L. Andrews
Norwich, UK
Andrey A. Demidov
Boston, USA
November 1998

Foreword

Natural photosynthesis depends on the efficiency of two ultrafast processes – energy transfer and electron transfer. Initial electron transfer takes place from the excited state of a 'special' pair of chlorophyll or bacteriochlorophyll molecules within a pigment-protein complex called a reaction center and eventually leads to the creation of a pH gradient across the biological membrane in which the complex sits. This pH gradient is the source of chemical energy for essentially all life on earth. Typically each reaction center is surrounded by pigment protein complexes containing a total of 200–300 chlorophyll molecules whose function is to absorb sunlight and transfer the excitation energy to the special pair. This light harvesting process is the key to photosynthesis proceeding at the optical turnover rate: for example in bright sunlight the special pair would receive about one direct excitation per second. The light harvesting complexes boost this rate to 200–300 per second by virtue of energy transfer with greater than 90% efficiency.

Of the two processes, electron and energy transfer, the former has been far more extensively investigated, despite the fact that the fundamental theory for energy transfer was proposed by Förster several years before the seminal work of Marcus on electron transfer [1,2]. Indeed the timescale of photosynthetic energy transfer within the light harvesting complexes was calculated to be about 10–13 s by Duysens as early as 1964. Perhaps this extraordinarily short timescale discouraged extensive investigation of the primary photosynthetic energy transfer steps. Certainly direct observation of the primary steps was inconceivable at the time. Likewise the structures of the photosynthetic light harvesting (or antenna) complexes were unknown. Even the idea that the pigment molecules were always associated with proteins was not universally accepted even into the 1970s.

The advent of femtosecond spectroscopy has radically changed the experimental perspective with experiments on the few tens of femtoseconds timescale becoming routine. Similarly heroic efforts by crystallographers have led to atomic resolution structures for both reaction center and light harvesting complexes, initiated by the work of Michel and Deisenhofer on the purple bacterial reaction center [3]. The structures of the light harvesting complexes, exemplified by the light harvesting 2 (LH2) complex of purple bacteria solved by McDermott *et al.* [4], and the Fenna – Matthews – Olson complex [5] raise many challenging questions about the mechanisms of energy migration in these systems. Four articles in this volume are devoted to a discussion of these issues and taken together provide a detailed survey of the current state of knowledge on photosynthetic light harvesting.

Valkunas, Trinkunas and Liuolia describe the theory and applications of excitation annihilation in molecular aggregates with particular emphasis on results from photosynthetic light harvesting complexes. Following localization of the excitation, measurement of the annihilation process allows determination of excitation diffusion coefficients in energy transfer systems.

Gnanakaran *et al.* provide a comprehensive discussion of the role of localization of excitation in energy transfer in reaction centers and antenna complexes. Their chapter also gives a very useful introduction to the optical properties of dimers and aggregates.

Van Grondelle and Somsen discuss photosynthetic light harvesting in the context of the structures of several chlorophyll-protein complexes. The different complexes display a variety of spatial organizations from highly ordered (as in LH2 of purple bacteria) to apparently disordered in photosystem I and the Light Harvesting Complex II of plants. As the authors note the connecting theme in these structures is arranging for the maximum number of possible pigments close to the site where electron transfer occurs. However, this arrangement must also allow for the vectorial electron transfer to the primary electron acceptor, which must successfully compete with the reduction of antenna molecules by the primary donor. Thus arrangements with large numbers of molecules roughly equidistant from the primary donor, and located at a separation greater than that of the primary electron acceptor are favored.

Savikhin, Buck and Struve describe the current understanding of the electronic structure and dynamics of the first photosynthetic protein to have its structure determined – the Fenna – Matthews – Olson protein. Strong inter-pigment coupling and many inequivalent sites make this a challenging system to study, but Savikhin *et al.* conclude that the laser-prepared excited states are highly localized by disorder.

Förster's derivation of the resonance inductive mechanism of energy transfer [1] fundamentally arises from the Coulombic interaction of transition densities on donor and acceptor molecules while Dexter demonstrated that electron exchange coupling could induce energy transfer between states with forbidden optical transitions [6]. Several chapters in this book extend these ideas. Juzeliunas and Andrews use quantum electrodynamics to present a description of energy transfer that is valid from small to large separations where the transfer mechanisms becomes radiative rather than radiationless. The chapter by Berberan-Santos *et al.* takes up the theory of radiative transport in detail. Surprisingly little effort has been devoted to numerical estimates of coupling between donor and acceptor, particularly when separations are so small that orbital overlaps are significant. Scholes explores this topic and clarifies a number of aspects of the distance-dependence of coupling between chromophores.

Energy transfer depends on the relative orientation of donor and acceptor. Van der Meer describes orientational effects on pairwise energy transfer while Demidov and Andrews develop models for the time dependence of the fluorescence polarization in multichromophore systems of increasing complexity. These models allow structural formation to be inferred from the experimental data. Building on the reasonably complete theoretical understanding of long range energy transfer, Dos Remedios and Moens describe the use of resonance energy transfer to provide structural and functional information on proteins. Many exciting developments have occurred in this field, including single molecule energy transfer studies perhaps leading the way to mechanistic studies of single proteins folding to their native state, for example.

The broad range of applications of fluorescence and fluorescence energy transfer to studies in molecular biology and biotechnology assure that resonance energy transfer will be a vital component of the new science and technology of the next millenium. The present book should provide an important source of the methods and concepts that will underpin the new science.

G. R. Fleming
Berkeley, California
September 1998

REFERENCES

1. Förster, Th. 1948. *Ann Physik* 2: 55.
2. Marcus, R. A. 1956 *J. Chem. Phys.* 24: 966.
3. Deisenhofer, J., Epp, E., Miki, K., Huber, R., Michel, H. 1984 *J. Mol. Biol.* 180: 385.
4. McDermott, G., Prince, S. M., Freer, A. A., Hawthornthwaite-Lawless, A. M., Papiz, M. Z., Cogdell, R. J. and Isaacs, N. W. 1995 *Nature* 374: 517.
5. Fenna, R. E., Matthews, B. W., Olson, J. M., Shaw, E. K. 1974 *J. Mol. Biol.* 84: 231.
6. Dexter, D. L. 1953 *J. Chem. Phys.* 21: 83.

1

Resonance energy transfer in proteins

Cristobal G. dos Remedios and Pierre D. J. Moens
The University of Sydney, Australia

1.1 INTRODUCTION

When it comes to spectroscopic examination of proteins, biophysicists often favor nuclear magnetic resonance because it yields an abundance of structural information. However, one has to be prepared to spend large amounts of time trying to figure out how to calculate protein structures. So, before you rush off and join their ranks, consider the disadvantages of NMR spectroscopy, for things are not as one-sided as they seem. NMR spectroscopists usually have to work with unreasonably high (millimolar) protein concentrations where many if not most proteins aggregate and/or become insoluble. NMR is normally limited to proteins of molecular weight of 20 000 Da or less, which excludes many, perhaps most, of the biologically important proteins. Neutral pH is not always the best solvent condition for NMR even though it is biologically the most significant. And, finally, NMR spectrometers are hideously expensive.

There are other ways of investigating protein structure and function. Fluorescence spectroscopy can provide structural and functional information and, as we shall see, the subdiscipline of fluorescence resonance energy transfer (FRET) spectroscopy is a particularly valuable tool for measuring distances – which can then interpret conformational changes as small as 1 Å. The development of resonance energy transfer owes much to Th. Förster and G. Weber, two founding fathers who gave us the theoretical and practical tools to perform energy transfer measurements. Resonance energy transfer spectroscopy can happily manage with micromolar protein concentrations, it has no problem

Resonance Energy Transfer. Edited by David L. Andrews and Andrey A. Demidov
© 1999 John Wiley & Sons Ltd

with large, even gigantic proteins, the experiments are relatively quick and easy to perform over a wide pH range, including physiological pH, and the fluorometers are much less expensive than NMR spectrometers.

1.2 SOME BASIC CONSIDERATIONS

Resonance energy transfer requires two probes, a fluorescent donor and an acceptor which need not be fluorescent. Irradiation of the donor fluorophore with light energy of an appropriate frequency effectively produces an oscillating dipole. This in turn may resonate with a dipole in an acceptor probe in the near field. This resonating dipole–dipole interaction involves a transfer of energy from a donor fluorophore to an acceptor chromophore. The process is normally regarded as nonradiative and so not involving the emission and absorption of photons. For any given donor–acceptor pair, the energy lost by the donor is gained by the acceptor. Resonance energy transfer is effective over distances ranging from < 10 Å to about 100 Å.

Fluorescent probes belong to the family of luminescent moieties, which absorb light energy at one wavelength and emit it at a longer wavelength. In fluorescence, absorption of light rapidly produces excitation from the ground singlet state (S_0) to a higher (S_1) state (see Fig. 1.1). A very rapid

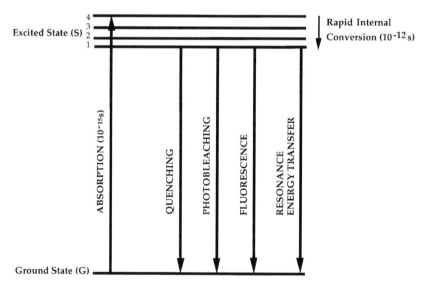

Figure 1.1 A simplified Jablonski diagram illustrating the electronic transitions involved in the excitation of a fluorophore and the pathways by which energy can be lost. The times in brackets indicate the orders of magnitude for the processes

(picosecond) decay to the lowest vibrational level in this state is followed by a slower return to the ground state, accompanied by photon emission. Return to the ground state generally occurs with a lifetime of the order of 10^{-8} s.

Fluorescence emission is only one of several potential pathways for the loss of excitation energy. Excited fluorophores can be deactivated by competing pathways whose rate constants include: (1) the fundamental emissive rate (k_F) of photons; (2) loss as heat by internal conversion (k_{IC}); (3) transfer to a quencher molecule by molecular collision or by forming a complex (k_Q); (4) loss by photodestruction or photobleaching (k_{PB}); and (5) the formation of a triplet state through intersystem crossing (k_{ISC}). In this chapter we will consider another possibility, namely (6) the rate constant for fluorescence resonance energy transfer (k_{FRET}). This involves a transfer of resonance energy from a donor to an acceptor separated by several atomic diameters. Coupling of the singlet electronic states of the acceptor to states in the donor is essential for the energy-transfer process. Each of the above six processes can occur in parallel, and measurement of the rate constant of one process can be used to determine the rate constants of any of the others. The most common procedure is to determine k_{FRET} by observing k_F in the presence and absence of an acceptor, but it is possible to observe the transfer process by recording k_{PB}. Bastiaens and Jovin recently combined quantitative confocal laser scanning microscopy and fluorescence lifetime imaging microscopy to observed a labeled protein kinase C beta I isozyme which was injected into living cells. By measuring the rate of photobleaching they achieved a novel determination of resonance energy transfer by imaging the photobleaching of the fluorescence using digital imaging microscopy [9].

The reciprocal relationship between the process rates and fluorescence lifetime is given by

$$1/\tau = k_{FRET} + k_{OP}, \qquad (1.1)$$

where τ is the fluorescence lifetime of a donor probe, k_{FRET} has already been defined, and k_{OP} is the rate constant for all other processes. From this relationship, one can see that an increase in FRET will result in a decrease in the lifetime of the donor.

A useful analogy suggested by Bob Clegg is to think of excitation energy first entering a "room" and then being able to leave via a number of possible different "doors," each of which represents a different and independent process. It is important to realize that, in resonance energy transfer experiments, it is common to assume that all of the energy is lost from the system via the door labeled "FRET." However, there may be a significant variation in the energy lost via the other doors. Under these circumstances there will be significant errors in the calculation of the donor–acceptor distance.

Each of the letters in the FRET acronym has significance. Although the first character might signify *Fluorescence* because of the requirement for a fluorescent (luminescent or light emitting) donor, as the source of energy transmitted to an acceptor, others (see [146]) argue against this because it is not the fluorescence that is transferred. Those who would otherwise wish to use the acronym RET for such reasons may prefer to think of FRET as *Förster* Resonance Energy Transfer. There is no requirement for the acceptor to be fluorescent, but when it is, we refer to it as sensitized fluorescence. *Resonance* is used because the acceptor must be able to resonate with the donor oscillator, and the term *Energy* is included because it is the electronic energy that resonates rather than the fluorescence. *Transfer* is important because FRET requires that the resonance occurs over distances which are much larger than atomic distances.

We think many readers will be surprised and impressed by the range of information that can be obtained using FRET spectroscopy. We hope that the recent data reviewed here will suggest useful new strategies to help solve research problems, particularly where subtle changes in structure must be detected. Finally, we review the contribution of FRET spectroscopy to clever devices such as biosensors – which may send you dashing for the telephone to call your stockbroker, perhaps next year!

Before proceeding, we want to declare that our interest in FRET centers on the contractile proteins of muscle cells (actin, myosin, the regulatory proteins – tropomyosin and the troponin complex) [5, 13, 38, 40, 66, 107, 110]. However, these comprise only a fraction of the total work on FRET involving protein structure. About 300 papers have been published since 1993, of which about half are devoted to proteins. About 25% of all those papers were concerned with analysing structure in two-dimensional membranes, about 20% were concerned with the role of FRET in funneling light into the reaction center of photosynthesis, and about 5% of the reports dealt with DNA/RNA structure. This chapter will concentrate on FRET in proteins.

1.3   A SHORT HISTORY OF FRET DETERMINATIONS

In 1947, Bücher and Kaspers [17] reported the now classic demonstration of the transfer of excitation energy in a protein. They showed that carboxymyoglobin is photochemically cleaved into myoglobin and CO, using monochromatic light with wavelengths ranging from 280 nm (where the energy is principally absorbed by aromatic amino acids) through to 546 nm (where the absorption is due to the carboxyheme group).

A seminal reference is Förster's 1948 paper [49], which was originally published in German but was subsequently translated into English [52]. This paper concentrated on homotransfer (between identical fluorophores) and introduced concepts such as R_0 (the Förster distance). In the following year [50], Förster introduced the notion of heterotransfer (between dissimilar donors and acceptors). Förster examined energy transfer from trypaflavine (as the donor) and rhodamine B (as the acceptor) in methanol, and, by a clever experimental device, he concluded that the transfer had occurred over a distance (58 Å) that was several times the molecular dimensions of the fluorophores. Bob Clegg [28] has recently pointed out that Förster's original classical derivation was first published in German in 1946 [48], but is little known because it was never translated into English.

Förster subsequently updated [51] his quantum-mechanical analysis of FRET and defined the elements that determine the efficiency of the transfer in what has now become known as the Förster equation. In this widely cited paper he refined a key parameter, the Förster distance (R_0), as the distance separating the donor and acceptor probes where the efficiency of transfer is 50%. To our knowledge, the first paper to demonstrate FRET in proteins was published by Weber and Teale in 1957 [156].

It took some time for the Förster theory to be incorporated into the biochemical literature, and it was probably his 1965 publication [51] that stimulated the subsequent rush of papers. In 1960, Stryer [134] had reported energy transfer in proteins and peptides, and then in 1965 he reported that almost 100% efficiency could be obtained for FRET from Tyr and Trp residues to a noncovalently bound ANS in apohemoglobin (from which the heme group was removed). Stryer's laboratory subsequently produced a string of papers, many with co-authors who have stayed in the field [56, 68, 135, 136, 142, 147, 157]. Stryer's papers covered a spectrum of applications, including proteins, membranes, and nucleic acids.

An excellent review of the quantum-mechanical approach to the theory of resonance energy transfer was published by Van Der Meer and his colleagues [146]. This book, which includes the correct form of the Förster equation for virtually every choice of parameter units – and much more – is an excellent source of both theory and practice of FRET spectroscopy and, in our opinion, it would be a tremendous help for prospective FRET spectroscopists. An elegant, more extensive and highly readable consideration of the derivation of the Förster equation was recently given by Clegg [28].

For many years, FRET determinations were clouded by concerns over the uncertainty of the value of the indeterminate orientation factor (κ^2). This parameter is a function of the relative orientations of the donor and acceptor oscillating dipoles. In principle, κ^2 can adopt values between zero and four. It is probable that these extremes can be disregarded and the range of donor–acceptor distances can be estimated from anisotropy measurements. However, the resulting range of acceptable distances is often so large as to render the distance values almost meaningless. This problem is the subject of a separate chapter in this volume (Chapter 4) and so it will not be discussed in any detail.

However, we wish to reassure readers that, in practice, virtually every reported FRET distance has assumed κ^2 to have a numerical value of 2/3, and that the FRET distances so obtained are in quite good agreement [113] with the crystallographic distances for the same proteins. Thus, although by the late 1970s many of the initial doubts about the Förster theory had begun to dissipate, residual fears persisted well into the 1990s, (e.g. see [40]).

Arguably the best steady state fluorescence instruments originated in Gregorio Weber's Urbana Illinois laboratory. Over the past 30 years, the design of fluorometers has gradually improved, and today's instruments are much more sensitive, versatile, and easy to operate. This is particularly true for the measurement of fluorescence lifetime. The first lifetime fluorometers were pulsed instruments limited by the duration of the flashlamps (about 1 ns). Developments in the design of lasers and Pockels cells have made it possible to perform multiple-phase modulation experiments which have improved the time resolution into the picosecond range. Today's instruments are interfaced to fast, inexpensive personal computers, which have made it easier to acquire, process, and calculate the final donor–acceptor distances. Finally, software developments such as the Globals Unlimited program produced by Enrico Gratton and his colleagues have added real power to the analysis of data from fluorescence spectra.

1.4   THE COMPONENTS OF THE FÖRSTER EQUATION

Förster derived his equation from quantum-mechanical considerations as well as on classical grounds. Clegg [28] describes in detail two classical derivations (based on oscillating dipoles) and compares them with the quantum-mechanical (coulombic interaction of donor and acceptor) derivation. Readers who wish to grasp the physical mechanism of FRET should consult this analysis and the related references therein.

The Förster equation defines the essential elements needed to calculate, from the following expression for the transfer efficiency, the distance between a donor and an acceptor:

$$E = 1/(1 + R^6/R_0^6), \qquad (1.2)$$

where

$$R_0^6 = (8.79 \times 10^{23})k^2 n^{-4} \Phi_d J_{da}, \qquad (1.3)$$

in which R_0 is the donor–acceptor distance (Å) when the efficiency of FRET is 50%; Φ_d is the quantum efficiency of the donor probe attached to the protein of interest; κ^2 is the Förster orientation factor; n is the refractive index of the solvent (which is generally assumed to be isotropic) and has values in the range 1.33–1.40 ([132] quotes the range 1.33–1.60); and J_{da} is the overlap of the donor emission spectrum with the absorption spectrum of the acceptor and is discussed in more detail below. The constants in Eqn (1.3) depend on the units employed. Where R_0 is expressed in Å, the absorbance units employed in determining the overlap integral J_{da} (see below) are in $M^{-1}cm^3$. The magnitude of the constant (8.79×10^{23}) will change if other units are employed. A detailed discussion of these constants is provided by Van Der Meer and his colleagues [146].

A graphical representation of the relationship between FRET efficiency (E) and the donor–acceptor distance (R) is shown in Fig. 1.2. Note that the

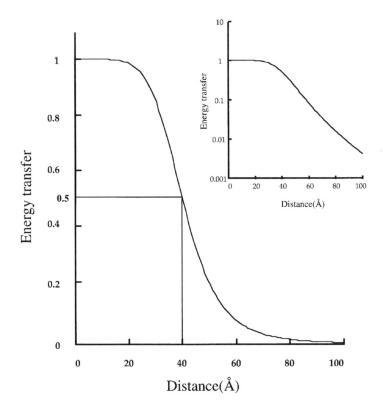

Figure 1.2 The relationship between donor–acceptor energy transfer and the distance (Å) that separates these probes. The distance corresponding to 0.5 transfer efficiency is defined as the Föster distance (R_0). The inserted figure (top right) describes the same data but with transfer efficiency shown on a log scale. Note that the slope of the transfer efficiency becomes flat at distances $\leqslant 0.5 \times R_0$. Also, the upper limit of the donor–acceptor distance can considerably exceed $1.5 \times R_0$, and its value depends on the precision of the determination of energy transfer

maximum efficiency in this relationship is reached more abruptly than its minimum. This feature suggests that, for any given precision in the determination of FRET efficiency, longer distances may be determined slightly more accurately than shorter distances. At low FRET efficiencies, the determination of the distance is limited by the precision with which the fluorescence intensities or lifetimes are measured. Normally, the cumulative errors of these measurements amount to about 1–3%. The insert in Fig. 1.2 shows the relationship between FRET efficiency and distance expressed on a semilogarithmic scale. Thus, if the precision of determinations of fluorescence intensity or lifetime is limited to about ± 1–3%, most estimates of the donor–acceptor distances will be limited within the range $0.5 < R < 1.5R_0$. However, if this precision can be significantly increased – for example, by greatly improving the signal-to-noise ratio – distances of $(1.8$–$1.9) \times R_0$ can be measured. This theoretic goal can be achieved with acceptors such as the lanthanide ions like Tb(III) (see below), where sensitized emission peaks occur in a low-noise region of the spectrum and so can be determined with great precision. Very few papers have attempted to capitalize on this feature of FRET determinations.

For efficient energy transfer, the quantum yield (Φ_d) of the donor and the absorption coefficient of the acceptor (ε_A) must be sufficiently large, and the donor spectrum must overlap the absorption spectrum of the acceptor. The orientation factor is a theoretic consideration here, but in practice there is probably sufficient motion of the probes to satisfy an assumed value of 2/3. Thus, practical difficulties come down to determination of the two parameters: Φ_d and J_{da}.

1.5   QUANTUM YIELD

In 1957, Weber and Teale [156] were the first to quantify this parameter. The quantum yield or quantum efficiency (Φ_d) of a fluorophore is defined as the number of light quanta emitted by a donor divided by the number of quanta absorbed by it. Thus Φ_d has a theoretic maximum value of 1.0, and several fluorophores have achieved this extreme value. On the other hand, Φ_d may be so small (< 0.01) that it is beyond the detection of modern instruments. Inclusion of this parameter in the Förster equation takes into account the many other pathways that lead to deactivation of the excitation energy, other than the emission of a photon. Most fluorometers deal with this parameter by recording the fluorescence intensity relative to a known reference standard, such as quinine sulfate or rhodamine. For an extensive listing of Φ_d values, see Van Der Meer *et al.* [146].

Haugland [67] provides a useful listing in which the absorption and emission spectra are ranked according to the product of the quantum yield and extinction

Table 1.1 A ranking of donor-acceptor pairs based on their Förster (R_0) distances (tryptophanyl data are listed separately in Table 2). Data in this table were extracted from Van Der Meer *et al.* [146] and supplemented with more recent publications. Note that where the same probe pair is associated with different R_0 values, the quantum yields and spectral overlaps of the donors differ because the donor is attached to different proteins

Förster Distance (R_0) Å	Donor Probe (locus)	Acceptor Probe (locus)	Donor Quantum Yield (Φ)	Overlap Integral (J) 10^{14} M^{-1} cm^{-1} nm^4	References
5	ANT-AMP (nucleotide)	Tb(III) (Ca site)	0.12	0.0014	[119]
7.7	Tb(III) (Ca site)	Pr(III) (Ca site)	0.49	0.00024	[142]
10.0–11.0	Eu(III) (Ca site)	Nd(III) (Ca site)	0.39–0.70	0.0014	[16]
9.2–11.4	ε-ADP (nucleotide)	Co(II) (metal ion)	0.40–0.47	0.00291–0.00307	[1]
11	EMal (Cys)	CPM (Cys)	0.15	1.10	[146]
21.9	DCl (Lys)	DDPM (Cys)	0.12	0.60	[101]
27.5–31.1	ε-ADP/ATP (nucleotide)	DDPM (Cys)	0.35–0.50	0.82–0.98	[1, 103, 105]
29	IAEDANS (Cys)	DDPM (Cys)	0.63	0.62	[31]
30.5	IAEDANS (Cys)	TNP (Lys)	0.64	0.821	[37]
32	FITC (Lys)	TNP-ATP (nucleotide)	0.4	1.76	[3]
33	MANT-dGDP (nucleotide)	sNBD (Lys)	0.29	2.319	[112]
34.9	ANS (noncovalent)	DABMI (Cys)	0.25	4.65	[85]
36–40.0	CPM (Cys)	5-IAF (Cys)	0.14	10.2–15.6	[62, 63]
37.5–45.5	ε-ADP (nucleotide)	5-IAF (Cys)	0.35–0.71	5.22–8.30	[83, 107]
38.9–43.8	IAEDANS (Cys)	DABMI (Cys)	0.53–0.63	5.02–8.35	[26, 102, 140]
40.3	IAEDANS (Cys)	TNP-ADP (nucleotide)	0.48	5.87	[37]
40.3	BFP	GFP	0.38	–	[70]
40.7–61.7	FITC (Lys)	TRITC (Lys)	0.34	29	[80]
40.9–48.1	IAEDANS (Cys)	FITC (Lys)	0.33–0.44	9.3–18.4	[74, 98]
43.8–51	ε-ADP (nucleotide)	TNP-ADP (nucleotide)	0.63–0.80	6.15–14	[37, 103]
45	CPM (Cys)	EMal (Cys)	0.42	10	[60]
46.2	IAEDANS (Cys)	IANBD (Cys)	0.63	1.01	[107]
48	CPM (Cys)	FMal (Cys)	0.42	17	[61]
51	ε-ADP (nucleotide)	TNP-ADP (nucleotide)	0.8	14	[37]

Table 1.1 (*contd.*)

Förster Distance (R_0) Å	Donor Probe (locus)	Acceptor Probe (locus)	Donor Quantum Yield (Φ)	Overlap Integral (J) 10^{14} M^{-1} cm^{-1} nm^4	References
52	IAEDANS (Cys)	5-IAF (Cys)	0.63	17.3	[138]
53.8	IAEDANS (Cys)	5-IAF (Cys)	0.48–0.63	1.05–30.1	[26, 83, 91, 138]
54	FITC (Lys)	EITC (Lys)	0.48	33.9	[19, 45]
56	FITC (Lys)	RITC (Lys)	1.0	–	[23]
56.1	IAEDANS (Cys)	IAE (Cys)	–	–	[54]
57	FMal (Cys)	EMal (Cys)	0.61	37	[61]
58	Tb (III)-DTPA cs 124 (Cys)	TMRIA (Cys)	0.7 (1.0 in D$_2$O)	–	[129]
58	5-IAF (Cys)	ErITC (Lys)	0.4	46	[3]
62	FITC (Lys)	ErITC (Lys)	0.4	46	[3]
64 (in H$_2$O) 70 (in D$_2$O)	Tb-DTPA cs124 (Cys)	Cy3	–	–	[128]
65.7	Tb(III)-cryptate	RB	1.0	4.51	[88]
90	TBP-Eu (III)	APC	0.3	490	[93]

Abbreviations: ε-ADP, 1-N(6)-ethenoadenosine-5′-diphosphate (in some references the triphosphate was used); APC, allophycocyanin; ANS, 1-anilinonaphthalene-8-sulfonic acid; ANT-AMP, anthraniloyl-2′-adenosine-5′-monophosphate; BFP, blue fluorescent protein; CPM, 7-diethylamino-3-(4′-maleimidophenyl)-4-methyl coumarin; Cy3 is the commercial name for an Amersham's carbocyanine dye; DABMI, 4-dimethylamino-phenylazophenyl-4′-maleimide; DCl, dansyl chloride; DDPM, N-(4-(dimethylamino)-3,5-dinitrophenyl)-maleimide; EITC, eosin isothiocyanate; EMal, eosin-5-maleimide; ErITC, erythrosin-5′-isothiocyanate; FITC, fluorescein-5′-isothiocyanate; FMal, fluorescein-5-maleimide; GFP, green fluorescent protein; IAE, 5-(iodoacetamido)eosin; IAEDANS, 5-((2-(iodoacetyl)amino)ethyl)amino)-naphthalene-1-sulfonic acid; 5-IAF, 5-(iodoacetamido)fluorescein; IANBD, 4-(N-(iodoacetoxy)ethyl-N-methylamino)-7-nitrobenz-2, 1, 3-oxadiazol; MANT-dGTP, 2′(3′)-O-(N-methylanthraniloyl)deoxy-guanosine-5′-triphosphate; RB, rhodamine B; RITC, rhodamine isothiocyanate; Ru(III), Tris(2,2′-bipyridyl)ruthenium; sNBD, succinimidyl-6-[(7-nitrobenz-2-oxa-1,3-diazol-4-yl)amino]hexanoate; TBP-Eu(III), a complex of Tb(III) with tribipyridine diamine; TNP, trinitrophenyl, the reaction product of trinitrobenzene sulfonate; Tb-DTPA-cs124, a chelate of Tb(III) with diethylenetriaminepentaacetic acid which is covalently attached to the organic chromophore carbostyril 124; TMRIA, 5-tetramethylrhodamine iodoacetamide; TNP-ADP, 2′,3′-O-(1, 4, 6-trinitrocyclohexadienylidine)adenosine-5′-diphosphate; TRITC, tetramethylrhodamine-5-isothiocyanate. Atomic symbols are used without abbreviation and their valence is shown in brackets.

coefficient of each probe. When this table is viewed with the FRET experiment in mind, it is an easy matter to select probes which have large quantum yields and a good overlap of the donor emission and acceptor absorption spectra. The Φ values for a wide range of probes are listed in Table 1.1.

1.6   DETERMINING THE SPECTRAL OVERLAP

J_{da} is defined as follows:

$$J_{da} = \int F_D(\lambda)\varepsilon_A(\lambda)\lambda^4 \, d\lambda \Big/ \int F_D(\lambda)d\lambda \quad (M^{-1} \, cm^3), \qquad (1.4)$$

where $F_D(\lambda)$ is the donor fluorescence per unit wavelength interval and $\varepsilon_A(\lambda)$ is the molar extinction coefficient of the acceptor at wavelength λ. The spectral overlap of the donor emission and acceptor absorption spectra (J_{da}) must be determined if the FRET efficiency is to be converted into a distance. This can be a difficult problem because, while the absorption spectrum can be determined with considerable accuracy, the fluorescence emission, expressed in arbitrary units, is often less accurately known. Weber [153] originally dealt with this question by assuming that the spectral bands followed Gaussian distributions.

Another approach [118] to the problem is to plot the absorption and fluorescence emission spectra on a wavenumber (reciprocal wavelength) scale. The donor emission spectrum can then be approximated to the mirror image of its absorption spectrum. This assumption is generally justified because it is known that, for any given oscillator, the two spectra are close to a mirror image, with their symmetry axis at the wavenumber corresponding to the energy gap between the thermally relaxed and excited states. The absorption and emission spectra can then be obtained graphically. This approximation is discussed on page 127 of the book by Pesce *et al.* [118]. The method is particularly useful for homotransfer between donor–acceptor pairs.

The method described by Van Der Meer *et al.* [146] is the one that we prefer. In this method, the fluorescence emission spectrum is integrated (preferably using the digital output of the fluorometer) and the integral is set to unity. The absorption spectrum can be reliably determined using a modern digital spectrophotometer, which generally has an RS232 port through which the absorption spectrum data can be easily downloaded to a personal computer. The peak value of the absorption spectrum is then set to one mole and the wavelength scales (nm) are aligned, thereby allowing the spectral overlap to be integrated.

1.7   STEADY STATE OR TIME-RESOLVED MEASUREMENTS?

This is really a rhetorical issue. In fact, FRET should be determined using *both* steady state and time-resolved methods. Each has advantages and disadvantages. Determinations of FRET from lifetime measurements are less prone to errors due to aggregation and insolubility than steady state measurements, because they are less affected by scattered light.

The great majority of FRET efficiency determinations are made from measurements of steady state fluorescence intensity, principally because the calculation of FRET efficiency is more straightforward. The main problems arising from these measurements are as follows: (i) there may be significant stray light from the exciting beam passing to the emission detectors (double grating monochromators largely avoid this problem); (ii) light coming from the exciting source monochromator and from the sample will, at least to some extent, be polarized – and this can influence intensity measurements unless the exciting polarizer is set to the perpendicular orientation and the emission polarized set to the "magic" angle (54.7°); (iii) the optical density of the sample should remain $\leqslant 0.05$ OD units to avoid effects due to secondary reabsorption of the fluorescence.

The relationship between the fluorescence intensity and the FRET efficiency (E) is as follows:

$$E = (1 - I_{DA}/I_D), \qquad (1.5)$$

where I_{DA} and I_D are the fluorescence intensities of the donor in the presence and absence of acceptor respectively. The major advantage of fluorescence lifetime measurements is that they are essentially uncomplicated by effects due to the concentration of the donor fluorophore. Traditionally, lifetime measurements were made using pulsed instruments, but this method can only be used with fluorophores which have τ values greater than a few nanoseconds. In 1969, Spencer and Weber [131] developed the cross-correlation phase fluorometer and this crucial improvement led the way to the development of the current instruments. The early single-phase determinations had the potential to create serious artifacts, but over the intervening period multiple-phase instruments have overcome these shortcomings. Today, most investigators employ the phase modulation method. The relationship between fluorescence lifetime and FRET efficiency is:

$$E = (1 - \tau_{DA}/\tau_D), \qquad (1.6)$$

where τ_{DA} and τ_D are the fluorescence lifetimes of the donor in the presence and absence of acceptor respectively.

1.8   RESONANCE ENERGY TRANSFER USING INTRINSIC AMINO ACIDS

The great majority of FRET experiments have been performed using proteins in which specific amino-acid side chains were modified with donor and acceptor probes, i.e. with extrinsic probes. However, it is possible to perform experiments using probes which are intrinsic, i.e. built into the proteins. There are two obvious advantages of using intrinsic probes. One is that there is no need to label the protein. The second is that by using the native structure to obtain the spectra, one need not be concerned about inducing structural perturbations due to the labeling procedures.

In principle, three natural amino acids can be used as intrinsic donors; tryptophan, tyrosine, and phenyl alanine. All three have weak Φ values. Trp has a larger quantum yield than Tyr and is therefore more useful as an intrinsic probe because it is easier to detect (see Table 1.2). The contribution of Tyr emission to the fluorescence intensity of a protein is amplified because

Table 1.2 Spectroscopic data for FRET probes involving at least one intrinsic fluorophore (Trp) and a range of acceptor probes attached to a variety of sites on proteins. (For details see text)

Förster Distance R_0 (Å)	Donor locus/ Acceptor locus	Acceptor probe	Quantum Yield (Φ) of donor	Spectral Overlap (J) 10^{-15} cm^3 M^{-1}	Reference
	Trp/porphyrin	porphyrin	0.13	300	[32]
	Trp/Ca	Tb(III)			[161]
3.4	Trp/Ca	Tb(III)	0.12	–	[90]
5–7	Trp/Trp	Trp	0.13	–	[132]
13–15	Tyr/Trp	Trp	0.03–0.05	–	[132, 154, 155, 158]
14	Trp/ATP	MANT-ATP	0.081	0.61	[46]
16–19	Phe/Trp	Trp	–	–	[115]
	Trp/ANT	ANT	0.16		[120]
17.8	Trp/Cys	IAEDANS	0.14	–	[133]
18	Trp/His	Ru(III)	0.20	1.02	[53]
18.6	Trp/Lys	DC	0.13		[44]
19.6	Trp/N-terminus	DC	0.13	0.436	[87]
21	Trp/Tyr	Tyr	0.13	–	[132]
21.2–22	Trp/Cys	IAEDANS	0.11–0.129	0.543–0.576	[42, 76, 86]
24	Trp/Cys	MIANS			[73]
24.1	Trp/Lys	Coumarin	0.08	4.82	[44]
25	Trp/Lys	5-IAF	0.13	1.3	[42]
25.4	Trp/flavonol	flavonol	0.11	16.4	[137]
28.6	Trp/Cys	IAEDANS			[29, 64]
29.4	Trp/hydrophobic	ANS	0.13	3.26	[24]

Abbreviations: ANS, 1-anilinonaphthalene-8-sulfonic acid; ANT, 2-amino-benzoic acid; DC, dansyl chloride; coumarin, 7-diethylamino-3-(4′-maleimidophenyl)-4-methyl coumarin; flavonol, 3-hydroxyflavone; 5-IAF, 5-(iodoacetamido)fluorescein; IAEDANS, 5-((2-(iodoacetyl)amino) ethyl)amino)-naphthalene-1-sulfonic acid; MANT-ATP, 2′(3′)-O-(N-methylanthraniloyl)adenosine-5′-triphosphate; MIANS, 2-(4′-maleimidylanilino)naphthalene-6-sulfonic acid.

Tyr residues usually occur more frequently in protein sequences. Further, Tyr fluorescence emission overlaps the emission spectrum of Trp, which complicates measurements of Trp fluorescence intensity. On the other hand, Phe has a very weak extinction and a small quantum yield, and so is rarely used as an intrinsic fluorophore. Naphthyl alanine is not a natural amino acid, but it has the advantage of having a relatively long fluorescence lifetime.

Quite recently, a new approach was described which sounds most promising. A fluorescent unnatural amino acid was introduced at known sites into the transmembrane neurokinin-2 (NK2) receptor protein by suppression of UAG nonsense codons with the aid of a chemically misacylated synthetic tRNA specifically designed for the incorporation of unnatural amino acids during heterologous expression in *Xenopus* oocytes. Fluorescence-labeled NK2 mutants containing a unique 3-*N*-(7-nitrobenz-2-oxa-1,3-diazol-4-yl)-2,3-di-aminopropionic acid (NBD-Dap) residue can be introduced at specific sites, and the resulting proteins were found to be functionally active in a native membrane environment. Microspectrofluorometry was used to measure FRET efficiency between the introduced fluorescent NBD probe and a fluorescently labeled NK2 heptapeptide antagonist. The calculated distances fixed the ligand in space and defined the structure of the receptor in a molecular model for NK2 ligand–receptor interactions [144]. This is the first report of the incorporation of a fluorescent unnatural amino acid into a protein in intact cells, as well as being the first measurement of distances between a G protein-coupled receptor and its ligand by FRET. The authors state that this method can be generally applied to the analysis of spatial relationships in integral membrane proteins such as receptors or channels, but clearly there are possibilities for introducing fluorescent labels into other proteins. The NBD probe can be matched with acceptors such as TNP–ATP, so that FRET can be determined without the need for covalent modification of the engineered NBD-labeled protein.

1.9   HOMOTRANSFER BETWEEN INTRINSIC PROBES

Homotransfer is the phenomenon whereby electronic energy is transferred between identical fluorophores. By definition, homotransfer requires more than one identical fluorophore in an oscillating system. With intrinsic probes, it arises because the emission spectrum of a fluorophore such as Trp overlaps, at least to a small extent, the absorption spectrum of a nearby Trp. In principle, Trp homotransfer can be either within a single protein or between proteins. As with all fluorescence spectroscopy in the UV and near-UV, a buffer blank must be subtracted from the protein spectrum to remove the water Raman contribution.

1.9.1   Trp–Trp

When Trp fluorescence resonance energy homotransfer does occur, the contribution of each Trp to the protein fluorescence is usually evaluated using molecular genetics techniques to systematically and conservatively replace each Trp with a non-Trp residue such as Cys or Phe. For example, the pore-forming domain of colicin contains three Trp residues, and the contribution of each was studied using steady state and multifrequency phase fluorometry. Site-directed mutants containing one or two Trps revealed that the fluorescence emission was dominated (53%) by one Trp and that mutual homotransfer FRET occurred between the other two [149]. In the case of 3 α-hydroxy steroid dehydrogenase, the three Trps present in the native protein were sequentially replaced by Tyr, to isolate the contribution of each Trp. All Trps contributed to the Trp fluorescence, but one contributed more than the others [79].

The contribution of each of the seven Trps of human carbonic anhydrase II has been determined by replacing each Trp with a Phe or Cys residue. Martensson *et al.*[92] found that two Trps dominated ($\sim 90\%$) the emission. They noted that no energy transfer was observed between two pairs of Trps which were close in the sequence (Trp16 to Trp5 and Trp192 to Trp209), and that the emission of a seventh Trp was effectively quenched by a nearby His residue. In this elegant series of experiments, they used 2D NMR spectra of ^{15}N-labeled proteins to confirm that the mutations from Trp to Phe or Cys residues had essentially no long-range effects on structure, and therefore that the perturbations of structure in the vicinity of each mutated Trp were small.

Trp environment may play a role in the determination of FRET efficiency. Time-resolved and steady state fluorescence spectroscopy can resolve the contributions of each of the intrinsic Trps. Eftink *et al.* [43] generated single-Trp mutants (Trp replaced by Phe) in the sub-units of the trp apo-repressor protein from *E. coli*, and concluded that homotransfer occurred between the Trps in the wild-type protein. They also used collisional quenching studies to show that one of the Trps buried in a hydrophobic region of the protein is immobile, and has a short (307 nm) emission maximum compared to the other Trps (315 nm). Exposed Trp residues in a hydrophilic environment have relatively long emission maxima (e.g. 330–350 nm) and thus may contribute disproportionately to FRET efficiency compared to those in a hydrophobic environment, particularly when there is little or no excitation of the acceptor at the shorter (305–310 nm) wavelengths. Reshetnyak *et al.* [122] have contributed to these analyses by showing that an emission spectrum containing multiple Trps can be deconvoluted into their hydrophobic and hydrophilic contributions.

1.9.2   Effect on κ^2

When there are many Trps in a protein sequence (for example, the seven Trps in carbonic anhydrase II), it is possible for energy to migrate from one Trp to another, including the Trp that was the initial donor. Runnels and Scarlata [125] recently reported that homotransfer strongly depolarizes fluorescence emission as a result of intermolecular excitation energy exchange between an initially excited, oriented molecule and a randomly oriented neighbor within a membrane. They concluded that in certain experiments homotransfer may well play an important role in randomizing the donor probe, and thus help to achieve a κ^2 value of 2/3.

1.9.3   Weber "red-edge" effect

In 1970, Weber and Shinitsky [155] first described the so-called Weber "red-edge" effect where energy migration that normally occurs between identical molecules fails to occur. For Trps, this happens when the protein is excited at wavelengths between 295 and 310 nm, and it can be a valuable way to avoid both Trp homotransfer and Tyr–Trp heterotransfer. In proteins where there is Trp homotransfer, the ratios of fluorescence polarization values at 310 nm to those at 295 nm are characteristically high (2.39–2.84). We recently studied the Trp fluorescence polarization of actin, which has four Trps, two of which are separated by only a few residues and might therefore be expected to undergo homotransfer. The ratio of the polarization at 310/295 nm in monomeric actin was 1.66, which is low and virtually excludes the possibility of Trp–Trp homotransfer in this protein (Moens, Helms, dos Remedios, and Jameson, unpublished data).

1.10   HETEROTRANSFER

Heterotransfer is simply defined as FRET between nonidentical fluorophores. It covers the great majority of FRET experiments reported in the literature. In this section, we will concentrate on those heterotransfer FRET experiments where at least one of the fluorophores is an amino-acid side chain. With heterotransfer the spectral separation between the donor absorption and acceptor emission is usually large. Consequently, reverse heterotransfer (from acceptor to a donor) is small enough to be neglected in most instances.

We distinguish two types of heterotransfer experiments, those where both donor and acceptor are different but are natural or intrinsic to the proteins, e.g. between amino-acid side chains (Phe–Trp and Tyr–Trp), and those where one of the fluorophores is an amino acid and the other is an extrinsic label. In all of these experiments, Trp has the appropriate spectral properties to enable it to play a central role in a wide range of FRET experiments.

1.10.1 Heterotransfer with two intrinsic probes

1.10.1.1 Tyr–Trp

Gregorio Weber [153] was one of the first to describe FRET between Tyr and Trp residues in a protein, together with a report of their range of R_0 values. Since then, a number of reports have used Tyr–Trp FRET to measure structural parameters in a range of proteins. They noted that the quantum efficiency of Tyr in phenol (0.2) is much larger than the quantum efficiency of Tyr in proteins (0.03–0.05) [153]. The excitation spectrum of Tyr overlaps that of Phe and Trp, and consequently it is difficult to isolate its contribution to the FRET process unless the protein of interest is deficient in either Phe or Trp residues. The maximum absorbance of Trp at 280 nm is about five times higher than that of Tyr. In addition, Trp has a quantum efficiency which is 50–100 times that of Tyr. Consequently, Trp emission tends to dominate the fluorescent emission of most proteins. The quantum efficiency of Phe is so small that in practice it can be ignored. Excitation at 280 nm essentially eliminates the fluorescence of Phe and excites both Tyr and Trp fluorescence. Excitation at 295–310 nm can eliminate the excitation of Tyr (the "red-edge" effect) although, as stated above, there is the possibility that Trp emission might excite Tyr residues which are very close. Like Tyr, free Trp (indole) has a quantum efficiency (0.42) which is significantly larger than that of Trp in proteins (0.04–0.3) [153, 154]. These lower quantum yields are associated with shorter fluorescence lifetimes [153].

The combination of small quantum yields and weak extinction coefficients for these residues produces R_0 values which are small compared to those achieved with extrinsic probes. These features may be very useful for examining small structural changes (conformational changes), but a small R_0 can be limiting when longer distances are needed. Assuming that $\kappa^2 = 2/3$, Steinberg [132] published a range of Förster distances from 5 Å to 7 Å.

Rischel *et al.* [123] investigated the structure of the molten globule state in horse heart apomyoglobin using FRET experiments. They converted the Tyr residue at position 146 into 3-nitro-Tyr, obtaining the distances between this side chain and the two Trp side chains from time-resolved Trp fluorescence decay curves. Since both Trp residues are located in the N-terminal α-helix and the modified Tyr residue is located in the C-terminal α-helix, these measurements provided information about this helix–helix distance.

Mely *et al.* [95] used FRET to refine the three-dimensional structure of peptides corresponding to the two zinc-saturated finger motifs of the nucleocapsid protein NCp7 of HIV-1. They determined a Tyr–Trp distance of 7–12 Å by fluorescence resonance energy transfer. They then refined this distance range using NMR to obtain a mean inter-chromophore distance of 7.3 ± 0.7 Å. This is a powerful and important confirmation of the ability of FRET to accurately determine distances.

1.10.1.2   Phe–Trp

Ostrowski *et al.* [115] used steady state and nanosecond pulse fluorometry to study Trp residues of human prostatic acid phosphatase. Fluorescence excitation spectra revealed that the excitation energy of Phe was transferred to Trp(s) in both the native and refolded forms, but not in the denatured form. Since the FRET efficiency was higher in the native enzyme than in the refolded structure (which had been unfolded in 6 M urea at pH 2.5), they concluded that the structure of the refolded enzyme must be subtly but definitely different from that of the native enzyme structure, and that local conformation of the active center was regenerated upon refolding. This is an example of a use of FRET to detect conformational changes without the need to actually report donor–acceptor distances.

1.10.2   Heterotransfer with one intrinsic probe

1.10.2.1   Trp-labeled His

The quencher Ru(III) can be attached to a His residue and thus a Trp–His distance can be obtained. In cytochrome c peroxidase, nearly all of the intrinsic fluorescence arises from one Trp, which is approximately 26 Å from the heme group of this protein and is known to be partially buried. The Trp fluorescence was determined with the acceptor, pentaamineruthenium(III), covalently attached to a specific His residue. The modified protein was shown by X-ray diffraction analysis to be structurally identical to the native protein. The Trp fluorescence intensity of the Ru–His complex is 17% less than from native protein. This result is consistent with the FRET efficiency expected from distance calculations [53]. This is another example of agreement between the distances determined by FRET and their corresponding structure from X-ray diffraction.

1.10.2.2   Trp-labeled Lys

FRET determinations and anisotropy decay measurements can be valuable tools for characterizing conformational dynamics of short peptides. Maliwal *et al.* [89] used Trp as a donor in RNase by replacing the wild type Phe 8, and introduced a 5-(dimethylamino)-1-naphthalenesulfonyl (dansyl) acceptor group at positions 1 or 18. The observed mean distances were in reasonable accord with the X-ray structure of RNase and, overall, their results suggest that the specific amino-acid sequence will significantly influence the relationship between distance distribution parameters and conformational dynamics in the case of short peptides.

FRET measurements have also been reported [44] using two synthetic donor–acceptor zinc finger peptides containing a single Trp as the donor and either a dansyl group or a 7-amino-4-methyl-coumarin-3-acetyl group as an acceptor, attached to the ε-amino side chain of a C-terminal Lys. Calculation of distance distributions revealed a shorter distance and suggested a unique conformation (i.e. a narrow distribution of distances) when metal was bound and a longer distance with greater conformational flexibility when metal ion was absent from these peptides. Thus, distance distributions can be valuable guides to conformational change.

1.10.2.3   Trp-labeled amino-terminus

In a similar series of experiments using the peptide mellitin, Lakowicz and his colleagues [87] determined the distance between a Trp located 19 residues from the N-terminus and a dansyl group located at the α-amino group of the N-terminus. They used frequency-domain spectroscopy to measure FRET efficiencies to determine the distribution of distances between these two loci free in solution, and when mellitin was complexed with calmodulin, troponin C or palmitoyloleoyl-L-alpha-phosphatidylcholine vesicles. A wide range (28.2 Å) of donor–acceptor distances was found for free mellitin, consistent with that expected for the random coil state. This distribution narrowed (to 8.2 Å) when the mellitin was complexed with calmodulin, and narrowed even further (4.9 Å) when it was bound to phosphatidylcholine vesicles.

1.10.2.4   Trp-labeled Cys

In an elegant series of experiments, Isaac *et al.* [76] combined recombinant DNA technology and FRET to investigate spatial relationships between segments of the Msx-1 homeodomain using an AEDANS acceptor probe attached to engineered Cys residues in different helices and the single invariant Trp. They found that the donor–acceptor pairs were 19 Å, 23 Å and 16 Å apart [76]. These experiments successfully detected subtle alterations in the conformation of this protein.

A similar strategy can be used with small proteins and peptides. Federwisch *et al.* [47] studied three mutants of the anaphylatoxin C5a containing substituted Trps (donor) and an additional Cys for the attachment of IAEDANS (acceptor). They established that the structural integrity and biological activity of the molecules were not affected by the substitutions, and they then carried out FRET determinations from a Trp to a Cys-directed probe some 50 residues toward the C-terminus. They observed a distance distribution of 24 ± 8 Å, compatible with the view of the C-terminal chain as a slightly stretched helix pointing away from the body of the molecule.

In another set of experiments, FRET and circular dichroism were combined to study the thermal denaturation of a wild type and a mutant staphylococcal nuclease. CD spectra suggested that residual structures remain in the denatured state. Steady state FRET from intrinsic Tyr residues to the single Trp was measured at different temperatures. The results revealed a substantial degree of energy transfer which would not have been expected in a random, denatured structure. The cysteine residue in an engineered mutant (Lys-78-Cys) was labeled with IAEDANS and the fluorescence decays of a specific Trp were determined. Both CD and FRET data confirmed that the thermally denatured state was both compact and heterogeneous.

A sizeable change in conformation was reported by O'Malley *et al.* [114], who measured the distance between a unique Trp and an acceptor (4-phenyl-azophenylmaleimide, PAPM) attached to the unique Cys of α-apohemoglobin. When this protein undergoes assembly and binds its β-chain, the change in FRET efficiency reveals an increase of 4 Å (18 Å to 22 Å) between the probes. More importantly, the distribution of FRET distances (20 Å to 8.5 Å) is reduced, indicating a decrease in the flexibility of the α-apohemoglobin chain within the assembled protein. Thus, FRET can be used to measure specific site-to-site conformational changes as well as flexibility.

Pouchnik *et al.* [120] synthesized two novel sulfhydryl-reactive fluorescent probes, 2-amino-benzoic acid, 2-(bromoacetyl)hydrazide (Br-ANT) and *N*-[2-[(bromoacetyl)amino]ethyl]-2-(methylamino)benzamide (Br-MANT), which contain an anthraniloyl and *N*-methyl anthraniloyl group, respectively. These probes react with cysteinyls via their bromoacetyl group. The cysteine adducts have absorption maxima (323 and 326 nm) and molar extinction coefficients (2100 and 2900 M^{-1} cm^{-1} respectively) which make them excellent acceptors for Trp. Other spectral properties ($\Phi = 0.16$ and 0.42; emission maxima 432 and 440 nm respectively) are very attractive from the FRET viewpoint. The emission maxima of ANT–Cys and MANT–Cys shift to shorter wavelengths with decreasing solvent polarity and both react selectively and stoichiometric-ally with the single cysteine residue of a model protein (a myosin light chain). These probes have long (9–10 ns) lifetimes (see Table 1.2).

1.10.2.5   Trp–ANS

The determination of FRET distances does not necessarily require the donor or acceptor probes to be covalently bound to a protein. The nucleotide and metal cation probes are obvious examples. However, there is another class of loci which cannot be regarded as specific loci (e.g. ATPase and cation-binding sites) but which nonetheless provide useful information. Apolar fluorescent probes such as 8-anilinonaphthalene-1-sulfonate (ANS) and 6-propionyl-2-dimethyl-aminonaphthaleneprodan) bind to hydrophobic pockets on proteins (as well as to membrane lipids) and these probes are good acceptors for Trp fluorescence. If there is a 1 : 1 stoichiometry of these hydrophobic probes to the protein of interest, and if there is a single Trp, then a precise distance can be determined. However, even if these conditions are not achieved (for example, where multiple ANS molecules bind to a protein) and even if there are multiple Trp donors, *changes* in FRET efficiency can nonetheless be measured. The contributions from each Trp can be investigated using collisional quenchers, as can the contributions of ANS by displacement with apolar eosin probes. Examples of changes in Trp emission in the presence of ANS using bovine serum albumin, phospholipase A2, ovalbumin and lysozyme are discussed by Chang *et al.* [24].

Watala *et al.* [152] used FRET to estimate the inter-chromophore distance between a membrane protein tryptophan and ANS molecules embedded in the membrane lipid bilayer. They found that when tPA binds to the membrane protein it undergoes a conformational change which exposes the Trp to the solvent and alters the lipid–protein interactions in platelet membranes. Thus, although the noncovalent ANS precludes the determination of specific distances, it is still possible to detect conformational changes.

FRET has been used to study the binding between ion channels and fluorescent-labeled drugs. A series of high-affinity probes was synthesized which bind to a specific domain of muscle L-type calcium channels. For example, a synthetic benzathiazepinone was labeled with the BODIPY (Molecular Probes Inc.) probe and time-resolved binding kinetics determined. These experiments demonstrate that FRET can provide valuable insights into the molecular pharmacology of ion channels [14].

1.10.2.6   Trp-nucleotides

With the availability of a variety of nucleotide analogs, FRET distances between Trp residues and ATP binding sites can now be calculated. MANT–ATP and ANT–ATP, containing anthraniloyl and *N*-methyl anthraniloyl groups respectively, are good acceptors of Trp emission because they have a good extinction coefficient ($\varepsilon = 5 \times 10^{-3}\,\mathrm{M^{-1}\,cm^{-1}}$), and their excitation maxima (about 350 nm) coincides reasonably well with the emission maximum of Trp (315–350 nm). Furthermore, their emission maxima (at about 440 nm) exhibit large Stokes shifts. MANT–ATP and ANT–ATP bind well to the myosin ATPase site, but they do not bind to other proteins such as actin which cannot tolerate modifications of the ribose moiety. The spectral characteristics of MANT–ATP and ANT–ATP (see Table 1.2) are similar to those of IAEDANS ($\lambda_{ex} = 336\,\mathrm{nm}$; $\lambda_{em} = 482$ nm).

Falson *et al.* [46] reported that a fraction of F1-ATPase, corresponding to a putative nucleotide-binding domain, can be overexpressed in *E. coli* and can bind either MANT–ATP or MANT–ADP on a mol/mol basis. The distance between a Trp residue located in the nucleotide-binding site and MANT–ATP or MANT–ADP was estimated by FRET to be 13 Å and 11 Å respectively. This suggests that the Trp is close to the polyphosphate moiety of the nucleotide, and it was tentatively assigned a position in the sequence based on a comparison with the H-ras p21 protein.

The situation becomes much more complex when there is more than one Trp donor in a protein. Burghardt and his colleagues [117] recently devised a way around the problem of multiple Trp donors and a single MANT–ATP acceptor probe. The active subfragment of myosin, subfragment-1, contains four Trps whose positions are known from the crystal structure of this fragment [121]. The spatial relationship between myosin and F-actin is critical to an understanding of the function of these two proteins, and it has been a matter of ongoing speculation [126]. Contrary to conventional experimental FRET design, the problem was analysed by choosing a donor–acceptor pair with a very short R_0. Other Trps were eliminated because they were outside the range of distances sensed by the R_0 of this probe pair. Using a precise set of controls, it was demonstrated that there is a conformational change in S-1 when F-actin binds, and with the assistance of its crystal structure it was shown that the direction of the conformational change (between myosin Trp505 and MANT–ATP) was consistent with the current concepts of conformational change in myosin. A more extensive discussion of nucleotide analogs is provided in a later section.

$Co(NH_3)_4$ ATP has an absorption maxima of 365 nm, and although it has a weak extinction coefficient, it is an excellent acceptor for Trp emission. We know of no reports which have employed the Trp / $Co(NH_3)_4$ ATP probe pair, which can be expected to have an R_0 of about 15 Å.

1.10.2.7   Trp-ligands

FRET measurements can be used to sense the binding of specific ligands such as PAMPS (poly 2-acrylamido-2-methylpropanesulfonate) to lysozyme. Xia *et al.* [159] used Trp residues on lysozymes as donors and a pyrene acceptor attached to PAMPS, showing that FRET occurs when lysozyme preferentially interacts with these pyrene sites. Picosecond time-resolved studies were carried out using the single Trp residue on the apo-protein of horseradish peroxidase (apoHRP) and an anionic porphyrin bound at its surface. These data revealed a FRET efficiency corresponding to a Trp–porphyrin distance of 15 Å [32]. Table 1.2 provides a useful guide to the use of Trp as an intrinsic probe in FRET experiments. Note that the precise value of the quantum yield and the spectral overlap of Trp with the listed acceptor probes may vary from protein to protein.

1.11 THE RANGE OF DISTANCES DETERMINED BY RESONANCE ENERGY TRANSFER

The Förster equation (Eqn (1.3)) precisely defines the properties of those parameters which influence the calculation of the Förster distance. These include the quantum yield (Φ) of the donor and the overlap (J_{da}) of the donor emission and acceptor absorption spectra. A reliable rule for estimating the upper and lower limits of donor–acceptor distances (R) is to limit the interpretation of R to $(1 \pm 0.5)R_0$.

1.11.1 The range of R_0 values available for protein probes

Van Der Meer *et al.* went to the trouble of compiling the Förster distances for about 270 donor–acceptor probe pairs, many directly applicable to proteins. Using their table of ranked Förster distances, it is possible to select a distance close to a distance you wish to measure and then look up the best probe pair to attach to the protein. The R_0 values listed in this chapter (Table 1.1) should be used to supplement the more extensive list provided by Van Der Meer *et al.* The combined information suggests that R_0 can range from a short 3.4 Å [90] to 90 Å [93], a more extensive range than reported by Van Der Meer *et al.*

1.11.2 Strategies for measuring short distances

The measurement of short distances (< 20 Å) is limited by two factors. It requires a probe pair with a short R_0 and it depends on the accuracy of the energy transfer determinations (FRET efficiencies). In practice, FRET measurements become unreliable when the inter-probe distance is less than 0.5 R_0, and therefore the measurement of short distances requires the selection of probe pairs with short R_0 values. These tend to involve the lanthanide ions and so are limited to cation-binding sites. For example, Van Der Meer *et al.* [146] list an R_0 of 7.7 Å using Tb(III) as the donor and Pr(III) as the acceptor [142]. Marriott *et al.* [90] reported an R_0 of 3.4 Å for Trp and Tb(III). As we discuss in a later section, these lanthanide ions are capable of displacing Ca ions. With these probes the problem of κ^2 is minimized because it is reasonable to assume that the ions precess rapidly compared to the long (millisecond) lifetime of Tb(III).

R_0 can be minimized by choosing donor–acceptor pairs with a small overlap integral, by selecting donors with low quantum yields and/or by using acceptors with small extinction coefficients. For non lanthanide donor–acceptor pairs, short R_0 values include 11 Å for eosin-5-maleimide (donor) and 7-diethyl-amino-3-(4'-maleimido-phenyl)-4-methyl coumarin (acceptor) and 14 Å for fluorescein-5 maleimide (donor) and CPM (acceptor) [146]. Table 1.1

demonstrates that the short R_0 values have overlap integrals which are 2–3 orders of magnitude smaller than probes with a long R_0. This strategy was used effectively by Burghardt and his colleagues, as described above [117].

When the distance between a donor and an acceptor is < 20 Å, the size of the probes themselves becomes a significant factor. Clearly, if both probes contain oscillating dipoles of dimensions 10–15 Å (measured along their longest chord), then a 20 Å measurement would have little accuracy, even though it might still be sensitive to changes in the donor–acceptor distance because of a high precision. Several approaches can be used to determine distances < 10 Å, but they require small probes and a judicious selection of spectral properties.

1.11.3   The minimum distance

For most aromatic donors and acceptors, separation by distances < 10 Å involves either strong interactions between their ground states or transfer by exchange interactions. This minimal distance depends on the molecular or atomic dimensions of the donor and acceptor – which are assumed to precess randomly, thereby satisfying the conditions for assuming that $\kappa^2 = 2/3$. As pointed out above, intrinsic donors and acceptors such as Tyr and Trp have been used to determine distances as short as 7 Å. However, as we will see below, free lanthanide ions such as Tb(III) and Eu(III) have the smallest dimensions of all the FRET probes. Their ionic radii are about 1 Å (assuming a coordination number of 8) and are comparable to Ca(II), which has a similar radius (assuming a coordination number of 7) [34]. The lower limit of R requires that there is no collision between the oscillating donor and acceptor dipoles and so, in principle, these probes can be used to determine distances as small as 3–4 Å.

1.11.4   Strategies for measuring long distances

The maximum distance that can be measured between a donor and an acceptor will also depend on elements in the Förster equation. Thus, the spectral overlap parameter (J) should be maximal and the quantum yield (Φ) should be large, preferably 1.0.

A guide to achieving long FRET distances is given in Table 1.1, where we have selected and arranged the R_0 values that are relevant to proteins, from Van Der Meer *et al.*, and supplemented them with information from the literature since 1994. Given that these parameters can be maximized, and remembering that the energy of the donor radiates into the surrounding space, and that this energy must resonate with the dipole of the acceptor, it is generally accepted that the maximum measurable distance (R) is approximately 100 Å. To our knowledge, no distances larger than 100 Å have been reported.

For proteins, the longest R_0 values include: 90 Å for the Eu(III) chelate tribipyridine diamine (donor) and allophycocyanin (a very large, 104 kD acceptor) [93]; 65.7 Å for Tb(III)-chelates (donor) and rhodamine B (acceptor) [88]; 62 Å for fluorescein-5-isothiocyanate (donor) and erythrocin-5'-isothiocyanate (acceptor) [3]; 56 Å for fluorescein-5-isothiocyanate (donor) and rhodamine isothiocyanate (acceptor) [23]; and 52 Å for 5-((2-(iodoacetyl)amino)ethyl)amino)-naphthalene-1-sulfonic acid (donor) and 5-(iodoacetamido) fluorescein (acceptor) [138]. The lanthanide ion donors with their very long R_0 values suggest that it may be possible to more accurately measure lanthanide emission when there are very low noise levels, and that this may make it possible to measure distances $\gg 100$ Å [128].

1.12   PRECISE LOCATION OF RESONANCE ENERGY TRANSFER PROBES

With intrinsic fluorophores such as Trp and Tyr, the delocalized electrons reside in their indole and phenyl rings, identifiable within protein structures available from protein databases. The precise location of extrinsic probes is not so simple because although we can be reasonably sure where the dipoles are located, the linkers to amino-acid side chains introduce an uncertainty which could only be resolved by a structural analysis of the protein labeled with the probe in question.

For many proteins, precise location of a probe may not be possible. Under these circumstances, one is left with an uncertainty equivalent to the radii of the probes, as well as the mirror image solution. We employed precisely this tactic in 1987 when the crystal structures of actin and myosin subfragment-1 were not yet determined [40]. Our modeled FRET distances turned out to have a precision of about ±5 Å but, more importantly, the consistency of the FRET distances convinced us that the FRET determinations (all of which had assumed that $\kappa^2 = 2/3$) were correct. Indeed, the FRET distances proved useful in tracing the protein backbone of the actin X-ray structure [81]. In many instances, the mirror image solution can be eliminated, since it will often be located deep within the protein. Since we know that the probe has reacted with a surface residue, it cannot be located deep within the protein structure and therefore this possibility may be disregarded.

1.13   PROPERTIES OF PROBES

1.13.1   The best probes?

A question commonly asked by those commencing FRET experiments is: Which probe should I use? To give an answer requires some information. Is it intended to use intrinsic or extrinsic probes? What loci will be labeled? What is the approximate distance to be measured?

Ideally, the probes should not perturb the native structure of a protein. This is best achieved by using intrinsic probes (such as Trp), metal ions (such as Cu) or fluorescent analogs of natural ligands (e.g. ε-ATP). Unfortunately, this may not be either possible or convenient because of multiple Trps, or the probe may be located in an unfortunate part of the structure. Whenever a specific distance is required, it will probably be necessary to attach a donor or acceptor probe to a particular locus in a protein structure.

A comprehensive summary of probes and their properties can again be found in the book by Van Der Meer *et al.* [146], where they are grouped according to their reactivities with specific groups (e.g. Cys-directed probes). This source provides tables which match donor and acceptor probe pairs according to the magnitude of their Förster distances (R_0); acceptors are matched to donors and vice versa.

There are several properties of donor or acceptor probes which determine their suitability for FRET. These include: (i) a suitably high quantum yield; (ii) a lack of interference from other fluorophores such as Trp which may also be present; (iii) an insensitivity to local changes in solvent environment; and (iv) a resistance to photobleaching.

1.13.2   Quantum yield

The quantum yield is a function of both the rate of radiative processes (k_R) and the rate of nonradiative processes (k_{NR}), i.e. $\Phi = k_R/(k_R + k_{NR})$. Quantum yield is difficult to predict because its value depends on the environment of the probes when bound to a protein. Take 1,5-IAEDANS, for example. This is one of the most commonly employed donor probes in the FRET literature. In Table 1.1 its quantum yield reported by numerous authors varies by as much as a factor of two, but the variations recorded in Table 7.3 of Van Der Meer *et al.* [146] are even greater (0.15–1.0). Some of this variation is due to different environments in different proteins, and some of it due to the fact that other ligands may bind close to the IAEDANS site and alter its quantum yield. Thus, unless it is planned to exactly replicate a published experiment, it is wise to re-determine the value of Φ for any particular donor.

Although it is usual, it is not essential to determine the quantum yield of the donor in order to calculate a donor–acceptor distance. Selvin [128] points out that the FRET efficiency from the fluorescence emission of the acceptor probe can be used to calculate the inter-probe distance, and in this case it is dependent *only* on the *acceptor quantum yield*, not on the donor quantum yield. There is a particular advantage in this strategy when studying sensitized emission of the lanthanide ions, but quantification of the FRET efficiency from changes in emission fluorescence can in principle be used with any probe pair.

1.13.3　pH resistance

The emission spectra of some fluorophores are highly sensitive to pH, while other are comparatively less so. This effect is due to the rearrangement of the π electrons produced by protonation of the probe. The fluorescence emission of fluorescein is sensitive over the range 5–8, i.e. over the pH range which can be tolerated by many proteins. Its quantum yield increases by about 50% when going from pH 6.5 to pH 8, and some fluorescein derivatives used to sense intracellular pH [67] are even more sensitive to pH. Rhodamine is comparatively insensitive to pH. Molecular Probes Inc. now markets fluorescein-like probes which are effectively insensitive to solvent pH.

1.13.4　Sensitivity to solvent polarity

Pyrene iodoacetamide reacts with a specific Cys residue (Cys374) in actin. Pyrene-labeled monomeric actin is very weakly fluorescent, but the quantum yield increases more than twentyfold when actin polymerizes. As a consequence, it has become a well-established tool for following the assembly of actin into filaments [30].

1.13.5　Resistance to photobleaching

Another "good" property of probes is their resistance to photobleaching. In this respect, rhodamine is better than fluorescein. This must be balanced against the higher quantum yield of fluorescein. All fluorophores can be destroyed by photodestruction. Some, such as Trp, are very susceptible and can be excited only a few times before they photobleach. Others, such as fluorescein, can be excited about 10 000 times and yet others several hundred thousand times. This bleaching involves the physical disruption of the aromatic structure. Photobleaching competes with other processes during the deactivation of a probe. For example, Ha *et al.* [65] used photobleaching to identify energy transfer between a single donor and a single acceptor. A FRET measurement can be made by observing photobleaching of either the donor or the acceptor probe [128].

1.13.6　Temperature sensitivity

In general, it is preferable to carry out FRET experiments at temperatures lower than room temperature. There are two reasons for this. First, the fluorescence intensity is higher at lower temperatures (because at higher temperatures the chances of other processes [k_{IC}, k_{ISC}, etc.] is increased). In addition, proteins are more susceptible to contaminating proteases and so will be more rapidly digested at elevated temperatures. Most modern fluorometers provide a capacity for temperature control. Further, it is wise to gently stir the sample to maintain an even temperature as well as to minimize photobleaching.

1.13.7   Water solubility of probes

Most fluorescent probes are soluble in nonpolar solvents and are usually dissolved at high (mM) concentrations in solvents such as dimethyl sulfoxide (DMSO) or dimethyl formamide (DMF). Probes which are water soluble can also be unstable and break down with prolonged storage. Figure 1.3 illustrates the structure of one of the most commonly used probes (1,5-IAEDANS) as well as two nonfluorescent acceptor probes, DABMI and DDPM.

1.13.8   Nucleotide probes

The number of available fluorescent nucleotide analogs has steadily increased over the past 25 years. Two classes of covalent modifications have been produced, namely modifications of the purine ring and modifications of the ribose.

The ATP analog ε-ATP was developed over 25 years ago by Leonard *et al.* [127]. It was the first of the ATP analogs to be widely used for FRET in proteins and its popularity stems from its high affinity for most ATP sites. An ε-ATP donor can be coupled with a variety of acceptors including: divalent cation acceptors such as Ni(II) and Co(II), which have R_0 values of 11 and 12 Å respectively; with the nonfluorescent acceptor DDPM ($R_0 = 31$ Å); and with Cys-directed probes such as IANBD ($R_0 = 41$ Å) – see [40] for a review. Formycin triphosphate (FTP) has short excitation and emission maxima and is a potential donor probe for other probes such as dansyl attached to Tyr69 in actin [7] and Tb(III). Unfortunately, it is no longer commercially available.

Modifications of the 2' or 3' positions (e.g. MANT – and ANT–ATP) have been described earlier. FRET determinations can be made using a MANT–ATP donor and a nonfluorescent acceptor (DDPM) attached to a specific Cys residue, although so far none has been reported. MANT–ATP and ANT–ATP were first synthesized by Hiratsuka [71] and are not yet commercially available. Trinitrophenyladenosine-5'-triphosphate (TNP-ATP) contains a Meisenheimer complex located at the 2' and 3' positions of the ATP ribose and is another useful ATP analog. It has a large extinction coefficient ($\varepsilon = 26 \times 10^{-3}\,M^{-1}\,cm^{-1}$) and a very broad (400–550 nm) absorption spectrum, making it an excellent acceptor probe for a large number of donor probes such as IAEDANS; see, e.g., [36] for a review. An R_0 value of 51 Å for the probe pair

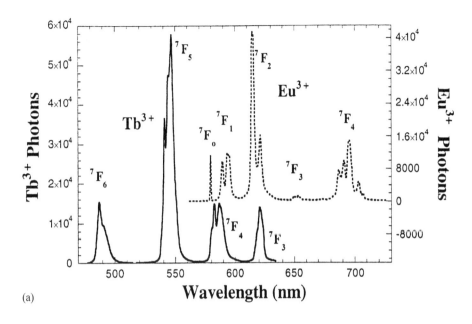

(a)

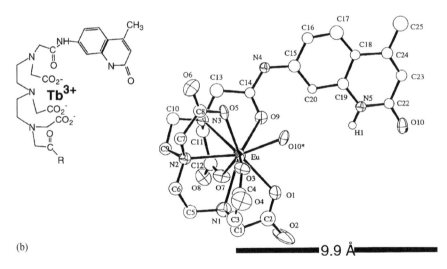

(b)

9.9 Å

Figure 1.3 (a) Steady state emission spectra of Tb(III) (solid line) and Eu(III) (dashed line) chelated to DTPA covalently linked to carbostryl-124. Note the baseline separation and hypersharp nature of the emission peaks. (b) The structure of the DTPA-cs-124 sensitizer of Tb (left) or Eu (right) ions. The lanthanide–sensitizer complex is approximately 12.8Å × 8.1Å × 8.3Å, as indicated by the 9.9 Å bar. This sensitizer is about the same size as many organic fluorophores. (Reproduced with permission from P. R. Selvin)

TNP–ATP and ε-ATP [36] demonstrates that these analogs can sense nucleo-tide–nucleotide distances in the range 25–75 Å. TNP–ATP is also fluorescent, and TNP derivatives of GTP are also available. The Amersham dyes Cy3 and Cy5 have been incorporated into analogs of ATP, and used in single molecule studies by Yanagida and his colleagues. Unfortunately, they too are not yet commercially available.

It is widely believed that the myosin S-1 "head" is responsible for substantial conformational changes when it interacts with F-actin. However, in 1984 Miki and Iio [99, 101] reported a conformational change of about 4 Å (33 Å to 29 Å) when they measured the distance between the C-terminus (Cys374) and the nucleotide-binding sites of actin. This change accompanied the self-assembly of actin monomers to F-actin. Small changes were also reported on changing the ligands at the nucleotide and divalent cation-binding sites (for a review of these changes, see dos Remedios and Moens [38]). Thus, nucleotide probes can be used to detect conformational changes. They are convenient to use, since they usually bind with high specificity and affinity and the unbound nucleotides can be simply removed by passage through a short column of Dowex-1.

1.14   LABELING SPECIFIC RESIDUES IN PROTEINS

The presence of a single cysteinyl in a protein does not guarantee that it will be labeled by a Cys-directed probe such as 1,5-IAEDANS. The γ-SH side chain may be partially or completely inaccessible and/or the added fluorophore may react with a His, Met or Tyr. Some of this lack of specificity can be avoided by using the maleimide derivative, which is less reactive with other side chains unless the pH is raised above 8. Maleimides such as pyrene maleimide have the advantage that they are essentially nonfluorescent until they react with a thiol. However, the specificity of a fluorescent probe is often difficult to predict. The sixth edition of Haugland's Handbook [67] provides an excellent introduction to the reaction specificities of amines, thiols, carboxylic acids, and glutamines, and extensive technical information is now available on a web site (`http://www.probes.com`).

Given this degree of uncertain specificity and predictability, and the need for a guide to protein labeling, we will describe the specific labeling in terms of our experience with a variety of side chains in muscle proteins including actin and myosin (Tables 1.3 and 1.4). In each instance, it is wise to take some precautions, as follows. (i) Dissolve the probe in an appropriate solvent (most hydrophobic labels canbe dissolved at high concentrations (200–400 $\times$ the protein concentra-tion) in dimethyl sulfoxide (DMSO), which we prefer to dimethyl formamide (DMF), since the latter often contains traces of reactive amines. The final concentration of solvent should not exceed 1–2% of the final volume. (ii) When adding the label, use a fine injector-type syringe (SGE or Hamilton) and slowly

Table 1.3 A summary of the common probes used to react with specific amino acid side chains. The wavelengths listed under Ex λ and Em λ represent the peak of the absorbance and emission spectra respectively. Where there is no Em value listed the probe is not fluorescent

Probe	Ex λ (max)	Em λ (max)	Comment	Reference
Cys-Directed				
Haloacetyl Probes			Unstable to light, especially in solution	[68]
1,5-IAEDANS	337 nm	480 nm	Common SH-probe, ~100 nm Stokes' shift, long lifetime	[75]
IAF	492 nm	515/516 nm	Proteins may be stereo-specific for 5- or 6-IAF; pH sensitive 5–8	[2, 150]
EIA	519 nm	540 nm	High quantum yield; acceptor for fluorescein	[141]
Br-ANT/MANT	323/326 nm	432/440 nm	ANT is smallest fluorescent organic probe	[120]
TMRI	555 nm	580 nm	Photostable; pH insensitive	[141]
IANBD	472 nm	536 nm	Sensitive to thiol environment	[82]
DABMI	460 nm	N/A	Smaller than DDPM; $\varepsilon = 3000\,M^{-1}\,cm^{-1}$	[140]
Pyrene iodoacetamide	339 nm	384 nm	Sensitive to solvent environment	[30]
Maleimide Probes				
Pyrene maleimide	338 nm	375 nm	Non-fluorescent until it reacts	[68]
DDPM	450 nm	N/A	Larger than DABMI; $\varepsilon = 24,800\,M^{-1}\,cm^{-1}$	[58]
MIANS	313 nm	420 nm	Reacts with hydrophobic cysteinyls	[72]
Lys-Directed Probes				
Dansyl-Cl	372 nm	557 nm (in water)	Sensitive to environment; large Stokes' shift	[4]
FITC	494 nm	519 nm	Water-soluble	[18]
TNBS	345 nm	N/A	Not absorbance until it reacts; water-soluble; $\varepsilon = 14,500\,M^{-1}\,cm^{-1}$	[37]
Tyr-Directed Probes				
FITC	495 nm	520 nm	Tyr requires prior nitration	[25]
Gln-Directed Probes				
Dansyl cadaverine	340 nm	520 nm	Labelled by transglutaminase	[139]
Rhodamine cadaverine	555 nm	580 nm	Labelled by transglutaminase	[139]
Eosin cadaverine	524 nm	544 nm	Labelled by transglutaminase	[139]
Fluorescein cadaverine	494 nm	518 nm	Labelled by transglutaminase	[139]

Table 1.3 (*contd.*)

Probe	Ex λ (max)	Em λ (max)	Comment	Reference
Nucleotide Probes				
ε-ATP	265 nm	411 nm		[127]
TNP-ATP	408 nm	–	Broad spectrum acceptor, non-fluorescent in water;	[68]
MANT-ATP	350 nm	440 nm	Better quantum yield than ANT	[71]
Non-Covalent Probes				
ANS	372 nm	480 nm	Non-fluorescent in water, used for membrane-bound proteins	[85]

Table 1.4   A list of the donor-acceptor pairs used to study structure and structural changes in the contractile protein, actin. This information has been arranged to facilitate the selection of a suitable donor-acceptor probe pair for Cys, Lys, Gln, Tyr, divalent cation and nucleotide sites on proteins. All data refer to FRET measurements made in monomeric actin and they avoid the complication of FRET between adjacent actin monomers in assembled actin filaments. These data are taken from recent reviews [39, 106] and have been supplemented with more recent data. For a complete list of abbreviations see Table 1 legend

Residue	Donor	Acceptor	R (Å)	Reference
Cys10[1]	IAEDANS (C10)	Co(II) (Me site)	14	[96]
	IAEDANS (C10)	DDPM (C374)	31	[106]
	Dansyl-Cl (Y69)	DABMI (C10)	27	[6]
	ε-ATP (ATP site)	DABMI (C10)	41	[97]
	FTP (ATP site)	DABMI (C10)	32	[6]
Gln41[2]	Dansyl (Q41)	DDPM (C374)	31	[111]
Lys61	ε-ATP (ATP site)	FITC (K61)	25	[99]
	IAEDANS (C374)	FITC (K61)	39	[4]
	FTP (ATP site)	FITC (K61)	21	[99]
	FITC (K61)	Co(II) (Me site)	19	[100]
Tyr69	IAEDANS (C374)	Dansyl-Cl (Y69)	39	[4]
	Dansyl-Cl (Y69)	DABMI (C374)	25	[101]
	FTP (ATP site)	Dansyl-Cl (Y69)	21	[7]
	Dansyl-Cl (Y69)	Co(II) (Me site)	11	[6]
Cys374[3]	ε-ATP (ATP site)	IANBD (C374)	33	[108]
	ε-ATP (ATP site)	DDPM (C374)	29	[105]
	IAEDANS (C374)	Co(II) (Me site)	23	[108]
	FTP (ATP site)	DABMI (C374)	30	[5]
	ε-ATP (ATP site)	NBD-Cl (C374)	29	[104]

Residue	Donor	Acceptor	R (Å)	Reference
ATP site	ε-ATP (ATP site)	IANBD (C374)	35	[27]
	ε-ATP (ATP site)	NBD-Cl (C374)	29	[104]
	FTP[4] (ATP site)	Tb(III) (Me site)	16	[5]
Divalent	Dansyl-Cl (Y69)	Co(II) (Me site)	11	[6]
cation site	FTP (ATP site)	Tb(III) (Me site)	16	[6]
	ε-ATP (ATP site)	Ni(II) (Me site)	< 10	[108]

Notes: 1 Labelling of G-actin at Cys10 required mild denaturing conditions such as 1M urea;
2 Enzymatic transfer of label using transglutaminase and a cadaverine-conjugated probe;
3 Cys374 is the most reactive of four Cys residues;
4 FTP (formycin-5'-triphosphate) will substitute for ATP in G-actin but it is not strictly an analog of ATP.

add the label over a period of 2–3 minutes, with the needle tip immediately above a rapidly rotating small magnetic stirbar. (iii) Check that there is a sufficiently high buffer concentration, because the pH may be lowered as the reaction proceeds. (iv) The labeling of Cys residues can be most effectively stopped by sedimenting the protein, but if this is not feasible, add a tenfold excess of *N*-acetyl cysteine. (v) Protect the label and the labeled protein from light.

1.14.1   Cys-directed probes

These are the most commonly reported FRET probes of protein structure. The halide group confers on the probe a strong reactivity with –SH groups of protein side chains (and any other –SH groups present). Trayer's group [10] reported that the bromoacetyls are more cysteinyl-specific than the iodoacetyl derivatives. Recently, Cremo's group [120] described new sulfhydryl-directed probes (linked by bromoacetyl) which are both small in size and avoid the problems caused by the large hydrophobic rings of more conventional probes. These Br-ANT and Br-MANT probes are particularly good acceptors for Trp. Cysteinyl-directed probes do not react with proteins unless the Cys side chains are reduced by reagents such as DTT, but care must be taken to remove these reducing, agents before labeling begins.

There are surprising differences in the reactivities of Cys residues in different proteins, e.g. 6-IAF reacts with the most reactive Cys (SH1) in myosin [2] while 5-IAF reacts with Cys374 (one of four Cys residues) in actin [141]. This stereo-specificity is difficult to predict and the selection of one or the other probe is probably best done empirically. The fluorescent properties of these isomers are practically identical. Fluorescein and rhodamine derivatives have been widely used to specifically label protein cysteinyls.

1,5-IAEDANS is by far the most popular of the haloacetyl Cys-directed probes. It is reasonably water soluble, it has a reasonably long fluorescence lifetime (10–20 ns) and it exhibits a large Stokes' shift – which is a real advantage for FRET experiments where it is commonly paired with fluorescein probes (see Table 1.1). While IAEDANS usually reacts with cysteinyls [36], there are proteins such as troponin-I in which it reacts with a Lys [97].

Maleimides, like the haloacetyls, also react with sulfhydryls. Some probes (e.g. DABMI: 4-[[[4-(dimethyl-amino)phenyl]azo]phenyl]maleimide) are not fluorescent and make excellent Cys-directed acceptors because the donor emission is not complicated by a contribution from the acceptor. Pyrene is very sensitive to solvent environment and its maleimide derivatives makes it react with Cys side chains. It has an unusually large extinction coefficient. Pyrene-labeled actin monomers greatly increase their fluorescence emission upon assembly into filaments, making this a popular probe of actin-binding proteins known to regulate actin microfilament assembly.

In actin, Cys10 is much less reactive than Cys374 and when it was first labeled [8] we used 1 M urea to partially unfold the proteins' to allow the probes to react with the normally buried Cys10. Subsequently, Drewes and Faulstich [41] showed that replacement of the native ATP with ADP made Cys10 much more reactive.

1.14.2   Lys-directed probes

Nearly all proteins contain Lys residues and usually there is more than one. Their ε-amino side chains are therefore a favourite labeling site for light microscopy and FACS scanning, where the attachment of multiple probes is an advantage. The opposite is true for FRET measurements. The moderately basic nature of these Lys side chains combined with the poor reactivity of acylating reagents such as isothiocyanates means that labeling reactions are best set at pH > 8.5–9.0. Further, these reagents tend to be unstable in water, particularly at high pH.

Papp *et al.* [116] labeled the ATPase proteins present in sarcoplasmic reticulum using either fluorescein isothiocyanate (FITC) or eosin isothiocyanate (EITC). Both labels react with the same Lys residues. By varying the proportions of the two labels in the reaction, they were able to vary the donor–acceptor ratio and demonstrate efficient energy transfer between adjacent ATPase proteins. This illustrates how science can progress in the face of seemingly impossible odds, since one would never *a priori* surmise that in a complex system such as isolated sarcoplasmic reticulum vesicles (where there are different proteins as well as lipids and other macromolecules) there would be any likelihood of specifically labeling one protein.

Dansyl chloride is also a good probe for Lys residues. A bonus with this label is the fact that it is hardly fluorescent until it reacts with amines. A less desirable property is its nonpolar nature, which affects its reactivity, since most proteins are dissolved in aqueous solvents. Dinitrophenyl sulfonyl chloride reacts with Lys as DNP-Lys. Trinitrobenzene sulfonate (TNBS) is a good reagent for Lys residues because it almost completely lacks an absorbance until it reacts with the target Lys. This means that the reaction rate can be followed in a spectrophotometer by monitoring the increase in OD of the solution at about 350 nm. An unusual application of multiple TNP acceptors was described as the "painted actin filament" experiment in which multiple Lys residues were reacted with TNBS which effectively coated the surface of the actin with TNP (actin contains 19 Lys residues). A donor-labeled myosin fragment was allowed to form an actomyosin complex and we looked for FRET to see if the myosin came within 45–50 Å of the actin filament surface. No FRET was observed and the reason for this became clear when, years later, the myosin S-1 crystal structure was "docked" with a model of F-actin. This model showed that the labeled myosin site was sufficiently far from the actin filament surface to prevent significant FRET.

1.14.3   Tyr-directed probes

Tyr residues react with FITC. Burtnick [18] was the first to preferentially react Tyr69 of actin with this probe. This reaction is carried out in two steps, a nitration reaction to form the ortho-amino-tyrosine, followed by a second reaction where the FITC reacts preferentially with the nitro-tyrosine. FITC is a good acceptor for 1,5-IAEDANS, and it is a good donor for acceptors such as rhodamine, eosin and erythrocin (see Tables 1.3 and 1.4). Dansyl chloride can be used to probe a specific Tyr residue in actin. Barden and Miki observed a small conformational change ($\sim$ 1 Å) by measuring the distance between Tyr69 and Cys374 [7].

1.14.4   Gln-directed probes

The labeling of glutamine is catalysed by transglutaminase isolated from guinea pig liver. The probes usually have a short (five carbon) spacer which enhances the reaction. Cadaverine derivatives of dansyl, fluorescein, eosin and tetramethylrhodamine are currently available. These can all be transferred to the glutamine side chain in the presence of transglutaminase. The method was first employed with actin by Takashi [138] and subsequently by other laboratories, including ours [111]. Like Lys residues, there are multiple Gln residues in actin and yet, under the control of the enzyme, a unique Gln is labeled. The selectivity of the transglutaminase depends on the solvent accessibility of a Gln to this enzyme. Some Gln residues may require a leaving group other than cadaverine.

The following generalizations can be made: (i) some probes are more reactive than others; (ii) not all so-called residue-specific probes are completely specific; (iii) the reaction specificity of amino-acid side chains depends on the solvent pH; (iv) the binding of proteins to their natural or imposed ligands or other proteins can change reactivities of side chains either by direct steric hindrance or by more remote allosteric effects; and (v) be prepared to try a probe and then test whether it specifically and uniquely labels the expected residue.

1.14.5   Determination of unique labeling of proteins

Protein databases now contain a very large number of amino-acid sequences and partial sequences, so it is increasingly likely that the sequence of any protein of interest is already known. The definitive identification of specific labeling is to isolate the peptide containing the label and then perform N-terminal sequencing. If the fluorophore is stable to acid conditions, then it will appear as an unidentified amino acid. It is an unfortunate fact that many proteins have blocked N-termini, which makes this strategy unlikely to succeed for intact proteins.

However, there are simpler ways to identify the labeled residue if the sequence of the protein is known. Limited enzymatic proteolysis of the labeled protein can be performed and the products then separated on highly cross-linked SDS PAGE gels. The labeled peptides can be recognized on a UV viewing box. Even better, the labeled peptides can be identified (and even quantified) using a Molecular Dynamics Inc. FluorImager, provided that the label can be excited by the standard 488 nm argon ion laser. PC software programs are available which can predict the sizes of the labeled peptides generated from uniquely labeled proteins, and usually this analysis is repeated with more than one enzyme. We refer readers to the Director of the Australian National Genomic Information Service (ANGIS); for more information, consult: `http://www.csu.edu.au/molecular/angis.html`.

A similar strategy can be employed in the absence of sequence information, because the peptides which are generated by the proteolysis are generally small (5–15 residues), so they can be separated by HPLC and then sequenced without incurring great expense. We recommend first performing an amino-acid composition analysis to determine the maximal number of residues that can react with the probe.

1.14.6   Nonspecific labeling

If a probe reacts with more than one locus on the protein, a small amount of label may react at a second (or third) site on the protein even if the labeling ratio is about 1 : 1. If an unintended donor site is closer to the acceptor than expected, its effect can dominate the transfer efficiency due to the sixth inverse power relationship between FRET efficiency and donor–acceptor distance. Such extraneous labeling is probably responsible for most of the reported differences between experimental FRET distances and those determined from X-ray diffraction analyses. On the other hand, if an extraneous site is further away from the acceptor or donor probe, then it will have little effect on the measurement. A useful quick check for the uniqueness of a particular donor is to perform a lifetime measurement. A unique lifetime usually (but does not always) indicates a unique labeled site.

1.14.7 Multiple acceptors

Some years ago, Taylor [141] took advantage of the known helical geometry of the monomers in the actin filament to show that when a donor-labeled actin monomer is co-polymerized with acceptor-labeled monomers, it is possible to calculate the radial coordinate of the labels, in this instance covalently attached to the Cys-374 residues. This is discussed in detail in Section 1.16, which deals with FRET in radially symmetrical systems.

1.14.8 Multiple donors

In multidonor, single-acceptor systems that contain different classes of donors, quenching may be used to assess the contribution of energy transfer from each class of donor. The tubulin–colchicine complex was used as a donor–acceptor system to show that two inaccessible tryptophanyls are at or near the colchicine-binding site [11]. Another example is the resonance energy transfer between four Trps and a single MANT acceptor in myosin, described above [126].

1.15 RESONANCE ENERGY TRANSFER EXPERIMENTS USING LANTHANIDE IONS

1.15.1 Luminescence

While it is true that the fluorescence emission from aromatic fluorophores is invariably derived from electric dipoles, this is not true for the lanthanide ions. Some of their transitions are of other types and cannot be involved in conventional FRET. As Selvin recently pointed out, strictly speaking, the emission from the lanthanide ions is also not fluorescence (a singlet–singlet transition) and is more properly referred to as luminescence, because it arises from high-spin to high-spin transitions [128]. In general, the emission spectra of these ions are not sensitive to solvent conditions, because their inner-shell electrons are the ones that undergo the transitions and are therefore inaccessible to the solvent. Further, the quantum yield of the lanthanide chelates is large, usually approximately 0.5 but potentially as large as 1.0 in D_2O.

1.15.2   Lanthanide chelates

Lanthanides such as Tb(III) and Eu(III) have very weak absorption cross-sections, but this defect can be overcome by forming chelates; for example, with salicylic acid-maleimide derivatives. The principal action of these chelates is to remove water, which normally quenches the lanthanide ion emission. A celebrated example of this sensitized fluorescence emission is the emission of Tb-chelate using rhodamine B as the acceptor [88]. The donor–acceptor pair has a Förster distance of 65.7 Å, where Tb(III) is assumed to have a quantum yield of 1.0 (in D_2O, slightly less in water) and an overlap integral of 4.51. Also, Tb(III) can undergo FRET to other lanthanides (e.g. Pr(III), Nd(III), Ho(III), and Er(III)) and so these ions can be used to accurately determine distances between Ca^{2+} sites over distances of 5–15 Å, since the lanthanide ions usually bind to the Ca^{2+} sites (see Table 1.1).

1.15.3   Long R_0 values

Lanthanide-cryptates can be used as very long-lived donor labels. Mathis [93] reported an extremely large R_0 value (~ 90 Å) for Eu(III) and allophycocyanin. This is the outcome of a very high overlap integral of the donor (Eu) and the acceptor (allophycocyanin) and a maximal quantum yield of the sensitized Eu(III) emission. Mathis used this chemistry to detect energy transfer between two antibodies (one antibody with the Eu chelate, the other with the allophycocyanin) bound to a single antigen (which in this case was prolactin).

Recently, Selvin [128] reported R_0 values for both Tb(III) and Eu(III) where their fluorescence (Fig. 1.4a) was sensitized by chelation to diethylenetriamine pentaacetic acid (DTPA). Sensitizers such as this can be covalently attached to the chromophore carbostyril (Fig. 1.4) and used to chelate Tb(III) or Eu(III) (Fig. 1.4b). The complex is referred to as DTPA-cs124. In D_2O, using Tb-DTPA-cs124 as the donor and the carbocyanine dye Cy3 as the acceptor, the R_0 value is 70 Å (64 Å in water). Using Eu-DTPA-cs124 as the donor and Cy5 as the acceptor, the R_0 value is also 70 Å (or 54 Å in water). Thus, with these long R_0 probes it should be possible to measure distances over the entire range from 30 Å to 110 Å using a single donor–acceptor pair.

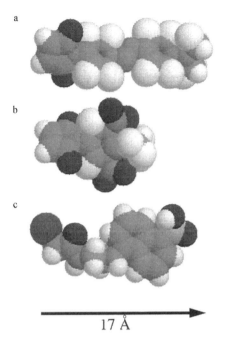

Figure 1.4   Space-filling models of two nonfluorescent acceptor probes, DABMI (a) and DDPM (b), together with the fluorescent probe, 1,5-IAEDANS (c). Structures were obtained by energy minimization using Cerius^2 software. All probes are shown in their fully extended configuration. The arrowhead points away from the points of covalent attachment of the probes to the Sγ sulfur of a cysteinyl side chain

1.15.4   Short R_0 values

At the other end of the scale (see Table 1.1), very short R_0 values of around 5 Å can be determined using lanthanide ions such as Tb(III) coupled to acceptors such as anthraniloyl-nucleotides. Anthraniloyl-2'-AMP has been used an a donor probe bound to the nucleotide site of *myo*-inositol monophosphatase, which also is known to bind lanthanide ions. This enzyme–nucleotide–lanthanide complex has been crystallized with Gd(III) and an ANT–Tb(III) distance of 10 Å was determined [119].

1.15.5   Lanthanides and κ^2

An appealing property of the lanthanides is that not only do they have multiple dipole moments (making their emission effectively isotropic), but one can also safely assume that they precess rapidly during their highly extended excited lifetimes (about 1.8 ms). This means that for a lanthanide donor–acceptor pair such as Tb(III):Nd(III), κ^2 will certainly be 2/3. Where only the donor is a lanthanide (i.e. where the emission is effectively isotropic), then κ^2 can vary in the range 1/3 to 4/3, which produces a distance uncertainly of no more than 12%. Emission spectra of a Tb-chelate and a Eu-chelate are illustrated in Fig. 1.4b (from [128]), as well as the decay of Tb luminescence with time. The structure of the lanthanide chelates is also reproduced.

Selvin *et al.* [129] recently reported experiments in which a distance of 72 Å was determined using a donor–acceptor pair with a Förster distance of 58 Å. They used a Tb(III)-chelate as the donor attached to the regulatory light chain of myosin, and 5-tetramethylrhodamine iodoacetamide (TMRI) as the acceptor attached to the most reactive cysteinyl on the myosin heavy chain. In this case, the anisotropy of the acceptor was high, indicating that the TMRI probe was restricted to a cone angle wobble of about 10°, but even in this case the maximum possible error was $\leqslant 10\%$. Given the very long lifetime of the Tb-chelate (about 1.8 ms), κ^2 is likely to be close to 2/3 because of the high probability that the probe will randomly precess during this extended excited state.

1.15.6 Lanthanide–Trp

In membrane protein channels, Tb(III) is usually employed as a donor probe of Ca(II)-binding sites [35]. In a study of the 33 kDa protein of photosystem 2 from plant chloroplasts [161], the emission of Tb(III) (no sensitizer present) was markedly enhanced when bound to this protein, due to nonradiative FRET between a Trp residue and the bound Tb(III). The average distance between the Trp and the Tb(III) was 10.5 Å. The use of pH titration indicated that Glu and/or Asp carboxyl side groups were involved in the Ca(II) binding. These experiments are appealing, because they involve minimum modification of the protein.

Calbindin D28K protein binds three moles of Tb(III). Veenstra *et al.* [148] determined which of its six EF-hand structures were responsible for Tb(III) binding by constructing three mutants, one lacking EF-hand 2, another lacking EF-hands 2 and 6, and a third containing only EF-hands 3 and 4. Binding properties were studied by FRET using Trp residues as donors, showing that two of the EF-hand domains were not essential for Ca binding, and that their deletion does not alter the pattern of Tb(III) binding. Using FRET from Trp residues, competition experiments using Ca ions demonstrated that one of the high-affinity Tb-binding sites had a greater affinity than the other two, and that it was filled first.

1.15.7 Lanthanide–lanthanide

In calbindin, FRET experiments from Tb(III) to Ho(III) have shown that two of the Tb-binding sites are close while the third site is more distant [148]. Inter-metal ion distances have also been determined using laser-induced luminescence of Eu(III) as a donor and Nd(III) as the acceptor. Both ions replace Ca in the calcium-binding protein, calmodulin. Eu(III) lifetimes (1–2 ms) were measured and the uncertainty in the calculated R values were primarily due to estimates of the quantum yield rather than the orientation factor, which in this case can safely be assumed to be 2/3. The R_0 values for this donor–acceptor pair at the four Ca sites were 10–11 Å and the distances were determined to be 11–12 Å. These results are in excellent agreement with those measured from the X-ray structures (10.9–12.4 Å) [16].

1.15.8 Lanthanides to map protein surfaces

In an unusual experimental twist, Yamamoto *et al.* [160] measured the electro-static potentials around specific and localized portions of protein surfaces. Diffusion-enhanced FRET was used to distinguish donors from freely diffusing acceptors, thus quantifying the sensitivity to the electrostatic potential around the donor with charged acceptors in a controled environment. FRET from an excited Tb(III) chelate of known electric charge was used to sense a series of acceptors of different charges but bearing the same chromophore group. The rate of energy transfer was calculated theoretically as well as experimentally (by time-resolved detection of Tb luminescence). The authors then studied the electrostatic conditions around two specific sites on the surface of actin mole-cules. A negative potential was found around the myosin-binding site, which was nearly neutralized by the addition of myosin S-1. On the other hand, the potential at the phalloidin-binding site was not significantly affected. This approach can be modified for the more general problem of mapping protein–protein interaction sites.

1.16 MEASUREMENTS IN RADIALLY SYMMETRICAL SYSTEMS

In radially symmetrical assemblies of proteins such as actin, myosin, and intermediate filaments, it is possible to precisely determine the radial coordinate of probes bound to specific loci. The first application of this technique exam-ined the dynamic equilibrium of actin polymerization [141, 151]. Three years later, Miki *et al.* [103, 106, 108] calculated the radial coordinates of Cys-10 and the nucleotide-binding site (NUC). Kasprzak *et al.* [84] subsequently took the method a step further by studying the changes in the radial coordinate for probes bound to actin when a second protein (in this case, the myosin S-1 head) interacts with F-actin. All of these findings led to the proposal of an actin filament model based on FRET data [113]. Some of the early FRET measurements overestimated the radial coordinates and, more recently, we have shown [110] that labeling the protein with large extrinsic probes can modify the assembly of the monomer and therefore underestimate FRET efficiency. This artifact can be limited by careful selection of the probes. Smaller probes such as IAEDANS and DABMI can be randomly assembled by phalloidin, while large probes such as 5-iodoacetamide fluorescein (5-IAF) and 5-eosin maleimide do not seem to randomly assemble, even in the presence of phalloidin. Thus, the addition of phalloidin appears to induce random co-assembly of monomers labeled with smaller probes and unlabelled monomers, but it fails to randomly assemble actins with the larger probes [110].

In a helically symmetrical system it is also possible to examine the effects of the binding of protein and other ligands on radial coordinates. For example, we recently showed that binding of myosin S-1 subfragments to actin filaments significantly increases the radial coordinates of probes attached to Cys-374 [109].

1.16.1 Determination of radial coordinates

Before determining the radial coordinate of a locus in a protein filament, some basic knowledge of the system is required. For example, the translation and rotation of the monomer in the helix can usually be obtained by electron microscopy. Armed with this information, the radii (r) can be calculated from the distances (R_n) that separate a donor and an acceptor on nth adjacent monomers, using the expression

$$R_n = [r^2(1 - \cos(n\theta))^2 + r^2 \sin^2(n\theta) + (nT)^2]^{1/2}, \qquad (1.7)$$

where θ and T are the rotation (in degrees) and the translation between two consecutive monomers, respectively. Care should be taken to select probe pairs with a sufficiently large R_0 (see Table 1.1) to allow FRET between adjacent monomers.

Each protein sub-unit must be separately labeled at a homologous (i.e. identical) site with either a donor or an acceptor probe, and the two populations of labeled monomers are then mixed to obtain different acceptor molar ratios. It should be insured that the probes do not perturb the ability of the monomers to assemble. If the labeling does disturb the assembly process, the result could be a polymer in which the donor and acceptor probes are non-randomly distributed, leading to an overestimation of the radial position of the probes [110].

In a randomly assembled polymer, the probability of a donor being surrounded by 0 to nth acceptor-labeled monomers is given by a simple polynomial function and depends on the acceptor molar ratio. The distance between the donor and the surrounding acceptors can be calculated for a given radius (Eqn 17.1) using

$$E_k = \frac{R_0^6}{R_0^6 + 1/\sum_N R_n^{-6}}, \tag{1.8}$$

where N is the number of acceptors for the kth arrangement of acceptors around a donor. Hence we can compute the FRET efficiency for each (kth) arrangement. The energy transfer efficiency in a filament (E_t) is the sum of the transfer efficiencies for each kth arrangement (E_k) multiplied by the probability of each arrangement (δ_k). The sum is taken over all possible arrangements (K) of acceptors around a donor and is given by

$$E_t = \sum_{k=0}^{K} \delta_k E_k. \tag{1.9}$$

When this equation is computed for each acceptor molar ratio (0, 0.01, ..., 1.0), one can build a series of theoretic curves of FRET efficiency versus acceptor molar ratio for different radii (Fig. 1.5).

The FRET efficiencies can now be determined and the data plotted for the different acceptor molar ratios together with the theoretical curve for each radius, to find which radius best fits the data. A proven experimental design involves the preparation of separate solutions of different acceptor molar ratios and one solution containing only donor-labeled monomers. The latter is then used to monitor the changes of the donor fluorescence following buffer modifications (addition of salt, changes of pH, addition of proteins, etc.) and to correct the donor fluorescence in the solution containing acceptor-labeled monomers.

If it is not possible to separately label the same amino-acid side chain with donor or acceptor probes, the radial coordinate can still be determined. Knowledge of the radial coordinate of one site (r) and of the intramonomer spatial relationship of that site with respect to the one sought can provide the radial coordinate of the unknown site (Fig. 10.6).

Miki *et al.* [103, 108] devised equations which solve this problem, calculating the radial coordinate of Cys-10 in F-actin knowing the radius of the nucleotide-binding site. The distances between a donor probe and an adjacent acceptor probe can be calculated using the following equations:

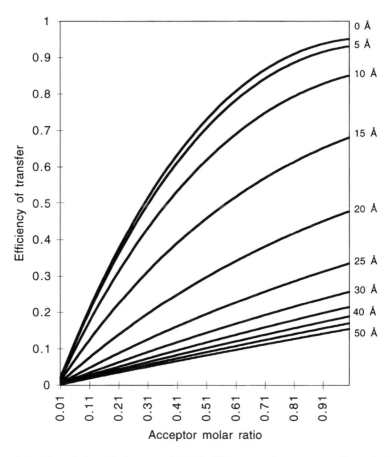

Figure 1.5   The relationship between FRET efficiency and acceptor molar ratio (concentration of acceptor-labeled monomers versus the total concentration of monomer in the solution), calculated for multiple radial coordinates (0–50 Å) of probes bound to a specific amino acid in a helically symmetrical polymer

$$R_n^2 = (ur \cos(n\theta) - r \cos \Phi)^2 + (ur \sin(n\theta) - r \sin \Phi)^2 + (nT + h)^2 \quad (1.10)$$

and

$$R_{\text{intra}}^2 = (ur - r \cos \Phi)^2 + r^2 \sin^2 \Phi + h^2, \quad (1.11)$$

where h, Φ, and R_{intra} are the height, angle, and intramonomer distance between the two probes bound to nonhomologous sites respectively; and ur is the unknown radius: n, T, and θ have already been defined.

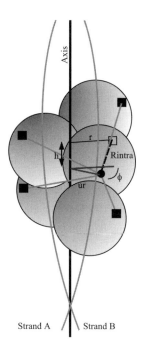

Figure 1.6 A schematic representation of a filament. The intramonomer spatial relationship between two nonhomologous residues (■ and ●) is determined by the intramonomer distance (R_{intra}), the angle (ϕ), and the height (h) between the two residues along the filament axis and their respective radial coordinates (r and ur). When all parameters are known except one radial coordinate, it is possible to calculate the distance (represented by the straight gray lines) separating the donor probe (●) to the acceptor probes (■) using Eqns 1.10 and 1.11. Then, from the efficiency of transfer between the donor on one monomer and four acceptors on adjacent monomers (two in the same strand as the donor and two in the opposite strand), one can calculate (Eqns 1.8 and 1.9) the unknown radial coordinate

1.16.2 $\kappa^2 = 2/3$ in F-actin

We are fortunate with the actin filament in that we can assume $\kappa^2 = 2/3$ with minimal risk of error. This is due to the fact that in a filament, we have multiple (up to four) acceptors around a donor, each acceptor having a different orientation. This results in the averaging of κ^2, thereby dramatically reducing the risk of it not being equal to 2/3. In 1992, Censullo *et al.* [21] used a "worst case" approach to the problem, using values for the minima and maxima of the average orientation factor rather than values considered to be statistically probable. In addition, they restricted the amount of freedom of the fluorophore by the use of a simple cone model having a single axis of rotation. They concluded that, due to the multiplicity of acceptors in the F-actin FRET experiments, less motion is required of the chromophores in order to apply $\kappa^2 = 2/3$.

Thus, for F-actin, the potentially troublesome nature of the orientation factor was eliminated and subsequently we were able to show [109] that there is a substantial (4 Å) change in the radial coordinate of the C-terminus of actin when myosin S-1 binds. However, the uncertain nature of the orientation factor should not deter the FRET investigation of helically symmetrical (or other) ordered arrays of proteins, since we know of no instance to date where the assumed value of κ^2 need be other than 2/3.

FRET experiments might also be used to obtain structural information about protein–protein distances in larger assemblies, such as tubulin–tubulin distances in microtubules. Finally, FRET can be employed to study nonhelical crystalline aggregates of proteins, such as the two-dimensional crystalline microcrystals of rhodopsin [157], sarcoplasmic reticulum Ca-ATPase [147], and Ln-induced sheets and tubes of actin [37]. There is no *a priori* reason why this analysis should not be used to measure distances in true protein crystals (e.g. Trp donors) where acceptor substrates can be diffused into the crystals and small ($\geqslant 1$ Å) structural changes detected.

1.17 COMPARISON WITH CRYSTALLOGRAPHIC DISTANCES

In his celebrated review, Stryer [135] demonstrated that molecular distances determined by FRET for tRNA essentially agree with those determined by X-ray crystallography. An empirical approach can be used with proteins other than tRNA. If the protein structure has already been determined from X-ray diffraction (for example, it is known for actin [81]), then we can model the structures of donor and acceptor probes and insert them into the crystal structure. Using conventional molecular graphics, we can directly determine the distances between regions of the probe.

We obtained generally good agreement between FRET distances and crystallographic dimensions for the side chains to which the FRET probes are attached [107]. In this comparison, the molecular dimensions of the probes must be taken in account. Seven loci were modeled and all but one (Cys10) agreed with the atomic coordinates. The data for this exceptional FRET locus was obtained by labeling actin which had been pretreated with 1 M urea, a procedure which probably distorted its structure.

We recently compared (unpublished observations) the crystallographic distances between the four Trp indole rings in actin and modeled acceptor probes (either DDPM or DABMI) attached to Cys374. The Trp–Cys distances determined by FRET agreed with the crystallographic distances to within 1–2 Å. A number of examples of good agreement between FRET distances and the corresponding distances in crystal structures of the same protein (e.g. [16, 53, 89, 95]) have been described in preceeding sections, and the number is steadily increasing to the point at which the consensus is that FRET reports of distances are substantially correct rather than potentially erroneous.

1.18   USING RESONANCE ENERGY TRANSFER TO CONSTRAIN MOLECULAR MODELS

At the beginning of this chapter, we compared FRET with NMR spectroscopy. In fact, there is little comparison when it comes to generating molecular models of small (< 20 kDa) proteins. Unfortunately, there are still many proteins that are of biological interest which are larger than 20 kDa and which have not yet been crystallized. For this group of proteins, FRET spectroscopy is a realistic proposition.

Dewitte *et al.* [33] developed an algorithm that enumerates all the possible conformations of a protein that satisfy a given set of distance restraints. They investigated the dependence of the number of conformations on pairwise distance restraints for several proteins including crambin, pancreatic trypsin inhibitor and ubiquitin. Knowledge of only one or two contacts was sufficient to restrict the number of candidate structures to approximately 1000 conformations. Pairwise r.m.s. deviations of atomic position comparisons between pairs of these 1000 structures revealed that they can be grouped into about 25 families. These results suggest a new approach to assessing alternative protein folds, given a very limited number of distance restraints available from several spectroscopic techniques, including FRET spectroscopy. Thus, FRET may turn out to be both a prospective technique and a sensor of conformational change in proteins whose structures are already known.

1.19   RESONANCE ENERGY TRANSFER WITH SINGLE FLUOROPHORES: NEW WAVE EXPERIMENTS

Throughout the preceeding discussion, the FRET technique has been applied to populations of donors and acceptors where only the *average* transfer efficiency is measured. Today, a single myosin molecule (or its subfragments) can be manipulated to pull on single actin filaments and the resulting force directly measured [77, 78]. The sensitivity of detectors is such that it is now possible to perform a range of solution-state experiments by quantifying the fluorescence intensity of a single molecule labeled with a donor and an acceptor. For example, single myosin "heads" have been observed and single ATP hydrolysis events have been visualized using fluorescent probes [55].

Some of these experiments have already been accomplished. Yanagida (Biomotron Project) in Osaka observed FRET between a donor and an acceptor on a single myosin head (S-1). Conformational changes were detected in monomers of F-actin directly observed in a FRET experiment where an individual doubly labelled actin monomer in a filament of actin interacts with unlabeled myosin. Another possible experiment is the determination of FRET efficiency between a donor-labeled monomer in an otherwise unlabeled assembly of actin monomers and a single acceptor-labeled myosin head. A similar set of experiments can be contemplated for other motor proteins such as kinesin [145].

Ha *et al.* [65] extended the sensitivity of FRET to single probe molecules using tetramethylrhodamine isothiocyanate as the donor and Texas Red as the acceptor linked to 10 or 20 bp lengths of DNA. FRET efficiency was determined from the photodestruction dynamics of either the donor or the acceptor. Where the acceptor was photobleached there was an immediate rise in the intensity of the donor, because it no longer transfers its energy to the acceptor. The inverse experiment, where the donor is photobleached, was also performed, and here a small residual acceptor fluorescence arose due to a small direct excitation of the acceptor – which was used to correct the data. Ha *et al.* performed conventional solution-state FRET experiments and observed a 65% efficiency with the labels on a 10 bp DNA sample, and 32% FRET efficiency for the 20 bp sample (where R_0 is 52 Å). In these single-molecule experiments, these authors observed large variations in efficiency, which could be explained by a number of assumptions, including deviations of κ^2 from its assumed value of 2/3. They acknowledged that experiments on dried samples can and should be extended to solution-state observations of these single probe pair experiments. Nevertheless, the experiment was a pioneering one.

1.20 INTRAMOLECULAR ENERGY TRANSFER IN PROTEINS BOUND TO MEMBRANES

Although this review of FRET and proteins does not attempt to cover the literature concerned with membranes, there are several recent examples where FRET has been used to demonstrate conformational changes in proteins associated with membrane systems. For a detailed analysis of FRET in two-dimensional membrane systems, readers should consult the original paper by Fung and Stryer [56]. A large number of papers have concerned FRET between membrane-bound proteins and acceptors in the surrounding lipid layer. A useful summary is available in Van Der Meer *et al.* [146].

Time-resolved FRET of proteins in membranes offers several advantages, including the ability to distinguish between specific and random protein–protein interactions. Runnels *et al.* [125] specifically labeled the N-terminus of melittin with fluorescein and monitored the monomer–tetramer equilibrium both in solution and in lipid bilayers. Their results suggested that homotransfer may be a promising method for the study of protein oligomerization in membranes.

Synapsin I is a nerve terminal phosphoprotein which interacts with synaptic vesicles and actin in a phosphorylation-dependent manner. Ceccaldi *et al.* [20] used FRET between purified components labeled with fluorescent probes, and showed that binding of synapsin I to actin can be demonstrated when synaptic vesicles are present in the medium. The interaction appears to be modulated by ionic strength and by synapsin I phosphorylation.

1.21   GREEN FLUORESCENT PROTEIN

Certain bioluminescent coelenterates such as the jellyfish *Aequorea victoria* employ a luminescent protein (aequorin) to excite a fluorescent protein, called the Green Fluorescent Protein (GFP), resulting in an energy transfer reaction which produces green light [22]. GFP is large (238 residues, 27.3 kD) and barrel-shaped (27×42 Å). It is a particularly stable protein with a core that contains a cyclized Ser65–Tyr66–Gly67 sequence, which provides the source of delocalized electrons needed for the fluorescence. The development of mutants of the wild-type GFP is increasingly being used as a marker of gene expression because its fluorescence enables a localization of the protein in bacterial, plant and animal (including mammalian) cells. The heterologous expression of GFP cDNA has also meant that it can be mutated to produce proteins with different fluorescent properties. Variants with more intense fluorescence or alterations in the excitation and emission spectra have been produced [69, 130]. These technical developments suggest that specific proteins can be tagged with a GFP probe of the appropriate emission wavelength and intensity. Further, they can be detected at very low concentrations using the new generation of biosensors that are currently under development (see below).

The wild-type GFP has two excitation peaks (λ_{ex} at 395 and 475 nm) and an emission maximum (λ_{em}) at 508 nm, but it is only weakly fluorescent. In addition, it dimerizes, photobleaches and has a number of idiosyncratic features, all of which have prompted a search for mutants with improved spectral properties. One mutant, the Blue Fluorescent Protein (BFP), is blue-shifted ($\lambda_{ex} = 432$ and 453 nm; $\lambda_{em} = 480$ nm; $\Phi = 0.72$) while others are slightly red-shifted ($\lambda_{ex} = 396$ and 504 nm; $\lambda_{em} = 514$ nm; $\Phi = 0.54$) [69]. These and other mutants have opened the way for FRET between BFP and the GFP mutants [143]. In one report, a modified and engineered GFP was linked by a 25-residue peptide to a BFP by concatenating their cDNAs, encoding a polyhis tag and a 25-residue linker. The R_0 for this pair was about 40 Å (see Table 1.1) [70]. In this covalently coupled donor–acceptor pair, the BFP mutant was excited at 368 nm and the GFP mutant emission monitored at 511 nm. Trypsin cleavage of the linker peptide produces a slow increase in blue emission which, in turn, can be used to monitor the activity of proteases. This technology could be applied to automated industrial enzyme production.

GFP and its derivatives have been used for an increasing array of experiments. It has been linked to myosin, allowing single molecules to be directly observed and manipulated, so that the turnover of individual ATP molecules by the myosin ATPase can be observed [78]. A complex of two GFP variants covalently linked by a calmodulin-binding sequence peptide has been used to detect Ca-dependent changes in living cells.

Persechini and his colleagues recently reported [124] the use of FRET with two GFP mutants linked by an 18-residue peptide. This peptide was capable of

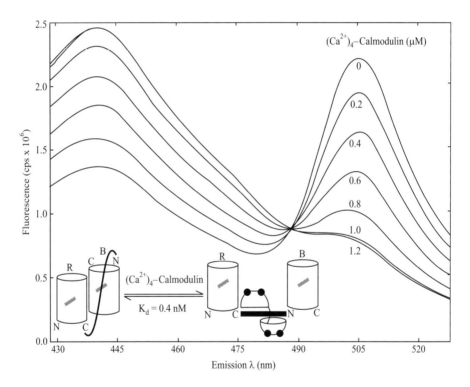

Figure 1.7 Fluorescence emission spectra of two mutants of the green fluorescent protein (GFP) joined by a peptide (GTSSGSSTGA) corresponding to the sequence of the calmodulin-binding domain of myosin light chain kinase (MLCK). FRET occurs between the donor BGFP (labeled B: λ_{ex} max = 382 nm; λ_{em} max 448 nm) and the acceptor RBFP (labeled R: with λ_{ex} max = 495 nm; λ_{em} max = 509 nm). Fluorophores are represented as shaded rectangles within the barrel-shaped GFPs. Binding of calmodulin (total present = 1 μM) to the MLCK peptide occurs (K_d = 0.4 nM) in the presence of free Ca ions and leads to a decrease in FRET efficiency over the Ca concentration range from 50 nM} through 1 μM (reproduced with permission from Romoser *et al.*)

binding calmodulin provided that free Ca ions were present. Their *in vitro* experiment is illustrated in Fig. 1.7, but they were able to perform the same experiment by microinjecting the two GFP mutants linked by the peptide together with small amounts of calmodulin. In this experimental design, the linker peptide is straightened when Ca induces the binding of calmodulin, and this in turn reduces the FRET efficiency between the previously adjacent GFP mutants. This system is capable of detecting the presence of Ca^{2+} (in the concentration range from 50 nM through 1 μM) and is likely to find a number of uses as a biosensor of free calcium ions *in vivo*. A potential limiting factor is the slowness of the response and the possibility that the injected calmodulin may perturb the cells' native calmodulin.

1.22 RESONANCE ENERGY TRANSFER AND BIOSENSORS: A NEW AND PROMISING TECHNOLOGY

Microchips are the basis for some interesting new technology, and combined with fluorescence signals the FRET technique holds the promise of significant advantages over electrical signals. Not only are optical signals immune to electronic interference, but the polychromatic nature of most fluoro-chemical assays provides potentially more useful data. A common difficulty with optical biosensors is their inability to routinely re-calibrate the optical and electronic components of the system throughout the life of the sensor.

With this in mind, Meadows and Schultz [94] constructed an optical fiber biosensor system for glucose. The biochemical assay is based on an homoge-neous singlet/singlet FRET affinity assay. The sensor probe indirectly measures glucose concentrations from the level of fluorescence quenching caused by competition between rhodamine-labeled concanavalin A (receptor) and fluor-escein-labeled dextran. The fluorescein signal was used as an indicator of glucose concentrations and the rhodamine signal for internal calibration. Using this sensor system, dextran concentrations of 0.05 μg/ml and glucose concentrations up to 16 mg/ml were detected with a time response of approx-imately 10 min.

Taylor and his colleagues [57] predicted that fluorescent protein biosensors will form the basis of a new generation of reagents that are capable of reporting specific molecular events in living cells. Creative designs of fluorescent protein biosensors will allow us to measure the molecular dynamics of macromolecules, metabolites, and ions in single cells. These designs have emerged from the fields of synthetic organic chemistry, biochemistry and molecular biology. Taylor *et al.* expect that advances in fluorescent probe design, computer-driven optical instrumentation, and software will permit us to engineer endogenous cellular components that act as functional reporters. These designs will shift the emphasis of molecular measurement from single cell to living tissues and even to the whole organism.

Another example of this technology is the multifrequency phase and modulation technique which is used with total internal reflection fluorescence to probe *in situ* the distribution of free and analyte-bound sites of a fluorescently labelled, silica-immobilized F(ab′) antibody fragment [15]. One can determine the affinity constant for the immobilized antibody, the effects of storage time and the excited-state fluorescence decay parameters, and follow time-dependent changes in antibody conformation using a silica chip.

Total internal reflection fluorescence was also used to analyse rabbit immunoglobulin G (IgG) nonspecifically adsorbed on thin nylon films. FRET measurements were recorded for two systems: (i) adsorbed fluorescein (FITC)-tagged IgG and rhodamine (TRITC) tagged nylon; and (ii) FITC-labeled IgG and TRITC-labeled anti-rabbit IgG. These experiments were used to examine the nylon/IgG/solution interactions as a function of buffer pH and washing conditions [59].

FRET can be used in bioengineering problems such as the monitoring of the purification of an enzyme. Substrates can be labeled with a donor and an acceptor which are deliberately placed close to each other so that the FRET efficiency is 100%. This system can then easily be bound to the surface of a biosensor cell and illuminated by an evanescent field. As the stages in enzyme purification gradually improve, the rate at which the donor–acceptor-labeled substrate is cleaved will be observed as an increase in fluorescence intensity (or lifetime) of the donor probe.

1.23 SHORTCOMINGS

1.23.1 Difficulties of specific labeling

We have already emphasized the requirement to specifically label a protein at a unique locus. This is usually difficult and is sometimes impossible to achieve, mainly because of the lack of absolute specificity of the labels for particular amino-acid side chains. Given the R^{-6} dependence of FRET efficiency on donor–acceptor separation, it is clear that the introduction of even a small molar fraction of an extraneous donor at a locus *closer* to the acceptor probe will dominate the transfer. As a consequence, the donor–acceptor distance will be seriously biased, making it appear much closer than in the absence of the extraneous label. Thus, nonspecific or unintended labeling must be avoided. Measurement of fluorescence lifetimes can be an advantage, because the extraneous label may be recognized by having a significantly longer or shorter lifetime. Note that these comments do not apply to FRET determinations which quantify the acceptor emission (see Section 1.15).

1.23.2   Artifacts and probes

The design of FRET experiments assumes that the probes required to measure transfer efficiency do not perturb the structure and function of the system being measured. In a few instances we know that the label does *not* alter the properties of the amino-acid side chain to which it is attached; pyrene labeling of actin is a good example [30]. But other cases may not be so free of this problem [110]. Considerable controversy has arisen because some investigators claim that the labeling of a particular Cys residue (e.g. Cys707) in myosin strongly modifies its ATPase activity [12]. Others claim that labeling of Cys374 on actin (the site most frequently labeled) distorts the ability of actin to activate myosin ATPase activity. Another example of a perturbation of protein structure is the modification of Cys10 in actin, which has consistently been shown to be significantly perturbed from its location in the crystal structure [113].

Many extrinsic probes are bulky and, consequently, not only does the presence of a large extrinsic label potentially perturb, but the probes themselves are often highly nonpolar. As a result, they may preferentially bind to the hydrophobic pockets of a protein rather than allowing them to precess about their attachment to the protein surface, thus generating a potential artifact. Wherever possible, the surface labeling of proteins by extrinsic probes should be done using small, water-soluble probes. Thus, for FRET experiments on a protein which has not been labeled and characterized previously, it is best to carry out functional tests to show that the labeling *per se* does not significantly alter the properties of the native protein.

1.23.3   Limited accuracy

Where FRET experiments are performed to determine a particular donor–acceptor distance, the accuracy of the measurement is typically about ± 1–2%. This value has several contributing components, including determination of the concentrations of the protein and the labeling ratio. Some of the inaccuracy comes from the background fluorescence that is present in many samples. Sensitized emission of lanthanide ions overcomes this latter difficulty, because of the sharp emission bands which can be determined with great accuracy, and because they occur in regions of the spectrum in which there is no contribution from acceptor probes.

1.23.4   The orientation factor κ^2

Uncertainty in the value of κ^2 in the Förster equation has traditionally been invoked as a possible explanation for deviations between observed and expected distances. We know of no instance in which κ^2 has been shown to be responsible for significant errors in donor–acceptor distances. However, this does not mean that κ^2 can be completely ignored, and a separate chapter of this volume (Chapter 4) has been devoted to this question.

1.24   THE FUTURE OF FRET

It is likely that it will soon be possible to model FRET probes into existing protein databases and to accurately position these probes so that conformational changes be more readily interpreted. Considerable advances have been made with the lanthanide chelates, including the development of new chelates with leaving groups capable of providing a high degree of confidence that they are attached to specific side chains. These lanthanide luminescence sensitizers promise to extend the upper range of FRET distances well beyond the current maximum of 100 Å, while the lanthanide–lanthanide donor–acceptor pairs have already extended the lower range.

In principle, intrinsic probes offer the best opportunity for determining conformational changes in proteins and an increasing effort will be made to take advantage of those probes which produce minimal perturbation of the native conformation. It is now possible to engineer a fluorescent probe into a specific site and this will significantly enhance the use of FRET *in vivo*. We envisage that the 'problem' of κ^2 will disappear. Finally, the single-donor and single-acceptor experiments now under way in several laboratories are likely to lead to completely new insights into the way in which we perceive structure and function in proteins.

1.25   SUMMARY

FRET determinations are capable of accurately measuring distances that separate donor and acceptor probes. By judicious selection of the donor–acceptor probe pair, it is possible to measure distances ranging from <5 Å through to >100 Å. The accuracy of most FRET measurements is about 1%, limiting the effective distance determinations to about $\pm 1.5 \times R_0$. At this stage, the best application of FRET measurements in proteins is for the determinations of *changes* in distance – these can be made with an accuracy of $\pm$ 1–2 Å. FRET data closely correspond to the crystallographic distances where known and this, along with other factors, suggests that in the great majority of protein FRET determinations it is reasonable to assume that $\kappa^2 = 2/3$.

Dedication

This chapter is dedicated to the memory of Gregorio Weber. His seminal contribution to the field of fluorescence resonance energy transfer spectroscopy was inimitable. His quiet charm and encouraging nature will be remembered by all who knew him.

Acknowledgments

The authors were supported by a project grant from the National Health & Medical Research Council of Australia, a U2000 postdoctoral fellowship from The University of Sydney, and a project grant from the Australian Research Council. We thank Robert Clegg, Enrico Gratton, Dave Jameson, Paul Selvin, and the members of the Muscle Research Unit for helpful discussions and comments. We thank Dr. Selvin for permission to reproduce Fig. 1.3 and Dr. Persechini for permission to reproduce Fig. 1.7.

References

1. Aguirre, R., S.-H. Lin, F. Gonsoulin, C.-K. Wang, and H. C. Cheung (1989). Characterization of the ethenoadenosine diphosphate binding site of myosin subfragment 1. Energetics of the equilibrium between two states of nucleotide. S1 and vanadate-induced global conformation changes detected by energy transfer. *Biochemistry* 28: 799–807.
2. Ajtai, K., P. J. Ilich, A. Ringler, S. S. Sedarous, D. J. Toft, and T. P. Burghardt 1992. Stereospecific reaction of muscle fiber proteins with the 5′ and 6′ iodoacetamido derivatives of tetramethyl rhodamine: only the 6′ isomer is mobile on the surface of S1. *Biochemistry* 31: 12 431–12 440.
3. Ambler, E., A. Abbott, and W. J. Ball Jr. 1992. Structural dynamics and oligomeric interactions of Na^+, K^+-ATPase as monitored using fluorescence energy transfer. *Biophys. J.* 61: 553–568.
4. Barden, J. A. 1985. Fluorescence energy transfer between Tyr-69 and Cys-374 in actin. *Biochem. Int.* 11: 583–589.
5. Barden, J. A., and C. G. dos Remedios 1984. The environment of the high-affinity cation binding site on actin and the separation between cation and ATP sites as revealed by proton NMR and fluorescence spectroscopy. *J. Biochem.* 96: 913–921.
6. Barden, J. A., and C. G. dos Remedios 1987. Fluorescence resonance energy transfer between sites in G-actin. The spatial relationship between Cys-10, Tyr-69, Cys-374, the high affinity metal and the nucleotide. *Eur. J. Biochem.* 168: 103–109.
7. Barden, J. A., and M. Miki 1986. The distance separating Tyr-69 from the high-affinity nucleotide and metal binding sites in actin. *Biochem. Int.* 12: 321–329.
8. Barden, J. A., M. Miki, and C. G. dos Remedios 1986. Selective labeling of Cys-10 on actin. *Biochem. Int.* 12: 95–101.
9. Bastiaens, P. I., and T. M. Jovin 1996. Microspectroscopic imaging tracks the intracellular processing of a signal transduction protein: fluorescent-labeled protein kinase C beta I. *Proc. Natl Acad. Sci., USA* 93: 8407–8412.

10. Bhandari, D. G., H. R. Trayer, and I. P. Trayer 1985. Resonance energy transfer evidence for two attached states of the actomyosin complex. *FEBS Lett.* 187: 160–166.

11. Bhattacharyya, A., B. Bhattacharyya and S. Roy 1993. A study of colchicine tubulin complex by donor quenching of fluorescence energy transfer. *Eur. J. Biochem.* 216: 757–761.

12. Bobkov, A. A., E. A. Bobkova, E. Homsher, and E. Reisler 1997. Activation of regulated actin by SH1-modified myosin subfragment 1. *Biochemistry* 36: 7733–7738.

13. Boey, W., W. Huang, B. Bennetts, J. Sparrow, and C. G. dos Remedios 1994. Fluorescence resonance energy transfer within the regulatory light chain of myosin. *Eur. J. Biochem.* 219: 603–610.

14. Brauns, T., Z. W. Cai, S. D. Kimball, H. C. Kang, R. P. Haughland, W. Berger, S. Berjukov, S. Hering, H. Glossmann, and J. Striessing 1995. Benzothiapinone binding domain of purified L-type calcium channels – direct labeling using a novel fluorescent diltiazem analogue. *Biochemistry* 34: 3461–3469.

15. Bright, F. V. 1993. Probing biosensor interfaces by multifrequency phase and modulation total internal reflection fluorescence (MPM–TIRF). *Appl. Spectrosc.* 47: 1152–1160.

16. Bruno, J., De W. Horrocks Jr., and K. Beckingham 1996. Characterization of Eu(III) binding to a series of calmodulin binding site mutants using laser-induced Eu(III) luminescence spectroscopy. *Biophys. Chem.* 63: 1–16.

17. Bücher, T., and J. Kaspers 1947. Photochemische Spaltung der Kohlenoxydmyoglobins durch ultraviolette Strahlung. *Biochim. Biophys. Acta* 1: 21–34.

18. Burtnick, L. 1984. Modification of action with fluorescein isothiocyanate. *Biochim. Biophys. Acta* 791: 57–62.

19. Carraway, K. L., J. G. Koland, and R. A. Cerione 1989. Visualization of epidermal growth factor (EGF) receptor aggregation in plasma membranes by fluorescence resonance energy transfer. Correlation of receptor activation with aggregation. *J. Biol. Chem.* 264: 8699–8707.

20. Ceccaldi, P. E., F. Benfenati, E. Chieregatti, P. Greengard, and F. Valtorta 1993. Rapid binding of synapsin-1 to F-actin and G-actin – a study using fluorescence resonance energy transfer. *FEBS Lett.* 329: 301–305.

21. Censullo, R., J. C. Martin, and H. C. Cheung 1992. The use of the isotropic orientation factor in fluorescence resonance energy transfer (FRET) studies of the actin filament. *J. Fluorescence* 2: 141–155.

22. Chalfie, M. 1995. Green fluorescent protein. *Photochem. Photobiol.* 62: 651–656.

23. Chan, S. S., D. J. Arndt-Jovin, and T. M. Jovin 1979. Proximity of lectin receptors on the cell surface measured by fluorescence energy transfer in a flow system. *J. Histochem. Cytochem.* 27: 56–64.

24. Chang, L. S., E. Y. Wen, J. J. Hung, and C. C. Chang 1994. Energy transfer from tryptophan residues of proteins to 8-anilinonaphalene-1-sulfonate. *J. Prot. Chem.* 13: 635–640.

25. Chantler, P. D., and W. B. Gratzer 1975. Effects of specific chemical modification of actin. *Eur. J. Biochem.* 60: 67–72.

26. Chantler, P. D., and T. Tao 1986. Interhead fluorescence energy transfer between probes attached to transitionally equivalent sites on the regulatory light chains of scallop myosin. *J. Mol. Biol.* 192: 87–99.

27. Cheung, H. C., and B. Liu 1984. Distance between nucleotide site and cysteine-373 of G-actin by resonance energy transfer measurements. *J. Muscle Res. Cell Motil.* 5: 65–80.

28. Clegg, R. M. 1996. Fluorescence resonance energy transfer. In *Fluorescence Imaging Spectroscopy and Microscopy*. X. F. Wang and B. Herman, editors. John Wiley, New York, pp. 179–251.

29. Clottes, E., and C. Vial 1996. Discrimination between the four tryptophan residues of MM-creatine kinase on the basis of the effect of N-bromosuccinimide on activity and spectral properties. *Arch. Biochem. Biophys.* 329: 97–103.

30. Cooper, J. A., S. B. Walker, and T. D. Pollard 1983. Pyrene actin: documentation of the validity of a sensitive assay for actin polymerization. *J. Muscle Res. Cell Motil.* 4: 253–262.

31. Dalbey, R. E., J. Weiel, and R. G. Yount 1983. Förster energy transfer measurements of thiol 1 to thiol 2 distances in myosin subfragment 1. *Biochemistry* 22: 4696–4706.

32. Das, T. K., and S. Mazumdar. 1995. Conformational substrates of apoprotein of horseradish peroxidase in aqueous solution – a fluorescence dynamics study. *J. Phys. Chem.* 99: 13 283–13 290.

33. Dewitte, R. S., S. W. Michnick, and E. Shakhnovich 1995. Exhaustive enumeration of protein conformations using experimental constraints. *Protein Sci.* 4: 1780–1791.

34. dos Remedios, C. G. 1977. Ionic radius selectivity of skeletal muscle membranes. *Nature* 270: 750–751.

35. dos Remedios, C. G. 1981. Lanthanide ion probes of calcium-binding sites on cellular membranes. *Cell Calcium* 2: 29–51.

36. dos Remedios, C. G., and R. Cooke 1984. Fluorescence energy transfer between probes on actin and probes on myosin. *Biochim. Biophys. Acta* 188: 193–205.

37. dos Remedios, C. G., and M. J. Dickens. 1978. Actin microcrystals and tubes formed in the presence of gadolinium ions. *Nature* 276: 731–733.

38. dos Remedios, C. G., and P. D. J. Moens 1995. Actin and the actomyosin interface: a review. *Biochim. Biophys. Acta* 1228: 99–124.

39. dos Remedios, C. G., and P. D. J. Moens 1995. Fluorescence resonance energy transfer spectroscopy is a reliable "ruler" for measuring structural changes in proteins – dispelling the problem of the unknown orientation factor. *J. Struct. Biol.* 115: 175–185.

40. dos Remedios, C. G., M. Miki, and J. A. Barden 1987. Fluorescence resonance energy transfer measurements of distances in actin and myosin. A critical evaluation. *J. Muscle Res. Cell Motil.* 8: 97–117.

41. Drewes, G., and H. Faulstich 1991. A reversible conformational transition in muscle actin is caused by nucleotide exchange and uncovers cysteine in position 10. *J. Biol. Chem.* 266: 5508–5513.

42. Dunn, B. M., C. Pham, L. Raney, D. Baayasekara, W. Gillespie, and A. Hsu 1981. Interaction of alpha-dansylated peptide inhibitors with porcine pepsin: detection of complex formation by fluorescence energy transfer and chromatography and evidence for a two-step binding scheme. *Biochemistry* 20: 7206–7211.

43. Eftink, M. R., G. D. Ramsay, L. Burns, A. H. Maki, C. J. Mann, C. R. Matthews, and C. A. Ghiron 1993. Luminescence studies with Trp repressor and its single tryptophan mutants. *Biochemistry* 32: 9189–9198.

44. Eis, P. S., and J. R. Lakowicz 1993. Time-resolved energy transfer measurements of donor–acceptor distance distributions and intramolecular flexibility of a CCHH zinc finger peptide. *Biochemistry* 32: 7981–7993.

45. Epe, B., K. G. Steinhauser, and P. Woolley 1983. Theory of measurement of Förster-type energy transfer in macromolecules. *Proc. Natl Acad. Sci., USA* 80: 2579–2583.

58   Resonance energy transfer in proteins

46. Falson, P., F. Penin, G. Divita, J. P. Lavergne, A. Dipietro, R. S. Goody, and D. C. Gautheron 1993. Functional nucleotide-binding domain in the F(1)F(0)-ATP synthase alpha-subunit from the yeast *Schizosaccaromyces-pombe*. *Biochemistry* 32: 10 387–10 397.
47. Federwisch, M., A. Wollmer, M. Emde, T. Stuhmer, T. Melcher, A. Klos, J. Kohl, and W. Bautsch 1993. Tryptophan mutants of human C5A anaphylatoxin – a fluorescence anisotropy decay and energy transfer study. *Biophys. Chem.* 46: 237–248.
48. Förster, T. 1946. Energiewanderung und Fluoreszenz (Energy transfer and fluorescence). *Naturwissenschafften* 6: 166–175.
49. Förster, T. 1948. Zwischenmolekulare Energiewanderung und Fluoreszenz (Intermolecular energy migration and fluorescence). *Ann. Phys.* 2: 55–75.
50. Förster, T. 1949. Experimentelle und theoretische Untersuchung des zwischenmolekularen Übergangs von Elektronenanregungsenergie (Experimental and theoretical study of the intermolecular transfer of excitation energy). *Z. Naturforsch.* 4A: 321–327.
51. Förster, T. 1965. Delocalized excitation and excitation transfer. In *Istanbul Lectures, Part III: Action of Light and Organic Crystals*. O. Sinanglu, editor. Academic Press, New York, pp. 93–137.
52. Förster, T. 1993. Intermolecular energy migration and fluorescence. In *Biological Physics*. E. V. Mielczarek, E. Greenbaum, and R. S. Knox, editors. American Institute of Physics, New York, pp. 183–221.
53. Fox, T., L. Ferreira-Rajabi, B. C. Hill, and A. M. English 1993. Quenching of intrinsic fluorescence of yeast cytochrome-C peroxidase by covalently-bound and noncovalently-bound quenchers. *Biochemistry* 32: 6938–6943.
54. Franzen, J. S., P. Marchetti, and D. S. Feingold 1980. Resonance energy transfer between catalytic sites of bovine liver uridine diphosphoglucose dehydrogenase. *Biochemistry* 19: 6080–6089.
55. Funatsu, T., Y. Harada, M. Tokunaga, K. Saito, and T. Yanagida 1995. Imaging of single fluorescent molecules and individual ATP turnovers by single myosin molecules in aqueous solution. *Nature* 374: 555–559.
56. Fung, B. K.-K., and L. Stryer 1978. Surface density determination in membranes by fluorescence energy transfer. *Biochemistry* 17: 5241–5248.
57. Giuliano, K. A., P. L. Post, K. M. Hahn, and D. L. Taylor 1995. Fluorescent protein biosensors – measurement of molecular dynamics in living cells. *Ann. Rev. Biophys. Biomolec. Struct.* 24: 405–434.
58. Gold, A. H., and H. L. Segal 1964. The amino acid sequence of a hexapeptide containing an essential sulfhydryl group of a rabbit muscle glyceraldehyde-3-phosphate dehydrogenase. *Biochemistry* 3: 778–782.
59. Grabbe, E. S. 1993. Total internal reflection fluorescence with energy transfer – a method for analyzing IgG adsorption on nylon thin films. *Langmuir* 9: 1574–1581.
60. Griep, M. A., and C. S. McHenry 1990. Dissociation of the DNA polymerase III holoenzyme beta 2 subunits is accompanied by conformational change at distal cysteines 333. *J. Biol. Chem.* 265: 20 356–20 363.
61. Griep, M. A., and C. S. McHenry 1992. Fluorescence energy transfer between the primer and the beta subunit of the DNA polymerase III holoenzyme. *J. Biol. Chem.* 267: 3052–3059.
62. Grossman, S. H. 1983. Resonance energy transfer between the active sites of myocardial-type creatine kinase (isozyme MB). *Biochemistry* 22: 5369–5375.
63. Grossman, S. H. 1989. Resonance energy transfer between the active sites of rabbit muscle creatine kinase: analysis by steady-state and time-resolved fluorescence. *Biochemistry* 28: 4894–4902.

64. Grossman, S. H., H. M. France, and J. R. Mattheis 1992. Heterogeneous flexibilities of the active site domains of homodimeric creatine kinase: effect of substrate. *Biochim. Biophys. Acta* 1159: 29–36.

65. Ha, T., Th. Enderle, D. F. Ogletree, D. S. Chemla, P. R. Selvin, and S. Weiss 1996. Probing the interaction between two single molecules: Fluorescence resonance energy transfer between a single donor and a single acceptor. *Proc. Natl Acad. Sci. USA* 93: 6264–6268.

66. Hambly, B. D., J. A. Barden, M. Miki, and C. G. dos Remedios 1986. Structural and functional domains on actin. *BioEssays* 4: 124–128.

67. Haugland, R. P. 1996. *Handbook of Fluorescent Probes and Research Biochemicals.* Molecular Probes Inc., Oregon.

68. Haugland, R. P., J. Yguerabide, and L. Stryer 1969. Dependence of the kinetics of singlet–singlet energy transfer on spectral overlap. *Proc. Natl Acad. Sci. USA* 63: 23–30.

69. Hein, R., and R. Y. Tsien 1996. Engineering green fluorescent protein for improved brightness, longer wavelengths and fluorescence resonance energy transfer. *Curr. Biol.* 6: 178–182.

70. Hein, R., D. C. Cubltt, and R. Y. Tsien 1995. Improved green fluorescence. *Nature* 373: 663–664.

71. Hiratsuka, T. 1983. A new ribose-modified analog of adenine and guanine nucleotides available as substrates for various enzymes. *Biochim. Biophys. Acta* 742: 496–508.

72. Hiratsuka, T. 1992. Movement of Cys-697 in myosin ATPase associated with ATP hydrolysis. *J. Biol. Chem.* 267: 14 941–14 948.

73. Hiratsuka, T. 1992. Spatial proximity of ATP-sensitive tryptophanyl residue(s) and Cys-697 in myosin ATPase. *J. Biol. Chem.* 267: 14 949–14 954.

74. Huang, K.-H, R. H. Fairclough, and C. R. Cantor 1975. Singlet energy transfer studies of the arrangement of proteins in the 30 S *Escherichia coli* ribosome. *J. Mol. Biol.* 97: 443–470.

75. Hudson, E. N., and G. Weber 1973. Synthesis and characterization of two fluorescent sulfhydryl reagents. *Biochemistry* 12: 4154–4161.

76. Isaac, V. E., L. Patel, T. Curran, and C. Abateshen 1995. Use of fluorescence resonance energy transfer to estimate intramolecular distances in the MSX-1 homeodomain. *Biochemistry* 34: 15 276–15 281.

77. Ishijima, A., H. Kojima, H. Higuchi, Y. Harada, T. Funatsu, and T. Yanagida 1996. Multiple- and single-molecule analysis of the actomyosin motor by nanometer piconewton manipulation with a microneedle: Unitary steps and forces. *Biophys. J.* 70: 383–400.

78. Iwane, A. H., T. Funatsu, Y. Harada, M. Tokunaga, O. Ohara, S. Morimoto, and T. Yanagida 1997. Single molecular assay of individual ATP turnover by a myosin–GFP fusion protein expressed *in vitro. FEBS Lett.* 407: 235–238.

79. Jez, J. M., B. P. Schlegel, and T. M. Penning 1996. Characterization of the substrate binding site in rat liver 3 alpha-hydroxysteroid / dihydrodiol dehydrogenase. The roles of tryptophans in ligand binding and protein fluorescence. *J. Biol. Chem.* 271: 30 190–30 198.

80. Johnson, D. A., J. G. Voet, and P. Taylor 1984. Fluorescence energy transfer between cobra alpha-toxin molecules bound to the acetylcholine receptor. *J. Biol. Chem.* 259: 5717–5725.

81. Kabsch, W., H. G. Mannherz, D. Suck, E. F. Pai, and K. C. Holmes 1990. Atomic structure of the actin: DNase I complex. *Nature* 347: 37–49.

82. Kaspryzk, P. G., P. M. Anderson, and J. J. Villafranca 1983. Fluorescence energy transfer experiments with *Escherichia coli* carbamoyl phosphate synthetase. *Biochemistry* 22: 1877–1882.
83. Kasprzak, A. A., P. Chaussepied, and M. F. Morales 1989. Location of a contact site between actin and myosin in the three-dimensional structure of the acto-S1 complex. *Biochemistry* 28: 9230–9238.
84. Kasprzak, A. A., R. Takashi, and M. F. Morales 1988. Orientation of the actin monomer in the F-actin filament: radial coordinate of glutamine-41 and effect of myosin subfragment-1 binding on the monomer orientation. *Biochemistry* 27: 4512–4522.
85. Kella, N. K. D., D. D. Roberts, J. A. Shafer, and I. J. Goldstein 1984. Fluorescence energy transfer studies on lima bean lectin. Distance between the subunit hydrophobic binding site and the thiol group essential for carbohydrate binding. *J. Biol. Chem.* 259: 4777–4781.
86. Lakey, J. H., D. Baty, and F. Pattus 1991. Fluorescence energy transfer distance measurements using site-directed single cysteine mutants. The membrane insertion of colicin A. *J. Mol. Biol.* 218: 639–653.
87. Lakowicz, J. R., I. Gryczynski, G. Laczko, W. Wiczk, and M. L. Johnson 1994. Distribution of distances between the tryptophan and the N-terminal residue of mellitin in its complex with calmodulin, troponin C, and phospholipids. *Protein Sci.* 3: 628–637.
88. Leder, R. O., S. L. Helgerson, and D. D. Thomas 1989. The transverse location of the retinal chromophore in the purple membrane by diffusion-enhanced energy transfer. *J. Mol. Biol.* 209: 683–701.
89. Maliwal, B. P., J. R. Lakowicz, G. Kupryszewski, and P. Rekowski 1993. Fluorescence study of conformational flexibility of RNase S-peptide – distance distribution, end-to-end diffusion, and anisotropy decays. *Biochemistry* 32: 12 337–12 345.
90. Marriott, G., W. R. Kirk, and K. Weber 1990. Absorption and fluorescence spectroscopic studies of the Ca-dependent lipid-binding proteins p36: the anexin repeat as the Ca-binding site. *Biochemistry* 29: 7004–7011.
91. Marsh, D. J. and S. Lowey 1980. Fluorescence energy transfer in myosin subfragment-1. *Biochemistry* 19: 774–784.
92. Martensson, L. G., P. Jonasson, P. O. Freskgard, M. Svensson, U. Carlsson, and B. H. Jonsson 1995. Contribution of individual tryptophan residues to the fluorescence spectrum of native and denatured forms of human carbonic anhydrase II. *Biochemistry* 34: 1011–1021.
93. Mathis, G. 1993. Rare earth cryptates and homogeneous fluoroimmunoassays with human sera. *Clin. Chem.* 39: 1953–1959.
94. Meadows, D. L., and J. S. Schultz 1993. Design manufacture and characterization of an optical fiber glucose affinity sensor based on an homogeneous fluorescence energy transfer assay system. *Analyt. Chim. Acta* 280: 21–30.
95. Mely, Y., N. Jullian, N. Morellet, H. Derocquigny, C. Z. Dong, E. Piemont, B. P. Roques, and D. Gerard 1994. Spatial proximity of the HIV-1 nucleocapsid protein zinc fingers investigated by time-resolved fluorescence and fluorescence resonance energy transfer. *Biochemistry* 33: 12 085–12 091.
96. Miki, M. 1987. The recovery of polymerizability of Lys-61-labelled actin and resonance energy transfer measurements. *Eur. J. Biochem.* 164: 229–235.
97. Miki, M. 1990. Resonance energy transfer between points in a reconstituted skeletal muscle thin filament. A conformational change of the thin filament in response to a change in Ca ion concentration. *Eur. J. Biochem.* 187: 155–162.

98. Miki, M. 1991. Detection of conformational changes in actin by fluorescence resonance energy transfer between tyrosine-69 and cysteine-374. *Biochemistry* 30: 10 878–10 884.

99. Miki, M., and T. Iio 1984. Fluorescence energy transfer measurements between the nucleotide binding site and Cys-373 in actin and its application to the kinetics of polymerization. *Biochim. Biophys. Acta* 790: 201–207.

100. Miki, M., and K. Mihashi 1978. Fluorescence energy transfer between ε-ADP at nucleotide binding site and N-(4-dimethyl-amino-3, 5-dinitrophenyl)-maleimide at Cys-373 of G-actin. *Biochim. Biophys. Acta* 533: 163–172.

101. Miki, M., and P. Wahl 1984. Fluorescence energy transfer between points in actosubfragment-1 rigor complex. *Biochim. Biophys. Acta* 790: 275–283.

102. Miki, M., and P. Wahl 1985. Fluorescence energy transfer between points in G-actin: the nucleotide binding site, the metal binding site and Cys-373 residue. *Biochim. Biophys. Acta* 828: 188–195.

103. Miki, M., J. A. Barden, and C. G. dos Remedios 1986. Fluorescence resonance energy transfer between the nucleotide binding site and Cys-10 in G-actin and F-actin. *Biochim. Biophys. Acta* 872: 76–82.

104. Miki, M., J. A. Barden, and C. G. dos Remedios 1986. The distance separating Cys-10 from the high affinity metal binding site in actin. *Biochem. Int.* 12: 807–813.

105. Miki, M., C. G. dos Remedios, and J. A. Barden 1987. The spatial relationship between the nucleotide site, Lys-61 and Cys-374 in actin. *Eur. J. Biochem.* 168: 339–345.

106. Miki, M., B. D. Hambly, and C. G. dos Remedios 1986. Fluorescence energy transfer between nucleotide binding sites in an F-actin filament. *Biochim. Biophys. Acta* 871: 137–141.

107. Miki, M., S. I. O'Donoghue, and C. G. dos Remedios 1992. Structure of actin observed by fluorescence resonance energy transfer spectroscopy. *J. Muscle Res. Cell Motil.* 13: 132–145.

108. Miki, M., J. A. Barden, B. D. Hambly, and C. G. dos Remedios 1986. Fluorescence energy transfer between Cys-10 residues in actin filaments. *Biochem. Int.* 12: 725–731.

109. Moens, P. D. J., and C. G. dos Remedios 1997. A conformational change in F-actin when myosin binds: fluorescence resonance energy transfer detects an increase in radial coordinate of Cys-374. *Biochemistry* 36: 7353–7360.

110. Moens, P. D. J., D. Yee, and C. G. dos Remedios 1994. Determination of the radial coordinate of Cys-374 in F-actin using fluorescence resonance energy transfer spectroscopy – effect of phalloidin on polymer assembly. *Biochemistry* 33: 13 102–13 108.

111. Moraczewska, J., P. D. J. Moens, H. Strzelecka-Golaszewska, and C. G. dos Remedios 1995. Fluorescence resonance energy transfer determination of the location of Gln-41 in G-actin, DNase I-G-actin and F-actin. *Biochem. J.* 317: 605–611.

112. Nomanbhoy, T. K., D. A. Leonard, D. Manor, and R. A. Cerione 1996. Investigation of the GTP-binding GTPase cycle of CDC42HS using extrinsic reporter group fluorescence. *Biochemistry* 35: 4602–4608.

113. O'Donoghue, S. I., B. D. Hambly, and C. G. dos Remedios 1992. Models of actin monomer and filament from fluorescence resonance-energy transfer. *Eur. J. Biochem.* 205: 591–601.

114. O'Malley, S. M., and M. J. McDonald 1994. Monitoring the effect of subunit assembly on the structural flexibility of human apohemoglobin by steady state fluorescence. *J. Prot. Chem.* 13: 561–567.

115. Ostrowski, W. S., R. Kuciel, F. Tanaka, and K. Yagi 1993. Fluorimetric analysis of native, urea-denatured and refolded prostatic acid phosphatase. *Biochim. Biophys. Acta* 1164: 319–326.
116. Papp, S., S. Pikula, and A. Martonosi 1987. Fluorescence energy transfer as an indicator of Ca^{2+}-ATPase interactions in sarcoplasmic reticulum. *Biophys. J.* 51: 205–220.
117. Park, S., K. Ajtai, and T. P. Burghardt 1996. Optical activity of a nucleotide-sensitive tryptophan in myosin subfragment 1 during ATP hydrolysis. *Biophys. Chem.* 63: 67–80.
118. Pesce, A. J., C.-G. Rosén, and T. L. Pasby 1971. *Fluorescence Spectroscopy. An Introduction for Biology and Medicine.* Marcel Decker, New York.
119. Pineda, T., M. J. Thorne, M. G. Gore, and J. E. Churchich 1996. Spectroscopic studies of myo-inositol monophosphatase with a novel fluorescent substrate. *Biochim. Biophys. Acta* 1292: 259–264.
120. Pouchnik, D. J., L. E. Laverman, F. Janiakspens, D. M. Jameson, G. D. Reinhart, and C. R. Cremo 1996. Synthesis and spectral characterization of sulfhydryl-reactive fluorescent probes. *Analyt. Biochem.* 235: 26–35.
121. Rayment, I., H. M. Holden, M. Whittaker, C. B. Yohn, M. Lorenz, K. C. Holmes and R. A. Milligan 1993. Structure of the actin–myosin complex and its implications for muscle contraction. *Science* 261: 58–65.
122. Reshetnyak, Y. K., and E. A. Burstein 1997. Assignment of log-normal components of fluorescence spectra of serine proteases to the clusters of tryptophan residues. *Biofizika* 41: 293–300.
123. Rischel, C., P. Thyberg, R. Rigler, and F. M. Poulsen 1996. Time-resolved fluorescence studies of the molten globule state of apomyoglobin. *J. Mol. Biol.* 257: 877–885.
124. Romoser, V. A., P. M. Hinkle, and A. Persechini 1997. Detection in living cells of Ca^{2+}-dependent changes in the fluorescence emission of an indicator composed of two green fluorescent protein variants linked by a calmodulin-binding sequence – A new class of fluorescent indicators. *J. Biol. Chem.* 272: 13 270–13 274.
125. Runnels, L. W., and S. F. Scarlata 1995. Theory and application of fluorescence homotransfer to mellitin oligomerization. *Biophys. J.* 69: 1569–1583.
126. Schroder, R. R., D. J. Manstein, W. Jahn, H. Holden, I. Rayment, K. C. Holmes, and J. A. Spudich 1993. 3-dimensional atomic model of F-actin decorated with Dictyostelium myosin-S1. *Nature* 364: 171–174.
127. Secrist, J. A. III., J. R. Barrio, and N. J. Leonard 1972. A fluorescent modification of adenosine triphosphate with activity in enzyme systems: 1, N^6-ethenoadenosine-5′-triphosphate. *Science* 175: 646–647.
128. Selvin, P. R. 1996. Lanthanide-based resonance energy transfer. *IEEE J. Selected Topics Quantum Elect.* "Lasers in Biology" 2: 1077–1087.
129. Selvin, P. R., E. Burmeister Getz, and R. Cooke 1998. Luminescence resonance energy transfer measurements on the power-stroke state of actomyosin. *Biophys. J.* 74: 2451–2458.
130. Sirokman, G., T. Wilson, and J. W. Hastings 1995. A bacterial luciferase reaction with a negative temperature coefficient attributable to protein–protein interaction. *Biochemistry* 34: 13 074–13 081.
131. Spencer, R. D., and G. Weber 1969. Measurement of subnanosecond lifetimes with a cross-correlation phase fluorimeter. *Ann. N. Y. Acad. Sci.* 158: 361–376.
132. Steinberg, I. Z. 1971. Long-range nonradiative transfer of electronic excitation energy in proteins and polypeptides. *Ann. Rev. Biochem.* 40: 83–114.

133. Steiner, R. F., D. Juminaga, S. Albaugh, and H. Washington 1996. A comparison of the properties of the binary and ternary complexes formed by calmodulin and troponin C with two regulatory peptides of phosphorylase kinase. *Biophys. Chem.* 59: 277–288.

134. Stryer, L. 1960. Energy transfer in proteins and polypeptides. *Radiat. Res.* Suppl. 2: 432–451.

135. Stryer, L. 1978. Fluorescence energy transfer as a spectroscopic ruler. *Ann. Rev. Biochem.* 47: 819–846.

136. Stryer, L., and R. P. Haughland 1967. Energy transfer: a spectroscopic ruler. *Proc. Natl Acad. Sci., USA* 58: 719–726.

137. Sytnik, A., and I. Litvinyuk 1996. Energy transfer to a proton-transfer fluorescence probe: tryptophan to a flavonol in human serum albumin. *Proc. Natl Acad. Sci., USA* 93: 12 959–12 963.

138. Takashi, R. 1979. Fluorescence energy transfer between subfragment 1 and actin points in the rigor complex of acto subfragment-1. *Biochemistry* 18: 5164–5169.

139. Takashi, R. 1988. A novel actin label. A fluorescent probe at glutamine 41 and its consequences. *Biochemistry* 27: 938–943.

140. Tao, T., M., Lamkin, and S. S. Lehrer 1983. Excitation energy transfer studies of the proximity between tropomyosin and actin in reconstituted skeletal muscle thin filaments. *Biochemistry* 22: 3059–3066.

141. Taylor, D. L., J. Reidler, J. A. Spudich, and L. Stryer 1981. Detection of actin assembly by fluorescence energy transfer. *J. Cell Biol.* 89: 362–367.

142. Thomas, D. D., W. F. Carlsen, and L. Stryer 1978. Fluorescence energy transfer in the rapid diffusion limit. *Proc. Natl Acad. Sci., USA* 75: 5746–5750.

143. Tsien, R. Y., B. J. Bacskai, and S. R. Adams 1993. FRET for studying intracellular signalling. *Trends Cell Biol.* 3: 242–245.

144. Turcatti, G., K. Nemeth, M. D. Edgerton, U. Meseth, F. Talabot, M. Peitsch, J. Knowles, H. Vogel, and A. Chollet 1996. Probing the structure and function of the tachykinin neurokinin-2 receptor through biosynthetic incorporation of fluorescent amino acids at specific sites. *J. Biol. Chem.* 271: 19 991–20 008.

145. Vale, R. D., T. Funatsu, D. W. Pierce, L. Romberg, Y. Harada, and T. Yanagida 1996. Direct observation of single kinesin molecules moving along microtubules. *Nature* 380: 451–453.

146. Van Der Meer, B. W., G. I. Coker, and S.-Y. Chen 1994. *Resonance Energy Transfer, Theory and Data.* VCH Publishers, New York.

147. Vanderkooi, J. M., A. Lerokomar, H. Nakamura, and A. Martonosi 1977. Fluorescence energy transfer between Ca^{2+} transport molecules in artificial membranes. *Biochemistry* 16: 1262–1267.

148. Veenstra, T. D., M. D. Gross, W. Hunziker, and R. Kumar 1995. Identification of metal binding sites in rat brain calcium-binding protein *J. Biol. Chem.* 270: 30 353–30 358.

149. Vos, R., Y. Engelborghs, J. Izard, and D. Baty 1995. Fluorescence study of the three tryptophan residues of the pore-forming domain of colicin A using multi-frequency phase fluorometry. *Biochemistry* 34: 1734–1743.

150. Wang, Y.-L., and D. L. Taylor 1980. Preparation and characterization of a new molecular cytochemical probe: 5-iodoacetamidofluorescein-labeled actin. *J. Histochem. Cytochem.* 28: 1198–1206.

151. Wang, Y.-L., and D. L. Taylor 1981. Probing the dynamic equilibrium of actin polymerisation by fluorescence energy transfer. *Cell* 27: 429–436.

152. Watala, C., T. Pietrucha, E. Dzieiatkowska, K. Gwozdinski, and C. S. Cierniewski 1993. Tissue-type plasminogen activator induces alterations in structure and

conformation of membrane proteins upon its interaction with human platelets. *Chem.–Biol. Interactions* 89: 115–127.

153. Weber, G. 1960. Fluorescence–polarization spectrum and electronic-energy transfer in proteins. *Biochem. J.* 75: 345–352.

154. Weber, G. 1960. Fluorescence–polarization spectrum and electronic-energy transfer in tyrosine, tryptophan and related compounds. *Biochem. J.* 75: 335–345.

155. Weber, G., and M. Shinitsky 1970. Failure of energy transfer between identical aromatic molecules on excitation at the long wavelength end of the absorption spectrum. *Proc. Natl Acad. Sci., USA* 65: 823–830.

156. Weber, G., and F. W. S. Teale 1957. Determination of the absolute quantum yield of fluorescent solutions. *Trans. Faraday Soc.* 53: 646–655.

157. Wu, C. W., and L. Stryer 1972. Proximity relationships in rhodopsin. *Proc. Natl Acad. Sci., USA* 69: 1104–1108.

158. Wu, P. G., E. James, and L. Brand 1993. Compact thermally-denatured state of a staphlococcal nuclease mutant from resonance energy transfer measurements. *Biophys. Chem.* 48: 123–133.

159. Xia, J. L., P. L. Dubin, Y. Morishima, T. Sato, and B. B. Muhoberac 1995. Quasielastic light scattering, electrophoresis, and fluorescence studies of lysozyme-poly (2-acrylamido-methylpropyl sulfate) complexes. *Biopolymers* 35: 411–418.

160. Yamamoto, T., S. Nakayama, N. Kobayashi, E. Munekata, and T. Ando 1994. Determination of electrostatic potential around specific locations on the surface of actin by diffusion-enhanced fluorescence resonance energy transfer. *J. Mol. Biol.* 241: 714–731.

161. Zhang, L. X., J. Wang, and H. G. Liang 1995. Investigation of the Ca^{2+}-binding sites on the 33 KDa protein in photosystem 2 using Tb^{3+} as a fluorescence probe. *Photosynthetica* 31: 203–208.

Unified theory of radiative and radiationless energy transfer

Gediminas Juzeliūnas[1] and David L. Andrews[2]

[1]Institute of Theoretical Physics and Astronomy, Lithuania
[2]University of East Anglia, UK

2.1 INTRODUCTION

The resonance transfer of energy between chemical species separated beyond wavefunction overlap has, until quite recently, commonly been regarded as being mediated by one of two distinct mechanisms: radiationless transfer, generally associated with the names of Perrin [46], Förster [23], Dexter [19] and Galanin [24, 25], and radiative transfer. The former applies over short distances and is characterized by an inverse sixth-power dependence on the separation between the donor and acceptor; the latter is a longer-range effect characterized by an inverse square law. Within the framework of a unified theory [4, 6, 7, 8, 10, 11, 26, 37, 38, 53], it has emerged that both are but limits of a more general mechanism which operates over all distances. This includes an intermediate range over which some degree of competition between the two traditional mechanisms might have been envisaged – and where it transpires that a third, previously hidden, interaction gains equal prominence. However, the new approach offers a number of other advantages beyond its greater compass. In particular, it can properly accommodate the dielectric influence of the medium across which energy transfer takes place [37], it lends itself to the rigorous analysis of energy transfer dynamics [38, 52], and it can be incorporated into stochastic theories of ensemble energy transport [15]. Further, it is a theory

Resonance Energy Transfer. Edited by David L. Andrews and Andrey A. Devidov.

which has successfully expedited the resolution of a number of serious conceptual and other problems latent in earlier treatments. It is the purpose of this chapter to summarize the key features of the modern theoretic development, to identify results in a form amenable to direct implementation, and to highlight the paradigm shift in the conceptualization of resonance energy transfer.

2.2 BACKGROUND

2.2.1 Theoretical framework

The framework within which the unified theory naturally emerges is quantum electrodynamics (QED) [28], best implemented in the molecular formulation largely due to Craig and Thirunamachandran [16]. This is a theory in which both matter and radiation are subject to quantum development, in contrast to the more familiar semiclassical approach where radiation is treated as a classical electromagnetic field. Quantum electrodynamics is in fact the only theory in which the photon concept has any legitimacy, despite the latter's invocation at some point in almost every semiclassical description. It is a theory in which retardation is also naturally accommodated, reflecting the finite speed of signal propagation. It is such retardation features, for example, which are responsible for modifying at mesoscopic distances the inverse sixth-power distance-dependence of the London potential (the attractive part of the 6–12 Lennard–Jones potential) to the correct and experimentally verified form given by the Casimir–Polder formula – in which the asymptotic behavior at large distances proves to be of inverse seventh-power form [16, 42]. Lastly, QED is a theory in which matter and radiation, treated on a common footing, together comprise a closed dynamical system, as illustrated in Fig. 2.1 for the simple case of photon absorption. Notwithstanding the theory's intrinsic logical appeal, it is the incorporation of retardation features in particular which vindicates the application to resonance energy transfer. It is this aspect which proves crucial in identifying the link between radiationless and radiative transport of excitation energy – it also clarifies their relationship to the classical model of dipole coupling [5].

2.2.2 Historical development

More than a half century has elapsed since the first pioneering attempts by Kikuchi [39], Fermi [21], Heitler and Ma [30] and Hamilton [27] to address by quantum electrodynamical methods the theory of resonance energy transfer. It was clearly not the intention in the earliest studies, concentrating on the longer-range (far zone) energy transfer, to forge any link between the so-called radiative and radiationless mechanisms. In the 1960s, Avery [10] and Gomberoff and

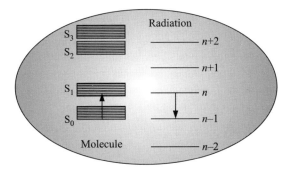

Figure 2.1 A representation of the absorption of light by a molecular system, emphasizing the quantum-electrodynamical view of the molecule plus the radiation as a single closed system. The molecule undergoes an upward transition from its ground to its first excited singlet state, while the radiation experiences a transition from a state with n photons to one with $n - 1$

Power [26] made the first attempts to extend the Förster theory of the short-range radiationless energy transfer [2, 23–25] to arbitrary transfer distances, to include long-range radiative transfer. Such a unified approach to radiative and radiationless energy transfer received a considerable boost of interest in the 1980s [4, 8, 9, 11, 17, 47] and 1990s [6, 7, 14, 18, 35, 52, 53]. Through such studies it has been demonstrated [35, 11, 8, 7, 6, 4] that, in the far zone, the unified mechanism for energy transfer equates to emission of a photon by a donor molecule and subsequent recapture of the photon by an acceptor, thus proving the equivalence of the so-called radiationless and radiative mechanisms. In other developments [44, 54, 18, 38, 52], the time evolution of the transfer dynamics has been explicitly considered within the framework of the unified theory, analysing in detail the transfer dynamics beyond the rate regime.

Another important raft of issues relates to the incorporation within the unified theory of the effects of the surrounding medium. Although a handful of sporadic attempts to accommodate medium effects in excitation transfer has appeared previously [7, 10, 17], it was the case until quite recently that most QED theories totally ignored the influence of such effects. The 1989 treatment by Craig and Thirunamachandran [17], incorporating effects of a third molecule in the energy transfer between a selected pair of molecules, led to a new discussion of the way to include dielectric characteristics. It was suggested from macroscopic arguments that the vacuum dielectric permittivity ϵ_0 entering the rate of excitation transfer *in vacuo* should be replaced by its medium counterpart ϵ to represent the screening. Nevertheless, in using this prescriptive approach other important medium effects, such as local fields, energy losses due to the absorbing medium, and influences on the character of the transfer rates in passing from the near to the far zone, were not considered. More recently, a QED theory was developed by us [37, 38] which systematically dealt with these issues by microscopically including the molecular medium. In contrast to the conventional QED theories in which energy transfer is cast in terms of the intermolecular propagation of virtual photons, the new theory has been formulated by invoking the concept of bath polaritons ("medium-dressed" photons) mediating the process. The medium effects have been shown to play an important role in making the unified theory self-consistent [35, 37, 38]. In fact, it was a systematic treatment of the surrounding medium that made it possible to solve the problem of potentially infinite ensemble rates of energy transfer associated with the far zone inverse square law, as will be discussed in detail in Section 2.5.

2.2.3 Physical basis

Before proceeding further, it is worth saying something about the electronic basis for these interactions. Although usually couched in terms of a coupling between molecules, the theory to be described here is valid for energy transfer between any two species with a distinct electronic integrity – for convenience and generality we can simply refer to them as the donor and acceptor. The theory thus embraces not only intermolecular transfer, but also transfer between chromophores within any larger structure such as a protein or other host structure – provided that these chromophores are not electronically coupled by a resonance structure such as a conjugated chain. As it is generally most convenient to consider electronic properties in terms of electric dipoles, quadrupoles, etc., we shall deal in the common multipolar (Power–Zienau–Woolley) form of QED. A key feature of this formulation is the exact cancelation of all intermolecular coulombic (longitudinal) interactions [16], so that any process involving two or more electrically neutral species invokes the creation and annihilation of virtual photons: in the jargon of particle physics, where QED is more often applied, we would say that these are the gauge bosons that mediate intermolecular interaction. If that sounds daunting, the calculations which it engenders in the case of resonance energy transfer nonetheless prove remarkably straightforward.

2.2.4 Layers of complexity

To develop a usable and general result for the rate of energy exchange in a real system, it is most instructive and convenient to progressively refine a working model by inclusion of salient detail. The plan in the following sections of this chapter is therefore to work as follows:

- The first stage (Section 2.3) is to define the basic concepts of the unified theory, subsequently providing the derivation of the transition matrix element for energy transfer between a pair of chromophores separated by an arbitrary distance R. This leads to identification of the radiative R^{-1} term featured in the quantum amplitude for the transition (the transition matrix element), along with the near-zone R^{-3} contribution. At this stage, neither the effect of the surrounding medium nor the vibrational structure for each of the transfer species is explicitly included.

- The next step (Section 2.4) is to include the effects of energy-level structure for each donor and acceptor. Accordingly, the transfer rate is represented in terms of the overlap integral between the donor fluorescence and the acceptor absorption spectra, establishing a connection with the Förster theory of radiationless energy transfer. The subsequent consideration of the range-dependence of the fluorescence depolarization illustrates the general theory.

- The third element (Section 2.5) is accommodation of the electronic influence of the absorbing molecular medium between the donor and acceptor sites. This is reflected in refractive and dissipative effects on the transfer of excitation energy, rectifying the otherwise anomalous R^{-2} dependence of the transfer rates between the selected pair at large separation. Using the corrected pair rates one can calculate (Section 2.5.2) the total rate of decay of an initially excited molecule due to the energy transfer to the surrounding medium. The contributions due to energy transfer in the far zone are then identified as the rate of spontaneous emission in the absorbing medium, the effects of the surrounding medium being incorporated on a fully microscopic basis.

- Finally, Section 2.6 analyses in detail the transfer dynamics for a pair of species in a dielectric medium. Starting from a general consideration of the time evolution, a connection is first established with the temporal basis of the previous sections that describe the process in terms of the energy transfer rates. Attention is then focused on situations that do not fit into the rate regime, and where different dynamical aspects are apparent.

2.3 THE BASIS OF THE UNIFIED THEORY

2.3.1 General formulation

In the multipolar (Power–Zienau–Woolley) formulation of QED [16, 50, 57], the Coulomb interaction between molecules is represented by the propagation of transverse virtual photons, and it is the coupling between the molecules and the quantized radiation field which is responsible not only for molecular absorption and emission, but also for intermolecular energy transfer. In this formalism, the Hamiltonian for the system can generally be written as

$$H = H_{\text{rad}} + \sum_{X} H_{\text{mol}}(X) + \sum_{X} H_{\text{int}}(X), \qquad (2.1)$$

where H_{rad} is the Hamiltonian for the radiation field and $H_{mol}(X)$ is the Hamiltonian for molecule X, the summations being over all molecules of the system. The coupling between the molecular subsystem and the quantized field is represented by a set of terms $H_{int}(X)$ that describe the interaction of the field with the individual molecules. For general purposes, it is sufficient to express the interaction terms in the electric dipole approximation, although the formalism that we employ is perfectly amenable to the incorporation of higher multipole terms [52]. Thus we write

$$H_{int}(X) = -\varepsilon_0^{-1} \boldsymbol{\mu}(X) \cdot \mathbf{d}^{\perp}(\mathbf{R}_X), \tag{2.2}$$

where $\boldsymbol{\mu}(X)$ is the electric dipole operator of the molecule X positioned at $\mathbf{R}_X$, and $\mathbf{d}^{\perp}(\mathbf{R}_X)$ is the electric displacement operator calculated at the molecular site. The latter displacement operator and the radiation Hamiltonian may be cast as [16]

$$\mathbf{d}^{\perp}(\mathbf{R}) = i \sum_{\mathbf{k},\lambda} \left(\frac{\hbar ck\varepsilon_0}{2V} \right)^{1/2} \mathbf{e}^{(\lambda)}(\mathbf{k}) \left\{ a^{(\lambda)}(\mathbf{k}) e^{i\mathbf{k}\cdot\mathbf{R}} - a^{(\lambda)^{\dagger}}(\mathbf{k}) e^{-i\mathbf{k}\cdot\mathbf{R}} \right\} \tag{2.3}$$

and

$$H_{rad} = \sum_{\mathbf{k},\lambda} a^{(\lambda)+}(\mathbf{k}) a^{(\lambda)}(\mathbf{k}) \hbar ck + e_{vac}, \tag{2.4}$$

where in each expression a sum is taken over radiation modes characterized by wave-vector $\mathbf{k}$ and polarization vector $\mathbf{e}^{(\lambda)}(\mathbf{k})$ (with $\lambda = 1, 2$); $a^{(\lambda)+}(\mathbf{k})$ and $a^{(\lambda)}(\mathbf{k})$ are the corresponding operators for creation and annihilation of a photon, V is the quantization volume, and e_{vac} is the energy of the photon vacuum.

2.3.2   Energy transfer between a donor and acceptor pair in vacuum

In this and the next sections, we shall analyse the energy transfer between a pair of species (to be referred to as donor and acceptor, labeled by D and A) without taking into account the influences of other molecules comprising the surrounding medium [4, 6, 7, 8, 9, 10, 11, 16, 18, 26]. In such a situation the general Hamiltonian Eqn 2.1 reduces to the following:

$$H = H^0 + V, \tag{2.5}$$

with

$$H^0 = H_{\text{rad}} + H_{\text{mol}}(D) + H_{\text{mol}}(A) \tag{2.6}$$

and

$$V = H_{\text{int}}(D) + H_{\text{int}}(A). \tag{2.7}$$

To represent the energy transfer from D to A, the initial and final state vectors are chosen to be the following eigenvectors of the zero-order Hamiltonian H^0:

$$|I\rangle = |D^*\rangle|A\rangle|0\rangle, \quad |F\rangle = |D\rangle|A^*\rangle|0\rangle, \tag{2.8}$$

the corresponding energies being

$$E_I = e_{D^*} + e_A + e_{\text{vac}}, \quad E_F = e_D + e_{A^*} + e_{\text{vac}}, \tag{2.9}$$

where $|0\rangle$ denotes the photon vacuum, $|D^*\rangle$ and $|D\rangle$ label the initial and final states of the donor, $|A\rangle$ and $|A^*\rangle$ are the corresponding state vectors of the acceptor (the asterisk referring to a molecule in an electronically excited state), and e_{D^*} and e_A (e_D and e_{A^*}) are the appropriate energies of the donor and acceptor in their initial (final) states. For generality, the state vectors of donor and acceptor are considered to implicitly contain vibrational contributions that are normally separable from the electronic parts on the basis of the Born–Oppenheimer principle, both for the ground and excited electronic molecular states. The vibronic sublevels will be explicitly included into the theory in the following section.

In passing, we note that the conventional (semiclassical) theories of radiationless energy transfer [2, 23, 25] do not consider photon states, and the energy transfer appears as a first-order process induced by an instantaneous Coulomb interaction. Such an approach is justified in the near zone, i.e. when the distance R of donor–acceptor separation is much less than the reduced wavelength $\lambda = \lambda/2\pi$, λ being the wavelength corresponding to the transfer energy. In the QED formalism employed here, the quantized electromagnetic field is treated on an equal footing to the molecular subsystem, both subsystems comprising a united dynamical system described by the full Hamiltonian H. Here the energy transfer emerges as a second-order process mediated by intermolecular propagation of virtual photons (see Fig. 2.2), and the theory is no longer restricted to the near zone of the separation distances R. The rate of excitation energy transfer, associated with the initial and final states of Eqn. 2.8, can be generally written using the Fermi Golden Rule (see Section 2.6 for more detail; and, for example, [21, 51]), as

$$W_{FI} == \frac{2\pi}{\hbar}|\langle F|T|I\rangle|^2\delta(E_I - E_F), \tag{2.10}$$

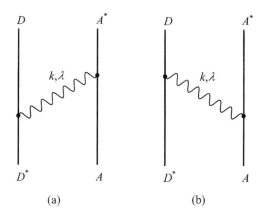

Figure 2.2   The two time-ordered diagrams for resonance energy transfer, time progressing upward. In both cases the virtual photon labeled k, λ mediates the transfer of energy from the initially excited donor D^* to the acceptor A

where T is the transition operator, given by

$$T = V + V \frac{1}{E_1 - H^0 + is} V + \ldots = T^{(1)} + T^{(2)} + \ldots \quad (S \to +0) \quad (2.11)$$

and the higher order terms can be neglected for our purposes. In the QED approach, the first-order term $T^{(1)} \equiv V$, representing photoabsorption and photoemission by individual molecules (Fig. 2.3), does not contribute to the transfer rate as given by Eqns 2.10 and 2.11. It is the second-order contribution

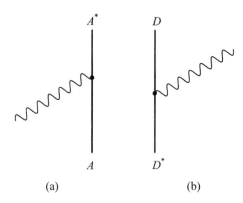

Figure 2.3   Time-ordered diagrams for (a) photoabsorption and (b) photoemission

$T^{(2)}$ that is the leading term responsible for the resonance energy transfer in question. Using Eqns 2.4–2.9, the second-order transition matrix element reads

$$\langle F|T^{(2)}|I\rangle = \frac{1}{\hbar}\sum_{\mathbf{k},\lambda}\sum_{q=1}^{2}\frac{\langle F|V|M_q(\mathbf{k},\lambda)\rangle\langle M_q(\mathbf{k},\lambda)|V|I\rangle}{(-1)^{q+1}cK - ck + is}, \qquad (2.12)$$

with

$$\hbar cK = e_{A^*} - e_A = e_{D^*} - e_D > 0 \qquad (2.13)$$

being the transfer energy. Here $|M_p(\mathbf{k},\lambda)\rangle$ $(q = 1, 2)$ denote the intermediate states in which both donor and acceptor are either in the ground $(q = 1)$ or in the excited $(q = 2)$ electronic states:

$$|M_1(\mathbf{k},\lambda)\rangle = |D\rangle|A\rangle|\mathbf{k},\lambda\rangle \qquad (2.14)$$

and

$$|M_2(\mathbf{k},\lambda)\rangle = |D^*\rangle|A^*\rangle|\mathbf{k},\lambda\rangle, \qquad (2.15)$$

the radiation field being promoted to a one-photon state:

$$|\mathbf{k},\lambda\rangle \equiv a^{(\lambda)^+}(\mathbf{k})|0\rangle. \qquad (2.16)$$

The two types of intermediate state correspond to the two possible sequences of transitions undergone by the donor and acceptor. In the first case $(q = 1)$, the transition $D^* \to D$ precedes the transition $A^* \to A^*$, as in Fig. 2.2a, whereas in the second case $(q = 2)$ one has the opposite ordering, as in Fig. 2.2b. The latter sequence represents an apparently anomalous situation in which the upward transition of the acceptor A is accompanied by the creation of a virtual photon, and the subsequent annihilation of the photon induces the downward transition by the donor $D^* \to D$. Nevertheless, both types of transition must be included in the theory according to the normal rules of time-dependent perturbation theory. The precision with which the law of energy conservation has to apply to the virtual photons is determined by the time–energy Uncertainty Principle: $\delta E \delta t \geqslant \hbar$, where $\delta t = R/c$ is the time necessary for a photon to cover the distance between the donor and acceptor, giving

$$\delta E \geqslant \frac{\hbar cK}{KR}. \qquad (2.17)$$

Consequently, the contribution due to intermediate states of the second type, Fig. 2.2b, is of essential importance in the near zone ($KR \ll 1$), for which the energy uncertainty δE greatly exceeds the energy transferred, $\hbar c K$. At large separations $(KR \gg 1)$, where a virtual photon mediating the interaction between the donor and acceptor exists for a time that appreciably exceeds the duration of the optical cycle, the photon acquires real character: $\delta E \ll \hbar c K$. In such a situation, the contribution associated with Fig. 2.2b diminishes to a point at which it can be considered negligible.

Substituting Eqns 2.2, 2.8, 2.9, and 2.14–2.16 into Eqn 2.12, one arrives at the following expression for the transition matrix element:

$$\langle F|T^{(2)}|I\rangle = \mu_{A_l}^{\text{full}} \, \theta_{lj}^{\text{vac}}(K, \mathbf{R}) \mu_{D_j}^{\text{full}}, \tag{2.18}$$

with

$$\theta_{lj}^{\text{vac}}(K, \mathbf{R}) = \frac{1}{\hbar\varepsilon_0^2} \sum_{\mathbf{k},\lambda} \left[\frac{\langle 0|d_l^\perp(\mathbf{R}_A)|\mathbf{k}, \lambda\rangle\langle \mathbf{k}, \lambda|d_j^\perp(\mathbf{R}_D)|0\rangle}{cK - ck + is} \right. $$
$$\left. - \frac{\langle 0|d_j^\perp(\mathbf{R}_D)|\mathbf{k}, \lambda\rangle\langle \mathbf{k}, \lambda|d_l^\perp(\mathbf{R}_A)|0\rangle}{cK + ck - is} \right] \quad (\mathbf{R} = \mathbf{R}_A - \mathbf{R}_D) \tag{2.19}$$

and

$$\boldsymbol{\mu}_D^{\text{full}} = \langle D|\boldsymbol{\mu}(D)|D^*\rangle, \qquad \boldsymbol{\mu}_A^{\text{full}} = \langle A^*|\boldsymbol{\mu}(A)|A\rangle, \tag{2.20}$$

where implied summation over the repeated Cartesian indices (l and j) is assumed, and where $\boldsymbol{\mu}_D^{\text{full}}$ and $\boldsymbol{\mu}_A^{\text{full}}$ are the transition dipole moments of the donor and acceptor, respectively. The superscript "full" indicates that the molecular state vectors entering the transition dipoles of Eqn 2.20 contain both electronic and vibrational contributions, as made explicit in the following section. Here also $\theta_{lj}^{\text{vac}}(K, \mathbf{R})$ is the tensor for the retarded dipole–dipole coupling between the donor and acceptor in vacuum: modifications to the tensor by the surrounding molecular medium will be considered in Section 2.5. Using Eqn 2.3 for the displacement operator and performing summation over the photon polarizations ($\lambda = 1, 2$), the tensor Eqn 2.19 reduces to

$$\theta_{lj}^{\text{vac}}(K, \mathbf{R}) = \sum_{\mathbf{k}} \frac{(\delta_{lj} - \hat{k}_l\hat{k}_j)k}{2\varepsilon_0 V} \left[\frac{e^{i\mathbf{k}\cdot\mathbf{R}}}{K - k + is'} - \frac{e^{-i\mathbf{k}\cdot\mathbf{R}}}{K + k - is'} \right] \quad (\mathbf{R} = \mathbf{R}_A - \mathbf{R}_D), \tag{2.21}$$

where $\hat{\mathbf{k}} = \mathbf{k}/k$ is the unit vector along the wave-vector $\mathbf{k}$, and $s' = s/c \to +0$ is the new infinitesimal. Replacing the sum over $\mathbf{k}$ by an integral and performing the angular integration, Eqn 2.21 can be written as

$$\theta_{lj}^{\text{vac}}(K, \mathbf{R}) = \left(-\nabla^2 \delta_{lj} + \nabla_l \nabla_j\right) \frac{1}{4\pi^2 \varepsilon_0} G(K, R), \tag{2.22}$$

where

$$G(K, R) = \int_0^\infty \frac{\sin(kR)}{R} \left[\frac{1}{K - k + \mathrm{i}s'} - \frac{1}{K + k - \mathrm{i}s'}\right] \mathrm{d}k \tag{2.23}$$

$$= \frac{1}{2\mathrm{i}R} \int_{-\infty}^\infty \left[\frac{\mathrm{e}^{\mathrm{i}kR}}{K - k + \mathrm{i}s'} - \frac{\mathrm{e}^{\mathrm{i}kR}}{K + k + \mathrm{i}s'}\right] \mathrm{d}k \tag{2.24}$$

is the Green function, the sign of the infinitesimal s' being reversed in the nonresonant term of the original integral in Eqn 2.23 to expand the integration contour to negative values of k in Eqn 2.24. The expanded contour of integration can be closed up by a large semicircle in the upper complex half-plane, subsequently calculating the residue at $k = K + \mathrm{i}s'$ to yield

$$G(K, R) = -\pi \frac{\exp(\mathrm{i}KR)}{R}. \tag{2.25}$$

It is noteworthy that the imaginary infinitesimal featured in the transition matrix element, Eqn 2.11 and the subsequent equations, emerges intrinsically from the time-dependent analysis of the problem, as will be demonstrated in Section 2.6. It is the presence of such an infinitesimal s' that ensures the correct bypassing of poles in Eqns 2.23 and 2.24. Thus, one automatically avoids analytic problems associated with the choice of the integration contour of the Green function, discussed previously [8].

Using Eqn 2.25, the electromagnetic tensor, Eqn 2.22, now takes the final form

$$\theta_{ij}^{\text{vac}}(K, \mathbf{R}) = \frac{K^3 \mathrm{e}^{\mathrm{i}KR}}{4\pi \varepsilon_0} \left[(\delta_{ij} - 3\hat{R}_i \hat{R}_j)\left(\frac{1}{K^3 R^3} - \frac{i}{K^2 R^2}\right) - (\delta_{ij} - \hat{R}_i \hat{R}_j)\frac{1}{KR}\right], \tag{2.26}$$

with $\hat{\mathbf{R}} = \mathbf{R}/R$. The above tensor, whose role in the first-order dispersion forces was much earlier established by Stephen [55], contains the R^{-3} term characteristic of the near zone, the radiative R^{-1} term operating in the far zone, as well as an R^{-2} contribution that plays an important role at critical retardation distances: $KR \simeq 1$. Note that the long- and short-range terms are characterized by different orientational dependencies: this factor will lead to the range dependence of the fluorescence anisotropy, to be discussed in Section 2.4.2.

2.4   SPECTRAL FEATURES

2.4.1   Connection with the Förster theory

According to the Born–Oppenheimer Principle, the molecular state vectors can be separated into electronic and vibrational parts, as

$$|\mathbf{D}^*\rangle = |\mathbf{D}^*_{el}\rangle|\boldsymbol{\varphi}^{(n)}_{D^*}\rangle, \qquad |\mathbf{D}\rangle = |\mathbf{D}_{el}\rangle|\boldsymbol{\varphi}^{(r)}_{D}\rangle, \tag{2.27}$$

$$|\mathbf{A}\rangle = |\mathbf{A}_{el}\rangle|\boldsymbol{\varphi}^{(m)}_{A}\rangle, \qquad |\mathbf{A}^*\rangle = |\mathbf{A}^*_{el}\rangle|\boldsymbol{\varphi}^{(p)}_{A}\rangle, \tag{2.28}$$

where the subscript "el" refers to the electronic part of the state vectors, the indices n, r, m, and p specifying the vibrational, rotational, etc. sublevels of the transfer species D and A. The transition dipole moments, Eqn 2.20, then split into electronic and vibrational contributions, as

$$\boldsymbol{\mu}^{full}_{D} = \langle\mathbf{D}_{el}|\boldsymbol{\mu}(\mathbf{D})|\mathbf{D}^*_{el}\rangle\langle\boldsymbol{\varphi}^{(r)}_{D}|\boldsymbol{\varphi}^{(n)}_{D^*}\rangle = \boldsymbol{\mu}_{D}\langle\boldsymbol{\varphi}^{(r)}_{D}|\boldsymbol{\varphi}^{(n)}_{D^*}\rangle, \tag{2.29}$$

$$\boldsymbol{\mu}^{full}_{A} = \langle\mathbf{A}^*_{el}|\boldsymbol{\mu}(\mathbf{A})|\mathbf{A}_{el}\rangle\langle\boldsymbol{\varphi}^{(p)}_{A^*}|\boldsymbol{\varphi}^{(m)}_{A}\rangle = \boldsymbol{\mu}_{A}\langle\boldsymbol{\varphi}^{(p)}_{A^*}|\boldsymbol{\varphi}^{(m)}_{A}\rangle, \tag{2.30}$$

where, according to the Condon Principle, the dipole operators of the donor and acceptor, $\boldsymbol{\mu}(D)$ and $\boldsymbol{\mu}(A)$ respectively, are assumed not to depend on the vibrational degrees of freedom, $\boldsymbol{\mu}_D$ and $\boldsymbol{\mu}_A$ being the appropriate electronic parts of the transition matrix elements.

Substituting the transition dipoles, Eqns 2.29 and 2.30, into Eqn 2.18, the full rate of donor–acceptor transfer reads, after performing the necessary averaging over the initial molecular states and summing over the final molecular states in Eqn 2.10,

$$W_{DA} = \frac{2\pi}{\hbar}\sum_{n,m,r,p}\boldsymbol{\rho}^{(n)}_{D^*}\boldsymbol{\rho}^{(m)}_{A}|\langle\boldsymbol{\varphi}^{(r)}_{D}|\boldsymbol{\varphi}^{(n)}_{D^*}\rangle\langle\boldsymbol{\varphi}^{(p)}_{A^*}|\boldsymbol{\varphi}^{(m)}_{A}\rangle|^2|\mu_{A_i}\mu_{D_j}|^2|\theta^{vac}_{ij}(K,\mathbf{R})|^2$$

$$\delta\left(e_{D^*_n} + e_{A_m} - e_{D_r} - e_{A^*_p}\right). \tag{2.31}$$

Here, $\boldsymbol{\rho}^{(n)}_{D^*}$ and $\boldsymbol{\rho}^{(m)}_{A}$ are the population distribution functions of the initial vibrational states of the donor and acceptor respectively, the vibrational indices also being included in the energies of the initial and final states that feature in the energy-conserving delta function. In analogy to the Förster theory [2, 25], the pair transfer rate, Eqn 2.31, can be expressed, using Eqn 2.26 for $\theta^{vac}_{ij}(K,\mathbf{R})$, in terms of the overlap integral between the donor and acceptor spectra:

$$W_{DA} = \frac{9}{8\pi c^2\tau_D}\int_0^\infty F_D(\omega)\sigma_A(\omega)\omega^2 g^{vac}(\omega,\mathbf{R})d\omega, \tag{2.32}$$

with

$$g^{vac}(\omega, \mathbf{R}) = \eta_3^2 \frac{c^6}{\omega^6 R^6} + (\eta_3^2 - 2\eta_1\eta_3) \frac{c^4}{\omega^4 R^4} + \eta_1^2 \frac{c^2}{\omega^2 R^2}. \tag{2.33}$$

In the above equations,

$$\sigma_A(\omega) = \frac{\pi\omega\mu_A^2}{3\varepsilon_0 c} \sum_{m,p} \rho_A^{(m)} |\langle \varphi_{A^*}^{(p)} | \varphi_A^{(m)} \rangle|^2 \delta\left(e_{A_p^*} - e_{A_m} - \hbar\omega\right) \tag{2.34}$$

and

$$F_D(\omega) = \frac{\omega^3 \tau_D \mu_D^2}{3\varepsilon_0 \pi c^3} \sum_{n,r} \rho_{D^*}^{(n)} |\langle \varphi_D^{(r)} | \varphi_{D^*}^{(n)} \rangle|^2 \delta(e_{D_n^*} - e_{D_r} - \hbar\omega), \tag{2.35}$$

are, respectively, the cross-section for the acceptor absorption and the donor emission spectra (the latter $F_D(\omega)$ normalized to unity), τ_D being the radiative lifetime of the donor. Here also

$$\eta_q = (\hat{\boldsymbol{\mu}}_A \cdot \hat{\boldsymbol{\mu}}_D) - q\left(\hat{\mathbf{R}} \cdot \hat{\boldsymbol{\mu}}_A\right)\left(\hat{\mathbf{R}} \cdot \hat{\boldsymbol{\mu}}_D\right) \qquad (q = 1, 3) \tag{2.36}$$

are the orientational factors, the carets referring to unit vectors. In passing we note that the orientational factor η_3 that characterizes the near-zone transfer is identical to the kappa whose square is the familiar short-range orientational factor; see, for example, Chapter 4.

The rate equation 2.32 accommodates contributions due to both radiation-less and radiative energy transfer. In the near zone, Eqn 2.35 reduces to the usual Förster rate for nonradiative energy transfer [2, 25]:

$$W_{DA}^{Först} = \frac{9c^4\eta_3^2}{8\pi\tau_D R^6} \int_0^\infty F_D(\omega)\sigma_A(\omega)\omega^{-4} \, d\omega, \tag{2.37}$$

characterized by an R^{-6} distance dependence and the orientational factor η_3^2 (kappa squared). In the far zone, Eqn 2.32 provides the radiative result $W_{DA} \propto \eta_1^2/R^2$. It is noteworthy that the two limiting cases differ not only in their distance dependence, but also in their orientational factors. This leads to a completely different transfer-induced fluorescence depolarization in these two cases, an issue to be discussed in detail in the following subsection. Note also that although the effects of retardation are contained in the pair-rate Eqn 2.32, the result has been derived without taking into account the influence of mole-cules other than the donor and acceptor: the effects of the intervening medium will be considered in Section 2.5.

We conclude this subsection with a remark concerning nonrigid systems that have fast rotational motion of the donor and acceptor. In such a situation, the factor $g^{\text{vac}}(\omega, \mathbf{R})$ that enters Eqn 2.32 should be replaced by its orientational average:

$$g_{av}^{\text{vac}}(\omega, \mathbf{R}) = \frac{2}{9} \left(\frac{3c^6}{\omega^6 R^6} + \frac{c^4}{\omega^4 R^4} + \frac{c^2}{\omega^2 R^2} \right). \tag{2.38}$$

The factor $g_{av}^{\text{vac}}(\omega, \mathbf{R})$ is then related to the excitation transfer function $A(K, R)$ introduced in [8, 4], as

$$g_{av}^{\text{vac}}(\omega, \mathbf{R}) = \left(\frac{4\pi\varepsilon_0 c^3}{3\omega^3} \right)^2 A(\omega/c, R). \tag{2.39}$$

2.4.2   Range-dependence of the fluorescence depolarization

In this subsection we shall consider the polarization character of the system fluorescence [25], applying the unified approach to determine the depolarization of fluorescence through acceptor decay following radiative or radiationless energy transfer. For the usual nonradiative (Förster) mechanism, the transfer rate depends only weakly on the average mutual orientation of the donors and acceptors (see Eqn 2.41 below for the average of the appropriate orientational factor). This is the reason for the well-known and considerable (1/25) reduction of fluorescence anisotropy following a single act of energy transfer in an isotropic or randomly oriented system [2, 25]. By contrast, in the radiative mechanism, the energy transfer between species with parallel transition dipoles is greatly preferred, leading to a smaller loss of polarization in a randomly oriented system (compare Eqn 2.40, in which the angle-dependent term is weighted by a factor of 7]. Consequently, the residual anisotropy following a single act of photon reabsorption is substantially (seven times) greater than in the case of nonradiative transfer.

Here, following [6], a general formula will be derived which connects and accommodates the above limiting cases, also providing results that are valid for intermediate distances where neither radiative nor radiationless transfer dominates. Note that the rotational depolarization is assumed to be negligible. In order to arrive at a formula that exhibits the effects of the relative donor–acceptor orientation in an ensemble, it is necessary to average the pair-rate, Eqn. 2.32, over the orientation of the radius vector $\mathbf{R}$, keeping a fixed mutual orientation between the donor and acceptor. We then obtain the following results for the rotational averages of the orientational factors featured in Eqn 2.33.

$$\overline{\eta_1^2} = \tfrac{1}{15}(7\cos^2\theta + 1), \qquad \overline{\eta_3^2} = \tfrac{1}{5}(\cos^2\theta + 3) \tag{2.40, 2.41}$$

and

$$\overline{\eta_1 \eta_3} = \tfrac{1}{3}\overline{\eta_3^2}, \tag{2.42}$$

where $\cos\theta = \hat{\boldsymbol{\mu}}_A \cdot \hat{\boldsymbol{\mu}}_D$. Substituting the above averages into Eqns 2.32 and 2.33, we thus obtain

$$\overline{W(R)} \propto (3y_6 + y_4 + 7y_2)\cos^2\theta + (9y_6 + 3y_4 + y_2), \tag{2.43}$$

with

$$y_n - \frac{c^{n-2}}{R^n} \int_0^\infty \omega^{(2-n)} \Gamma_D(\omega) \upsilon_A(\omega) \mathrm{d}\omega \quad (n = 2, 4, 6). \tag{2.44}$$

Therefore, the properly normalized function for the orientational distribution of excited acceptors is given by

$$f(\theta, R) = \frac{3}{10} \frac{\left(3 + \widetilde{K^2}R^2 + 7\widetilde{K^4}R^4\right)\cos^2\theta + \left(9 + 3\widetilde{K^2}R^2 + \widetilde{K^4}R^4\right)}{3 + \widetilde{K^2}R^2 + \widetilde{K^4}R^4}, \tag{2.45}$$

with

$$\widetilde{K^n} = \frac{\displaystyle\int_0^\infty (\omega/c)^n F_D(\omega)\sigma_A(\omega)\omega^{-4}\,\mathrm{d}\omega}{\displaystyle\int_0^\infty F_D(\omega)\sigma_A(\omega)\omega^{-4}\,\mathrm{d}\omega}. \tag{2.46}$$

Of special interest is the fluorescence anisotropy, defined by

$$r = \frac{I_{II} - I_\perp}{I_{II} + 2I_\perp}, \tag{2.47}$$

where I_{II} and $I_\perp$ are components of the fluorescence intensity polarized, respectively, parallel to and perpendicular to the polarization of the excitation light. In the case where fluorescence occurs directly from the molecule which absorbs the incident light (the donor), the anisotropy is designated r_0; where fluorescence occurs following a single-step transfer of energy to another molecule (the acceptor), the anisotropy is designated r_1. The value of r_0, if intramolecular relaxation of the donor produces no change of electronic state, has its theoretical maximum of 0.4 [2, 25].

Here it is the result for r_1 which is of principal interest; the fluorescence anisotropy following a chain of energy transfer events can be directly calculated from this result. In terms of r_0, the acceptor anisotropy r_1 can be expressed as

$$r_1 = \langle P_2(\cos \theta) \rangle r_0, \qquad (2.48)$$

where $P_2(\cos \theta) = \frac{1}{2}(3\cos^2 \theta - 1)$ is the second-order Legendre polynomial, and the angular brackets denote the orientational average,

$$\langle P_2(\cos \theta) \rangle = \frac{1}{2} \int_0^\pi P_2(\cos \theta) f(\theta)\sin \theta \, d\theta. \qquad (2.49)$$

Substituting Eqn 2.45 into Eqn 2.49, we obtain the following most general result for the fluorescence anisotropy:

$$r_1(R) = \frac{r_0}{25} \frac{7\widetilde{K^4}R^4 + \widetilde{K^2}R^2 + 3}{\widetilde{K^4}R^4 + \widetilde{K^2}R^2 + 3}. \qquad (2.50)$$

The above equation is valid for arbitrary separations R. As shown in Eqns 2.55 and 2.56 below, the familiar near- and far-zone results are the asymptotes of this formula. It is to be pointed out that, in general, the residual anisotropy depends not only on the transfer distance, but also on the shapes of the spectral lines through the averages $\widetilde{K^2}$ and $\widetilde{K^4}$ featured in Eqn 2.50. However, as the widths of the absorption and emission lines are considerably less than the photon frequency, Eqn 2.50 can be rewritten without significant loss of generality as

$$r_1(R) = \frac{r_0}{25} \frac{7\widetilde{K}^4 R^4 + \widetilde{K}^2 R^2 + 3}{\widetilde{K}^4 R^4 + \widetilde{K}^2 R^2 + 3}, \qquad (2.51)$$

where $\widetilde{K}$ is the averaged value of K calculated by use of Eqn 2.46 with $n = 1$. In the case in which the absorption and emission lines are of Gaussian shape,

$$F_D(\omega) \propto \omega^3 \exp\left[(\omega - \omega_D)^2/2\sigma^2\right], \qquad \sigma_A(\omega) \propto \omega \exp\left[(\omega - \omega_A)^2/2\sigma^2\right], \quad (2.52)$$

we have

$$\widetilde{K^2} = \widetilde{K}^2\left[1 + (\sigma/\tilde{\omega})^2/2\right], \qquad \widetilde{K^4} = \widetilde{K}^4\left[1 + 3(\sigma/\tilde{\omega})^2 + 3(\sigma/\tilde{\omega})^4/4\right].$$
$$(2.53, 2.54)$$

This means that if, for instance, the ratio $\sigma/\tilde{\omega}$ is equal to 0.1, the error made using the relationship of Eqn 2.51 instead of the exact result, Eqn 2.50, is less than a few percent. Further, both formulas provide the correct asymptotes at small and large distances, as follows.

$\tilde{K}R \ll 1$. Here, Eqns 2.50 and 2.51 reproduce the usual Galanin result,

$$r_1 \equiv r_1^{\text{nonrad}} = r_0/25. \tag{2.55}$$

$\tilde{K}R \gg 1$. Here we arrive at the result for the fluorescence depolarization associated with radiative energy transfer from the donor to an acceptor in the far zone:

$$r_1 \equiv r_1^{\text{rad}} = 7r_0/25. \tag{2.56}$$

The distance-dependence of the relative anisotropy r_1/r_0 calculated according to Eqn 2.51 is presented in Fig. 2.4. One can see the anisotropy rise to significant values at distances much less than those normally associated with radiative energy transfer. For instance, with a donor–acceptor separation of $R = 1.5/\tilde{K} = 0.75\lambda/\pi$, the relative anisotropy r_1/r_0 attains the value of 3/25, which is considerably higher than the result for radiationless transfer, as follows from Eqn 2.51.

In connection with multiphoton fluorescence energy transfer, any microscopically disordered system exhibits the same sevenfold increase in fluorescence

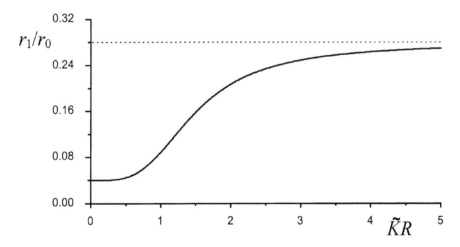

Figure 2.4 The distance-dependence of relative fluorescence anisotropy, displaying the increase associated with the onset of significant retardation on progressing from "radiationless" transfer at short distances to long-range "radiative" transfer. The abscissa values $\tilde{K}R$ signify the donor–acceptor distance R divided by the "reduced wavelength" $\lambda/2\pi$, where λ is the wavelength corresponding to the transferred energy

anisotropy as the donor–acceptor distance increases from the near-zone to the far-zone range. However, the detailed dependence on the relative orientations of the participating donor transition moments adds considerable complexity to the results, even in the two-photon case [3].

2.5   REFRACTION AND DISSIPATION

In deriving the pair rate, Eqn 2.32, the influence of molecules other than the selected donor and acceptor has not been taken into account. Therefore, straightforward application of such pair rates to any ensemble can lead to erroneous conclusions, the most dramatic fact being the prediction of a potentially infinite donor decay rate, calculated as the sum of contributions to all surrounding acceptors (by analogy with the Förster theory of radiationless energy transfer [2, 25]). The problem arises because the pair rates, Eqn 2.35, behave as R^{-2} in the far zone, whereas for a uniform distribution the number of acceptors in a shell centered on the donor grows as R^2. In this section, contributions due to other molecules are systematically included into the pair rates, providing *inter alia* a resolution to the above problem.

2.5.1   Influence of the molecular medium on the pair rates

Let us return to the general Hamiltonian, Eqn 2.1, which describes an ensemble of molecules (atoms or chromophore groups) coupled with the quantized radiation field. To deal with excitation transfer between a selected pair of molecules D and A, we shall now divide the full system into two parts. One subsystem comprises the selected pair of species D and A; another subsystem (to be referred to as the polariton bath) contains the quantized electromagnetic field and the remaining molecules that constitute the surrounding medium. Note that the molecules of the medium may, but do not necessarily, differ in type from D and A. With regard to the chosen partitioning of the system, the full Hamiltonian of Eqn 2.1 splits into a zero-order Hamiltonian and interaction operator, as

$$H = H^0 + V, \tag{2.57}$$

with

$$H^0 = H_{\text{bath}} + H_{\text{mol}}(D) + H_{\text{mol}}(A), \tag{2.58}$$

$$V = H_{\text{int}}(D) + H_{\text{int}}(A), \tag{2.59}$$

and

$$H_{\text{bath}} = H_{\text{rad}} + \sum_{X \neq D,A} [H_{\text{mol}}(X) + H_{\text{int}}(X)], \tag{2.60}$$

where the latter H_{bath} is the "bath" Hamiltonian that contains the radiation Hamiltonian H_{rad}, as well as contributions due to all molecules other than the donor and acceptor, the interaction Hamiltonians $H_{\text{int}}(X)$ being given by Eqns 2.2 and 2.3.

The partitioning of Eqns 2.57–2.59 has the same form as Eqns 2.5–2.7 employed for energy transfer *in vacuo*, subject to the replacement $H_{\text{rad}} \rightarrow H_{\text{bath}}$. Consequently, one can readily modify the previous Eqns 2.8–2.20 to suit the present situation. Specifically, the state vector $|0\rangle$ featured in the initial conditions, Eqn 2.8, now represents the ground state of the polariton bath (a combined system of the radiation field and the molecular medium), e_{vac} being the corresponding zero-point energy of the bath. Next, the transition matrix element, Eqn 2.12, and the state vectors for the intermediate states, Eqns 2.14 and 2.15, modify as follows:

$$\langle F|T^{(2)}|I\rangle = \frac{1}{\hbar} \sum_{\sigma} \sum_{q=1}^{2} \frac{\langle F|V|M_q(\sigma)\rangle\langle M_q(\sigma)|V|I\rangle}{(-1)^{q+1}cK - \Pi_\sigma + is}, \tag{2.61}$$

$$|M_1(\sigma)\rangle = |D\rangle|A\rangle|\sigma\rangle, \tag{2.62}$$

and

$$|M_2(\sigma)\rangle = |D^*\rangle|A^*\rangle|\sigma\rangle, \tag{2.63}$$

where

$$\hbar\Pi_\sigma = e_\sigma - e_{\text{vac}}, \tag{2.64}$$

is the excitation energy of the bath (i.e. the difference in energies between its excited and ground states). Here the index σ refers to the excited states of the bath that are accessible though a single action of the interaction operator V on the ground-state vector $|0\rangle$. Accordingly, the energy transfer is now regarded as being mediated by the elementary excitations of the bath (virtual polaritons) rather than by virtual photons of the "pure" electromagnetic field. The two types of intermediate states, Eqns 2.62 and 2.63, again correspond to the two time-orderings (Fig. 2.5) showing two different patterns for mediation of the intermolecular coupling by a medium-dressed photon (a bath polariton).

By analogy with Eqns 2.18 and 2.19, the transition matrix element can now be represented as

$$\langle F|T^{(2)}|I\rangle = \mu_{A_l}^{\text{full}}\theta_{lj}(K,\mathbf{R})\mu_{D_j}^{\text{full}}, \tag{2.65}$$

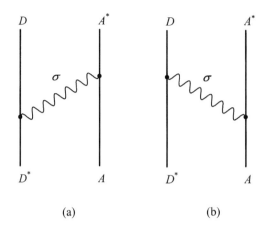

Figure 2.5   Time-ordered diagrams for polariton-mediated energy transfer

where

$$\theta_{lj}(K,\mathbf{R}) = \frac{1}{\hbar\varepsilon_0^2}\sum_\sigma \left[\frac{\langle 0|d_l^\perp(\mathbf{R}_A)|\sigma\rangle\langle\sigma|d_j^\perp(\mathbf{R}_D)|0\rangle}{cK - \Pi_\sigma + is} - \frac{\langle 0|d_j^\perp(\mathbf{R}_D)|\sigma\rangle\langle\sigma|d_l^\perp(\mathbf{R}_A)|0\rangle}{cK + \Pi_\sigma - is} \right]$$

(2.66)

is the tensor for the retarded dipole–dipole coupling between a pair of molecules within the medium: here the influences of the material medium arise through the detailed form of the eigenstates and the eigenenergies of the bath that enter Eqn 2.66. The eigenvalue problem can be bypassed in calculating the tensor $\theta_{lj}(K,\mathbf{R})$ by invoking the Green function formalism [37]. As an alternative, explicit summation over the normal modes σ can be performed in Eqn 2.66; this can be achieved by expanding (in terms of the normal modes of the dielectric medium [33, 34]) the local displacement operators $d_l^\perp(\mathbf{R}_A)$ and $d_l^\perp(\mathbf{R}_D)$ that enter Eqn 2.66. By either method, one arrives at the same result, Eqn 2.68, presented below. For instance, applying the Green function technique, the tensor of Eqn 2.66 is found to be [37]

$$\theta_{lj}(K,\mathbf{R}) = \frac{1}{n^2}\left(\frac{n^2+2}{3}\right)^2\sum_\mathbf{k} \frac{\left(\delta_{lj} - \hat{k}_l\hat{k}_j\right)k}{2\varepsilon_0 V}\left[\frac{e^{i\mathbf{k}\cdot\mathbf{R}}}{nK - k + is'} - \frac{e^{-i\mathbf{k}\cdot\mathbf{R}}}{nK + k - is'} \right],$$

(2.67)

where $n \equiv n(cK)$ is the refractive index of the medium given by Eqns 2.69–2.70 below. One can now repeat steps similar to those for the vacuum case,

Eqns 2.21–2.26, to yield the following final result for the retarded tensor of the dipole–dipole coupling in the dielectric medium:

$$\theta_{lj}(K,\mathbf{R}) = \frac{1}{n^2}\left(\frac{n^2+2}{3}\right)^2 \theta_{lj}^{\text{vac}}(nK,\mathbf{R}). \tag{2.68}$$

The tensor of Eqn 2.68 accommodates a screening contribution n^{-2} and a local field (Lorenz) factor $(n^2+2)/3$. In addition, the result contains the tensor $\theta_{lj}^{\text{vac}}(nK,\mathbf{R})$ with the same structure as the vacuum tensor, Eqn 2.26, the argument $y = nK$ now, however, being scaled by the refractive index n, as given by

$$n^2 = 1 + \frac{\alpha\rho/\varepsilon_0}{1 - \alpha\rho/3\varepsilon_0}, \tag{2.69}$$

where $\rho = N/V$ in turn is the number of molecules X per unit volume, and α is the molecular polarisability (calculated at the transfer frequency $\omega = cK$):

$$\alpha \equiv \alpha_X(cK) = \frac{1}{3\hbar}\sum_{m,p}\rho_X^{(m)}\left[\frac{\mu_X^2|\langle\varphi_{X^*}^{(p)}|\varphi_X^{(m)}\rangle|^2}{e_{X_p^*} - e_{X_m} - cK - \mathrm{i}s} + \frac{\mu_X^2|\langle\varphi_{X^*}^{(p)}|\varphi_X^{(m)}\rangle|^2}{e_{X_p^*} - e_{X_m} + cK + \mathrm{i}s}\right]. \tag{2.70}$$

In this representation, the quantity s is considered to be a small though a finite parameter that reflects the natural widths of each molecular line: replacement of an infinitesimal s by its finite counterpart to can be justified on rigorous dynamical grounds, as discussed in Section 2.6 (in which connection, the width s is labeled by another letter, η). Here also the indices m and p label the sublevels (vibrational, etc.) of the ground and excited state manifolds for the species X comprising the medium, μ_X is the electronic part of the transition dipole, as in Eqn 2.30, $e_{X_p^*} - e_{X_m}$ represent the excitation energies, and $\rho_X^{(m)}$ are the population distributions of the vibrational sublevels in the ground electronic state of the molecule X. It is noteworthy that an arbitrary number of vibrational sublevels can be accommodated for each molecule X of the medium in the framework of the formalism applied [33, 34, 37]. This includes *inter alia* a situation in which the molecular sublevels form a dense (quasi-continuum) set. In such a case, the molecular polarizability, Eqn 2.70, acquires an imaginary part in the absorbing areas of the spectrum, making the refractive index, Eqn 2.68, a complex quantity:

$$n = n' + \mathrm{i}n''. \tag{2.71}$$

The derivation of the tensor represented by Eqns 2.67 and 2.68 is based on a microscopic theory [33, 34, 37] in which the molecules of the medium are considered to be all of the same type, regularly placed to form a simple cubic lattice, and characterized by the same isotropic polarizabilities $\alpha^X \equiv \alpha$. Such a model may also describe a common situation in which nonisotropic species are randomly oriented in their sites. The resulting Eqn 2.68 seems, however, not to be sensitive to the possible lack of translational symmetry as well, as long as the vibrational widths of the spectral lines exceed the characteristic energies of resonance coupling between the molecules comprising the medium. It is therefore expected that the transition matrix element given by Eqns 2.65 and 2.68, should adequately describe the transfer of energy in a variety of amorphous media constituted of randomly situated and oriented molecules, and characterized by some energetic disorder as well. For such systems, the quantity α that enters Eqn 2.69 is to be understood as an averaged polarizability for all of the species X that constitute the medium:

$$\alpha = \bar{\alpha}^X = N^{-1} \sum_X \alpha^X, \tag{2.72}$$

N being the total number of molecules in the system. The subsequent analysis is also consistent with such a definition of the polarizability α.

The full rate of donor–acceptor transfer is again given by Eqn 2.31, subject to replacement of the tensor for the electromagnetic coupling in the vacuum by the tensor for the coupling in the medium, as $\theta_{lj}^{\mathrm{vac}}(K,\mathbf{R}) \to \theta_{lj}(K,\mathbf{R})$. Using Eqn 2.68 for $\theta_{ij}(K,\mathbf{R})$, the pair rate can be expressed, as in Eqn 2.32, in terms of the overlap integral between the donor and acceptor spectra:

$$W_{DA} = \frac{9}{8\pi c^2 \tau_D} \int_0^\infty F_D(\omega)\sigma_A(\omega)\omega^2 g(\omega, \mathbf{R}) e^{-2n'' \omega R/c} \, d\omega, \tag{2.73}$$

with $n \equiv n(\omega)$ and

$$g(\omega, \mathbf{R}) = |n|^2 \left| \eta_3 \left[\left(\frac{c}{n\omega R}\right)^3 - i\left(\frac{c}{n\omega R}\right)^2 \right] - n_1 \frac{c}{n\omega R} \right|^2$$

$$= \frac{1}{|n|^4} \left\{ \eta_3^2 \frac{c^6}{\omega^6 R^6} + 2\eta_3^2 n'' \frac{c^5}{\omega^5 R^5} + \left[\eta_3^2 |n|^2 - 2\eta_1\eta_3 (n'^2 - n''^2) \right] \right.$$

$$\times \frac{c^4}{\omega^4 R^4} + 2\eta_1\eta_3 n'' |n|^2 \frac{c^3}{\omega^3 R^3} + \eta_1^2 |n|^4 \frac{c^2}{\omega^2 R^2} \right\}, \tag{2.74}$$

where the quantities

$$\sigma_A(\omega) = \frac{\pi \omega \mu_A^2}{3\varepsilon_0 c} \frac{1}{n'} \left| \frac{n^2 + 2}{3} \right|^2 \sum_{m,p} \rho_A^{(m)} \left| \left\langle \varphi_{A^*}^{(p)} | \varphi_A^{(m)} \right\rangle \right|^2 \delta(e_{A_p^*} - e_{A_m} - \hbar\omega) \tag{2.75}$$

and

$$F_D(\omega) = \frac{\omega^3 \tau_D \mu_D^2}{3\varepsilon_0 \pi c^3} n' \left| \frac{n^2 + 2}{3} \right|^2 \sum_{n,r} \rho_{D^*}^{(n)} \left| \left\langle \varphi_D^{(r)} \middle| \varphi_{D^*}^{(n)} \right\rangle \right|^2 \delta\left(e_{D_n^*} - e_{D_r} - \hbar\omega\right), \quad (2.76)$$

can be identified (see Section 2.5.2), respectively, as the absorption cross-section of the acceptor and the emission spectrum of donor, both quantities being corrected for the electronic influence of the medium. The latter $F_D(\omega)$ is again normalized to unity in the sense

$$\int F_D(\omega)d\omega = 1. \quad (2.77)$$

The normalization constant τ_D – to be explicitly presented in Eqn 2.94 below – represents the radiative lifetime of the donor in the absorbing medium, as will be demonstrated in the following subsection. In this way, dielectric influences of the material medium feature in the pair rate, Eqn 2.73, through the refractive modifications of the spectral functions $F_D(\omega)$ and $\sigma_A(\omega)$, as well as through the factors $g(\omega, \mathbf{R})$ and $e^{-2n''\omega R/c}$. The latter exponential factor represents the Beer's law losses in the absorbing medium. This factor will be demonstrated to play a vital role at large separations between the transfer species, providing a physically sensible total rate of energy transfer to all the surrounding acceptors (see Section 2.5.2).

The pair rate, Eqn 2.73, can be presented meaningfully as the following sum of three terms:

$$W_{DA} = W_{DA}^{\text{Först}} + W_{DA}^{I} + W_{DA}^{\text{Far zone}}. \quad (2.78)$$

The first term,

$$W_{DA}^{\text{Först}} = \frac{9c^4 \eta_3^2}{8\pi\tau_D R^6} \int_0^\infty F_D(\omega)\sigma_A(\omega)|n|^{-4}\omega^{-4}e^{-2n''\omega R/c} \, d\omega, \quad (2.79)$$

represents the familiar Förster rate of energy transfer which is characterized by an R^{-6} distance dependence. Such a rate is the dominant contribution in the near zone ($KR \ll 1$) where the exponential factor $e^{-2n''\omega R/c}$ is close to unity and can therefore be disregarded. Consequently, the spectral integral of Eqn 2.79 is weighted by the factor $|n|^{-4}$ in the near zone, in agreement with the standard theory of radiationless energy transfer [2, 25]. Note that the local field-factors featured in the coupling tensor, Eqn 2.68, are now contained in the spectral functions $F_D(\omega)$ and $\sigma_A(\omega)$. The third term in Eqn. 2.78,

$$W_{DA}^{\text{Far zone}} = \frac{9\eta_1^2}{8\pi\tau_D R^2} \int_0^\infty F_D(\omega)\sigma_A(\omega)e^{-2n''\omega R/c} \, d\omega, \quad (2.80)$$

behaving as R^{-2}, dominates in the far zone ($KR \gg 1$). It is characterized by the overlap integral between the donor emission and acceptor absorption spectra, weighted by the Beer's law factor. The rate, Eqn 2.80, can be identified as the rate of radiative (far-zone) energy transfer involving spontaneous emission by a donor, propagation of the emitted photon through the absorbing medium, followed by its absorption at the acceptor. The factors $F_D(\omega)$, $e^{-2n''\omega R/c}$ and $\sigma_A(\omega)$ characterize the corresponding processes.

Lastly, the middle term of the pair-rate equation 2.78, W_{DA}^I, due to the remaining terms in the function of Eqn 2.74, becomes important at intermediate distances where $KR \simeq 1$. In general, this intermediate contribution contains not only the usual R^{-4} term [4, 8, 10, 11, 17, 26], but also additional terms in odd powers of R (i.e. R^{-3} and R^{-5}). However, in the case of a weakly absorbing medium ($n'' \ll n'$) one can disregard the latter odd rank terms to arrive at the following approximate expression:

$$W_{DA}^I = \frac{9c^2\left(\eta_3^2 - 2\eta_1\eta_3\right)}{8\pi\tau_D R^4} \int_0^\infty F_D(\omega)\sigma_A(\omega)|n|^{-2}\omega^{-2}e^{-2n''\omega R/c}d\omega. \tag{2.81}$$

For any such weakly absorbing medium, the function $g(\omega, \mathbf{R})$ that enters the pair rate, Eqn 2.73, can be written approximately as

$$g(\omega, \mathbf{R}) = |n|^2 g^{\mathrm{vac}}(|n|\omega, \mathbf{R}), \tag{2.82}$$

$g^{\mathrm{vac}}(|n|\omega, \mathbf{R})$ being its vacuum counterpart defined by Eqn 2.33, but with the argument ω now scaled by the modulus of the complex refractive index $|n|$. Finally, one can readily extend Eqn 2.51, which characterizes the range-dependence of fluorescence anisotropy in vacuum, for the depolarization that is taking place in the medium. Specifically, in the case of a weakly absorbing medium, one then has

$$r_1(R) = \frac{r_0}{25} \frac{7|n|^4 \widetilde{K}^4 R^4 + |n|^2 \widetilde{K}^2 R^2 + 3}{|n|^4 \widetilde{K}^4 R^4 + |n|^2 \widetilde{K}^2 R^2 + 3}, \tag{2.83}$$

i.e. the wave-vector $\widetilde{K}$ appears to be scaled by $|n|$, modifying to some extent the characteristic scale of distances. As such, the position at which the onset of significant retardation modifies the fluorescence anisotropy can be still closer than the vacuum formula would suggest.

2.5.2 Spontaneous emission as far-zone energy transfer

Consider the decay of an excited state of donor D through energy transfer to the surrounding species X. By summing up all appropriate pair rates, the full decay rate is

$$\Gamma_D = \sum_{X \neq D} W_{DX}, \tag{2.84}$$

where W_{DX} is the rate of excitation transfer between a pair of molecules D and X, as explicitly presented in the previous subsection (with $X \equiv A$). Using Eqn 2.78 for the pair rates, the decay rate of Eqn 2.84 can be cast as

$$\Gamma_D = \tilde{\Gamma}_D^{\text{Först}} + \Gamma_D^{\text{Far zone}}, \tag{2.85}$$

with

$$\tilde{\Gamma}_D^{\text{Först}} = \sum_{X \neq D} \left(W_{DX}^{\text{Forst}} + W_{DX}^I \right), \tag{2.86}$$

and

$$\Gamma_D^{\text{Far zone}} = \sum_{X \neq D} W_{DX}^{\text{Far zone}} \tag{2.87}$$

The former decay rate $\tilde{\Gamma}_D^{\text{Först}}$ contains contributions due to the Förster pair rates modified by the intermediate terms W_{DX}^I, the tilde over Γ reflecting such a modification. The constituent pair rates have been explicitly presented in Eqns 2.79 and 2.81 of the previous subsection.

In what follows, we shall concentrate on the latter decay rate $\Gamma_D^{\text{Far zone}}$ associated with far-zone (radiative) energy transfer. The far-zone transfer may be viewed [4, 10, 11, 35, 36] as spontaneous emission of a photon followed by its subsequent recapture by a distant acceptor. Adopting such a concept, we shall regard the contribution $\Gamma_D^{\text{Far zone}}$ as the rate of spontaneous emission in the absorbing medium. The approach is in a certain sense related to absorber theory [20, 45, 56] in which spontaneous emission is seen to be the result of direct interaction between the emitting atom and "the Universe," the latter acting as a perfect absorber at all emitted frequencies. In our situation, the surrounding medium does indeed act as a perfect absorber, even at extremely low concentrations of the absorbing species (or, alternatively, for an almost transparent condensed medium), as long as the system dimensions are large enough to insure eventual recapture of the emitted photon. For such a weakly absorbing medium, the rate $\Gamma_D^{\text{Far zone}}$ will be demonstrated to reproduce exactly the familiar rate [31, 33, 34, 40, 43] for the spontaneous emission in a transparent dielectric.

To obtain the proper decay rate $\Gamma_D^{\text{Far zone}}$ using Eqn 2.87, the pair transfer rates $W_{DX}^{\text{Far zone}}$ should not only reflect effects due to retardation, but also incorporate influences of the surrounding medium, as in the previous section. Following [35, 36], we shall demonstrate that $\Gamma_D^{\text{Far zone}}$ does indeed represent the rate of spontaneous emission in the absorbing medium. For this purpose, we shall substitute Eqn 2.80 for the pair rate $W_{DA}^{\text{Far zone}}$ (with $A \equiv X$) into Eqn 2.87, to yield

$$\Gamma^{\text{Far zone}} = \frac{9}{8\pi\tau_D} \int_0^\infty F_D(\omega) \sum_X R^{-2}\eta_1^2\sigma_X(\omega)e^{-2n''\omega R/c}\,d\omega, \tag{2.88}$$

with $\mathbf{R} \equiv \mathbf{R}_{XD}$. The decay rate $\Gamma_D^{\text{Far zone}}$ is built up of a large number of the pair rates operating predominantly in the far zone. Hence the summation over the molecules X can be changed to an integral over the radius vector $\mathbf{R}$, giving

$$\Gamma^{\text{Far zone}} = \frac{1}{\tau_D} \int_0^\infty d\omega \, F_D(\omega)\sigma(\omega) \int_0^\infty d\,\text{Re}^{-2n''\omega R/c} \tag{2.89}$$

$$= \frac{c}{2\tau_D} \int_0^\infty d\omega \, F_D(\omega)\sigma(\omega)/\omega n'', \tag{2.90}$$

where in Eqn 2.89 orientational averaging has been carried out $(\eta_1^2 \to \langle\eta_1^2\rangle = 2/9)$, and the cross-section of the molecular absorption $\sigma_X(\omega)$ has been replaced by its ensemble average, as given by

$$\sigma(\omega) \equiv \bar{\sigma}_x(\omega) = \frac{\omega}{c}\frac{1}{n'}\left|\frac{n^2+2}{3}\right|^2\frac{\bar{\alpha}_x''}{\varepsilon_0}. \tag{2.91}$$

The relationship of Eqn 2.91 has been written exploiting Eqn 2.75 for $\sigma_A(\omega)$ (with $A \equiv X$), with constraints due to the energy-conserving delta functions expressed in terms of the imaginary part of the molecular polarizability $\bar{\alpha}_X'' \equiv \alpha''$ using Eqn 2.70 for α_X. The imaginary part α'' is in turn related to the complex refractive index n, given by Eqn 2.69, as

$$n'' = \frac{1}{2n'}\left|\frac{n^2+2}{3}\right|^2\frac{\alpha''\rho}{\varepsilon_0}, \tag{2.92}$$

so that Eqn 2.91 reduces to

$$\sigma(\omega) \equiv \bar{\sigma}_x(\omega) = n''2\omega/c\rho. \tag{2.93}$$

In writing Eqn 2.92, use has been made of the generalized definition, Eqn 2.72, for the molecular polarizability α.

Equation 2.93 can be identified as the usual relationship between the imaginary part of the complex refractive index and the absorption cross-section. In this way, the quantity $\sigma(\omega)$ defined by Eqn 2.75 is indeed seen to represent the cross-section of molecular absorption in the lossy medium. Substituting Eqn 2.93 into Eqn 2.90, using Eqn 2.76 for $F_D(\omega)$ and the normalization condition Eqn 2.77, one finds

$$\Gamma^{\text{Far zone}} = \tau_D^{-1} = \sum_{n,r} (\tau_D^{rn})^{-1} \rho_{D^*}^{(n)}, \tag{2.94}$$

with

$$(\tau_D^{rn})^{-1} = \frac{\omega^3 \mu_D^2}{3\hbar\varepsilon_0\pi c^3} n' \left| \frac{n^2+2}{3} \right|^2 \left| \langle \varphi_D^{(r)} | \varphi_{D^*}^{(n)} \rangle \right|^2. \tag{2.95}$$

The rate $\Gamma^{\text{Far zone}} = \tau_D^{-1}$ as given by Eqns 2.94 and 2.95 represents the full rate of spontaneous emission by the donor, and so involves summation over the final levels and averaging over the initial levels of the donor, labeled respectively by r and n, $\rho_{D^*}^{(n)}$ being the population distribution of the vibrational levels of donor in the initially excited electronic state. The constituent terms $(\tau_D^{rn})^{-1}$ represent the partial rates of spontaneous emission associated with downward transitions of the donor between the specific levels n and r, where in each of these terms the refractive index is to be calculated at the appropriate frequency $(e_{D_n^*} - e_{D_r})/\hbar$.

The emission rates, Eqn 2.95, manifestly accommodate contributions due to the absorbing dielectric medium, including the local field factor. It is noteworthy that the present analysis is based on a microscopic QED theory [35, 37], the relationship Eqn 2.95 supporting previous phenomenological methods [12, 13] used to introduce local field corrections to the rates of spontaneous emission in an absorbing medium. In the limit where $\alpha'' \to 0$, $n'' \to 0$, the far-zone rate given in Eqns 2.94 and 2.95 reduces smoothly to the usual result for spontaneous emission in a transparent medium [31, 33, 34, 40, 43], there being a vanishing contribution due to the decay in the near and intermediate zones: $\tilde{\Gamma}_D^{\text{Först}} \to 0$. In other words, the present analysis reproduces in full the rate of spontaneous emission in transparent dielectrics $n'' = 0$, including *inter alia* the case of free space: $n'' = 0$, $n' = 1$. Here, free space is be to viewed as a limit in which the density of the absorbing species goes to zero, while the size of the system goes to infinity, so that the emitted photon is eventually recaptured somewhere in the system. It is noteworthy that in order to arrive at a sensible result for $\Gamma_D^{\text{Far zone}} = \tau_D^{-1}$, such as that given by Eqns 2.94 and 2.95, the influences of the absorbing medium are necessarily to be reflected in the pair transfer rates comprising the decay rate equation 2.87. In fact, it is the exponential factor $\exp(-2n''\omega R/c)$, which represents absorption losses at the intervening medium, that helps avoid the potentially infinite decay rate $\Gamma_D^{\text{Far zone}}$ (given by Eqn 2.87) due to the R^{-2} factor featured in the constituent pair rates, Eqn 2.80. Note also that the same result, Eqns 2.94 and 2.95, can be reproduced for the rate of spontaneous emission τ_D^{-1} in terms of another microscopic method that involves calculation of the quantum flow from the emitting molecule D in the absorbing medium [36].

2.6   DYNAMICS OF ENERGY TRANSFER BETWEEN A PAIR OF MOLECULES IN A DIELECTRIC MEDIUM

In previous sections, the transfer dynamics has been described in terms of well-defined rates for intermolecular energy transfer. It is the purpose of the current section to pursue in more detail temporal aspects of the energy transfer between a pair of species in a dielectric medium, through explicit consideration of the QED time evolution. A distinctive aspect of the approach developed in [38] is that it affords a combined analysis of rate and nonrate regimes, in the context of examining the influence of the dielectric medium on a microscopic basis. The theory is built on the foundation established in the previous section. Again, the approach exploits the concept of energy transfer mediated by bath polaritons. The theory also makes use of the microscopically derived tensor, Eqn 2.68, for the retarded and medium-dressed dipole–dipole coupling, now with regard to the dynamical behavior. The present section not only extends consideration beyond the rate description, but also re-examines conditions for that regime itself. That leads to incorporation of an energy renormalization for both of the ground and excited states of the transfer species, due to the interaction of these species with the molecules that belong to the medium, and also with each other. That is a feature not reflected in direct application of the ordinary Fermi Golden Rule.

   The section is organized as follows. In the next subsection the Heitler–Ma method [1, 28–30] for describing the quantum time evolution is first outlined, and subsequently reformulated to suit our current purposes, technical details of the derivation being placed in two appendices. Consequently, the basic equations for time evolution acquire a form that is more symmetric with respect to the initial and final states. Subsection 2.6.2 concentrates on the transfer dynamics between a pair of molecules in the molecular medium, starting from general considerations and leading to an analysis of both the rate regime and beyond. Note that the nonrate regime features in situations that lack an intrinsic density of molecular states for the participating species.

2.6.1   General description of time evolution

Consider the quantum dynamics of a system with a time-independent Hamiltonian that is separable as the sum of a zero-order Hamiltonian H^0 and an interaction term V (such as that defined by Eqn 2.57 in the previous section), where the eigenvectors of H_0 include, *inter alia*, both the initial state $|I\rangle$ and the final state $|F\rangle$ for the process. For reasons which will become apparent later, we shall commence work in the Schrödinger representation rather than the more common interaction representation. The state vector of the system then evolves at positive times from the state $|I\rangle$ at $t = +0$ as

$$S(t)|I\rangle = \Theta(t)e^{-iHt/t}|I\rangle \qquad (2.96)$$

$$= -\frac{1}{2\pi i}\int_{-\infty}^{+\infty} d\varepsilon\, e^{-i\varepsilon t/\hbar}(\varepsilon - H + i\eta)^{-1}|I\rangle \quad (\eta \to +0), \qquad (2.97)$$

$\Theta(t)$ being the unit step (Heaviside) function. Strictly, the quantity η is a positive infinitesimal, yet for finite times it may be considered to be a finite quantity that obeys the following: $\eta t/\hbar \ll 1$ (i.e. η should be kept much less than the inverse lifetime for the excited states). Under this condition, introduction of a finite η does not influence the quantum dynamics governed by Eqn 2.97. Retention of a finite value for η plays an important role in smoothing spectral lines. This makes the refractive index given by Eqns 2.69 and 2.70 a complex quantity in absorbing areas of the spectrum, comprising contributions due to densely spaced molecular sublevels of vibrational or other origin for each electronic transition. Note that the smoothing parameter η was labeled s in the previous sections.

The Heitler–Ma method [1, 28–30] may now be employed, giving (see Appendix A)

$$\langle F|(\varepsilon - H + i\eta)^{-1}|I\rangle = \frac{U_{FI}(\varepsilon)}{(\varepsilon - E_F + i\eta)(\varepsilon - E_1 + \frac{1}{2}i\hbar\Gamma_I(\varepsilon) + i\eta)}. \qquad (2.98)$$

Here $U_{FI}(\varepsilon) \equiv \langle F|U(\varepsilon)|I\rangle$ and $\Gamma_I(\varepsilon) \equiv \langle I|\Gamma(\varepsilon)|I\rangle$ are, respectively, the matrix elements of the off-diagonal transition operator $U(\varepsilon)$ and the diagonal damping operator $\Gamma(\varepsilon)$, both determined by the following recurrence relation:

$$U(\varepsilon) - \frac{i}{2}\hbar\Gamma(\varepsilon) = V + V(\varepsilon - H^0 + i\eta)^{-1}U(\varepsilon). \qquad (2.99)$$

For present purposes it is more convenient to represent the above in a non-recursive format, as

$$[U(\varepsilon) - \frac{i}{2}\hbar\Gamma(\varepsilon)]|I\rangle = \left[V + VP_I(\varepsilon - H^0 - P_IVP_I + i\eta)^{-1}P_IV\right]|I\rangle, \qquad (2.100)$$

where the projection (idempotent) operator,

$$P_I = 1 - |I\rangle\langle I|, \qquad (2.101)$$

identifies the exclusion of contributions by the initial state in the perturbation expansion of Eqn 2.100. Recasting the transition matrix element in a form in which the perturbational contribution by the final state also no longer explicitly features, one arrives at (see Appendix B)

$$U_{FI}(\varepsilon) = \frac{\varepsilon - E_F + i\eta}{\varepsilon - E_F - \frac{1}{2}\hbar\Gamma'_F + i\eta} U'_{FI}, \qquad (2.102)$$

where the newly defined quantities on the right, U'_{FI} and Γ'_F, both have implicit ε-dependence and are given by

$$U'_{FI} \equiv \langle F | \left[V + VP_I P_F (\varepsilon - H^0 - P_I P_F V P_I P_F + i\eta)^{-1} P_I P_F V \right] | I \rangle, \quad (2.103)$$

$$-\frac{i}{2}\hbar\Gamma'_F \equiv \langle F | \left[V + VP_I P_F (\varepsilon - H^0 - P_I P_F V P_I P_F + i\eta)^{-1} P_I P_F V \right] | F \rangle, \tag{2.104}$$

with

$$P_F = 1 - |F\rangle\langle F|. \qquad (2.105)$$

Finally, calling on Eqns 2.97, 2.98 and 2.102, one finds the following probability amplitude for the transition $|I\rangle \rightarrow |F\rangle$:

$$\langle F | S(t) | I \rangle = -\frac{1}{2\pi i} \int_{-\infty}^{+\infty} d\varepsilon \, U'_{FI} \, e^{-i\varepsilon t/\hbar} \left(\varepsilon - E_F + \frac{i}{2}\hbar\Gamma'_F + i\eta\right)^{-1}$$
$$\left(\varepsilon - E_I + \frac{i}{2}\hbar\Gamma_I + i\eta\right)^{-1}, \qquad (2.106)$$

which is an exact result. Here, the presence of both Γ_I and Γ'_F in the energy denominators explicitly accommodates the damping corrections and energy renormalization of the initial and final states. Consequently, the transfer amplitude as presented above has a form that is obviously more symmetric with respect to the initial and final states than would result from direct substitution of Eqn 2.98 into Eqn 2.97. Still, there is some dissymmetry with respect to these states, reflected by the prime on Γ'_F. The retention of this dissymmetry will be of vital importance in the case of sharp energy levels for the donor and acceptor, i.e. where the participating transfer species lack an intrinsic density of molecular states; this aspect is to be considered in subsection 2.6.2.2.

2.6.2   Transfer dynamics

The dynamical system of interest has been defined by the Hamiltonian, Eqns 2.57–2.60. For the representation of energy transfer between the donor and acceptor molecules in the dielectric medium, the initial and final state vectors and their energies are again (as in the previous section) considered to have the form of Eqns 2.8 and 2.9, which characterize the energy transfer *in vacuo*, where the state vector $|0\rangle$ is again to be understood as the ground state of the polariton bath (the combined system of the radiation field and the molecular medium), e_{vac} being the corresponding zero-point energy of such a bath. Because of the two-center character of the interaction operator, Eqn 2.59, it is convenient to carry out the corresponding partitioning in Eqns 2.100 and 2.104, writing

$$-\frac{i}{2}\hbar\Gamma_I(\varepsilon) = \Delta e_{D^*} + \Delta e_A - \frac{i}{2}\hbar\gamma_{D^*} - \frac{i}{2}\hbar\Gamma_{D^*A}, \qquad (2.107)$$

$$-\frac{i}{2}\hbar\Gamma'_F(\varepsilon) = \Delta e_D + \Delta e_{A^*} - \frac{i}{2}\hbar\gamma_{A^*} - \frac{i}{2}\hbar\Gamma'_{DA^*}, \qquad (2.108)$$

Here, one center contributions, denoted by a single index D (or A), are due to the terms that contain only one operator V_D (or V_A) in the perturbation expansions of Eqns 2.100 and 2.104. Such contributions have already been separated into real energy shifts and imaginary damping terms in the above equations. For instance, Δe_{D^*} and γ_{D^*} represent, respectively, the bath-induced level shift (energy renormalization) and the damping factor for the excited molecular state $|D^*\rangle$, there being no imaginary (damping) contributions for the ground molecular states $|D\rangle$ and $|A\rangle$. Each such energy renormalization (Δe_{D^*}, Δe_A, Δe_D, and Δe_{A^*}) embodies not only the radiative (Lamb) shift [16, 28, 42], but the also the contribution due to the dispersion interaction between the donor D (or acceptor A) and the molecular medium. Note that the dispersion energy appears now in the second order of perturbation, rather than the usual fourth order [16, 48, 49], as the coupling of the radiation field with the medium has already been included in the zero-order Hamiltonian H^0, given by Eqns 2.58 and 2.60. Here, we shall not consider the explicit structure of these energy shifts, which are to be treated as the parameters of the theory. The remaining (complex) quantities Γ_{D^*A} and Γ'_{DA^*} are two-center contributions resulting from cross-terms (containing both V_D and V_A) that emerge in the perturbation expansions of Eqns 2.100 and 2.104.

By making use of Eqns 2.107 and 2.108, the probability amplitude for the energy transfer, Eqn 2.106, becomes

$$\langle F|\tilde{S}(t)|I\rangle = -\frac{1}{2\pi i\hbar}\int_{-\infty}^{+\infty} d\omega$$

$$\frac{U'_{FI}(\omega)e^{-i(\omega-\omega_{A^*})t}}{\left(\omega - \omega_{A^*} + (i/2)\gamma_{A^*} + (i/2)\Gamma'_{DA^*} + i\eta'\right)\left(\omega - \omega_{D^*} + (i/2)\gamma_{D^*} + (i/2)\Gamma_{D^*A} + i\eta'\right)}$$
$$(\eta' \to +0),$$

$$(2.109)$$

where

$$\omega = (\varepsilon - e_D - e_A - \Delta e_D - \Delta e_A - e_{\text{vac}})/\hbar \qquad (2.110)$$

is a new variable, and

$$\omega_{D^*} = (e_{D^*} + \Delta e_{D^*} - e_D - \Delta e_D)/\hbar, \qquad \omega_{A^*} = (e_{A^*} + \Delta e_{A^*} - e_A - \Delta e_A)/\hbar$$
$$(2.111, 2.112)$$

are the excitation frequencies of the donor and acceptor. The frequencies ω_{D^*} and ω_{A^*} incorporate level shifts for both the ground and excited molecular states. Finally, in Eqn 2.109 transformation has been carried out to a modified interaction representation, as

$$\langle F|\tilde{S}(t)|I\rangle = \langle F|S(t)|I\rangle\exp(-\mathrm{i}(E_F + \Delta E_F)t/\hbar), \qquad (2.113)$$

the term "modified" referring to renormalization by the medium of the final state energy $E_F = e_D + e_{A^*}$ by the amount $\Delta E_F = \Delta e_D + \Delta e_{A^*}$.

Now, we turn our attention to the transition matrix element $U'_{FI}(\omega)$, which in the present study will be represented through an effective second-order contribution, as

$$U'_{FI}(\omega) \approx U^{(2)}_{FI}(\omega) = \mu^{\text{full}}_{A_l}\theta'_{lj}(\omega/c, \mathbf{R})\mu^{\text{full}}_{D_j}, \qquad (2.114)$$

with

$$\theta'_{lj}(\omega/c, \mathbf{R}) = \frac{1}{\hbar\varepsilon_0^2}\sum_{\sigma}\left[\frac{\langle 0|d_l^\perp(\mathbf{R}_A)|\sigma\rangle\langle\sigma|d_j^\perp(\mathbf{R}_D)|0\rangle}{\omega - \prod_{\sigma} +\mathrm{i}\eta'} + \frac{\langle 0|d_j^\perp(\mathbf{R}_D)|\sigma\rangle\langle\sigma|d_l^\perp(\mathbf{R}_A)|0\rangle}{\omega - \omega_{D^*} - \omega_{A^*} - \prod_{\sigma} +\mathrm{i}\eta'}\right]$$
$$(\mathbf{R} = \mathbf{R}_A - \mathbf{R}_D),$$
$$(2.115)$$

where implied summation over the repeated Cartesian indices (l and j) is assumed, μ^{full}_D and μ^{full}_A being the transition dipoles given by Eqn 2.20. As in the previous section, here $\hbar\prod_{\sigma} = e_\sigma - e_{\text{vac}}$ is the excitation energy of the bath, the index σ denoting excited (single polariton) states of the bath accessible from the ground state $|0\rangle$ by single action of the local displacement operator $\mathbf{d}^\perp(\mathbf{R}_X)(X = D, A)$. Within the range of frequencies ω close to molecular transition frequencies ω_{D^*} and ω_{A^*}, the energy denominator $(\omega - \omega_{D^*} - \omega_{A^*} - \prod_{\sigma} +\mathrm{i}\eta$ may be replaced by $(-\omega - \prod_s +\mathrm{i}\eta)$ in the nonresonant term of Eqn 2.115.[†]

† It follows from the time–energy Uncertainty Principle that retention of the original form (Eqn 2.115) is important only for times which are less than the inverse molecular transition frequency, $\omega_{D^*}^{-1}$, generally on the femtosecond timescale.

Consequently, the above tensor of Eqn 2.115 reduces to that relating to retarded dipole–dipole coupling in the medium:

$$\theta'_{l,j}(\omega/c, \mathbf{R}) \approx \theta_{l,j}(\omega/c, \mathbf{R}), \tag{2.116}$$

the latter tensor, $\theta_{l,j}(\omega/c, \mathbf{R})$, being defined by Eqn 2.66. Using Eqns 2.68 and 2.26, the tensor $\theta_{l,j}(\omega/c, \mathbf{R})$ reads explicitly

$$\theta_{l,j}(\omega/c, \mathbf{R}) = n\left(\frac{n^2+2}{3}\right)^2 \frac{\omega^3 e^{in\omega R/c}}{4\pi c^3 \varepsilon_0} \left[(\delta_{ij} - 3\hat{R}_i\hat{R}_j)\left(\frac{c^3}{n^3\omega^3 R^3} - \frac{ic^2}{n^2\omega^2 R^2}\right) \right.$$
$$\left. - (\delta_{ij} - \hat{R}_i\hat{R}_j)\frac{c}{n\omega R}\right], \tag{2.117}$$

where $n \equiv n(\omega)$ is the complex relative index given by Eqns 2.69 and 2.70.

It is noteworthy that one can relate the matrix element $U_{FI}^{(2)}(\omega)$ (given by Eqns 2.114 and 2.117) to the transition matrix element considered in the previous section (see Eqn 2.65), as

$$\langle F|T^{(2)}|I\rangle \equiv U_{FI}^{(2)}(\omega_{D^*}), \tag{2.118}$$

where $U_{FI}^{(2)}(\omega)$ is to be taken at the emission frequency $\omega = \omega_{D^*} = cK \approx \omega_{A^*}$. In the context of time evolution, it is important to retain the ω-dependence featured in the exponential phase factor $\exp(in\omega R/c)$ of the matrix element $U_{FI}^{(2)}(\omega)$. This leads to appearance of a time lag in the initial arrival of the excitation at the acceptor A, due to the finite speed of signal propagation. The remainder of the transition element $U_{FI}^{(2)}(\omega)$, together with other ω-dependent parameters that enter Eqn 2.109, will at this stage be evaluated at the resonant frequency, $\omega = \omega_{A^*} \approx \omega_{D^*}$. Linearizing the exponent,

$$n(\omega)\omega R/c \approx [n(\omega_{A^*})\omega_{A^*}R/c] + [(\omega - \omega_{A^*})R/v_g], \tag{2.119}$$

with

$$\frac{1}{v_g} = \frac{d}{d\omega}\left(\frac{n\omega}{c}\right)\bigg|_{\omega=\omega_{A^*}}, \tag{2.120}$$

and re-defining the origin of time $\tau = (t - R/v_g)$, the transfer amplitude, Eqn 2.109, takes the form

$$\langle F|\tilde{S}(t)|I\rangle = -\frac{1}{2\pi\hbar i}\int_{-\infty}^{+\infty} d\omega\, U_{FI}^{(2)}(\omega_{A^*})e^{-i(\omega-\omega_{A^*})\tau}$$

$$\left(\omega - \omega_{A^*} + \frac{i}{2}\gamma_{A^*} + i\eta'\right)\left(\omega - \omega_{D^*} + \frac{i}{2}\gamma_{D^*} + i\eta'\right)^{-1} \tag{2.121}$$

$$= \hbar^{-1}U_{FI}^{(2)}(\omega_{A^*})\Theta(\tau)\frac{e^{-\frac{1}{2}\gamma_{A^*}\tau} - e^{\left[-\frac{1}{2}\gamma_{D^*} + i(\omega_{A^*}-\omega_{D^*})\right]\tau}}{(\omega_{A^*} - \omega_{D^*}) + \frac{1}{2}i(\gamma_{D^*} - \gamma_{A^*})}, \tag{2.122}$$

which takes account of damping for both species D and A. Here the two-center contributions Γ_{D^*A} and Γ'_{DA^*} are, for the present, omitted; the physical basis of this approximation will be clarified in due course.

It is worth noting that the radiative group velocity v_g featured in Eqn 2.122, via the shift of the origin of time, describes the delay of the initial arrival of the excitation at acceptor A, whereas the phase velocity $v_\phi = c/n$, which enters the exponential factor $\exp(in\omega R/c)$ of the transition matrix element $U_{FI}^{(2)}(\omega_{A^*})$ (with $n \equiv n(\omega_{A^*})$), characterizes the changes of optical phase with distance. Note also that incorporation of the time lag in the above manner implies that the refractive index, and hence also the group velocity, takes real values. Nonetheless, the general result to follow (Eqn 2.123) for the transfer rates holds both for lossless and absorbing media.

2.6.2.1   Transfer rates

Let us consider first the case in which the spectral widths of the species participating in the transfer exceed the magnitude of the corresponding transition matrix elements. The overall migration is then incoherent, described as a multistep process involving uncorrelated events of excitation transfer between the molecules of the system. With regard to the selected pair D–A, by omitting the relaxation terms γ_{D^*} and γ_{A^*} in Eqn 2.122, and for times in excess of the transit time R/v_g, the resultant rate of the excitation transfer reads

$$W_{FI} = \frac{d}{dt}|\langle F|\tilde{S}(t)|I\rangle|^2 = \frac{2\pi}{\hbar}|U_{FI}^{(2)}(\omega_{D^*})|^2\delta(\hbar\omega_{D^*} - \hbar\omega_{A^*}), \tag{2.123}$$

This provides the Fermi Golden Rule exploited previously, subject to the replacement of $U_{FI}^{(2)}(\omega_{D^*})$ by $\langle F|T^{(2)}|I\rangle$, using the relationship of Eqn 2.118. The full pair transfer rate W_{DA} is subsequently obtained by means of the standard procedure involving averaging over initial and summing over final molecular sublevels, as in the previous sections.

A new feature that arises in the present dynamical context is that the excitation frequencies ω_{D^*} and ω_{A^*} have now been modified (renormalized) through interaction of the transfer species D and A with the molecules of the surrounding medium. The mutual interaction of D with A may also be taken into account by retaining the omitted terms Γ_{D^*A} and Γ'_{DA^*} in Eqn 2.121. That alters the molecular excitation frequencies ω_{D^*} and ω_{A^*} featured in the energy conservation δ-function of Eqn 2.123, by the amounts $-\mathrm{Im}\,\Gamma_{D^*A}/2$ and $-\mathrm{Im}\,\Gamma'_{DA^*}/2$, respectively: these represent changes in the excitation energy of each transfer species due to its interaction with the other. The effects of such corrections decrease with distance, and over the separations of interest where R is greater than typical intermolecular distances within the medium, they contribute negligibly.

At this juncture, a remark should be made concerning some asymmetry of the formalism with regard to the initial and final states, as reflected by the prime on Γ'_{DA^*}. The rate regime generally implies the presence of a dense structure of (usually vibrational) molecular energy levels within the electronic manifolds of D and A. Hence the apparent asymmetry in question vanishes, as either inclusion or exclusion of the individual states (such as $|I\rangle$ or $|F\rangle$) in the intermediate-state summation does not significantly alter the quantities Γ_{D^*A} and Γ'_{DA^*}. It is a different story in the case where there is no intrinsic density of molecular states for the participating species, as is to be considered next.

2.6.2.2   Nonrate regime

Suppose now that each of the ground and exited state manifolds of D and A is characterized by only one molecular sublevel, so that the subsystem D–A may be treated as a pair of two-level species. Ignoring contributions from states with two or more mediating bath excitations (polaritons), the exchange of energy between D and A now occurs exclusively through intermediate states in which both transfer species are either in their ground or excited states, the bath being in a one-polariton excited state. Under these conditions, the quantities Γ_{D^*A} and Γ'_{DA^*} introduced in Eqns 2.107 and 2.108 are as follows:

$$-\frac{\mathrm{i}}{2}\Gamma_{D^*A} = \left\{\left[U_{FI}^{(2)}(\omega)\right]^2/\hbar^2\right\}(\omega - \omega_{A^*} + \frac{\mathrm{i}}{2}\gamma_{A^*} + \mathrm{i}\eta')^{-1}, \qquad (2.124)$$

$$\Gamma'_{DA^*} = 0, \qquad (2.125)$$

where use has been made of Eqns 2.100 and 2.104. Substituting these results for Γ'_{DA^*} and Γ_{D^*A} into the general dynamical equation 2.109, the probability amplitude becomes

$$\langle F|\tilde{S}(t)|I\rangle = -\frac{1}{2\pi\hbar\mathrm{i}}\int_{-\infty}^{+\infty}\mathrm{d}\omega\,U_{FI}^{(2)}(\omega)\mathrm{e}^{-\mathrm{i}(\omega-\omega_{A^*})t}\left\{\left(\omega - \omega_{A^*} + \frac{\mathrm{i}}{2}\gamma_{A^*} + \mathrm{i}\eta'\right)\right.$$
$$\left.\left(\omega - \omega_{D^*} + \frac{\mathrm{i}}{2}\gamma_{D^*} + \mathrm{i}\eta'\right) - \left[U_{FI}^{(2)}(\omega)/\hbar\right]^2\right\}^{-1}.$$

$$(2.126)$$

To illustrate the precise form of the time evolution for one specific application, one finds for the case of identical species ($\omega_{D^*} = \omega_{A^*}, \gamma_{D^*} = \gamma_{A^*}$), and without regard to the delay time R/ν_g, the following:[†]

$$|\langle F|\tilde{S}(t)|I\rangle|^2 = \tfrac{1}{2}[\cosh(\gamma_{DA}t) - \cos(2\Omega_{DA}t)]e^{-\gamma_{D^*}t}, \qquad (2.127)$$

where the transfer frequency Ω_{DA} and the inverse time γ_{DA}, respectively, represent the real and imaginary parts of the transition matrix element:

$$\hbar^{-1}U_{FI}^{(2)}(\hbar\omega_{A^*}) = \Omega_{DA} - \frac{i}{2}\gamma_{DA}. \qquad (2.128)$$

Equation 2.127 has a form that is familiar from the case of energy transfer between molecules *in vacuo* [22, 32, 41], although the parameters $\gamma_{DA}, \gamma_{A^*}$, and Ω_{DA} here display the influence of the medium. Note that although in writing Eqns 2.124–2.128 the transfer species D and A have been modeled as two-level systems, the formulation still allows each of the surrounding molecules to possess an arbitrary number of energy levels, thus accommodating the cases of both absorbing and lossless media.

The result given by Eqn 2.127 represents an oscillatory, to- and-fro exchange of excitation, accompanied by damping. That type of dynamical behavior is a direct consequence of the absence of a density of final states, a feature which obviously makes the rate description inadequate. Nonetheless, a distinction should be drawn between the short-range reversible Rabi-type oscillatory behavior, which does not represent any real flow of energy from D to A, and the long-range behavior. In the latter case, the excitation energy of the donor is irreversibly passed to the acceptor. Under such circumstances it is appropriate to introduce transfer probabilities (rather than rates), as will be shown below.

In the long-range limit, the contribution $[U_{FI}^{(2)}]^2$, associated with the coupling between D and A, may legitimately be omitted in the denominator of the integrand in Eqn 2.126. The system then follows the same time evolution as described through the earlier Eqn 2.122, where in the current consideration D and A are not necessarily identical two-level species. The transfer dynamics given by Eqn 2.122 reflects both the initial arrival of excitation at the acceptor, commencing from time $t = R/\nu_g$, and subsequent decay of the resulting excited state of A. The rate of the latter decay may be considered to be the same as that for an individual acceptor in the dielectric medium (i.e. γ_{A^*}), since at large distances the remaining influence of the donor is minimal. Accordingly, the total transfer probability P may be defined as the probability for irreversible trapping of the excitation by the acceptor. Integrating the population-weighted rate of decay of the excited state of A, we obtain, for P,

---

[†] Such a time delay leading to the effects of multiple delay in to- and-fro exchange of excitation, has been investigated by Milonni and Knight [44], who analysed the transfer dynamics between a pair of species *in vacuo*.

$$P = \int_{R/v_g}^{\infty} |\langle F|\tilde{S}(t)|I\rangle|^2 \gamma_{A^*} \, dt \tag{2.129}$$

$$= (\Omega_{DA}^2 + \gamma_{DA}^2)\gamma_{D^*}^{-1} \frac{(\gamma_{D^*} + \gamma_{A^*})}{(\omega_{A^*} - \omega_{D^*})^2 + (\gamma_{D^*} + \gamma_{A^*})^2/4}, \tag{2.130}$$

which is in agreement with the previous far-zone result for the transfer of energy between a pair of molecules *in vacuo* [54]. In passing, we note that the individual rates of the excited-state decay $\gamma_{D^*} \equiv \Gamma_D$ and $\gamma_{A^*} \equiv \Gamma_A$, featured in the above equations, have been explicitly analysed in Section 2.5.2. Calling on Eqns 2.114, 2.116, 2.117 and 2.128, the long-range result of Eqn 2.130 assumes the following form in the case of a nonabsorbing medium:

$$P = \frac{9}{8\pi} \langle \sigma_A \rangle \left[(\hat{\boldsymbol{\mu}}_D \cdot \hat{\boldsymbol{\mu}}_A) - (\hat{\boldsymbol{\mu}}_D \cdot \hat{\mathbf{R}})(\hat{\boldsymbol{\mu}}_A \cdot \hat{\mathbf{R}}) \right]^2 / R^2, \tag{2.131}$$

with

$$\langle \sigma_A \rangle = \frac{1}{n} \left(\frac{n^2 + 2}{3} \right)^2 \frac{\mu_A^2 \omega_{A^*}}{3\varepsilon_0 \hbar c} \left[\frac{(\gamma_{D^*} + \gamma_{A^*})/2}{(\omega_{A^*} - \omega_{D^*})^2 + (\gamma_{D^*} + \gamma_{A^*})^2/4} \right] \tag{2.132}$$

Here, in addition to the appearance of the refractive pre-factors, the influence of the medium extends to the excitation frequencies ω_{D^*} and ω_{A^*}, as well as to the decay parameters γ_{A^*} and γ_{B^*}. The above $\langle \sigma_A \rangle$ may be identified as the isotropic absorption section of the acceptor, $\sigma_A(\omega)$, averaged over the normalized emission spectrum of the donor, $I_{D^*}(\omega)$:

$$\langle \sigma_A \rangle = \int_{-\infty}^{+\infty} \sigma_A(\omega) I_{D^*}(\omega) d\omega, \tag{2.133}$$

with $\sigma_A(\omega)$ and $I_{D^*}(\omega)$ given by

$$\sigma_A(\omega) = \frac{1}{n} \left(\frac{n^2 + 2}{3} \right)^2 \frac{\mu_A^2 \omega_{A^*}}{3\varepsilon_0 \hbar c} \left[\frac{\gamma_{A^*}/2}{(\omega - \omega_{A^*})^2 + (\gamma_{A^*}/2)^2} \right] \tag{2.134}$$

and

$$I_{D^*}(\omega) = \frac{1}{\pi} \left[\frac{\gamma_{D^*}/2}{(\omega - \omega_{D^*})^2 + (\gamma_{D^*}/2)^2} \right]. \tag{2.135}$$

Finally, one obtains, for the orientationally averaged probability,

$$\bar{P} = \langle \sigma_A \rangle / 4\pi R^2, \qquad (2.136)$$

which is the ratio of the spectrally averaged isotropic absorption cross-section to the spherical surface at distance $R, 4\pi R^2$. In the case of an absorbing medium, an exponential decay factor of the form $\exp(-2n''\omega R/c)$ would also feature in the above expression.

2.7   CONCLUSION

The unified theory of resonance energy transfer described in this chapter represents a seamless union of mechanisms that were previously considered to be entirely separate. It has resolved a number of thorny issues that were obscured or skirted by earlier theories and, in fully fledged form, it naturally accommodates all of the optical, dynamical and electronic influences of the host medium on the characteristics of donor–acceptor transfer. In particular, this theory has established the following:

- There is no competition between "radiationless" and "radiative" mechanisms for energy transfer. Both are manifestations of a single interaction which operates over the full range of distances beyond wavefunction overlap. This interaction additionally includes other, previously hidden, forms of interaction, whose distance dependence is characterized by intermediate power laws. With this theory, there is no longer any need to model systems in which energy migrates across both short- and long-range distances in terms of distinct mechanisms.
- The shift in the inverse power which is apparent as distance increases reflects a powerful interplay between the governing principles of relativistic retardation and quantum uncertainty, as they relate to the propagation of the signal mediating the energy transfer. At short distances the overwhelming quantum uncertainty is reflected in the completely virtual nature of the mediating photons; over long distances the uncertainty is small and the photons acquire unequivocally real character. The polarization of acceptor fluorescence has been shown to display a distance dependence which dramatically exhibits this behavior.
- The pair transfer rates provided by the unified theory have been integrated into the context of an ensemble treatment that is directly amenable to experimental application. The effect of spontaneous emission by donor species within an absorbing ensemble in all respects properly corresponds to the far-zone limit of unified energy transfer. Finally, the apparent rate divergence associated with transfer from a single donor within an infinite ensemble of acceptor species has been resolved by proper accommodation of the effect of the ensemble on the characteristics of (virtual) medium-

dressed photons that mediate the resonance energy transfer between the molecules.

Appendix A: Heitler–Ma method for analysis of the transition operator

Consider the following Green's operator:

$$\zeta(\varepsilon - H) = (\varepsilon - H + i\eta)^{-1}, \tag{A.1}$$

where the full Hamiltonian H is divided into the zero-order Hamiltonian H^0 and the interaction operator V as in Eqn 2.57. Let $N(\varepsilon)$ be the diagonal part of $\zeta(\varepsilon - H)$ (in the representation of the eigenvectors of H^0). Then the operator $\zeta(\varepsilon - H)$ may written as

$$\zeta(\varepsilon - H) = [1 + \zeta(\varepsilon - H_0)U(\varepsilon)]N(\varepsilon), \tag{A.2}$$

where $U(\varepsilon)$ is an off-diagonal operator. Multiplying Eqn A.2 by $\varepsilon - H$, one finds that

$$N^{-1}(\varepsilon) = \varepsilon - H + U(\varepsilon) - V\zeta(\varepsilon - H_0)U(\varepsilon). \tag{A.3}$$

The diagonal operator $N(\varepsilon)$ may be represented as

$$N(\varepsilon) = \left[\varepsilon - H_0 + \frac{i}{2}\hbar\Gamma(\varepsilon)\right]^{-1}, \tag{A.4}$$

$\Gamma(\varepsilon)$ being another diagonal matrix, so that one has

$$U(\varepsilon) - \frac{i}{2}\hbar\Gamma(\varepsilon) = V + V\zeta(\varepsilon - H_0)U(\varepsilon). \tag{A.5}$$

Calling on Eqns A.2, A.4, and A.5, one arrives at

$$\zeta(\varepsilon - H)|I\rangle = \frac{[1 + \zeta(\varepsilon - H_0)U(\varepsilon)]|I\rangle}{\varepsilon - E_I + i\hbar\Gamma_I(\varepsilon)/2 + i\eta} \tag{A.6}$$

The last two equations, together with the definition in Eqn A.1, provide the required results, Eqns 2.98 and 2.99 of the main text. Finally, the relationship A.6 can be represented in a nonrecursive format using Eqn 2.100, as

$$\zeta(\varepsilon - H)|I\rangle = \frac{[1 + P_I\zeta(\varepsilon - H_0 - P_I V P_I)P_I V]|I\rangle}{\varepsilon - E_I + i\hbar\Gamma_I(\varepsilon)/2 + i\eta}. \tag{A.7}$$

Appendix B: Modified approach to the transition operator

Using Eqn 2.100 of the main text, one finds for the transition operator that one has

$$\langle F|U(\varepsilon)|I\rangle = \langle X|P_I V|I\rangle, \tag{B.1}$$

with

$$\langle X| = \langle F|\left[1 + P_1 V P_1(\varepsilon - H_0 - P_I V P_I + i\eta)^{-1}\right]. \tag{B.2}$$

The above ket vector may be rewritten as

$$\langle X| = \langle F|(\varepsilon - H_0 + i\eta)(\varepsilon - H_0 - P_I V P_I + i\eta)^{-1},$$

giving

$$\langle X| = (\varepsilon - E_F + i\eta)\langle F|\zeta(\varepsilon - H_0 - P_I V P_I) \tag{B.3}$$

Next, in complete analogy to Eqn A.7 for the bra vectors, the above ket vector $\langle F|\zeta(\varepsilon - H_0 - P_I V P_I)$ is

$$\langle F|\zeta(\varepsilon - H_0 - P_I V P_I) = \frac{\langle F|[1 + V P_I P_F \zeta(\varepsilon - H_0 - P_F P_I V P_I P_F)]}{\varepsilon - E_F + i\hbar\Gamma'_F/2 + i\eta}, \tag{B.4}$$

with Γ'_F and P_F as in Eqns 2.104 and 2.105. Lastly, calling on Eqns B.1, B.3, and B.4, one arrives at the required result, Eqn 2.102, presented in the main text.

ACKNOWLEDGEMENT

One of us (G.J.) wishes to acknowledge the Alexander von Humboldt Foundation for financial support through a European Fellowship at the University of East Anglia, February-March 1998.

References

1. Agarwal, G. S. 1974. Quantum statistical theories of spontaneous emission and their relation to other approaches. In *Springer Tracts in Modern Physics*. G. Höhler, editor. Springer-Verlag, Berlin, Vol. 70.
2. Agranovich, V. M., and M. D. Galanin 1982. *Electronic Excitation Energy Transfer in Condensed Matter*. North-Holland, Amsterdam.

3. Allcock, P., and D. L. Andrews 1998. Two-photon fluorescence: resonance energy transfer. *J. Chem. Phys.* 108: 3089–3095.
4. Andrews, D. L. 1989. A unified theory of radiative and radiationless molecular energy transfer. *Chem. Phys.* 135: 195–201.
5. Andrews, D. L., and P. Allcock 1995. Bimolecular photophysics. *Chem. Soc. Revs* 24: 259–265.
6. Andrews, D. L., and G. Juzeliūnas 1991. The range dependence of fluorescence anisotropy in molecular energy transfer. *J. Chem. Phys.* 95: 5513–5518.
7. Andrews, D. L., and G. Juzeliūnas 1992. Intermolecular energy transfer: retardation effects. *J. Chem. Phys.* 96: 6606–6612.
8. Andrews, D. L., and B. S. Sherborne 1987. Resonant excitation transfer: a quantum electrodynamical study. *J. Chem. Phys.* 86: 4011–4017.
9. Andrews, D. L., Craig, D. P., and T. Thirunamachandran 1989. Molecular quantum electrodynamics. *Int. Rev. Phys. Chem.* 8: 339–383.
10. Avery, J. S. 1966. Resonance energy transfer and spontaneous photon emission. *Proc. Phys. Soc. (London)* 88: 1–8.
11. Avery, J. S. 1984. Use of the S-matrix in the relativistic treatment of resonance energy-transfer. *Int. J. Quant. Chem.* 25: 79–96.
12. Barnett, S. M., Huttner, B., and R. Loudon 1992. Spontaneous emission in absorbing dielectric media. *Phys. Rev. Lett.* 68: 3698–3701.
13. Barnett, S. M., Huttner, B., Loudon, R., and R. Matloob 1996. Decay of excited atoms in absorbing dielectrics. *J. Phys. B – Atom. Molec. Opt. Phys.* 29: 3763–3781.
14. Berberan-Santos, M. N., Nunes Pereira, E. J., and J. M. G. Martinho 1995. Stochastic theory of molecular radiative transport. *J. Chem. Phys.* 103: 3022–3028.
15. Berberan-Santos, M. N., Nunes Pereira, E. J., and J. M. G. Martinho 1997. Stochastic theory of combined radiative and nonradiative transport. *J. Chem. Phys.* 107: 10 480–10 484.
16. Craig, D. P., and T. Thirunamachandran 1998. *Molecular Quantum Electrodynamics.* Dover Publications, New York.
17. Craig, D. P., and T. Thirunamachandran 1989. Third-body mediation of resonance coupling between identical molecules. *Chem. Phys.* 135: 37–48.
18. Craig, D. P., and T. Thirunamachandran 1992. An analysis of models for resonant transfer of excitation using quantum electrodynamics. *Chem. Phys.* 167: 229–240.
19. Dexter D. L. 1953. A theory of sensitized luminescence of solids. *J. Chem. Phys.* 21: 836–850.
20. Durrant, A. V. 1980. A derivation of optical field quantization from absorber theory. *Proc. R. Soc. London* A 370: 41–59.
21. Fermi, E. 1932. Quantum theory of radiation. *Rep. Mod. Phys.* 4: 87–132.
22. Ficek, Z., Tanas, R., and S. Kielich 1986. Cooperative effects in the spontaneous emission from two non-identical atoms. *Opt. Acta* 33: 1149–1160.
23. Förster, Th. 1948. Zwischenmolekulare Energiewanderung und Fluoreszenz. *Ann. Phys.* 6: 55–75.
24. Galanin, M. D. 1953. On the question about influence of concentration on the solution fluorescence. *Zh. Eksp. Theoret. Fiz.* 28: 485–495.
25. Galanin, M. D. 1996. *Luminescence of Molecules and Crystals.* Cambridge International Science Publishing, Cambridge.
26. Gomberoff, L., and E. A. Power 1966. The resonance transfer of excitation. *Proc. Phys. Soc. (London)* 88: 281–284.
27. Hamilton, J. 1949. Damping theory and the propagation of radiation. *Proc. Phys. Soc. (London)* 62: 12–18.

28. Healy, W. P. 1982. *Non-relativistic Quantum Electrodynamics.* Academic Press, London, p. 138.

29. Heitler, W. 1954. *The Quantum Theory of Radiation.* Oxford University Press, New York, p. 163.

30. Heitler, W., and S. T. Ma 1949. Quantum theory of radiation damping for discrete states. *Proc. R. Irish Acad.* 52, Sect. A: 109–125.

31. Ho, S. T., and P. Kumar 1993. Quantum optics in dielectrics: macroscopic electromagnetic field and medium operators for a linear dispersive lossy medium – a microscopic derivation of the operators and their commutation relations. *J. Opt. Soc. Am.* B 10: 1620–1636.

32. Hutchinson, D. A., and H. F. Hameka 1964. Interaction effects on lifetimes of atomic excitations. *J. Chem. Phys.* 41: 2006–2011.

33. Juzeliūnas, G. 1995. Molecule–radiation and molecule–molecule processes in condensed media – a microscopic QED theory. *Chem. Phys.* 198: 145–158.

34. Juzeliūnas, G. 1996. Microscopic theory of quantization of radiation in molecular dielectrics – normal mode representation of operators for local and averaged (macroscopic) fields. *Phys. Rev. A* 53: 3543–3558.

35. Juzeliūnas, G. 1997. Spontaneous emission in absorbing dielectrics: a microscopic approach. *Phys. Rev. A* 55: R4015–R4018.

36. Juzeliūnas, G. 1998. Microscopic analysis of spontaneous emission in absorbing dielectrics. *J. Luminescence* 76 & 77: 666–669.

37. Juzeliūnas, G., and D. L. Andrews 1994. Quantum electrodynamics of resonance energy transfer. *Phys. Rev. B* 49: 8751–8763.

38. Juzeliūnas, G., and D. L. Andrews 1994. Quantum electrodynamics of resonance energy transfer. II. Dynamical aspects. *Phys. Rev. B* 50: 13 371–13 378.

39. Kikuchi, S. 1930. Über die Fortpflanzung von Lichtwellen in der Heisenberg-Paulischen Formulierung der Quantenelektrodynamik. *Z. Phys.* 66: 558–571.

40. Knoester, J., and S. Mukamel 1989. Intermolecular forces, spontaneous emission, and superradiance in a dielectric medium: polariton-mediated interactions. *Phys. Rev. A* 40: 7065–7080.

41. Lehmberg, R. H. 1970. Radiation from an *N*-atom system. II. Spontaneous emission from a pair of atoms. *Phys. Rev. A* 2: 889–896.

42. Milonni, P. W. 1994. *The Quantum Vacuum: an Introduction to Quantum Electrodynamics.* Academic Press, San Diego.

43. Milonni, P. W. 1995. Field quantization and radiative processes in dispersive dielectric media. *J. Mod. Opt.* 42: 1991–2004.

44. Milonni, P. W., and P. L. Knight 1974. Retardation in the resonance interaction of two identical atoms. *Phys. Rev. A* 10: 1096–1108.

45. Pegg, D. T. 1979. Absorber theory approach to the dynamic Stark effect. *Ann. Phys. (N.Y.).* 118: 1–17.

46. Perrin, F. 1932. Théorie quantique des transferts d'activation entre molécules de même espèce. Cas des solutions fluorescentes. *Ann. Phys. (Paris)* 17: 283–314.

47. Power, E. A., and T. Thirunamachandran 1983. Quantum electrodynamics with non-relativistic sources. III. Intermolecular interactions. *Phys. Rev. A* 28: 2671–2675.

48. Power, E. A., and T. Thirunamachandran 1993. A new insight into the mechanism of intermolecular forces. *Chem. Phys.* 171: 1–7.

49. Power, E. A., and T. Thirunamachandran 1993. Quantum electrodynamics with non-relativistic sources. V. Electromagnetic field correlations and interactions between molecules in either ground or excited-states. *Phys. Rev. A* 47: 2539–2551.

50. Power, E. A., and S. Zienau 1959. Coulomb gauge in non-relativistic quantum electrodynamics and the shape of spectral lines. *Phil. Trans. R. Soc. London* A 251: 427–454.
51. Rodberg, L. S., and Thaler R. M. 1967. *Introduction to the Quantum Theory of Scattering*. Academic Press, New York, Chapter 8.
52. Scholes, G. D., and D. L. Andrews 1997. Damping and higher multipole effects in the quantum electrodynamical model for electronic energy transfer in the condensed phase. *J. Chem. Phys.* 107: 5374–5384.
53. Scholes, G. D., Clayton, A. H. A., and K. P. Ghiggino 1992. On the rate of radiationless intermolecular energy transfer. *J. Chem. Phys.* 97: 7405–7413.
54. A. A. Serikov and Yu. M. Khomenko 1978. Molecular excitation transfer by reabsorption. *Physica* 93C: 383–392.
55. Stephen, M. J. 1964. First-order dispersion forces. *J. Chem. Phys.* 40: 669–673.
56. Wheeler, J. A., and R. P. Feynman 1949. Classical electrodynamics in terms of direct interparticle action. *Rev. Mod. Phys.* 21: 425–433.
57. Woolley, R. G. 1971. Molecular quantum electrodynamics. *Proc. R. Soc. London* A 321: 557–572.

3

Dynamics of radiative transport

Mário N. Berberan-Santos, Eduardo J. Nunes Pereira, and
José M. G. Martinho

Instituto Superior Técnico, Portugal

3.1 INTRODUCTION

Radiative transfer, i.e. the transfer of energy mediated by real (as opposed to virtual) photons, is ubiquitous in Nature. For instance, the first step of photosynthesis consists of radiative transfer from the Sun's photosphere to the chlorophyll molecules of green plant leaves. The donor and acceptor are in this case 150 million kilometers apart, and the process takes eight minutes to be completed. Radiative transfer is also of importance in astrophysics, plasmas and in atomic and molecular luminescence, and it plays an important role in solar concentrators, discharge and fluorescent lamps, scintillation counters, and lasers.

The type of radiative transfer to be discussed in this chapter consists of the emission of a photon by an electronically excited molecule, with subsequent absorption by an identical ground-state molecule. It involves distances much smaller than those of the above example, and consequently occurs on much shorter timescales, usually determined by molecular excited-state lifetimes and not by photon propagation times.

In assemblies of like atoms or molecules, one elementary process of radiative transfer leads to another, until one of two things happens: (a) the excitation energy is irreversibly lost through a nonradiative path (internal conversion, intersystem crossing, quenching, etc.), or (b) the photon escapes from the sample (Fig. 3.1). This repeated radiative transfer is known by a number of names: radiative transport, radiative migration, self-absorption, reabsorption,

Resonance Energy Transfer. Edited by David L. Andrews and Andrey A. Demidov.
© 1999 John Wiley & Sons Ltd

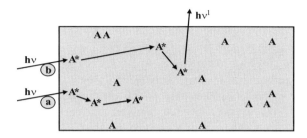

Figure 3.1   A schematic representation of the two kinds of photon trajectory. In case (a), the absorbed photon, of energy *hν*, never exits the sample, because the trajectory ends in a nonradiative decay process. In case (*b*), and after several reabsorptions, a photon of energy *hν'* escapes from the enclosure.

radiation imprisonment, and radiation trapping. Its importance depends on many factors: the extent of spectral overlap between absorption and emission, absorption strength, fluorescence quantum yield, concentration, cell size and shape, excitation and detection geometries, etc. In the molecular case, it plays a particularly important role in solutions of highly fluorescent compounds with a good absorption–emission spectral overlap, whether concentrated or in large volumes. When present, radiative transport affects the measured fluorescence decays and spectra, as well as the fluorescence polarization. These observables are then functions of the excitation and emission wavelengths, concentration, and excitation and detection geometries.

When preparing this review, it was realized that the theoretical and experimental studies of radiative transport at a microscopic level have been conducted thus far in two almost noncommunicating areas, one concerned with atomic resonance radiation, mainly in the gas phase, and the other concerned with molecular fluorescence, mainly in the condensed phases. While there are aspects that are specific to each field, many concepts are common. Indeed, some remarkably similar results have been independently obtained in the two fields. In discussing molecular processes, mention of the relevant results obtained in the atomic field is therefore appropriate.

3.2   OVERVIEW OF ATOMIC AND MOLECULAR RADIATIVE TRANSPORT

3.2.1   Atomic resonance radiation

The first experimental study of radiative transport was made in 1912 by R. W. Wood [67], the discoverer of resonance radiation, with mercury vapor. He observed that, upon continuous excitation of a given part of the vapor, 253.7 nm atomic mercury resonance radiation was emitted from a volume larger than that irradiated. By interposing a quartz window, he confirmed that the spread was due not to atomic motion, but rather to radiation scattering, by means of repeated absorptions and re-emissions. In 1923, Compton [23] put forward the first theory of radiative transport, considered by him as formally identical to gaseous diffusion: denoting the (base e) absorption coefficient of the medium by α, and by τ the average interval between absorption and emission of a quantum by an atom, the mean free path of the quanta is $1/\alpha$ and the effective average speed of the diffusing photons would be $1/(\alpha\tau)$. From these results, it followed that there was a "tendency of resonance radiation to remain imprisoned within a gas for a time which may be enormous in comparison with the time τ of imprisonment within individual atoms," in qualitative agreement with previous experimental observations. It was concluded that this effect should be important whenever the mean free path of the radiation was smaller than the dimensions of the apparatus (discharge tube). Compton's diffusion theory was later refined by Milne [43].

In 1927, Zemansky [68] carried out the first quantitative study of the phenomenon, again on the mercury 253.7 nm resonance line. He was able to show that the effective decay time of the radiation emitted in the central part of a discharge tube could be a thousand times greater than the intrinsic atomic state lifetime. An essential discrepancy between the experimental results and the Compton–Milne theory predictions was nevertheless found. Zemansky drew attention to several possible causes, including the effect of the line shape. However, he incorrectly assumed that "the radiation composing the whole absorption line diffused as a whole through the vapour," a mean absorption coefficient being thus applicable.

In 1932, Kenty [34] attempted to take into account the variation of the absorption over the frequency spectrum of the emission, by considering a Doppler line profile. He arrived at the surprising conclusion that, for an infinite medium, the average diffusion coefficient was infinite. This is indeed so, on account of the overwhelming contribution of the few photons originating at the band wings (that supposedly extend to infinite frequencies). This showed the general inapplicability of the diffusion equation to the radiative transport problem, an aspect that was, however, only clearly recognized in 1947, by Holstein [30, 31] and, independently, by Biberman [12]. Both derived a Boltzmann-type integro-differential equation for radiative transport, whose solution for a cutoff experiment was obtained by Holstein as a sum of time exponentials. An approximate solution, valid for sufficiently long times, was then given in terms of the slowest decaying exponential, whose argument was numerically obtained for two idealized enclosure geometries (infinite slab and infinite cylinder) and various spectral line shapes [30, 31]. Reasonable agreement with part of Zemansky's results was obtained. A refined experimental study of the mercury

253.7 nm line by Alpert, McCoubrey, and Holstein [2] yielded still better agreement, but also disclosed several complications not accounted for by the theory, namely the increasing contribution of collisional broadening with density.

Later studies of radiative transport in atomic gases have mainly dealt with refinements and experimental tests of approximate solutions to the Holstein–Biberman equation, taking into account aspects such as hyperfine structure [2, 62], simultaneous Doppler and Lorentz (collisional) broadening [62], Heisenberg (natural) broadening [32], incomplete frequency redistribution [50, 52, 54, 55], and inhomogeneous broadening [39]. Other works include approximate calculation of the full solution to the Holstein–Biberman equation, i.e. of the parameters in the sum of time exponentials (up to 20 terms), for the infinite slab [44, 60], for the infinite cylinder [45, 51, 61] and for the sphere [45], and the calculation of approximate solutions in the low-opacity region [20, 44], for which Holstein's long-time limit is of little use.

More recently, Wiorkowski and Hartmann developed and applied to experimental results [21, 64–66] a multiple scattering approach that allows calculation of the time dependence of the fluorescence, in response to an excitation pulse, in terms of an infinite series of unknown coefficients. The equivalence between this treatment and that based on the Holstein–Biberman equation has been shown [25, 36]. The multiple scattering approach is physically appealing, mathematically simpler and more flexible with respect to refinements. The unknown parameters of the theoretical decay law can be evaluated by Monte Carlo simulation [21, 64–66]. Lai, Liu, and Ma [36] have recently pointed out that, given the equivalence of the Holstein–Biberman equation and the multiple scattering method, the solution of this last treatment should asymptotically converge to the long-time exponential decay of the Holstein–Biberman equation. They then obtained formulas similar to those of the multiple scattering method directly from the Holstein–Biberman equation. Combining both types of solutions, it was shown that a single formula, with a finite number of terms, was able to reproduce both the short- and long-time behavior. The same result can also be used to obtain the steady state response of the system in terms of a finite sum [38].

3.2.2   Molecular fluorescence

Molecular radiative transport studies, theoretical or experimental, have not relied explicitly on the Holstein–Biberman equation. In fact, the only approximate solution of this equation known for many years, based on the fundamental mode, is often not appropriate to the molecular case. The typical molecular absorption coefficients and molecular absorption–emission overlaps are much smaller than those of atoms. Also, the fluorescence quantum yield is always smaller than unity. These three unfavorable factors render radiative transport much less efficient in molecular ensembles than in atomic ensembles. As a consequence, the average macroscopic decay times of molecular ensembles seldom attain values more than three or four times the intrinsic lifetime – while for atoms, values one to three orders of magnitude higher than the intrinsic lifetime are common. In this way, molecular fluorescence decays under the influence of radiative transport are usually nonexponential, and show no signs of the fundamental mode – which manifests itself only at very late times, when the intensity is negligible.

In the molecular field, motivation for the study of radiative transport *per se* came from two applied subjects: scintillation counting [13, 15, 35] and analytic fluorescence spectrometry [37, 42, 49]. Particularly in the last case, reabsorption was, and still is, an unwanted complication in determination of the true fluorescence spectrum, quantum yield and lifetime. The first attempt to seriously model self-absorption was that of Birks [14]. He proposed an admittedly simple kinetic model to account for the increase of the macroscopic decay time. In this model, excited molecules were classified according to generation, and a common self-absorption probability was used for all photons. This implied an exponential decay. Later, Birks [17] recognized that the self-absorption probabilities of the successive generations were in general different, but failed to notice that the theoretical decay would no longer be exponential. Birks's approximate treatment was also presented in his influential reviews of molecular luminescence [16, 18, 19]. For this reason, and also because of the inherent simplicity, it became widely used; see, e.g., [5, 7, 24, 28, 53, 58].

Notwithstanding this, Kilin and Rozman [35], following a kinetic approach similar to that used later by Birks [16, 17, 19], had in 1959 obtained not only the correct (nonexponential) time dependence, rediscovered in 1985 by Wiorkowski and Hartmann [65] for the atomic case, but also the influence of self-absorption on the emission spectrum. However, their work remained little known until it was revived and extended [40, 41, 46]. Almost simultaneously, a similar approach, although lacking the emission wavelength dependence, was presented and applied to experimental results [33, 56, 57]. A stochastic theory of radiative transport was afterwards presented [10], allowing the calculation of all observables (including the time dependence of the polarization) from known parameters [10, 47]. This approach was recently refined and extended to include the contribution of nonradiative transport [11].

3.3   THE HOLSTEIN–BIBERMAN EQUATION

3.3.1   Delta-pulse excitation

Although derived with regard to the resonance radiation transport problem, this equation – slightly generalized – is equally appropriate for the molecular case. Consider a given atomic or molecular ensemble, in a convex enclosure of otherwise arbitrary shape. Suppose that a small fraction of the ensemble is excited at time zero by a delta pulse, generating N_0 excited atoms or molecules, with a spatial distribution characterized by the number density $n(\vec{r}, t)$. Then, the time evolution of the number density of excited species $n(\vec{r}, t)$ will be given by

$$\frac{\partial n(\vec{r}, t)}{\partial t} = -\Gamma n(\vec{r}, t) + k_r \int_V f(\vec{r}, \vec{r}')n(\vec{r}', t)\mathrm{d}\vec{r}', \tag{3.1}$$

which is the Holstein–Biberman equation [12, 30]. In Eqn 3.1, the integration goes over the whole volume of the enclosure; Γ is the reciprocal atomic or molecular lifetime,

$$\Gamma = k_r + k_{nr} = \frac{1}{\tau_0}, \tag{3.2}$$

and k_r and k_{nr} are the radiative and nonradiative decay constants, respectively. In the atomic case, k_{nr} is due to quenching and is usually negligible. The function $f(\vec{r}, \vec{r}')$ is the average probability that a photon emitted by an excited atom or molecule at point $\vec{r}'$ will be absorbed by a ground-state atom or molecule in a unit volume element around point $\vec{r}$,

$$f(\vec{r}, \vec{r}') = \int_0^\infty F(v)k(v)\frac{1}{4\pi|\vec{r} - \vec{r}'|^2}\exp[-k(v)|\vec{r} - \vec{r}'|]\mathrm{d}v, \tag{3.3}$$

where $F(v)$ is the normalized emission spectrum, in the atomic case usually proportional to $k(v)$, the latter being the (base e) absorption coefficient of the medium.

Implicit in this derivation is: (i) neglect of the time of flight of the photons in comparison with the intrinsic lifetime τ_0. This is usually a very good approximation, breaking down only for very extended media of low opacity, where retardation effects are significant. Other usually valid approximations contained in Eqn 3.1 are (ii) neglect of boundary effects such as reflection or wall quenching, (iii) supposition of isotropic emission, (iv) assumption of spatial homogeneity of ground state species, and (v) neglect of material transport of excitation (diffusion, convection). Finally, (vi) complete frequency redistribution is assumed, i.e. a single emission frequency distribution $F(v)$ is valid for all atoms or molecules. This last approximation is not always valid, especially for atoms, and suitable extensions of Eqn 3.1 have been carried out [50, 52, 54, 55].

It is useful to define the spatial distribution function of excited species, $p(\vec{r}, t)$,

$$p(\vec{r}, t) = \frac{n(\vec{r}, t)}{\displaystyle\int_V n(\vec{r}, t)\mathrm{d}\vec{r}} = \frac{n(\vec{r}, t)}{N(t)}, \qquad (3.4)$$

where $N(t)$ is the total number of excited atoms or molecules in the enclosure at time t. Using Eqn 3.4, Eqn 3.1 can be rewritten as

$$\frac{\partial p(\vec{r}, t)}{\partial t} = k_r \left[\int_V f(\vec{r}, \vec{r}')p(\vec{r}', t)\mathrm{d}\vec{r}' - p(\vec{r}, t) \int_V \int_V f(\vec{r}, \vec{r}')p(\vec{r}, t)\mathrm{d}\vec{r}\,\mathrm{d}\vec{r}' \right]. \qquad (3.5)$$

From Eqn 3.5 and, on physical grounds, one sees that a stationary spatial distribution is attained for long times, obeying, irrespective of the initial distribution,

$$p_s(\vec{r}) = \frac{\displaystyle\int_V f(\vec{r}, \vec{r}')p_s(\vec{r}')\mathrm{d}\vec{r}'}{\displaystyle\int_V \int_V f(\vec{r}, \vec{r}')p_s(\vec{r}')\mathrm{d}\vec{r}\,\mathrm{d}\vec{r}'}. \qquad (3.6)$$

Equation 3.6 can be rewritten as

$$p_s(\vec{r}) = \frac{\alpha_s(\vec{r})}{\bar{\alpha}_s}, \qquad (3.7)$$

where the numerator is the average probability that a photon, originating from an excited-state population distributed according to the stationary spatial distribution, will be absorbed in a unit volume element around point $\vec{r}$; the denominator is the average probability that a photon originating from an excited-state population distributed according to the same distribution will be absorbed.

From Eqns 3.1 and 3.4, and using the definition of $N(t)$, one obtains

$$-\frac{\mathrm{d}N}{\mathrm{d}t} = \left[\Gamma - k_r \int_V \int_V f(\vec{r}, \vec{r}')p(\vec{r}, t)\mathrm{d}\vec{r}\,\mathrm{d}\vec{r}' \right] N = [\Gamma - k_r\bar{\alpha}(t)]N, \qquad (3.8)$$

where $\bar{\alpha}(t)$ is the probability that a photon originating from an excited-state population distributed according to the spatial distribution $p(\vec{r}, t)$ will be absorbed. Integration of Eqn 3.8 gives

$$N(t) = N_0 \exp(-\Gamma t)\exp\left[k_r \int_0^t \bar{\alpha}(u)\mathrm{d}u \right], \qquad (3.9)$$

and hence the normalized (i.e. scaled to one at $t = 0$) decay of intensity of the radiation emitted by the enclosed ensemble, integrated over all directions, $\rho(t)$, is given by

$$\rho(t) = \frac{1 - \bar{\alpha}(t)}{1 - \bar{\alpha}_s(0)} \exp(-\Gamma t) \exp\left[k_r \int_0^t \bar{\alpha}(u) du \right]. \tag{3.10}$$

It follows, both from Eqn 3.10 and from the stabilization of the average absorption probability with time, that a single exponential decay will result for sufficiently long times, with a lifetime given by

$$\tau_s = \frac{\tau_0}{1 - \bar{\alpha}_s \Phi_0}, \tag{3.11}$$

where Φ_0 is the intrinsic emission quantum yield, $\Phi_0 - k_r / \Gamma - k_r \tau_0$.

In this way, to obtain the long-time limit solution of the radiative problem, one needs only to compute the average absorption probability corresponding to the long-time stationary distribution. This distribution can in turn be numerically obtained from Eqn 3.6, for each particular geometry, ground state density and line shape.

It may be noted that the maximum possible value for τ_s is attained for unit absorption probability in Eqn 3.11,

$$\tau_{s, max} = \frac{1}{k_{nr}}. \tag{3.12}$$

The general solution of Eqn 3.1 can be written as [30]

$$n(\vec{r}, t) = \sum_i n_i(\vec{r}) e^{-\beta_i t}, \tag{3.13}$$

where the $n_i(\vec{r})$ are the stationary (but not necessarily positive for all values of $\vec{r}$) solutions of Eqn 3.1, and the β_i are the corresponding eigenvalues. The eigenvalue equation is obtained by insertion of Eqn 3.13 into Eqn 3.1, and takes the form

$$\Gamma n_i(\vec{r}) - k_r \int_V f(\vec{r}, \vec{r}') n_i(\vec{r}') d\vec{r}' = \beta_i n_i(\vec{r}). \tag{3.14}$$

Integrating both sides of Eqn. 3.14 over space, one obtains, for the eigenvalues,

$$\beta_i = \Gamma - k_r \bar{\alpha}_i, \tag{3.15}$$

where the generalized absorption probability $\bar{\alpha}_i$ is

$$\bar{\alpha}_i = \int_V \int_V f(\vec{r}, \vec{r}') p_i(\vec{r}') d\vec{r} \, d\vec{r}', \tag{3.16}$$

and the generalized distribution function $p_i(\vec{r})$ is

$$p_i(\vec{r}) = \frac{n_i(\vec{r})}{\int_V n_i(\vec{r}) d\vec{r}} = \frac{n_i(\vec{r})}{N_i}. \tag{3.17}$$

Insertion of Eqn 3.15 into the eigenvalue equation 3.14, rewritten in terms of the generalized distribution functions $p_i(\vec{r})$, shows that these functions are solutions of Eqn 3.6, i.e. they are indeed stationary distributions. Like the stationary solutions $n_i(\vec{r})$, these generalized distributions may take negative values, which are again devoid of direct physical meaning, except for the long-time solution, $p_s(\vec{r})$, that belongs to the smallest eigenvalue, $\beta_s = 1/\tau_s$, which is always positive.

From Eqns 3.13–3.17, the time-dependent distribution function can be written as

$$p(\vec{r}, t) = \frac{\sum_i \left(\int_V n_i(\vec{r}) d\vec{r} \right) p_i(\vec{r}) e^{k_r \bar{\alpha}_i t}}{\sum_i \left(\int_V n_i(\vec{r}) d\vec{r} \right) e^{k_r \bar{\alpha}_i t}} = \frac{\sum_i N_i p_i(\vec{r}) e^{k_r \bar{\alpha}_i t}}{\sum_i N_i e^{k_r \bar{\alpha}_i t}} = \frac{\sum_i c_i p_i(\vec{r}) e^{k_r \bar{\alpha}_i t}}{\sum_i c_i e^{k_r \bar{\alpha}_i t}}, \tag{3.18}$$

the coefficients c_i being the fraction of the excited population associated with the ith eigenvalue,

$$c_i = \frac{N_i}{N_0} \quad (i = 1, 2, \ldots). \tag{3.19}$$

Using $\bar{\alpha}(t)$ as defined for Eqn 3.8, one obtains, from Eqn 3.18,

$$\bar{\alpha}(t) = \frac{\sum_i c_i \bar{\alpha}_i e^{k_r \bar{\alpha}_i t}}{\sum_i c_i e^{k_r \bar{\alpha}_i t}}, \tag{3.20}$$

and inserting it in Eqn 3.10, this yields the sought-for solution:

$$\rho(t) = \exp(-\Gamma t) \sum_i a_i \exp(k_r \bar{\alpha}_i t) = \sum_i a_i \exp(-\beta_i t) = \sum_i a_i \exp(-t/\tau_i), \tag{3.21}$$

where

$$a_i = \frac{(1 - \bar{\alpha}_i)c_i}{\sum_i (1 - \bar{\alpha}_i)c_i} \qquad (3.22)$$

and

$$\tau_i = \frac{\tau_0}{1 - \bar{\alpha}_i \Phi_0}. \qquad (3.23)$$

The average decay time $\bar{\tau}$ is therefore

$$\bar{\tau} = \sum_i b_i \tau_i, \qquad (3.24)$$

with

$$b_i = \left(\frac{1 - \bar{\alpha}_i}{1 - \bar{\alpha}_i \Phi_0} c_i \right) \bigg/ \left(\sum_i \frac{1 - \bar{\alpha}_i}{1 - \bar{\alpha}_i \Phi_0} c_i \right). \qquad (3.25)$$

In this way, $\bar{\tau}$ reduces to τ_s only when $b_s \simeq 1$.

In most experiments, the measured decay corresponds to a given direction or solid angle, and not to the integrated signal over the surface of the enclosure. In such a case, and unless the excitation process and the enclosure geometry are such as to preserve a high symmetry (e.g. excitation along the central axis of a long cylinder), the decay will be a function of the direction of observation. Considering, for instance, that the recorded signal corresponds to the photons reaching a given point $\vec{r}_b$ at the system's boundary, the decay will be

$$\rho(t, \vec{r}_b) \propto \int_V [1 - \alpha_b(\vec{r})] n(\vec{r}, t) d\vec{r} = \sum_i N_i (1 - \bar{\alpha}_{bi}) e^{-\beta_i t}, \qquad (3.26)$$

where $\alpha_b(\vec{r})$ is the average probability that a photon emitted by an excited atom or molecule at point $\vec{r}$, and toward $\vec{r}_b$, will be absorbed by a ground-state atom or molecule before reaching the enclosure's surface at $\vec{r}_b$:

$$\alpha_b(\vec{r}) = \int_0^{|\vec{r}_b - \vec{r}|} \int_0^\infty F(v)k(v)\exp[-k(v)x] dv\, dx. \qquad (3.27)$$

Equation 3.26 becomes, after normalization,

$$\rho(t, \vec{r}_b) = \sum_i \alpha_{bi} e^{-\beta_i t}, \tag{3.28}$$

where

$$a_{bi} = \frac{(1 - \bar{\alpha}_{bi})c_i}{\sum_i (1 - \bar{\alpha}_{bi})c_i}, \tag{3.29}$$

and is similar in form to the overall decay, Eqn. 3.21.

A further case of interest is the direction- and frequency-resolved decay. This experimental situation is much less common in atomic than in molecular spectroscopy, probably on account of the much smaller emission spectral width. Such a decay is obtained from Eqn 3.26 by just removing the integration over frequencies carried out in Eqn 3.27:

$$\rho(t, v, \vec{r}_b) = \sum_i a_{bi}(v) e^{-\beta_i t}, \tag{3.30}$$

where

$$a_{bi}(v) = \frac{[1 - \bar{\alpha}_{bi}(v)]c_i}{\sum_i [1 - \bar{\alpha}_{bi}(v)]c_i}, \tag{3.31}$$

and where

$$\bar{\alpha}_{bi}(v) = \int_V \alpha_b(\vec{r}, v) n_i(\vec{r}) \mathrm{d}\vec{r}, \tag{3.32}$$

with

$$\alpha_b(\vec{r}, v) = \int_0^{|\vec{r}_b - \vec{r}|} k(v) \exp[-k(v)x] \mathrm{d}x. \tag{3.33}$$

3.3.2   Continuous excitation (photostationary state)

For decays in response to temporal profiles other than delta excitation, and still under the assumption of nonsaturating conditions, the decay is obtained from linear response theory as the convolution of the excitation profile with the delta response function (e.g. [8] and references therein). For continuous excitation, the steady state intensity is in this way shown to be proportional to $\sum_i a_i / \beta_i$, and we obtain, for the overall intensity I,

$$\frac{I}{I_0} = \frac{\Phi}{\Phi_0} = \sum_i c_i \frac{1 - \bar{\alpha}_i}{1 - \bar{\alpha}_i \Phi_0}, \tag{3.34}$$

where I_0 is the intensity expected in the absence of radiative transport, and Φ is the macroscopic emission quantum yield.

As might be expected, radiative transport causes a decrease in the overall intensity only when the intrinsic emission quantum yield is smaller than one. In such a case, the decrease may be strong, even for moderate optical densities, owing to the nonlinear dependence on Φ_0. In this regard, Eqn 3.34 can be made clearer by rewriting it as

$$\frac{\Phi}{\Phi_0} = \sum_i c_i (1 - \bar{\alpha}_i) \left[1 + \bar{\alpha}_i \Phi_0 + (\bar{\alpha}_i \Phi_0)^2 + (\bar{\alpha}_i \Phi_0)^3 + \ldots \right], \tag{3.35}$$

where the first term in brackets corresponds to the escape probability for the ith eigenmode, whose photons may suffer zero, one, two, three, etc. reabsorptions before escape.

Similarly, for the direction-resolved case, the steady state intensity, obtained from Eqn 3.30, is given by

$$\frac{I_b}{I_{0b}} = \frac{\Phi_b}{\Phi_0} = \sum_i c_i \frac{1 - \bar{\alpha}_{bi}}{1 - \bar{\alpha}_i \Phi_0}. \tag{3.36}$$

Now, some $\bar{\alpha}_{bi}$ may be negative, in such a way that, for some directions, the apparent quantum yield of emission will be larger than the intrinsic one. Of course, a compensation will exist, so that the integrated yield over all directions will be according to Eqn 3.34, as follows from Eqn 3.36.

Finally, for the direction- and frequency-resolved cases, one obtains for the steady state

$$I_b(\nu) = I_{0b} F(\nu) \sum_i c_i \frac{1 - \alpha_{bi}(\nu)}{1 - \bar{\alpha}_i \Phi_0}. \tag{3.37}$$

This equation can be used to account for the deformation of spectral shape caused by radiative transport. In the atomic case, this effect is known as self-reversal [22, pp. 62–65] (Fig. 3.2). In the molecular case, there is a strong decrease in the blue side of the emission, the only side to significantly overlap absorption (Fig. 3.3).

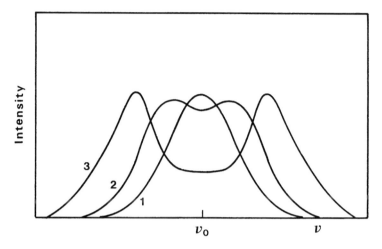

Figure 3.2   A schematic representation of atomic resonance line self-reversal. Curve 1 is the intrinsic absorption and emission spectrum. Increase of the atomic concentration leads to a progressive deformation of the emission spectrum (curves 2 and 3) to the point of yielding two apparently distinct lines.

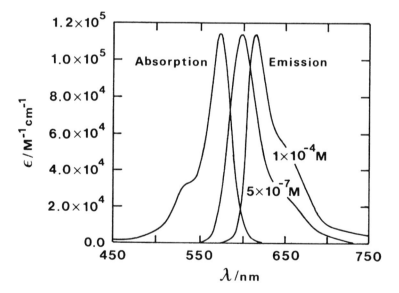

Figure 3.3   Absorption and fluorescence spectra (1 cm cell, right-angle geometry, excitation wavelength 530 nm) of the dye rhodamine 101 in acidified ethanol. Fluorescence spectra are normalized at the maximum. The emission spectrum of the concentrated solution shows a noticeable decrease of intensity on its blue side.

As mentioned above, with the exception of the long-time solution (corresponding to the eigenvalue β_s), all eigenfunctions take negative values at some points, and cannot thus be identified with physical distributions. This has been considered a drawback of the Holstein solution, being based on modes of difficult physical interpretation [21, 25, 36, 38, 64–66]. In fact, the negative values are essential to account for the possibility of rise times and increased apparent quantum yields of emission in some directions of space. These, in turn, have a clear meaning (see Section 3.5). Nevertheless, the original solution of the Holstein–Biberman equation, as given by Holstein, suffers from some real limitations: (i) the eigenvalues and eigenfunctions are not easily calculated; and (ii) it cannot be used to obtain the polarization of the emitted radiation. The approaches described in the next two sections overcome these difficulties.

3.4 MULTIPLE SCATTERING REPRESENTATION

3.4.1 Connection with the Holstein–Biberman equation

As mentioned in Section 3.2, Lai, Liu, and Ma [36] showed that the solution to the Holstein–Biberman equation could be given in a form alternative to that of Holstein's [30]. Rewriting Eqn 3.1 as

$$\frac{\partial n(\vec{r}, t)}{\partial t} = (-\Gamma + k_r \mathbf{L}) n(\vec{r}, t), \tag{3.38}$$

where the operator $\mathbf{L}$ is

$$\mathbf{L}[n(\vec{r}, t)] = \int_V f(\vec{r}, \vec{r}') n(\vec{r}', t) \mathrm{d}\vec{r}', \tag{3.39}$$

the symbolic solution of Eqn 3.38 is therefore

$$n(\vec{r}, t) = \exp(-\Gamma t + k_r t \mathbf{L}) n(\vec{r}, 0) = \exp(-\Gamma t) \exp(k_r t \mathbf{L}) n(\vec{r}, 0), \tag{3.40}$$

where the exponential of an operator $\mathbf{A}$ is defined as

$$\exp(\mathbf{A}) = 1 + \mathbf{A} + \frac{\mathbf{A}\mathbf{A}}{2} + \frac{\mathbf{A}\mathbf{A}\mathbf{A}}{3!} + \dots. \tag{3.41}$$

Equation 3.40 thus becomes [36]

$$n(\vec{r}, t) = \mathrm{e}^{-\Gamma t} \sum_{i=0}^{\infty} n_i(\vec{r}) \frac{(k_r t)^i}{i!}, \tag{3.42}$$

with

$$n_i(\vec{r}) = \mathbf{L}^i n(\vec{r}, 0). \tag{3.43}$$

It follows from Eqn 3.43 that

$$n_{i+1}(\vec{r}) = \mathbf{L} n_i(\vec{r}). \tag{3.44}$$

From Eqn 3.42, the spatial distribution function $p(\vec{r}, t)$ can be computed as

$$p(r, t) = \left[\sum_i n_i(\vec{r}) \frac{(k_r t)^i}{i!} \right] \Big/ \left[\sum_i N_i \frac{(k_r t)^i}{i!} \right], \tag{3.45}$$

and the time-dependent absorption probability as

$$\bar{\alpha}(t) = \left[\sum_i N_{i+1} \frac{(k_r t)^i}{i!} \right] \Big/ \left[\sum_i N_i \frac{(k_r t)^i}{i!} \right], \tag{3.46}$$

hence, from Eqn 3.10, and in agreement with [25],

$$\rho(t) = e^{-\Gamma t} \sum_{i=0}^{\infty} \frac{N_i - N_{i+1}}{N_0 - N_1} \frac{(k_r t)^i}{i!}. \tag{3.47}$$

It is seen that numerical values for this solution to the Holstein–Biberman equation can be obtained more easily than for that based on the eigenvalues. However, the asymptotic behavior toward a single exponential is not apparent from Eqn 3.47, and so computation of the solution for long times may demand a prohibitive number of terms. A remedy for this situation was proposed in [36] (see Eqn 3.66 below).

3.4.2   Kinetic model

The terms of the expansion given in Eqn 3.47 admit a simple physical interpretation: each corresponds to a given generation of excited atoms or molecules. More specifically, the ith term corresponds to the ith generation. It is in fact possible to directly obtain an equation with the same time-dependence as Eqn 3.47 from a kinetic model that considers the various generations, without invoking the Holstein–Biberman equation [35, 41, 65]. To do so, it is convenient to consider the following kinetic scheme:

$$I(t) \quad \overline{\alpha}_1 \, k_r \qquad \overline{\alpha}_2 \, k_r \qquad \overline{\alpha}_{n-1} \, k_r \qquad \overline{\alpha}_n \, k_r$$

$$\xrightarrow{\quad\quad} A_1^* \xrightarrow{\quad\quad} A_2^* \xrightarrow{\quad\quad} \cdots \xrightarrow{\quad\quad} A_n^* \xrightarrow{\quad\quad} \cdots$$

$$\Big| (1\text{-}\overline{\alpha}_1)k_r\text{+}k_{nr} \Big| (1\text{-}\overline{\alpha}_2)k_r\text{+}k_{nr} \quad \Big| (1\text{-}\overline{\alpha}_{n\text{-}1})k_r\text{+}k_{nr} \Big| (1\text{-}\overline{\alpha}_n)k_r\text{+}k_{nr}$$

where $N_1(t)$ is the number of first-generation molecules, directly excited by the pulse $I(t)$, $N_2(t)$ is the number of second-generation molecules, excited by absorption of the photons emitted by the first-generation molecules, etc., and $\bar{\alpha}_i$ is the average absorption probability of photons emitted by ith generation molecules. For delta-pulse excitation, one obtains [35, 41, 65]:

$$N_i(t) = N_0 a_i \frac{(k_r t)^{i-1}}{(i-1)!} \exp(-\Gamma t), \tag{3.48}$$

with

$$a_i = \prod_{j=1}^{i-1} \bar{\alpha}_j \quad (a_1 = 0). \tag{3.49}$$

The intensity due to N_i that will reach the boundary, at a given point $\vec{r}_b$, and for a given emission wavelength λ, will be

$$I_{bi}(\lambda, t) = k_r [1 - \alpha_{bi}(\lambda)] N_i(t). \tag{3.50}$$

Hence the total emission intensity is

$$I_b(\lambda, t) = \sum_{i=1}^{\infty} I_{bi}(\lambda, t) = k_r N_0 \sum_{i=1}^{\infty} [1 - \alpha_{bi}(\lambda)] \left(\prod_{j=1}^{i-1} \bar{\alpha}_j \right) \frac{(k_r t)^{i-1}}{(i-1)!} \exp(-\Gamma t), \tag{3.51}$$

and the decay will be

$$\rho_b(\lambda, t) = e^{-\Gamma t} \sum_{i=1}^{\infty} \frac{1 - \alpha_{bi}(\lambda)}{1 - \alpha_{b0}(\lambda)} a_i \frac{(k_r t)^{i-1}}{(i-1)!}, \tag{3.52}$$

while for the emission integrated over wavelengths and directions the decay is

$$\rho(t) = e^{-\Gamma t} \sum_{i=1}^{\infty} \frac{1 - \bar{\alpha}_i}{1 - \bar{\alpha}_1} a_i \frac{(k_r t)^{i-1}}{(i-1)!}. \tag{3.53}$$

This equation is identical to Eqn 3.47, as will be shown in Section 3.5. However, a physically clearer approach, without the involvement of unknown parameters, is desirable. We describe this in the next section.

3.5   STOCHASTIC APPROACH

3.5.1   Formulation and delta-pulse response

Consider the absorption of a photon at $t = 0$, according to a given spatial distribution $P_1(\vec{r})$. The excited molecule generated at time $t = 0$ will relax to the ground state, yielding a probability $p_n(\lambda, t)$ that, between t and $t + dt$, a photon with wavelength λ will hit the boundary at point $\vec{r}_b$ and thus leave the sample (neglecting reflection). This probability can be written as

$$p_b(\lambda, t) = \sum_{n=1}^{\infty} p_{bn}(\lambda, t), \tag{3.54}$$

where $p_{bn}(\lambda, t)$ is the probability that a photon with wavelength λ will cross the boundary at a point $\vec{r}_b$, between t and $t + dt$, after exactly n absorption–emission events. This probability can in turn be written as

$$p_{bn}(\lambda, t) = f_{bn}(\lambda) g_n(t), \tag{3.55}$$

where $f_{bn}(\lambda)$ is the probability that a photon with wavelength λ will hit the boundary at point $\vec{r}_b$ (thus leaving the sample), after exactly n absorption–emission events, and $g_n(t)$ is the probability that an nth-generation molecule will emit a photon between t and $t + dt$, given that it will emit one. Assuming that the photon propagation time is negligible, this probability (the normalized density function) is given by (see Appendix A)

$$g_n(t) = \Gamma \frac{(\Gamma t)^{n-1}}{(n-1)!} e^{-\Gamma t}. \tag{3.56}$$

The probability $f_{bn}(\lambda)$ is

$$f_{bn}(\lambda) = \frac{1}{4\pi} \Phi_0 F(\lambda) \int_V [1 - \alpha_b(\vec{r}, \lambda)] P_n(\vec{r}) d\vec{r}, \tag{3.57}$$

where $\alpha_b(\vec{r}, \lambda)$ is given by Eqn 3.33 and $P_n(\vec{r})$, the probability that an nth-generation photon will be emitted at $\vec{r}$, is

$$P_n(\vec{r}) = \int_V \int_V \cdots \int_V f(\vec{r}, \vec{r}_{n-1}) \, \Phi_0 f(\vec{r}_{n-1}, \vec{r}_{n-2}) \cdots$$
$$\Phi_0 f(\vec{r}_2, \vec{r}_1) P_1(\vec{r}_1) \mathrm{d}\vec{r}_{n-1} \, \mathrm{d}\vec{r}_{n-2} \cdots \mathrm{d}\vec{r}_1, \tag{3.58}$$

with $f(\vec{r}, \vec{r}')$ given by Eqn 3.3. Equation 3.58 can be written as a recurrence relation:

$$P_{n+1}(\vec{r}) = \Phi_0 \int_V f(\vec{r}, \vec{r}_n) P_n(\vec{r}_n) \mathrm{d}\vec{r}_n. \tag{3.59}$$

From it, one obtains the spatial distribution function $p_n(\vec{r})$ of the nth generation,

$$p_n(\vec{r}) = \frac{P_n(\vec{r})}{\int_V P_n(\vec{r}) \mathrm{d}\vec{r}} = \frac{\int_V f(\vec{r}, \vec{r}_{n-1}) p_{n-1}(\vec{r}_{n-1}) \mathrm{d}\vec{r}_{n-1}}{\int_V \int_V f(\vec{r}, \vec{r}_{n-1}) p_{n-1}(\vec{r}_{n-1}) \mathrm{d}\vec{r}_{n-1} \, \mathrm{d}\vec{r}}, \tag{3.60}$$

which shows the equivalence between this approach and that based on the Holstein–Biberman equation (compare with Eqns 3.6 and 3.44).

The normalized decay law will thus be

$$\rho_b(\lambda, t) = \frac{p_b(\lambda, t)}{p_b(\lambda, 0)} = \mathrm{e}^{-\Gamma t} \sum_{n=1}^{\infty} \frac{\int_V [1 - \alpha_b(\vec{r}, \lambda)] P_n^0(\vec{r}) \mathrm{d}\vec{r}}{\int_V [1 - \alpha_b(\vec{r}, \lambda)] P_1^0(\vec{r}) \mathrm{d}\vec{r}} \frac{(k_r t)^{n-1}}{(n-1)!}, \tag{3.61}$$

where $P_i^0(\vec{r})$ denotes $P_i(\vec{r})$ when $\Phi_0 = 1$. Using Eqn 3.60,

$$\rho_b(\lambda, t) = \mathrm{e}^{-\Gamma t} \sum_{n=1}^{\infty} \frac{1 - \alpha_{bn}(\lambda)}{1 - \alpha_{b1}(\lambda)} \frac{\int_V P_n^0(\vec{r}) \mathrm{d}\vec{r}}{\int_V P_1^0(\vec{r}) \mathrm{d}\vec{r}} \frac{(k_r t)^{n-1}}{(n-1)!}; \tag{3.62}$$

Taking into account Eqns 3.59 and 3.60 and the definition of $\bar{\alpha}_i$ (see Eqn 3.16) one has

$$\frac{\int_V P_n^0(\vec{r}) \mathrm{d}\vec{r}}{\int_V P_1^0(\vec{r}) \mathrm{d}\vec{r}} = \prod_{i=1}^{n-1} \bar{\alpha}_i \tag{3.63}$$

and Eqn 3.62 becomes identical to Eqn 3.52,

$$\rho_b(\lambda, t) = e^{-\Gamma t} \sum_{n=1}^{\infty} \frac{1 - \alpha_{bn}(\lambda)}{1 - \alpha_{b1}(\lambda)} \prod_{i=1}^{n-1} \bar{\alpha}_i \frac{(k_r t)^{n-1}}{(n-1)!}. \tag{3.64}$$

The emission integrated over wavelengths and directions is

$$\rho(t) = e^{-\Gamma t} \sum_{n=1}^{\infty} \frac{\int_V P_n^0(\vec{r}) d\vec{r} - \int_V \int_V f(\vec{r}, \vec{r}') P_n^0(\vec{r}) d\vec{r}\, d\vec{r}'}{\int_V P_1^0(\vec{r}) d\vec{r} - \int_V \int_V f(\vec{r}, \vec{r}') P_1^0(\vec{r}) d\vec{r}\, d\vec{r}'} \frac{(k_r t)^{n-1}}{(n-1)!}$$

$$= e^{-\Gamma t} \sum_{n=1}^{\infty} \frac{1 - \bar{\alpha}_n}{1 - \bar{\alpha}_1} \frac{\int_V P_n^0(\vec{r}) d\vec{r}}{\int_V P_1^0(\vec{r}) d\vec{r}} \frac{(k_r t)^{n-1}}{(n-1)!}. \tag{3.65}$$

Using Eqn 3.63 again, Eqn 3.65 becomes identical to Eqn 3.53.

Knowing that a stationary distribution is reached for the higher generations $(\bar{\alpha}_n \to \bar{\alpha}_s)$, one obtains

$$\rho(t) = e^{-\Gamma t} \sum_{n=1}^{\infty} \frac{1 - \bar{\alpha}_n}{1 - \bar{\alpha}_1} \prod_{i=1}^{n-1} \bar{\alpha}_i \frac{(k_r t)^{n-1}}{(n-1)!}$$

$$\simeq e^{-\Gamma t} \sum_{n=1}^{m} \frac{1 - \bar{\alpha}_n}{1 - \bar{\alpha}_1} \prod_{i=1}^{n-1} \bar{\alpha}_i \frac{(k_r t)^{n-1}}{(n-1)!} + e^{-\Gamma t} \sum_{n=m+1}^{\infty} \frac{1 - \bar{\alpha}_S}{1 - \bar{\alpha}_1} \left(\prod_{i=1}^{m} \bar{\alpha}_i \right) \bar{\alpha}_S^{n-m-1} \frac{(k_r t)^{n-1}}{(n-1)!}$$

$$= e^{-\Gamma t} \sum_{n=1}^{m} \frac{1 - \bar{\alpha}_n}{1 - \bar{\alpha}_1} \prod_{i=1}^{n-1} \bar{\alpha}_i \frac{(k_r t)^{n-1}}{(n-1)!} + e^{-\Gamma t} \frac{1 - \bar{\alpha}_S}{1 - \bar{\alpha}_1} \left(\prod_{i=1}^{m} \frac{\bar{\alpha}_i}{\bar{\alpha}_S} \right) \sum_{n=m+1}^{\infty} \bar{\alpha}_S^{n-m-1} \frac{(k_r \bar{\alpha}_S t)^{n-1}}{(n-1)!}$$

$$= e^{-\Gamma t} \left[\sum_{n=1}^{m} \frac{1 - \bar{\alpha}_n}{1 - \bar{\alpha}_1} \left(\prod_{i=1}^{n-1} \bar{\alpha}_i \right) \frac{(k_r t)^{n-1}}{(n-1)!} - \frac{1 - \bar{\alpha}_S}{1 - \bar{\alpha}_1} \left(\prod_{i=1}^{m} \frac{\bar{\alpha}_i}{\bar{\alpha}_S} \right) \sum_{n=m+1}^{\infty} \frac{(k_r \bar{\alpha}_S t)^{n-1}}{(n-1)!} \right]$$

$$+ \frac{1 - \bar{\alpha}_S}{1 - \bar{\alpha}_1} \left(\prod_{i=1}^{m} \frac{\bar{\alpha}_i}{\bar{\alpha}_S} \right) e^{(k_r \bar{\alpha}_S - \Gamma) t}, \tag{3.66}$$

which shows that a long-time exponential is asymptotically attained. The approximation made in Eqn 3.66 is of course better the higher m is. A similar approximation holds for Eqn 3.64. The usefulness of this approach has been demonstrated [36].

Some experimental results [48] and their comparison with theoretic predictions are now presented. The systems to be discussed are solutions of the fluorescent dye rhodamine 101 in acidified ethanol or ethanol–methanol (9 : 1 v/v) ($\Phi_0 = 0.9$ and $\tau_0 = 4.3$ ns at room temperature), whose absorption and emission spectra are depicted in Fig. 3.3. The enclosure under consideration is a typical 1 cm × 1 cm fluorescence cell. Two common detection geometries are considered: front-face (30°/60°) and right-angle (see Fig. 3.4).

The fluorescence decay of a 10^{-3} M solution of rhodamine 101, recorded with the front-face geometry, is shown in Fig. 3.5. The decay is nonexponential and the average lifetime is almost the double of the intrinsic lifetime. The fit to the theoretical decay, Eqn. 3.64, is quite good. In Fig. 3.6 a comparison is made between the experimental and calculated average lifetimes (from the equation, with the amplitudes evaluated by Monte Carlo simulation [47]). Again, there is a general agreement between experiment and theory [48]. The observed trend with emission wavelength deserves a comment: the average lifetime is lower in the overlap region than in the red side of the emission. This occurs because "blue" photons are either emitted close to the boundary, at early times, or are reabsorbed, later to appear as "red" photons.

The fluorescence decay of a 5×10^{-4} M solution of rhodamine 101 shown in Fig. 3.7, recorded with right-angle geometry, displays a significant rise time. The fit to the theoretical decay, Eqn. 3.64, is quite good. In Fig. 3.8 a comparison is made between the experimental and calculated average lifetimes (again with amplitudes evaluated by Monte Carlo simulation [47]). It is seen that the agreement between experiment and theory is good [48]. The observed trend

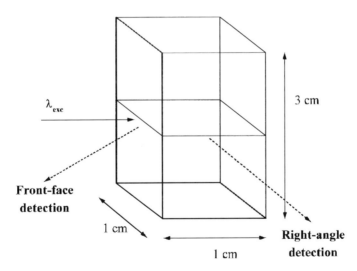

Figure 3.4 Experimental excitation and emission geometries, also reproduced in the Monte Carlo simulations. The enclosure size and shape correspond to a typical fluorescence cell. (Reprinted with permission from E. J. Nunes Pereira, M. N. Berberan-Santos, and J. M. G. Martinho, *J. Chem. Phys.* 104 (1996) 8950. Copyright 1996 American Institute of Physics.)

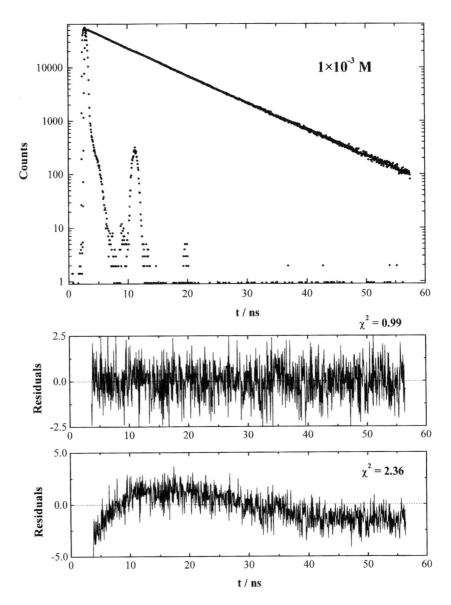

Figure 3.5   The fluorescence decay of a 10^{-3} M solution of rhodamine 101 in ethanol, at room temperature, front-face geometry. Excitation wavelength 294 nm; emission wavelength 620 nm. The average decay time is 8.4 ns. The fit to Eqn 3.64, with the intrinsic lifetime fixed at the dilute solution value of 4.34 ns, is quite good (upper residuals plot and reduced chi-squared value), while the fit to a single exponential is poor (lower residuals plot and reduced chi-squared value.)

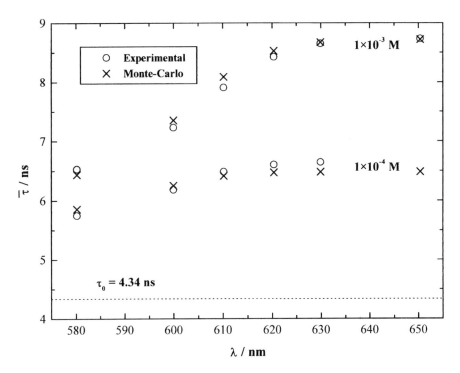

Figure 3.6   A comparison of experimental and predicted average lifetimes of rhodamine 101 for the front-face geometry, as a function of concentration and emission wavelength. Excitation wavelength 294 nm.

with emission wavelength is, however, opposite to that observed in front-face: the average lifetime is higher in the overlap region than in the red side of the emission. This occurs because "blue" photons, often absorbed, on average suffer more reabsorptions than "red" photons, on their way to the detector. The rise time, only observed in the blue [47], precisely reflects this waiting time for the arrival of "blue" photons. Holstein's solution of the Holstein–Biberman equation, in terms of eigenfunctions, provides for the rise times through negative amplitudes for some eigenmodes, at some points of space, as discussed in Section 3.3.2.

The physical picture can be made clearer by considering the spatial distribution of several generations of excited molecules, obtained by Monte Carlo simulation [47], as shown in Figs 3.9 and 3.10. As can be seen, the distribution of first-generation molecules is dictated by both the direction of the excitation beam and the opacity of the medium at the excitation wavelength, which controls its penetration. Excitation "diffusion" is more or less rapid according to the opacity in the spectral overlap region. For the front-face geometry

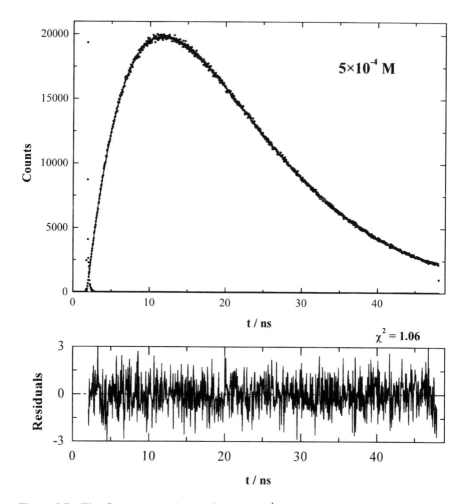

Figure 3.7 The fluorescence decay of a 5×10^{-4} M solution of rhodamine 101 in ethanol, at room temperature; right-angle geometry. Excitation wavelength 300 nm; emission wavelength 576 nm. There is a pronounced rise time. The fit to Eqn 3.64, with the intrinsic lifetime fixed at the dilute solution value of 4.34 ns, is quite good (residuals plot and reduced chi-squared value).

(Fig. 3.9), excited molecules are initially close to the boundary point for which emission is recorded, and then diffuse away from it. On the contrary, for the right-angle geometry (Fig. 3.10), excited molecules are initially far from the boundary point for which emission is recorded, but become closer over time. Only for very high generations (and whose contribution to the decay is very small), is the distribution close to the stationary distribution, corresponding to

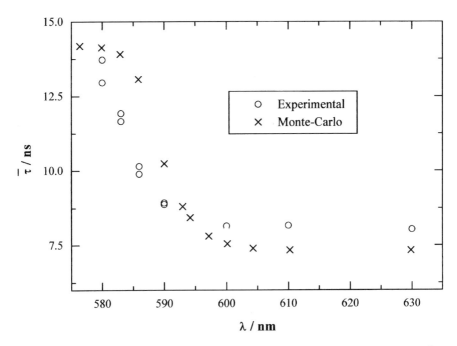

Figure 3.8   A comparison of experimental and predicted average lifetimes of 10^{-4} M solutions of rhodamine 101 for the right-angle geometry, as a function of the emission wavelength. Excitation wavelength 300 nm.

the fundamental mode of the Holstein–Biberman equation. This limiting distribution is centered in the middle of the cell and has approximately radial symmetry. Some tenth-generation distributions (see Plates 1 and 2) are already close to it.

3.5.2   Continuous excitation (photostationary state)

From Eqns 3.64 and 3.65 one may also obtain the steady state intensities,

$$\frac{I_b(\lambda)}{I_{0b}} = F(\lambda) \sum_{n=1}^{\infty} [1 - \alpha_{bn}(\lambda)] \left(\prod_{i=1}^{n-1} \bar{\alpha}_i \right) \Phi_0^{n-1}, \tag{3.67}$$

$$\frac{I_b}{I_{0b}} = \frac{\Phi_b}{\Phi_0} = \sum_{n=1}^{\infty} (1 - \bar{\alpha}_{bn}) \left(\prod_{i=1}^{n-1} \bar{\alpha}_i \right) \Phi_0^{n-1}, \tag{3.68}$$

$$\frac{I}{I_0} = \frac{\Phi}{\Phi_0} = \sum_{n=1}^{\infty}(1 - \bar{\alpha}_n)\left(\prod_{i=1}^{n-1}\bar{\alpha}_i\right)\Phi_0^{n-1} = (1 - \bar{\alpha}_1) + (1 - \bar{\alpha}_2)\,\Phi_0\bar{\alpha}_1 + \ldots . \quad (3.69)$$

These equations may be compared with Eqns 3.34, 3.36, and 3.37. Note also that they can be converted into finite sums by means of approximations similar to that carried out for Eqn. 3.66.

3.5.3   Fluorescence anisotropy

It is well known that nonradiative transport decreases the ensemble fluorescence anisotropy. For a pair of randomly oriented and nonrotating molecules, and for the Förster dipolar mechanism, Galanin calculated in 1950 [1, 27] that the acceptor fluorescence anisotropy is only 4% of that of the donor. This result was later shown to be in fact the zero-time value of the indirectly excited molecule anisotropy [9]. In any event, neglect of the contribution of indirectly excited molecules to the ensemble anisotropy is a good and frequently made approximation in nonradiative transport studies, where theoretic efforts concentrate on calculation of the survival probability of directly excited molecules. This calculation is difficult on account of the possibility of excitation return. Nevertheless, good approximations ([6] and references therein), extensively tested experimentally (see, e.g., [5, 29]) are available.

The situation with radiative transport is, in a sense, the opposite of that with nonradiative transport. In fact, and owing to its long-ranged nature, the return of excitation has negligible probability. On the other hand, the contribution of indirectly excited molecules to the overall anisotropy is considerable, and cannot be neglected: the radiative mechanism has a higher orientational selectivity than the nonradiative one [3, 4]. For the purposes of computing the effect of radiative transport on fluorescence anisotropy, we consider only results for directions contained in the horizontal plane (including the usual front-face and right-angle geometries), for which the anisotropy of fluorescence takes the highest value. We further suppose that molecular rotational motion is negligible during the lifetime and that the exciting photons carry vertical polarization.

We start with the calculation of the depolarization due to radiative transfer of electronic excitation energy. To conform to usage, we write as r_0 the anisotropy of first- generation molecules (the fundamental anisotropy), implying that $r_1 = r_0$. The anisotropy of second-generation molecules, indirectly excited by reabsorption, will be

$$r_2 = \beta r_0, \quad (3.70)$$

where β is the depolarization factor $(\beta < 1)$. In contrast to nonradiative transport, the probability of excitation return to the original molecule is negligible, and therefore the anisotropy of fluorescence of molecules belonging to the nth generation is obtained by repeated application of Eqn 3.70:

$$r_n = \beta^{n-1} r_0 \qquad (n = 1, 2, \ldots). \tag{3.71}$$

A quantum-electrodynamical calculation of the depolarization factor β, by Andrews and Juzeliūnas [4], gave $\beta = 0.28$ (an identical value is obtained by classical electrodynamics [10]; see Appendix B). This value may be compared to that of the *nonradiative* dipole–dipole transfer mechanism, which is $\beta = 0.04$ [1, 4, 9, 27]. The polarization retained after one transfer is thus seven times greater for the radiative case – precluding, as mentioned above, the common approximation in nonradiative transport of neglecting the contribution of higher order generations.

For excitation with vertically polarized light, the definition of anisotropy is

$$r = \frac{I_\parallel - I_\perp}{I_\parallel + 2I_\perp}, \tag{3.72}$$

where the parallel and perpendicular intensities are measured for at right-angles to the excitation, and contained in the horizontal plane. The denominator of Eqn 3.72 is usually proportional to the intensity emitted in all directions. An alternative measure of linear polarization is the polarization p,

$$p = \frac{I_\parallel - I_\perp}{I_\parallel + I_\perp}, \tag{3.73}$$

where the denominator is the intensity emitted in the direction of measurement. In most fluorescence experiments, anisotropy is a more useful parameter than polarization, because the denominator is proportional to the intensity of decay, and simpler expressions result. When several incoherent sources are present (e.g. from a mixture of fluorescent compounds), both polarization and anisotropy can be expressed as a sum of contributions, the weight of each being the fraction of the intensity emitted in all directions (anisotropy), or the fraction of the intensity emitted in the direction of measurement (polarization).

When radiative transport is present, the denominator of Eqn 3.72 is no longer proportional to the intensity decay. In fact, the symmetry of the emitting ensemble is lowered, and a complicated positional pattern of polarizations emerges. Both anisotropy and polarization become local quantities (i.e. relative to the measurement point $\vec{r}_b$). From an experimental point of view, Eqn 3.72 can still be used. However, from a theoretic viewpoint, information is limited to the decay at a given boundary point (Eqn 3.64), which is proportional to $I_\parallel(t) + I_\perp(t)$ and not to $I_\parallel(t) + 2I_\perp(t)$ (both measured at $\vec{r}_b$). Polarization, as given by Eqn 3.73, is therefore of more direct meaning. Nevertheless, given that anisotropy is the parameter used in the absence of radiative transport, it is important to obtain a generalized, albeit local anisotropy, that will reduce to the usual result in the limiting situation of negligible radiative transport. To do so, one takes into account the relation between the local anisotropy and local polarization,

$$p = \frac{3r}{2 + r},$$

(3.74)

or

$$r = \frac{2p}{3 - p}.$$

(3.75)

The total polarization is first obtained:

$$p_b(\lambda, t) = \sum_{n=1}^{\infty} a_{bn}(\lambda, t) p_n,$$

(3.76)

where p_n is the polarization of the nth generation. From Eqns 3.71 and 3.74,

$$p_n = \frac{3\beta^{n-1} r_0}{2 + \beta^{n-1} r_0}.$$

(3.77)

The fractional contribution $a_{bn}(t)$ is

$$a_{bn}(\lambda, t) = \frac{I_{bn}(\lambda, t)}{I_b(\lambda, t)} = \left\{ [1 - \alpha_{bn}(\lambda)] \left(\prod_{i=1}^{n-1} \bar{\alpha}_i \right) \frac{(k_r t)^{n-1}}{(n-1)!} \right\} \Bigg/ \left\{ \sum_{n=1}^{\infty} [1 - \alpha_{bn}(\lambda)] \left(\prod_{i=1}^{n-1} \bar{\alpha}_i \right) \frac{(k_r t)^{n-1}}{(n-1)!} \right\}.$$

(3.78)

The polarization is therefore

$$p_b(\lambda, t) = \left\{ \sum_{n=1}^{\infty} [1 - \alpha_{bn}(\lambda)] \left(\prod_{i=1}^{n-1} \bar{\alpha}_i \right) \frac{(k_r t)^{n-1}}{(n-1)!} \frac{3\beta^{n-1} r_0}{2 + \beta^{n-1} r_0} \right\} \Bigg/ \left\{ \sum_{n=1}^{\infty} [1 - \alpha_{bn}(\lambda)] \left(\prod_{i=1}^{n-1} \bar{\alpha}_i \right) \frac{(k_r t)^{n-1}}{(n-1)!} \right\},$$

(3.79)

and, finally, the anisotropy is obtained from Eqn 3.75:

$$r_b(\lambda, t) = r_0 \left\{ \sum_{n=1}^{\infty} [1 - \alpha_{bn}(\lambda)] \left(\prod_{i=1}^{n-1} \bar{\alpha}_i \right) \left(\frac{1}{2 + \beta^{n-1} r_0} \right) \frac{(k_r \beta t)^{n-1}}{(n-1)!} \right\} \Big/$$
$$\left\{ \sum_{n=1}^{\infty} [1 - \alpha_{bn}(\lambda)] \left(\prod_{i=1}^{n-1} \bar{\alpha}_i \right) \left(\frac{1}{2 + \beta^{n-1} r_0} \right) \frac{(k_r t)^{n-1}}{(n-1)!} \right\}. \tag{3.80}$$

By a reasoning similar to that for Eqn 3.66, it may be shown that for long times the anisotropy becomes

$$r_b(\lambda, t) \simeq r_0 \exp[-\bar{\alpha}_S (1 - \beta) k_r t]. \tag{3.81}$$

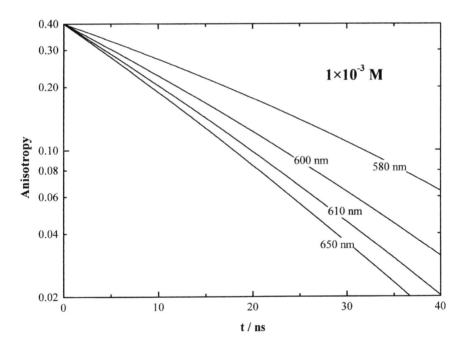

Figure 3.9 Theoretical anisotropy decays of 10^{-3} M rhodamine 101 in a rigid medium, front-face geometry, as a function of emission wavelength. Excitation wavelength 570 nm. (Reprinted with permission from E. J. Nunes Pereira, M. N. Berberan-Santos, and J. M. G. Martinho, *J. Chem. Phys.* **104** (1996) 8950. Copyright 1996 American Institute of Physics.)

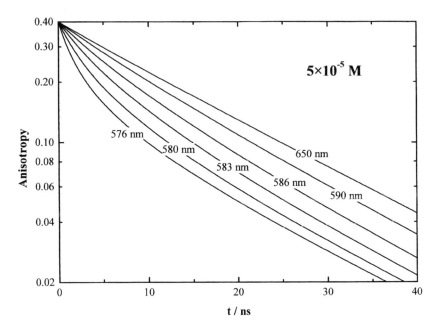

Figure 3.10   Theoretical anisotropy decays of 5×10^{-5} M rhodamine 101 in a rigid medium, right-angle geometry, as a function of emission wavelength. Excitation wavelength 300 nm. (Reprinted with permission from E. J. Nunes Pereira, M. N. Berberan-Santos and J. M. G. Martinho, *J. Chem. Phys.* 104 (1996), 8950, Copyright 1996 American Institute of Physics.)

Similarly, the polarization will then be

$$p_b(\lambda, t) \simeq \frac{3}{2} r_0 \exp[-\bar{\alpha}_S(1 - \beta)k_r t]. \tag{3.82}$$

It should be stressed that the reabsorption probabilities of Eqns 3.79–3.82 must be computed with an absorption probability whose orientational dependence is that of a radiating electric dipole, and not that of an isotropic emitter, because it is now assumed that molecular rotation is frozen during the lifetime. However, the results are expected not to greatly differ.

Theoretical results for the anisotropy decay of rhodamine 101 in a rigid medium, computed from Eqn 3.80, are now presented. In Fig. 3.9 the decay is displayed for a 10^{-3} M solution and with front-face geometry. It is faster for "red" photons than for "blue" ones, because the latter are on average "younger", i.e. of a lower generation, than the "red" ones, and thus more polarized. In Fig. 3.10 the decay for a 5×10^{-5} M solution, with right-angle geometry, is displayed. Conversely, this is faster for "blue" photons than for "red" ones, because the latter are on average "older", i.e. of a higher generation, than the "red" ones, and thus less polarized.

3.6   COMBINED RADIATIVE AND NONRADIATIVE TRANSPORT

3.6.1   Importance of nonradiative transport

The contribution of nonradiative energy transport has been completely neglected up to now. Nonradiative transport will be present whenever the average distance between molecules is smaller or of the order of the Förster radius for self-transfer. Because some of the parameters that favor radiative transport (such as high absorption–emission spectral overlap and a high molecular radiative constant) also favor nonradiative transport by the dipole–dipole mechanism, the Förster radius for self-transfer tends to be significant. It is therefore important to discuss the effects of nonradiative transport on the macroscopic observables such as the fluorescence intensity decay, quantum yield and anisotropy decay.

In a system in which both mechanisms are operative and, neglecting coherence, the excitation will perform a series of short-range hops by the nonradiative mechanism, alternating with long-distance jumps by the radiative one. What conclusions can be drawn from this picture? First, it is well known that the decay law is unaffected by pure nonradiative transport [26]. In this way, each series of short hops will not change the decay probability of that sub-ensemble. Secondly, because the hops are performed locally, the excitation spread during the lifetime does not exceed a few Förster radii, and cannot change significantly the spatial distribution of the generations considered in the radiative model. An interesting consequence of the nonradiative hops is the efficient randomization of the orientation of the emitting dipole. In this way, and depending on the importance of the nonradiative mechanism, the assumption of isotropic emission may be appropriate even in cases in which molecular rotation is insignificant during the excited-state lifetime. All of these considerations lead to the conclusion that nonradiative transport leaves the decay law and quantum yield practically unchanged.

3.6.2   Fluorescence anisotropy

As regards the anisotropy decay, the contribution of nonradiative transport may be quite important. Invoking again the model of series of short hops alternating with long jumps, a strong depolarization is expected for each series of hops. Nonradiative transport will therefore contribute to the anisotropy decay – when significant, it may even be the dominant mechanism. It is thus of interest to obtain an expression for the combined effect of radiative and nonradiative transport. To do so, we try to modify the anisotropy decay for pure radiative transport (Eqn 3.80). Following the nth radiative step, an $(n + 1)$th-generation molecule is excited (at a certain time t_{n+1}). Owing to the nonradiative hops, there is a probability $G(t - t_{n+1})$ that the excitation will remain in that molecule. If the next radiative jump $(n + 1 \rightarrow n + 2)$ occurs from the initially excited molecule, one may still apply the depolarization equation (3.70). If, on the other hand, the radiative jump originates from an indirectly excited molecule by the nonradiative mechanism, total depolarization is expected. In this way, the emission probability for polarized emission will be, for each radiative step, $g(t)G(t)$, where $g(t)$ is given by Eqn A.2 and $G(t)$ is the probability that the excitation is in the directly excited molecule, when nonradiative transport is operative. The function $G(t)$ is given by several theoretic treatments. For three-dimensional rigid media, a simple but accurate formula is the so-called HHB approximation [6, 29]:

$$G(t) = \exp\left(-0.8452\, N_F \sqrt{\frac{\pi \Gamma t}{2}}\right), \tag{3.83}$$

where N_F is the average number of molecules contained in a sphere of radius equal to the Förster radius for self-transfer. In this way, the probability of the emission of polarized radiation by nth-generation molecules will be

$$g'_n(t) = \underbrace{gG \otimes gG \otimes \ldots \otimes gG}_{nx}, \tag{3.84}$$

while the population of that generation will continue to be proportional to $g_n(t) = \Gamma[(\Gamma t)^{n-1}/(n - 1)!]e^{-\Gamma t}$ (see Appendix A). Hence, Eqn 3.71 is replaced by

$$r_n(t) = \beta^{n-1} r_0 \frac{g'_n(t)}{g_n(t)} \tag{3.85}$$

and Eqn 3.79 and 3.80 by

$$p_b(\lambda, t) = \left\{ \sum_{n=1}^{\infty} [1 - \alpha_{bn}(\lambda)] \left(\prod_{i=1}^{n-1} \bar{\alpha}_i \right) \left(\frac{3(\beta \Phi_0)^{n-1} r_0 g'_n(t)}{2 + \frac{g'_n(t)}{g_n(t)} \beta^{n-1} r_0} \right) \right\} \Bigg/ \tag{3.86}$$

$$\left\{ \sum_{n=1}^{\infty} [1 - \alpha_{bn}(\lambda)] \left(\prod_{i=1}^{n-1} \bar{\alpha}_i \right) \Phi_0^{n-1} g_n(t) \right\},$$

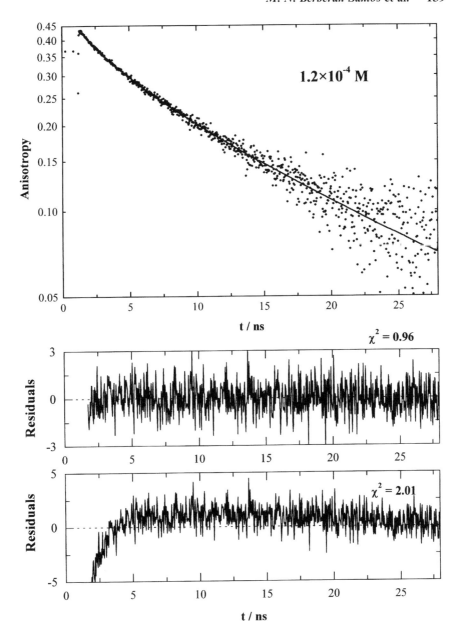

Figure 3.11 The fluorescence anisotropy decay of 1.2×10^{-4} M rhodamine 101 in acidified ethanol–methanol (9 : 1 v/v) at 100 K, right-angle geometry. The fit to Eqn 3.87, with due allowance for nonradiative transport, is quite good (upper residuals plot and reduced chi- squared value), while the fit to Eqn 3.80, where only radiative transport is considered, is clearly worse (lower residuals plot and reduced chi-squared value.)

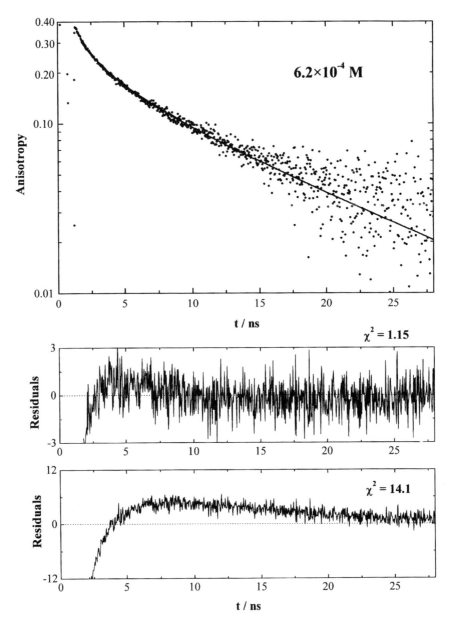

Figure 3.12 The fluorescence anisotropy decay of 6.2×10^{-4} M rhodamine 101 in acidified ethanol–methanol (9 : 1 v/v) at 100 K, right-angle geometry. The fit to Eqn 3.87, with due allowance for nonradiative transport, is good (upper residuals plot and reduced chi-squared value), while the fit to Eqn 3.80, where only radiative transport is considered, is much worse (lower residuals plot and reduced chi-squared value.)

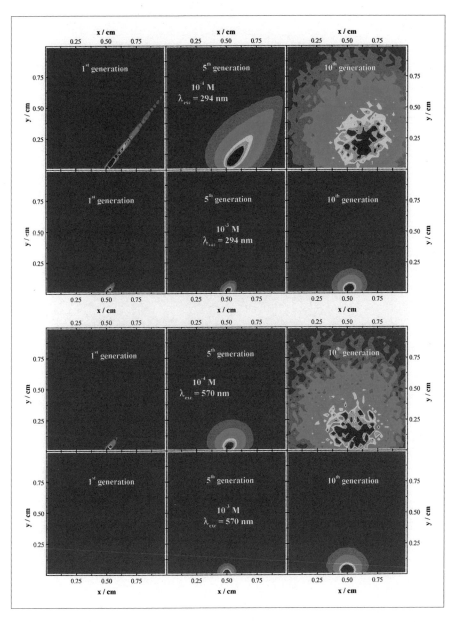

Plate 1 Spatial distribution functions for the first, fifth and tenth generations of excited molecules in a front-face geometry, obtained by Monte Carlo simulation. Reprinted with permission from E.J. Nunes Pereira, M.N. Berberan-Santos and J.M.G. Martinho, *J. Chem. Phys.* 104 (1996), 8950. Copyright 1996 American Institute of Physics.

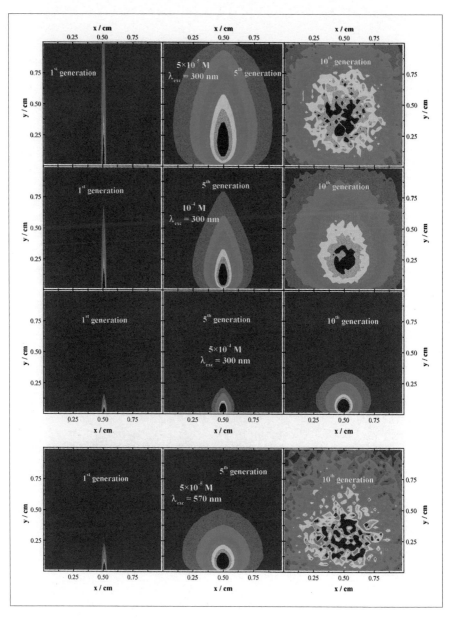

Plate 2 Spatial distribution functions for the first, fifth and tenth generations of excited molecules in a right-angle geometry, obtained by Monte Carlo simulation. Reprinted with permission from E.J. Nunes Pereira, M.N. Berberan-Santos and J.M.G. Martinho, *J. Chem. Phys.* 104 (1996), 8950. Copyright 1996 American Institute of Physics.

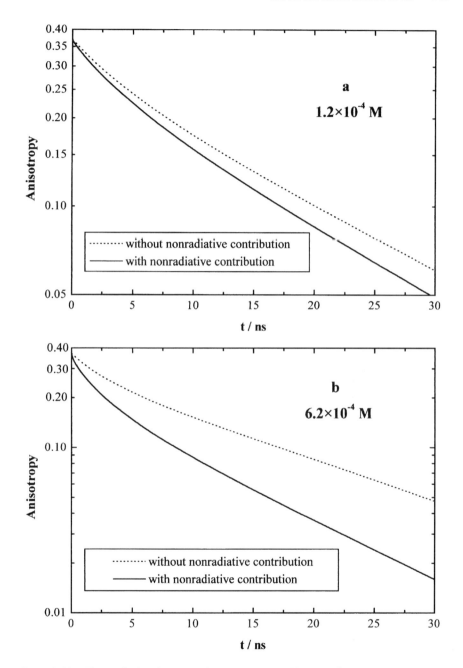

Figure 3.13   Theoretical anisotropy decays corresponding to Figs 3.11 and 3.12, but explicitly showing the effect of nonradiative transport.

$$r_b(\lambda, t) = r_0 \left\{ \sum_{n=1}^{\infty} [1 - \alpha_{bn}(\lambda)] \left(\prod_{i=1}^{n-1} \bar{\alpha}_i \right) \left(\frac{(\beta \Phi_0)^{n-1} g_n'(t)}{2 + \frac{g_n'(t)}{g_n(t)} \beta^{n-1} r_0} \right) \right\} \Bigg/$$

$$\left\{ \sum_{n=1}^{\infty} [1 - \alpha_{bn}(\lambda)] \left(\prod_{i=1}^{n-1} \bar{\alpha}_i \right) \left(\frac{\Phi_0^{n-1} g_n(t)}{2 + \frac{g_n'(t)}{g_n(t)} \beta^{n-1} r_0} \right) \right\}. \qquad (3.87)$$

It must be remarked that the above considerations and formulas are not completely general. A unified treatment of the problem of combined radiative and nonradiative transport that includes the continuous variation from the r^{-6} interaction to the r^{-2} interaction [3] is wanting. We have considered only the combined effect of the (extreme) radiative and nonradiative (dipole–dipole) processes. Nevertheless, this is expected to be valid for most situations.

Some experimental time-resolved and steady state anisotropies for rhodamine 101 in an ethanol–methanol (9 : 1 v/v) glass at 100 K, for the right-angle geometry (cell edge detection), are now presented and briefly discussed. In Figs 3.11 and 3.12 the anisotropy decays of 1.2×10^{-4} M and 6.2×10^{-4} M solutions are displayed. In both cases, the fit to Eqn 3.87 (with due allowance for nonradiative transport) is good, while neglect of the non-radiative transport contribution leads to significantly worse fits, deviations being higher for the highest concentration. The effect of nonradiative transport is shown, in Fig. 3.13, to be especially important for the 6.2×10^{-4} M solution.

3.7 CONCLUSION

An integrated view of atomic and molecular radiative transport theories has been presented, stressing the similarities between the two fields. It has been shown that the two main treatments of radiative transport dynamics, one based on the Holstein–Biberman equation and the other on a stochastic formulation, are essentially equivalent. However, the latter appears to be physically clearer and more versatile – allowing, for instance, for consideration of the combined effect of radiative and nonradiative transport. The applicability of the theoretic results presented has, in several cases, been illustrated by comparison with experimental results.

Appendix A: Probability of emission of a photon between t and t + dt by an nth-generation molecule

In order to compute $g_n(t)$, it suffices to consider that the instant of emission for a given realisation of an n-step process is (neglecting at first the photon propagation time between molecules)

$$t = \sum_{i=1}^{n} \Delta t_i, \qquad (A.1)$$

where Δt_i is the waiting time for the ith excited molecule involved in the sequence. Now the Δt_i are independent random variables with the common density function

$$g(\Delta t) = \Gamma e^{-\Gamma \Delta t}. \qquad (A.2)$$

Owing to the independence of the Δt_i, the random variable t has a density function given by the repeated convolution of Eqn A.2,

$$g_n(t) = \underbrace{g \otimes g \otimes \ldots \otimes g}_{n\times} = \Gamma \frac{(\Gamma t)^{n-1}}{(n-1)!} e^{-\Gamma t}, \qquad (A.3)$$

which is Eqn 3.56. The propagation time is neglected in this derivation. Its consideration is unnecessary for samples of a few centimeters, if the intrinsic decay lifetimes are of at least some nanoseconds. In such cases the decay times will be two or more orders of magnitude longer than the propagation times of individual hops ($1/c \sim 3$ ps/mm).

Appendix B: Depolarization factor for radiative transfer according to classical electrodynamics

Consider Fig. B.1. The depolarization factor β is given by [59, 63]

$$\beta = \frac{3\langle \cos^2 \omega \rangle - 1}{2}, \qquad (B.1)$$

where ω is the angle formed by the transition moments of the donor and of the acceptor, and corresponds to the rotation of the transition dipole when the energy transfer occurs. This rotation can be thought to occur in two steps: first, the donor's transition moment rotates by an angle χ, becoming coincident with

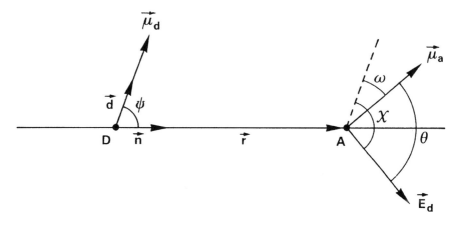

Figure B.1 A schematic representation of donor (D) and acceptor (A) relevant parameters for the computation of the depolarization factor: $\vec{\mu}_d$ and $\vec{\mu}_a$ are the donor and acceptor transition moments, $\vec{d}$ and $\vec{n}$ are the unit vectors along the transition moment of the donor and along the direction joining D and A, and $\vec{E}_d$ is the electric field of the donor at the acceptor. (Reprinted with permission from M. N. Berberan-Santos, E. J. Nunes Pereira, and J. M. G. Martinho, *J. Chem. Phys.* 103 (1995) 3022. Copyright 1995 American Institute of Physics.)

the direction of the electric field of the donor at the acceptor; it then rotates again by an angle θ so as to coincide with the direction of the acceptor's transition moment. Because these two angles are independent (χ depends only on the orientation of the donor; θ depends only on the orientation of the acceptor; and the donor and acceptor have uncorrelated orientations) Eqn B.1 can be rewritten as a product of two Soleillet factors [59, 63]:

$$\beta = \frac{3\langle\cos^2\omega\rangle - 1}{2} = \frac{3\langle\cos^2\chi\rangle - 1}{2} \times \frac{3\langle\cos^2\theta\rangle - 1}{2}. \qquad \text{(B.2)}$$

The electric field of the donor at the acceptor is, for the radiative zone,

$$\vec{E}_d = C\left[(\vec{n}\cdot\vec{d})\vec{n} - \vec{d}\right] = C(\cos\psi\,\vec{n} - \vec{d}), \qquad \text{(B.3)}$$

C being a distance-dependent factor. Now the absorption probability is proportional to

$$|\vec{E}_d \cdot \vec{\mu}_a|^2 \propto |\vec{E}_d|^2\cos^2\theta = \sin^2\psi\,\cos^2\theta. \qquad \text{(B.4)}$$

Because the orientational distribution function for the donor is

$$f(\psi) = \sin \psi, \qquad \psi \in \left[0, \frac{\pi}{2}\right] \tag{B.5}$$

and given Eqn B.4, the ψ orientational distribution for pairs with excited acceptors will be

$$g(\psi) = \frac{3}{2} \sin^3 \psi. \tag{B.6}$$

On the other hand,

$$\cos \chi = \frac{\vec{d} \cdot \vec{E}_d}{|\vec{E}_d|} = -\sin \psi; \tag{B.7}$$

hence

$$\langle \cos^2 \chi \rangle = \langle \sin^2 \psi \rangle = \int_0^{\frac{\pi}{2}} \sin^2 \psi \, g(\psi) d\psi = \frac{4}{5}. \tag{B.8}$$

In the same way, the θ distribution function for pairs with excited acceptors will be

$$g(\theta) = \frac{3}{2} \cos^2 \theta \sin \theta \tag{B.9}$$

and therefore

$$\langle \cos^2 \theta \rangle = \int_0^{\pi} \cos^2 \theta \, g(\theta) d\theta = \frac{3}{5}. \tag{B.10}$$

Finally, from Eqns B.2, B.8, and B.10, the depolarization factor is

$$\beta = \frac{7}{10} \times \frac{2}{5} = \frac{7}{25} = 0.28. \tag{B.11}$$

References

1. Agranovich, V. M., and M. D. Galanin 1982. *Electronic Excitation Energy Transfer in Condensed Matter*. North-Holland, Amsterdam.
2. Alpert, D., A.O. McCoubrey, and T. Holstein 1949. Imprisonment of resonance radiation in mercury vapor. *Phys. Rev.* 76: 1257–1259.
3. Andrews, D. L., and P. Allcock 1995. Bimolecular photophysics. *Chem. Soc. Rev.* 24: 259–265.
4. Andrews, D. L., and G. Juzeliūnas 1991. The range dependence of fluorescence anisotropy in molecular energy transfer. *J. Chem. Phys.* 95: 5513–5518.
5. Anfinrud, P. A., D. E. Hart, J. F. Hedstrom, and W. S. Struve 1986. Fluorescence depolarization of rhodamine 6G in glycerol: a photon-counting test of three-dimensional excitation transport theory. *J. Phys. Chem.* 90: 2374–2379.
6. Baumann, J., and M.D. Fayer 1985. Excitation transfer in disordered two-dimensional and anisotropic three-dimensional systems: effects of spatial geometry on time-resolved observables. *J. Chem. Phys.* 85: 4087–4107.
7. Baumann, J., G. Calzaferri, and T. Hugentobler 1985. Self-absorption and re-emission in wavelength-dependent fluorescence decay. *Chem. Phys. Lett.* 116: 66–72.
8. Berberan-Santos, M. N., and J. M. G. Martinho 1992. A linear response approach to kinetics with time-dependent rate coefficients. *Chem. Phys.* 164: 259–269.
9. Berberan-Santos, M. N., and B. Valeur 1991. Fluorescence depolarization by electronic energy transfer in donor–acceptor pairs of like and unlike chromophores. *J. Chem. Phys.* 95: 8048–8055.
10. Berberan-Santos, M. N., E. J. Nunes Pereira, and J. M. G. Martinho 1995. Stochastic theory of molecular radiative energy transport. *J. Chem. Phys.* 103: 3022–3028.
11. Berberan-Santos, M. N., E. J. Nunes Pereira, and J. M. G. Martinho 1997. Stochastic theory of combined radiative and non-radiative energy transport. *J. Chem. Phys.* 107: 10 480–10 484.
12. Biberman, L. M. 1947. On the theory of the diffusion of resonance radiation. *Zh. Eksperim. i Teor. Fiz.* 17: 416–426 [*Sov. Phys. JETP* 19, 584, 1949].
13. Birks, J. B. 1953. *Scintillation Counters*. Pergamon Press, London.
14. Birks, J. B. 1954. Energy transfer in organic phosphors. *Phys. Rev.* 94: 1567–1573.
15. Birks, J. B. 1964. *The Theory and Practice of Scintillation Counting*. Pergamon Press, Oxford.
16. Birks, J. B. 1970. *Photophysics of Aromatic Molecules*. John Wiley, London.
17. Birks, J. B. 1974. The influence of reabsorption and defects on anthracene crystal fluorescence. *Mol. Cryst. Liq. Cryst.* 28: 117–129.
18. Birks, J. B. 1975. Photophysics of aromatic molecules – a postscript. In *Organic Molecular Photophysics*. J. B. Birks, editor. John Wiley, London, Vol. 2.
19. Birks, J. B., and I. H. Munro 1967. The fluorescence lifetimes of aromatic molecules. *Prog. React. Kinet.* 4: 239–303.
20. Blickensderfer, R. P., W. H. Breckenridge, and J. Simons 1976. Diffusion theory of imprisonment of atomic resonance radiation at low opacities. *J. Phys. Chem.* 80: 653–659.
21. Braun, M., H. Liening, B. Storr, and P. Wiorkowski 1985. Comparison of experiments and theory in time resolved fluorescence spectroscopy. *Opt. Commun.* 53: 221–224.
22. Calvert, J. G., and J. N. Pitts Jr. 1966. *Photochemistry*. John Wiley, New York.
23. Compton, K. T. 1923. Some properties of resonance radiation and excited atoms. *Phil. Mag.* 45: 750–760.

24. Dhami, S., A. J. de Mello, G. Rumbles, S. M. Bishop, D. Phillips, and A. Beeby 1995. Phthalocyanine fluorescence at high concentration: dimers or reabsorption effect? *Photochem. Photobiol.* 61: 341–346.
25. Falecki, W., W. Hartmann, and R. Bocksch 1991. New aspects of Holstein's treatment of radiation trapping. *Opt. Commun.* 83: 215–219.
26. Förster, Th. 1948. Zwischenmolekulare Energiewanderung und Fluoreszenz. *Ann. Phys. (Leipzig)* 2: 55–75.
27. Galanin, M. D. 1996. *Luminescence of Molecules and Crystals.* Cambridge International Science Publishing, Cambridge.
28. Hammond, P. R. 1979. Self-absorption of molecular fluorescence, the design of equipment for measurement of fluorescence decay, and the decay times of some laser dyes. *J. Chem. Phys.* 70: 3884–3894.
29. Hart, D. E., P. A. Anfinrud, and W. S. Struve 1987. Excitation transport in solution: a quantitative comparison between GAF theory and time-resolved fluorescence profiles. *J. Chem. Phys.* 86: 2689–2696.
30. Holstein, T. 1947. Imprisonment of resonance radiation in gases. *Phys. Rev.* 72: 1212–1233.
31. Holstein, T. 1951. Imprisonment of resonance radiation in gases. II. *Phys. Rev.* 83: 1159–1168.
32. Huennekens, J., and A. Gallagher 1983. Radiation diffusion and saturation in optically thick Na vapor. *Phys. Rev. A.* 28: 238–247.
33. Kawahigashi, M., and S. Hirayama 1989. Microscopic fluorescence decay measurements on thin liquid films and droplets of concentrated dye solutions. *J. Luminescence* 43: 207–212.
34. Kenty, C. 1932. On radiation diffusion and the rapidity of escape of resonance radiation from a gas. *Phys. Rev.* 42: 823–842.
35. Kilin, S. F., and I. M. Rozman 1959. Effect of reabsorption on the fluorescence lifetimes of organic substances. *Opt. Spectrosc.* 6: 40–44.
36. Lai, R., S. Liu, and X. Ma 1993. Theoretical treatment of resonance radiation imprisonment: new approach to Holstein's equation. *Z. Phys. D* 27: 223–228.
37. Lipsett, F. R. 1967. The quantum efficiency of luminescence. *Progr. Dielectrics* 7: 217–319.
38. Ma, X., and R. Lai 1994. Theoretical treatment of radiation trapping: steady-state conditions and quenching experiment. *Phys. Rev. A* 49: 787–793.
39. Malyshev, V. A., and V. L. Shekhtman 1978. Trapping of radiation in an activated condensed medium. *Sov. Phys. – Solid State* 20: 1684–1691.
40. Martinho, J. M. G., and J. M. R. d'Oliveira 1990. Influence of radiative transport on energy transfer. *J. Chem. Phys.* 93: 3127–3132.
41. Martinho, J. M. G., A. L. Maçanita, and M. N. Berberan-Santos 1989. The effect of radiative transport on fluorescence emission. *J. Chem. Phys.* 90: 53–59.
42. Melhuish, W. H. 1961. Quantum efficiencies of fluorescence of organic substances: effect of solvent and concentration of the fluorescent solute. *J. Phys. Chem.* 65: 229–235.
43. Milne, E. A. 1926. The diffusion of imprisoned radiation through a gas. *J. London Math. Soc.* 1: 40–51.
44. Molisch, A. F., B. P. Ohery, and G. Magerl 1992. Radiation-trapping in a plane-parallel slab. *J. Quant. Spectrosc. Radiat. Transfer* 48: 377–396.
45. Molisch, A. F., B. P. Ohery, W. Schupita, and G. Magerl 1993. Radiation-trapping in cylindrical and spherical geometries. *J. Quant. Spectrosc. Radiat. Transfer* 49: 361–370.

46. Nunes Pereira, E. J., M. N. Berberan-Santos, and J. M. G. Martinho 1995. Uni-dimensional simulation of radiative transport of electronic energy. *J. Luminescence* 63: 259–271.

47. Nunes Pereira, E. J., M. N. Berberan-Santos, and J. M. G. Martinho 1996. Molecular radiative transport. II. Monte-Carlo simulation. *J. Chem. Phys.* 104: 8950–8965.

48. Nunes Pereira, E. J., M. N. Berberan-Santos, M. Vincent, J. Gallay, A. Fedorov, and J. M. G. Martinho. Molecular radiative transport. III. Experimental intensity decays. *J. Chem. Phys.*, in press.

49. Parker, C. A. 1968. *Photoluminescence of Solutions*. Elsevier, Amsterdam.

50. Parker, G. J., W. N. G. Hitchon, and J. E. Lawler 1993. Radiation trapping simulations using the propagator function method: complete and partial frequency redistribution. *J. Phys. B* 26: 4643–4662.

51. Payne, M. G., and J. D. Cook 1970. Transport of resonance radiation in an infinite cylinder. *Phys. Rev. A* 2: 1238–1248.

52. Payne, M. G., J. E. Talmage, G. S. Hurst, and E. B. Wagner 1974. Effect of correlations between absorbed and emitted frequencies on the transport of resonance radiation. *Phys. Rev. A* 9: 1050–1069.

53. Pines, D., and D. Huppert 1987. Time-resolved fluorescence depolarization measurements in mesoporous silicas. The fractal approach. *J. Phys. Chem.* 91: 6569–6572.

54. Post, H. A. 1986. Radiative transport at the 184.9-nm Hg resonance line. I. Experiment and theory. *Phys. Rev. A* 33: 2003–2016.

55. Romberg, A., and H.-J. Kunze 1988. Experimental investigation of the radiative transport of the resonance lines of sodium and lithium. *J. Quant. Spectrosc. Radiat. Transfer* 39: 99–107.

56. Sakai, Y., M. Kawahigashi, T. Minami, T. Inoue, and S. Hirayama 1989. Deconvolution of non-exponential emission decays arising from reabsorption of emitted light. *J. Luminescence* 42: 317–324.

57. Scully, A. D., A. Matsumoto, and S. Hirayama 1991. A time-resolved fluorescence study of electronic excitation energy transport in concentrated dye solutions. *Chem. Phys.* 157: 253–269.

58. Selanger, K. A., J. Falnes, and T. Sikkeland 1977. Fluorescence lifetime studies of rhodamine 6G in methanol. *J. Phys. Chem.* 81: 1960–1963.

59. Soleillet, P. 1929. Sur les paramètres caractérisant la polarisation partielle de la lumière dans les phénomènes de fluorescence. *Ann. Phys. (Paris)* 12: 23–97.

60. van Trigt, C. 1969. Analytically solvable problems in radiative transfer. I. *Phys. Rev.* 181: 97–114.

61. van Trigt, C. 1976. Analytically solvable problems in radiative transfer. IV. *Phys. Rev. A* 13: 726–733.

62. Walsh, P. J. 1959. Effect of simultaneous Doppler and collision broadening and of hyperfine structure on the imprisonment of resonance radiation. *Phys. Rev.* 116: 511–515.

63. Weber, G. 1953. Rotational brownian motion and polarization of the fluorescence of solutions. *Adv. Protein Chem.* 8: 415–459.

64. Wiorkowski, P. 1988. Monte Carlo calculations of multiple scattering of resonance photons: phase matrix algorithm and its application to 3-dimensional radiative transfer and self reversal of spectral lines. *Z. Phys. D* 10: 417–424.

65. Wiorkowski, P., and W. Hartmann 1985. Investigation of radiation imprisonment: Application to time resolved fluorescence spectroscopy. *Opt. Commun.* 53: 217–220.

66. Wiorkowski, P., and W. Hartmann 1988. Investigation of radiation trapping in time resolved fluorescence spectroscopy. *Z. Phys. D* 9: 209–214.

67. Wood, R. W. 1912. Selective reflexion, scattering and absorption by resonating gas molecules. *Phil. Mag. J. Sci.* 23: 689–714.
68. Zemansky, M. W. 1927. The diffusion of imprisoned resonance radiation in mercury vapor. *Phys. Rev.* 29: 513–523.

4

Orientational aspects in pair energy transfer

B. Wieb van der Meer
Western Kentucky University, USA

4.1 INTRODUCTION

Resonance energy transfer (RET) depends not only on the distance between the donor and the acceptor, but also on the relative orientations of the donor, the acceptor and the line connecting their centers. This chapter is about the orientational aspects of RET. Orientations of what, exactly? Which orientations are important? Specifically, the rate of radiationless energy transfer is proportional to an orientation factor, κ^2 (see Chapters 1 and 2), which depends on three orientations: that of the emission transition moment of the donor, the absorption transition moment of the acceptor, and the line connecting the centers of the donor fluorophore and the acceptor chromophore (see Fig. 4.1). We can introduce unit vectors: $\hat{d}$ along the emission transition moment of the donor, $\hat{a}$ along the absorption transition moment of the acceptor, and $\hat{r}$ along the line connecting their centers – that is, along the separation vector. The angle between $\hat{d}$ and $\hat{a}$ is θ_T, that between $\hat{d}$ and $\hat{r}$ is θ_D, and that between $\hat{a}$ and $\hat{r}$ is θ_A (see Fig. 4.2). There are three common ways of expressing κ^2 in angles:

$$\kappa^2 = (\cos\theta_T - 3\cos\theta_D\cos\theta_A)^2, \tag{4.1}$$

$$\kappa^2 = (\sin\theta_D\sin\theta_A\cos\varphi - 2\cos\theta_D\cos\theta_A)^2, \tag{4.2}$$

$$\kappa^2 = \cos^2\omega(1 + 3\cos^2\theta_D), \tag{4.3}$$

where φ is the angle between the projections of $\hat{d}$ and $\hat{a}$ on a plane perpendicular to $\hat{r}$, and ω is the angle between $\hat{a}$ and the electric dipole field due to the

Resonance Energy Transfer. Edited by David L. Andrews and Andrey A. Demidov.
© 1999 John Wiley & Sons Ltd

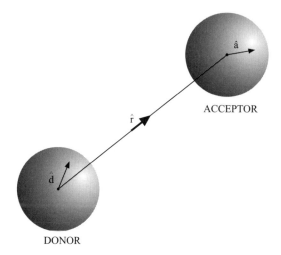

Figure 4.1   The unit vectors $\hat{d}, \hat{a}$, and $\hat{r}$: $\hat{d}$ lies along the emission transition moment of the donor, $\hat{a}$ along the absorption transition moment of the acceptor, and $\hat{r}$ along the separation vector, from the center of the donor fluorophore to the center of the acceptor chromophore

donor transition moment. The angles appearing in these three equations are illustrated in Fig. 4.2. The oscillating electric field produced by the emission dipole of the donor lies along the vector $3\hat{r} \cos \theta_D - \hat{d}$; a unit vector along this direction is $\hat{e}_D = \{3\hat{r} \cos \theta_D - \hat{d}\}/\sqrt{1 + 3 \cos^2 \theta_D}$.

An exercise that I strongly recommend to anybody interested in RET but not yet quite familiar with κ^2 is to take a notebook and draw $\hat{d}$, $\hat{a}$, and $\hat{r}$ in the following way. Open the notebook at a point at which both pages are blank (see Fig. 4.2). At the "hinge" of the notebook draw two dots not too close together, one for the location of the donor and one for that of the acceptor. The arrow from the donor to the acceptor is $\hat{r}$. Take the length of this arrow as your unit. Then draw $\hat{d}$, at the same length, on the left (or right) page as an arrow starting from the donor "dot," and draw two copies of $\hat{a}$, both on the right (or left) page parallel to each other, one starting at the acceptor "dot" and the other at that of the donor. Measure with a protractor θ_D and θ_A. Predict the value of θ_T if one of the pages were to be partly turned and held at a certain angle φ (see Fig. 4.2). Then actually perform the rotation through the value of φ that you had chosen (or turn through $180° - \varphi$, if the notebook was completely open) and check θ_T – by making a cut in the page through which the protractor can be inserted. Repeat this process of predicting and checking until you can do it quickly and accurately. Also perform the construction of $3\hat{r} \cos \theta_D - \hat{d}$, as shown at the bottom right of Fig. 4.2. (Project $\hat{d}$ on to $\hat{r}$, measure this projection, and mark off three times this distance, to yield the vector $3\hat{r} \cos \theta_D$. Add to this

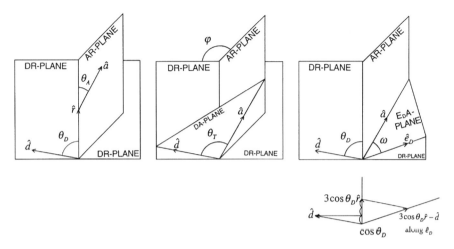

Figure 4.2   The polar angles $\theta_D, \theta_A, \theta_T$, and ω, and the azimuthal angle φ : θ_D is between $\hat{d}$ and $\hat{r}$, θ_A is between $\hat{a}$ and $\hat{r}$, θ_T is between $\hat{a}$ and $\hat{d}$, ω is between $\hat{a}$ and the electric field associated with the dipole emission transition moment, and φ is between the projections of $\hat{a}$ and $\hat{d}$ on a plane perpendicular to $\hat{r}$. The DR-plane is defined by $\hat{d}$ and $\hat{r}$, the AR-plane by $\hat{a}$ and $\hat{r}$, the DA-plane by $\hat{a}$ and $\hat{d}$, and the E_DA-plane by the electric donor field and $\hat{a}$. The bottom part of the panel on the left shows the construction of $3\hat{r}\cos\theta_D - \hat{d}$ (see text)

vector the vector $-\hat{d}$, to obtain $3\hat{r}\cos\theta_D - \hat{d}$.) The planes formed by $\hat{d}$ and $\hat{r}$, by $\hat{a}$ and $\hat{r}$, by $\hat{a}$ and $\hat{d}$, and by $\hat{e}_D$ and $\hat{a}$ are called the DR-plane, the AR-plane, the AD-plane, and the E_DA-plane, respectively. These planes are also shown in Fig. 4.2.

The highest value for κ^2 is 4, which occurs if both $\hat{d}$ and $\hat{a}$ are parallel (or antiparallel) to $\hat{r}$. The lowest value is 0, which refers to a situation in which $\hat{a}$ (or $\hat{d}$) is perpendicular to the electric field produced by the donor (or by the acceptor). The case $\kappa^2 = 4$ can only occur in a few ways, but $\kappa^2 = 0$ can be realized in an infinite number of ways (see Fig. 4.3). Consequently, the probability of finding a value near 4 is much lower than that of finding a κ^2 near 0 (see Table 4.6 below). Since the average of the cosine-squared of a polar angle is $\frac{1}{3}$, and both ω and θ_D are independent polar angles, it follows from Eqn 4.3 that the average value of κ^2 is $\frac{2}{3}$, if $\hat{d}, \hat{a}$, and $\hat{r}$ are randomly distributed. It is tempting, therefore, to replace κ^2 by $\frac{2}{3}$. It is completely justified to do just that, if the chromophores are spherical (lanthanides, for example, [16, 19]) or behave like a sphere because they rapidly diffuse through all possible orientations [8]. Replacement of κ^2 by $\frac{2}{3}$ leads to significant errors if the donor and/or acceptor are not rapidly diffusing or if they diffuse through a restricted set of orientations (see Sections 4.3 and 4.4 below).

Equation 4.3 for κ^2 depends on only two variables, ω and θ_D. This form of κ^2 refers to the electric dipole field produced by the oscillating electric field of the donor. κ^2 is proportional to the square of this field, which yields the factor $1 + 3 \cos^2 \theta_D$, and it is also proportional to $\cos^2 \omega$, where ω is the angle between this field and the acceptor transition moment. This means that if we pick a point for the location of the acceptor with respect to a donor placed at the origin, then alignment of the acceptor orientation $\hat{a}$ with the donor's electric field lines in the DR-plane (passing through that point) will yield the highest κ^2 value that can be achieved in that point, as is illustrated in Fig. 4.3. And if that point has polar coordinates r (the distance to the origin) and θ_D (the angle between $\hat{d}$ and the line connecting donor and acceptor), then the highest value for κ^2 is $1 + 3 \cos^2 \theta_D$. The $\cos^2 \omega$-proportionality implies that κ^2 is zero whenever $\hat{a}$ is along a line perpendicular to an electric field line; that is, at a tangent to an equipotential. This property is also illustrated in Fig. 4.3.

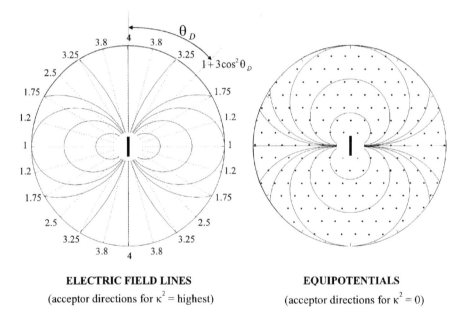

ELECTRIC FIELD LINES   **EQUIPOTENTIALS**

(acceptor directions for κ^2 = highest)   (acceptor directions for κ^2 = 0)

Figure 4.3   Electric field lines and equipotentials of the field associated with the the donor emission moment. The donor emission transition moment is along the heavy bar at the center. For an acceptor located at a point with polar coordinates r (the distance to the origin) and θ_D (the angle between $\hat{d}$ and the line connecting donor and acceptor) the highest value for κ^2 is $1 + 3 \cos^2 \theta_D$ and the lowest is zero. Since the equipotentials are curved planes, $\kappa^2 = 0$ can be obtained for an $\hat{a}$ in the DR-plane (shown), but also for an $\hat{a}$ perpendicular to the DR-plane (shown as dots) or other planes perpendicular to the field lines (not shown)

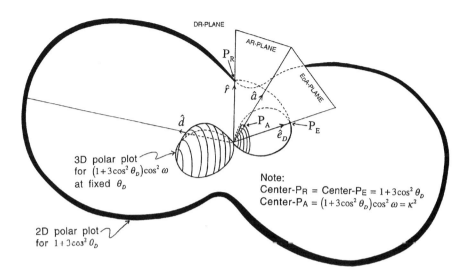

Figure 4.4 Construction of the value of κ^2 (see text)

Equation 4.3 also allows for a geometrical construction of the value of κ^2. This procedure is illustrated in Fig. 4.4. Figure 4.4 shows a paramecium-like two-dimensional polar plot for the factor $1 + 3 \cos^2 \theta_D$, in which the distance from the center to a point on this curve along a line that makes an angle θ_D with $\hat{d}$ is equal to $1 + 3 \cos^2 \theta_D$ (the distance between the center and P_R). This distance is measured off along a line through the center in the direction of $\hat{e}_D$ (the distance between the center and P_E), forming the basis of a three-dimensional polar plot of $(1 + \cos^2 \theta_D)\cos^2 \omega$, also shown in Fig. 4.4. In this three-dimensional plot θ_D is fixed and ω, the polar angle between $\hat{a}$ and $\hat{e}_D$, is varied. The line starting at the center along $\hat{a}$ will intersect the three-dimensional polar plot at a certain point P_A. The distance from this point to the center is equal to $(1 + \cos^2 \theta_D)\cos^2 \omega$, which is the value of κ^2.

It is clear from Eqns 4.1–4.3 and the definitions of the angles used in these equations that κ^2 does not change its value if we perform the following operations:

1. Flip the donor transition moment ($\hat{d} \rightarrow -\hat{d}$).
2. Flip the acceptor transition moment ($\hat{a} \rightarrow -\hat{a}$).
3. Let the donor and acceptor trade places ($\hat{r} \rightarrow -\hat{r}$).
4. Interchange the donor and acceptor ($\hat{d} \leftrightarrow \hat{a}$).

Often, $\hat{d}$ fluctuates rapidly around an orientation called the donor axis and specified with the unit vector $\hat{d}^x$, and, similarly, $\hat{a}$ fluctuates around $\hat{a}^x$. If these fluctuations take place on a timescale that is short compared to the transfer time (the inverse of the average rate of transfer), κ^2 can be replaced by $\langle \kappa^2 \rangle$, the dynamic average of the orientation factor [8, 21]. This average depends on d, a, Θ_T, Θ_D and Θ_A. The parameter d is the axial depolarization factor for the donor transition moment,

$$d \equiv \langle d_D^x \rangle = \langle \tfrac{3}{2} \cos^2 \psi_D - \tfrac{1}{2} \rangle = \int_0^{2\pi} (\tfrac{3}{2} \cos^2 \psi_D - \tfrac{1}{2}) \sin \psi_D F_D(\psi_D) d\psi_D, \quad (4.4)$$

where ψ_D is the angle fluctuating within a potential $V_D(\psi_D)$ between the donor transition moment and its axis, and $F_D(\psi_D) = \exp(-V_D(\psi_D)/k_B T)(k_B T =$ Boltzmann constant $\times$ absolute temperature) is the equilibrium distribution function for this angle. Similarly, the axial depolarization factor for the acceptor transition moment is

$$a \equiv \langle d_A^x \rangle = \langle \tfrac{3}{2} \cos^2 \psi_A - \tfrac{1}{2} \rangle = \int_0^{2\pi} (\tfrac{3}{2} \cos^2 \psi_A - \tfrac{1}{2}) \sin \psi_A F_A(\psi_A) d\psi_A, \quad (4.5)$$

where ψ_A is the angle fluctuating within a potential $V_A(\psi_A)$ between the donor transition moment and its axis, and $F_A(\psi_A) = \exp(-V_A(\psi_A)/k_B T)$ is the equilibrium distribution function for this angle. Note that it is assumed that the distribution functions F_D and F_A do not depend on the azimuthal angles γ_D and $\gamma_A (\gamma_D \{\gamma_A\}$ is the angle between the projection of $\hat{d}(\hat{a})$ on a plane perpendicular to $d^x \{\hat{a}^x\}$ and a specified axis in this plane). The transfer depolarization factor, x, is

$$x \equiv d_T^x = \tfrac{3}{2} \cos^2 \Theta_T - \tfrac{1}{2}, \quad (4.6)$$

where Θ_T is the angle between the donor axis and the acceptor axis: Θ_D is the angle between the donor axis and the separation vector, while Θ_A is that between the acceptor axis and the separation vector. The depolarization factors d, a, and x vary between $-\tfrac{1}{2}$ and 1, and can be obtained from fluorescence depolarization experiments [8] (see Section 4.3 below). The dynamical average of κ^2 is given by

$$\langle \kappa^2 \rangle = ad(\kappa_x)^2 + d(1-a)\cos^2 \Theta_D + a(1-d)\cos^2 \Theta_A - \tfrac{1}{3}d - \tfrac{1}{3}a + \tfrac{2}{3}, \quad (4.7)$$

with

$$\kappa_x = \cos \Theta_T - 3 \cos \Theta_D \cos \Theta_A. \quad (4.8)$$

4.2 κ^2 AND PROBABILITY

If the fastest rate of rotation in the system is much slower than the slowest rate of transfer, and also significantly less than k_D, the rate of direct (nontransfer) decay from the excited donor state, then the static averaging regime applies [3], and the intensity emitted by the donor at time t, after excitation with a flash of light, is proportional to

$$\int_0^\pi \int_0^\pi \exp(-(k_T + k_D)t)F_D(\psi_D)F_A(\psi_A)\sin \psi_D \sin \psi_A \, d\psi_D \, d\psi_A, \qquad (4.9)$$

assuming that the donor–acceptor distance r is constant during times of the order of $1/k_D$. The rate of transfer, k_T, is equal to $\kappa^2 k_D \frac{3}{2}(R_0'/r)^6$, where R_0' is the Förster distance for a value of κ^2 equal to $\frac{2}{3}$. On the other hand, if the largest k_T is tiny compared to k_D and also very small in relation to the smallest rate of rotation in the system, then the dynamic averaging condition applies and the donor intensity is proportional to

$$\exp(-(\langle k_T \rangle + k_D)t) \qquad (4.10)$$

where $\langle k_T \rangle$ is equal to $\langle \kappa^2 \rangle k_D \frac{3}{2}(R_0'/r)^6$ and $\langle \kappa^2 \rangle$ is given by Eqn 4.7.

In the dynamic averaging regime it is possible to deduce the uncertainty in the distance resulting from assuming an average value for κ^2, if some information on probabilities is available. There are three probability functions that are directly related to κ^2: the range probability, $P_R(0 \to \kappa^2)$, the probability distribution, $p(\kappa^2)$, and the apparent distance probability, $Q(r'/r)$. Here r' is the apparent distance – that is, the distance obtained assuming that $\kappa^2 = \frac{2}{3}$ – and r is the actual distance between the donor and acceptor. Specifically, $P_R(0 \to \kappa^2)$ is the probability of finding a value for κ^2 in the range from 0 to κ^2, $p(\kappa^2)d\kappa^2$ is the probability of encountering a value of the orientation factor lying in the interval κ^2 to $\kappa^2 + d\kappa^2$, and $Q(r'/r)d(r'/r)$ is the probability of finding the ratio of the apparent distance over the actual distance between r'/r and $(r'/r) + d(r'/r)$. It is clear that, in general, these three probability functions depend on the value of $\langle \kappa^2 \rangle$ in some unknown way. However, there are a number of special cases, specified in Table 4.1, in which specific information is available about these functions and/or their applications.

Table 4.1 Special cases in the dynamic averaging regime for donor and acceptor

Case	Value	Meaning
Linear (L)	1	Chromophore has purely linear transition moment
Planar (P)	$-\frac{1}{2}$	Chromophore is degenerated in a plane or rotating rapidly in a plane
Isotropic (I)	0	Chromophore has isotropically degenerated transition moment or rapidly rotating through all orientations

Table 4.2 Donor–acceptor combinations of degeneracies (L = linear, P = planar, I = isotropic), range of κ^2 values, most probable κ^2 value, and least probable κ^2 value in a statistical random ensemble of orientations. In this ensemble the average κ^2 equals $\frac{2}{3}$ in all cases

Case	$\hat{d}$	$\hat{a}$	Range of κ^2 values	Most probable κ^2 value	Least probable κ^2 value
L–L	1	1	0–4	0	4
L–P	1	$-\frac{1}{2}$	0–2	$\frac{1}{2}$	2
P–L	$-\frac{1}{2}$	1	0–2	$\frac{1}{2}$	2
L–I	1	0	$\frac{1}{3}-\frac{4}{3}$	$\frac{1}{3}$	$\frac{4}{3}$
I–L	0	1	$\frac{1}{3}-\frac{4}{3}$	$\frac{1}{3}$	$\frac{4}{3}$
P–P	$-\frac{1}{2}$	$-\frac{1}{2}$	$\frac{1}{4}-\frac{5}{4}$	Not known	$\frac{5}{4}$
P–I	$-\frac{1}{2}$	0	$\frac{1}{3}-\frac{5}{6}$	$\frac{5}{6}$	$\frac{1}{3}$
I–P	0	$-\frac{1}{2}$	$\frac{1}{3}-\frac{5}{6}$	$\frac{5}{6}$	$\frac{1}{3}$
I–I	0	0	$\frac{2}{3}-\frac{2}{3}$†	$\frac{2}{3}$†	$\frac{2}{3}$†

† In this case there is only one value for κ^2: $\kappa^2 = \frac{2}{3}$.

Table 4.3 Combinations of degeneracies (L = linear, P = planar, I = isotropic) and expressions for the range probability $P_R(0 \rightarrow \kappa^2)$

Case	$\hat{d}$	$\hat{a}$	$P_R(0 \rightarrow \kappa^2)$
L–L	1	1	$\begin{cases} \sqrt{\dfrac{\kappa^2}{3}}\ln(2+\sqrt{3}) &, \ 0 \leqslant \kappa^2 \leqslant 1 \\ \sqrt{\dfrac{\kappa^2-1}{3}}+\sqrt{\dfrac{\kappa^2}{3}}\ln\left(\dfrac{2+\sqrt{3}}{\sqrt{\kappa^2}+\sqrt{\kappa^2-1}}\right), & 1 \leqslant \kappa^2 \leqslant 4 \end{cases}$
L–P	1	$-\frac{1}{2}$	$\begin{cases} 1-\displaystyle\int_0^1\sqrt{1-\dfrac{2\kappa^2}{1+3x^2}}dx &, \ 0\leqslant\kappa^2\leqslant\frac{1}{2} \\ 1-\displaystyle\int_{\sqrt{\frac{1}{3}(2\kappa^2-1)}}^1\sqrt{1-\dfrac{2\kappa^2}{1+3x^2}}dx, & \frac{1}{2}\leqslant\kappa^2\leqslant2 \end{cases}$
P–L	$-\frac{1}{2}$	1	Same as L–P case
L–I	1	0	$\sqrt{\kappa^2-\frac{1}{3}}$
I–L	0	1	$\sqrt{\kappa^2-\frac{1}{3}}$
P–P	$-\frac{1}{2}$	$-\frac{1}{2}$	Not known
P–I	$-\frac{1}{2}$	0	$\sqrt{\frac{5}{3}-2\kappa^2}$
I–P	0	$-\frac{1}{2}$	$\sqrt{\frac{5}{3}-2\kappa^2}$
I–I	0	0	Kappa-squared only assumes the value $\kappa^2=\frac{2}{3}$

It is important to realize that there is no practical difference between certain forms of rapid motion and degeneracy of transition moments. For example, if the donor emission transition moment rapidly diffuses through all possible orientations, it appears as though its transition moment is isotropically degenerate. The special cases in Table 4.1 refer to isotropic (fast and complete rotation in all directions), planar (fast and complete rotation in a plane) and linear (no rotation) degeneracies. Planar degeneracy is approximately realized in naphthyl [10, 19, 20]. Close to isotropic degeneracy is found in the trivalent lanthanide ion [1, 13–12, 25]. Table 4.2 presents the range of kappa-squared values, the most probable, and the least probable kappa-squared value in a statistical random ensemble of orientations for the various combinations of degeneracies. Tables 4.3, 4.4, and 4.5 list expressions for the three probability functions.

Above, we have given information related to the three probability functions without explanation. We will now focus on the special case $a = d = 1$ (or the L–L degeneracy case) and explain in detail why the probability functions have the form given above. In this case all orientations of $\hat{d}$, $\hat{a}$, and $\hat{r}$ are equally

Table 4.4 Combinations of degeneracies (L = linear, P = planar, I = isotropic) and expressions for $p(\kappa^2)$

Case	$\hat{d}$	$\hat{a}$	$p(\kappa^2)$
L–L	1	1	$\begin{cases} \dfrac{1}{2\sqrt{3\kappa^2}}\ln(2+\sqrt{3}) & , \quad 0 \leqslant \kappa^2 \leqslant 1 \\[2ex] \dfrac{1}{2\sqrt{3\kappa^2}}\ln\left(\dfrac{2+\sqrt{3}}{\sqrt{\kappa^2}+\sqrt{\kappa^2-1}}\right), & 1 \leqslant \kappa^2 \leqslant 4 \end{cases}$
L–P	1	$-\frac{1}{2}$	$\begin{cases} \displaystyle\int_0^1 \dfrac{\mathrm{d}x}{\sqrt{1+3x^2}\sqrt{1+3x^2-2\kappa^2}} & , \quad 0 \leqslant \kappa^2 \leqslant \frac{1}{2} \\[2ex] \displaystyle\int_{\sqrt{\frac{1}{3}(2\kappa^2-1)}}^1 \dfrac{\mathrm{d}x}{\sqrt{1+3x^2}\sqrt{1+3x^2-2\kappa^2}}, & \frac{1}{2} \leqslant \kappa^2 \leqslant 2 \end{cases}$
P–L	$-\frac{1}{2}$	1	Same as L–P case
L–I	1	0	$p(\kappa^2) = \dfrac{1}{2\sqrt{\kappa^2-\frac{1}{3}}}$
I–L	0	1	Same as L–I case
P–P	$-\frac{1}{2}$	$-\frac{1}{2}$	Not known
P–I	$-\frac{1}{2}$	0	$\dfrac{1}{\sqrt{\frac{2}{3}-2\kappa^2}}$
I–P	0	$-\frac{1}{2}$	Same as P–I case
I–I	0	0	Kappa-squared only assumes the value $\kappa^2 = \frac{2}{3}$

Table 4.5 Combinations of degeneracies (L = linear, P = planar, I = isotropic) and expressions for $Q(\rho)\rho = r'/r$

Case	$\hat{d}$	$\hat{a}$	$Q(\rho)$
L–L	1	1	$\begin{cases} \dfrac{\sqrt{2}}{\rho^2}\ln\left(\dfrac{\sqrt{2}\left(\frac{2}{\sqrt{3}}+1\right)\rho^3}{1+\sqrt{1-\frac{3}{2}\rho^6}}\right), & \left(\frac{1}{6}\right)^{\frac{1}{6}} \leqslant \rho \leqslant \left(\frac{2}{3}\right)^{\frac{1}{6}} \\[4mm] \dfrac{\sqrt{2}}{\rho^2}\ln\left(2+\sqrt{3}\right), & \rho \geqslant \left(\frac{2}{3}\right)^{\frac{1}{6}} \end{cases}$
L–P	1	$-\frac{1}{2}$	$\begin{cases} \displaystyle\int_{\sqrt{\frac{1}{3}(\frac{4}{3}\rho^{-6}-1)}}^{1} \dfrac{4\rho^{-5}\,\mathrm{d}x}{\sqrt{1+3x^2}\sqrt{1+3x^2-\frac{4}{3}\rho^{-6}}}, & \left(\frac{1}{3}\right)^{\frac{1}{6}} \leqslant \rho \leqslant \left(\frac{4}{3}\right)^{\frac{1}{6}} \\[4mm] \displaystyle\int_{0}^{1} \dfrac{4\rho^{-5}\,\mathrm{d}x}{\sqrt{1+3x^2}\sqrt{1+3x^2-\frac{4}{3}\rho^{-6}}}, & \rho \geqslant \left(\frac{4}{3}\right)^{\frac{1}{6}} \end{cases}$
P–L	$-\frac{1}{2}$	1	Same as L–P case
L–I	1	0	$Q(\rho) = \dfrac{2\sqrt{3}}{\rho^2\sqrt{2-\rho^6}}$
I–L	0	1	Same as L–I case
P–P	$-\frac{1}{2}$	$-\frac{1}{2}$	Not known
P–I	$-\frac{1}{2}$	0	$Q(\rho) = \dfrac{4\sqrt{3}}{\rho^2\sqrt{5\rho^6-4}}$
I–P	0	$-\frac{1}{2}$	Same as P–I case
I–I	0	0	$\rho = r'/r$ only assumes the value $\rho = 1$

probable. It is convenient to start with the form of Eqn 4.3 for κ^2, which is depicted in Fig. 4.5. The range probability, $P_R(0 \rightarrow \kappa^2)$, called the "probability distribution" in [6], can be derived from this figure – it is related to the area under a curve similar to the ones shown in Fig. 4.5. Since θ_D and ω are polar angles, the area under the curve for a given value of κ^2 is the probability of finding κ^2 between 0 and that value. For κ^2 smaller than 1, $P_R(0 \rightarrow \kappa^2)$ is equal to the integral of $\kappa^2/(3x^2+1)$ between $x = \cos\theta_D = 0$ and $x = 1$: if κ^2 is larger than 1, the curve $\kappa^2 = $ constant will intersect the line $y = \cos\omega = 1$ at $x = \sqrt{(\kappa^2-1)/3}$. Therefore, in that case $P_R(0 \rightarrow \kappa^2)$ is given by $\sqrt{(\kappa^2-1)/3}$ plus the integral of $\kappa^2/(3x^2+1)$ between $x = \sqrt{(\kappa^2-1)/3}$ and $x = 1$. These integrations yield the following expression for $P_R(0 \rightarrow \kappa^2)$:

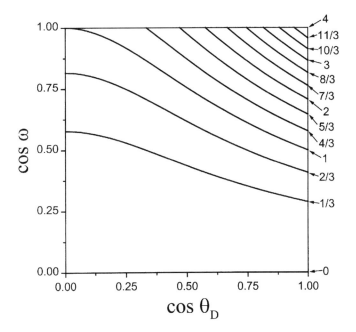

Figure 4.5   Lines of constant κ^2 in the $\cos \omega - \cos \theta_D$-plane. The family of curves corresponds to $\kappa^2 = N/3$, where $N = 0, 1, 2, \ldots, 12$

$$P_R\left(0 \rightarrow \kappa^2\right) = \begin{cases} \sqrt{\dfrac{\kappa^2}{3}} \ln\left(2 + \sqrt{3}\right) & , \quad 0 \leqslant \kappa^2 \leqslant 1, \\[3mm] \sqrt{\dfrac{\kappa^2 - 1}{3}} + \sqrt{\dfrac{\kappa^2}{3}} \ln\left(\dfrac{2 + \sqrt{3}}{\sqrt{\kappa^2} + \sqrt{\kappa^2 - 1}}\right), & 1 \leqslant \kappa^2 \leqslant 4. \end{cases} \tag{4.11}$$

This "range probability" is plotted in Fig. 4.6. The difference between $P_R(0 \rightarrow x_1)$ and $P_R(0 \rightarrow x_2)$ represents the probability of finding κ^2 between x_1 and x_2 in a random ensemble. Such probabilities for a series of intervals are listed in Table 4.6. The probability density or frequency distribution, $p(\kappa^2)$, is obtained by differentiating $P_R(0 \rightarrow \kappa^2)$ with respect to κ^2. The $p(\kappa^2)\mathrm{d}\kappa^2$ is the probability of encountering a value of the orientation factor lying in the interval κ^2 to $\kappa^2 + \mathrm{d}\kappa^2 : p(\kappa^2)$ is given by

$$p\left(\kappa^2\right) = \begin{cases} \dfrac{1}{2\sqrt{3}\kappa^2} \ln\left(2 + \sqrt{3}\right) & , \quad 0 \leqslant \kappa^2 \leqslant 1, \\[3mm] \dfrac{1}{2\sqrt{3}\kappa^2} \ln\left(\dfrac{2 + \sqrt{3}}{\sqrt{\kappa^2} + \sqrt{\kappa^2 - 1}}\right), & 1 \leqslant \kappa^2 \leqslant 4. \end{cases} \tag{4.12}$$

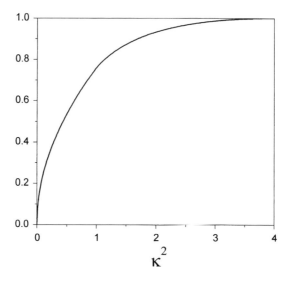

Figure 4.6   A plot of the "range probability" $P_R(0 \to \kappa^2)$, i.e. the probability of finding a value for κ^2 in the range 0 to κ^2, versus the orientation factor κ^2, for $a = d = 1$. The plot is for a statistically random ensemble of independent orientations of the transition moments and of the line connecting each donor-acceptor pair (L–L case)

Table 4.6   Probabilities of finding κ^2 values within a given interval in a statistically random ensemble of orientations. The numbers are accurate to the nearest 0.0001

Range of κ^2	Probability
$0-\frac{1}{3}$	0.4390
$\frac{1}{3}-\frac{2}{3}$	0.1818
$\frac{2}{3}-1$	0.1395
$1-\frac{4}{3}$	0.0848
$\frac{4}{3}-\frac{5}{3}$	0.0522
$\frac{5}{3}-2$	0.0357
$2-\frac{7}{3}$	0.0250
$\frac{7}{3}-\frac{8}{3}$	0.0174
$\frac{8}{3}-3$	0.0118
$3-\frac{10}{3}$	0.0075
$\frac{10}{3}-\frac{11}{3}$	0.0040
$\frac{11}{3}-4$	0.0012

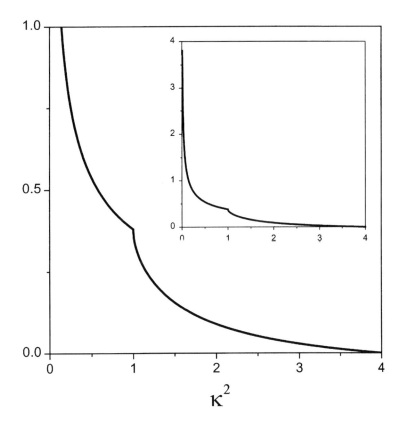

Figure 4.7 A plot of the probability density $p(\kappa^2)$ versus the orientation factor κ^2, for $a = d = 1$: $p(\kappa^2)d\kappa^2$ is the probability of encountering a value of the orientation factor lying in the interval κ^2 to $\kappa^2 + d\kappa^2$. The plot is for a statistically random ensemble of independent orientations of the transition moments and of the line connecting each donor–acceptor pair (L–L case)

From the plot of $p(\kappa^2)$ versus κ^2, shown in Fig. 4.7, it is obvious that the most probable value for κ^2 is zero. The expectation value or average value, which can be calculated by integrating $\kappa^2 p(\kappa^2)$ over the interval 0 to 4, is $\frac{2}{3}$ as expected. Expressions for $p(\kappa^2)$ and $P_R(0 \to \kappa^2)$ have been published by Dale *et al.* [8] and Tompa and Englert [12]. Both papers have typographical errors in these formulas. Dale *et al.* have published a correction in [8].

Haas *et al.* [10] have introduced the distribution function $Q(r'/r)$, which could be called the "apparent distance probability." Here r' is the apparent distance – that is, the distance obtained assuming that $\kappa^2 = \frac{2}{3}$ – and r is the actual distance between the donor and acceptor. Note that there is a direct relation between r'/r and κ^2, because $r'^6/\frac{2}{3}$ is equal to r^6/κ^2, so that

$$\left(\frac{r'}{r}\right)^6 = \frac{\frac{2}{3}}{\kappa^2}. \tag{4.13}$$

Since $p(\kappa^2)\mathrm{d}\kappa^2$ and $Q(r'/r)\mathrm{d}(r'/r)$ both denote the fraction of the population of donor–acceptor pairs whose orientational factor falls in the range of κ^2 to $\kappa^2 + \mathrm{d}\kappa^2$, we have

$$Q(r'/r)\mathrm{d}(r'/r) = p(\kappa^2)\mathrm{d}\kappa^2. \tag{4.14}$$

With Eqns 4.13 and 4.14, one can derive expressions for $Q(r'/r)$ from the expressions of $p(\kappa^2)$ for all the degeneracy cases listed in Table 4.1. For example, in the L–L case, one finds that

$$Q(r'/r) = \begin{cases} \sqrt{2}\left(\frac{r}{r'}\right)^4 \ln\left(\dfrac{\left(\frac{r'}{r}\right)^3 \left(\sqrt{6}+\frac{3}{2}\sqrt{2}\right)}{1+\sqrt{1-\frac{3}{2}\left(\frac{r'}{r}\right)^6}}\right), & \left(\frac{1}{6}\right)^{1/6} \le \frac{r'}{r} \le \left(\frac{2}{3}\right)^{1/6} \\[4ex] \sqrt{2}\left(\frac{r}{r'}\right)^4 \ln\left(2+\sqrt{3}\right), & \frac{r'}{r} > \left(\frac{2}{3}\right)^{1/6} \end{cases} \tag{4.15}$$

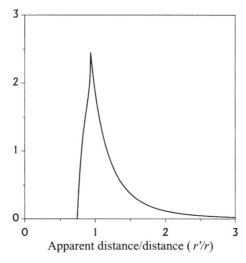

Apparent distance/distance (r'/r)

Figure 4.8   A plot of the apparent distance probability $Q(r'/r)$ versus r'/r, where r' is the apparent distance obtained when assuming $\kappa^2 = \frac{2}{3}$, and r is the actual distance between the donor and acceptor. $Q(r'/r)\mathrm{d}(r'/r)$ denotes the fraction of the population of donor–acceptor pairs whose ratio of apparent distance over actual distance falls in the range of r'/r to $(r'/r) + \mathrm{d}(r'/r)$. The plot is for a statistically random ensemble of independent orientations of the transition moments and of the line connecting each donor–acceptor pair (L–L case). If the motions of the transition moments are restricted or there is degeneracy, the peak of $Q(r'/r)$ is narrower and closer to unity [23]

where, as above, it is assumed that all orientations of the transition moments and the distance vector ($\hat{r}$) are equally probable. This apparent distance probability function is plotted in Fig. 4.8.

4.3 κ^2 AND ANISOTROPY

In the dynamic averaging regime, it is useful to derive a $\langle \kappa^2 \rangle_{\text{MIN}}$ and a $\langle \kappa^2 \rangle_{\text{MAX}}$ from anisotropy measurements, such that the average κ^2 is larger than or equal to $\langle \kappa^2 \rangle_{\text{MIN}}$ and smaller than or equal to $\langle \kappa^2 \rangle_{\text{MAX}}$. This procedure, developed by Dale and coworkers [2, 4, 6–9], is based on the following assumptions:

1. The donor emission moment, which is along the unit vector $\hat{d}$, fluctuates rapidly around $\hat{d}^X$ (a unit vector, called the "donor axis").
2. The acceptor absorption moment, which is along the unit vector $\hat{a}$, fluctuates rapidly around $\hat{a}^X$ (a unit vector, called the "acceptor axis").
3. The probability that the donor emission moment is along $\hat{d}$ depends only on the angle ψ_D between $\hat{d}$ and $\hat{d}^X$, but not on the azimuthal angle.
4. The probability that the acceptor absorption moment is along $\hat{a}$ depends only on the angle ψ_A between $\hat{a}$ and $\hat{a}^X$, but not on the azimuthal angle.
5. The angle θ_T between the momentary donor emission moment and the momentary acceptor absorption moment fluctuates around Θ_T, the angle between the donor axis and the acceptor axis (this is a consequence of assumptions 1 and 2).
6. The angle χ_D between the donor absorption and emission transition moments is unique if a single transition (for example, $S_2 \leftarrow S_0$) is excited. If, on the other hand, excitation occurs into overlapping transitions, χ_D can assume a range of values.
7. The same is true for χ_A, the angle between the acceptor absorption and emission transition moments.

The fluctuations of the transition moments around their "axes" may occur on a range of timescales. For example, a dipole could wobble around an axis on a picosecond timescale, and that axis could vibrate around another direction on a nanosecond timescale. Only those fluctuations concern us here that are fast compared to the average rate of transfer. We can average κ^2 over the corresponding range of orientations. The average κ^2 will depend on "depolarization factors," which are (averages of) second-rank Legendre polynomials of a cosine of a polar angle θ between 0 and π. That is, the depolarization factors have the form

$$d_\theta = \tfrac{3}{2} \cos^2 \theta - \tfrac{1}{2} \Rightarrow -\tfrac{1}{2} \leqslant d_\theta \leqslant 1, \tag{4.16}$$

or

$$\langle d_\theta \rangle = \langle \tfrac{3}{2} \cos^2 \theta - \tfrac{1}{2} \rangle \Rightarrow -\tfrac{1}{2} \leqslant \langle d_\theta \rangle \leqslant 1, \tag{4.17}$$

where the brackets denote a weighted average over the range of angles covered by θ during times small compared to the average transfer time. Note that $\hat{d}$ (the unit vector along the donor dipole) and d_θ (the depolarization factor) are very similar in appearance, but have completely different meanings. The relevant depolarization factors are

$$\langle d_{\chi D} \rangle = \langle \tfrac{3}{2} \cos^2 \chi_D - \tfrac{1}{2} \rangle, \qquad \langle d_{\chi A} \rangle = \langle \tfrac{3}{2} \cos^2 \chi_A - \tfrac{1}{2} \rangle, \tag{4.18, 4.19}$$

$$d = \langle d_D^X \rangle = \langle \tfrac{3}{2} \cos^2 \psi_D - \tfrac{1}{2} \rangle, \qquad a = \langle d_A^X \rangle = \langle \tfrac{3}{2} \cos^2 \psi_A - \tfrac{1}{2} \rangle \tag{4.20, 4.21}$$

$$\langle d_T \rangle = \langle \tfrac{3}{2} \cos^2 \theta_T - \tfrac{1}{2} \rangle, \qquad x \equiv d_T^X = \tfrac{3}{2} \cos^2 \theta_T - \tfrac{1}{2}. \tag{4.22, 4.23}$$

The angle brackets in Eqns 4.18 and 4.19 are significant if excitation into overlapping transitions occurs, but can be deleted if a single transition is excited. The depolarization factors listed above are related to anisotropies. The relevant anisotropy parameters are:

$r_{0D} = $ the "time-zero" value of the donor emission anisotropy,
$r_{0A} = $ the "time-zero" value of the acceptor emission anisotropy,
$r_{fD} = $ the fundamental anisotropy of the donor, in the absence of motion,
$r_{fA} = $ the fundamental anisotropy of the acceptor, in the absence of motion,
$r_{0T} = $ the "initial" transfer anisotropy – that is, the emission anisotropy of the acceptor resulting from exciting the donor in its $S_1 \leftarrow S_0$ band followed by transfer into $S_1 \leftarrow S_0$ of the acceptor,

where the terms "time-zero" and "initial" must be seen in the light of a Bürkli–Cherry plot [2, 23].

Soleillet's theorem [8] allows one to relate the depolarization factors to anisotropies. This theorem says that, if a series of depolarizing events intervenes between absorption and emission and if these events are independent of each other and exhibit cylindrical symmetry, then the fluorescence anisotropy is equal to 0.4 times the product of the corresponding depolarization factors. Applying Soleillet's theorem yields the following relations:

$$r_{fD} = 0.4 \langle d_{\chi D} \rangle, \qquad r_{fA} = 0.4 \langle d_{\chi A} \rangle, \tag{4.24, 4.25}$$

$$r_{0D} = r_{fD} d^2, \qquad r_{0A} = r_{fA} a^2, \qquad r_{0T} = dxa. \tag{4.26–4.28}$$

Note that the experimentally obtained factors d^2 and a^2 are in the range 0 to 1, whereas d and a are in the range -0.5 to 1.0, so that d and a can be either positive or negative if d^2 and a^2 are in the range 0.0 to 0.25. Similarly, x can be either positive or negative if d and/or a are of indeterminate sign. These sign ambiguities may be resolved if independent structural information is available (for example, if X-ray diffraction data suggest that a certain angle in a protein is close to a given value).

The average value of the orientation factor is given by Eqn 4.7, which allows one to find the range of possible κ^2 values. This possibility is obvious for the extreme situations: for $d = a = 0$, $\langle \kappa^2 \rangle$ can only assume the value $\frac{2}{3}$, whereas for $d = a = 1$, $\langle \kappa^2 \rangle$ is between 0 and 4. In general, there are two possible cases: case 1, x is not known, and case 2, x is known.

4.3.1 Case 1: unknown x

In the case that the transfer depolarization is not known, there are six possible candidates for maxima and minima, derived in [22]:

$$\kappa_A^2 = \tfrac{2}{3} + \tfrac{2}{3}a + \tfrac{2}{3}d + 2ad, \qquad \kappa_M^2 = \tfrac{2}{3} + \tfrac{1}{6}a + \tfrac{1}{6}d - ad + \tfrac{1}{2}|a - d|, \quad (4.29, 4.30)$$

$$\kappa_L^2 = \tfrac{2}{3} + \tfrac{1}{6}a + \tfrac{1}{6}d - ad - \tfrac{1}{2}|a - d|, \qquad \kappa_H^2 = \tfrac{2}{3} - \tfrac{1}{3}a - \tfrac{1}{3}d + ad, \quad (4.31, 4.32)$$

$$\kappa_T^2 = \tfrac{1}{9}(1 - a)(1 - d) + \tfrac{4}{9}\sqrt{(1 - a)(1 - d)(1 + 2a)(1 + 2d)}, \qquad (4.33)$$

$$\kappa_P^2 = \tfrac{2}{3} - \tfrac{1}{3}a - \tfrac{1}{3}d. \qquad (4.34)$$

The relative orientations of the donor and acceptor axes and the separation vector for these extrema are as follows:

For κ_A^2: the axes of both the donor and the acceptor are *aligned* with the separation vector.

For κ_M^2: the axes of the *most* prolate distribution (donor or acceptor) is aligned with the separation vector, while the other (acceptor or donor) is perpendicular to the separation vector. In other words, if $d > a$, the donor axis is aligned with the separation vector and the acceptor axis is perpendicular to it, and vice versa if $a > d$.

For κ_L^2: the axes of the *least* prolate distribution (donor or acceptor) is aligned with the separation vector, the other (acceptor or donor) is perpendicular to the separation vector.

For κ_H^2: the donor and acceptor are in an "*H-configuration*"; that is, their axes are parallel to each other but perpendicular to the separation vector.

For κ_T^2: the axes of both donor and acceptor are *tilted* at some angle from the separation vector, but are coplanar with this vector.

For κ_P^2: the separation vector, the donor axis and the acceptor axis are all *mutually perpendicular*.

In any point in the d–a plane one of these extrema is the maximum and another is the minimum, but it depends on the values of d and a which one of the six extrema is the maximum and which is the minimum. Accordingly, the d–a plane is divided into different regions, as defined in Table 4.7 [22].

It should be noted that anisotropy (or polarization) data can also be employed for calculating apparent distance probabilities. This approach has been taken by Haas and coworkers, who have shown that the peak of $Q(r'/r)$ becomes sharper and approaches unity, if the donor and acceptor polarization values come closer to zero [10].

Table 4.7   Regions in the (d, a)-plane with different maxima and minima for the case that the transfer depolarization is not known

For $0 \leqslant d \leqslant 1, 0 \leqslant a \leqslant 1$	$\kappa_P^2 \leqslant \langle \kappa^2 \rangle \leqslant \kappa_A^2$
For $-\frac{1}{2} \leqslant d \leqslant 0, \frac{1}{2} - d \leqslant a \leqslant 1$	$\kappa_H^2 \leqslant \langle \kappa^2 \rangle \leqslant \kappa_M^2$
For $\frac{1}{2} \leqslant d \leqslant 1, \frac{1}{2} - d \leqslant a \leqslant 0$	$\kappa_H^2 \leqslant \langle \kappa^2 \rangle \leqslant \kappa_M^2$
For $-\frac{1}{2} \leqslant d \leqslant -\frac{1}{3}, -\dfrac{1 + 3d}{3 + 5d} \leqslant a \leqslant \dfrac{1}{2} - d$	$\kappa_T^2 \leqslant \langle \kappa^2 \rangle \leqslant \kappa_M^2$
For $-\frac{1}{3} \leqslant d \leqslant 0, \dfrac{1 + 3d}{2 + 2d} \leqslant a \leqslant \dfrac{1}{2} - d$	$\kappa_T^2 \leqslant \langle \kappa^2 \rangle \leqslant \kappa_M^2$
For $0 \leqslant d \leqslant \frac{1}{2}, -\dfrac{1 + 3d}{3 + 5d} \leqslant a \leqslant \dfrac{1 - 2d}{2d - 3}$	$\kappa_T^2 \leqslant \langle \kappa^2 \rangle \leqslant \kappa_M^2$
For $\frac{1}{2} \leqslant d \leqslant 1, -\dfrac{1 + 3d}{3 + 5d} \leqslant a \leqslant \dfrac{1}{2} - d$	$\kappa_T^2 \leqslant \langle \kappa^2 \rangle \leqslant \kappa_M^2$
For $-\frac{1}{3} \leqslant d \leqslant 0, 0 \leqslant a \leqslant \dfrac{1 + 3d}{2 + 2d}$	$\kappa_L^2 \leqslant \langle \kappa^2 \rangle \leqslant \kappa_M^2$
For $0 \leqslant d \leqslant \frac{1}{2}, \dfrac{1 - 2d}{2d - 3} \leqslant a \leqslant 0$	$\kappa_L^2 \leqslant \langle \kappa^2 \rangle \leqslant \kappa_M^2$
For $-\frac{1}{2} \leqslant d \leqslant -\frac{1}{3}, 0 \leqslant a \leqslant -\dfrac{1 + 3d}{3 + 5d}$	$\kappa_A^2 \leqslant \langle \kappa^2 \rangle \leqslant \kappa_M^2$
For $0 \leqslant d \leqslant 1, -\frac{1}{2} \leqslant a \leqslant -\dfrac{1 + 3d}{3 + 5d}$	$\kappa_A^2 \leqslant \langle \kappa^2 \rangle \leqslant \kappa_M^2$
For $-\frac{1}{2} \leqslant d \leqslant -\frac{1}{3}, \dfrac{1 + 3d}{2 + 2d} \leqslant a \leqslant 0$	$\kappa_L^2 \leqslant \langle \kappa^2 \rangle \leqslant \kappa_H^2$
For $-\frac{1}{2} \leqslant d \leqslant 0, -\frac{1}{2} \leqslant a \leqslant \dfrac{1 - 2d}{2d - 3}$	$\kappa_L^2 \leqslant \langle \kappa^2 \rangle \leqslant \kappa_H^2$
For $-\frac{1}{2} \leqslant d \leqslant -\frac{1}{3}, \dfrac{1 - 2d}{2d - 3} \leqslant a \leqslant \dfrac{1 + 3d}{2 + 2d}$	$\kappa_T^2 \leqslant \langle \kappa^2 \rangle \leqslant \kappa_H^2$
For $-\frac{1}{3} \leqslant d \leqslant 0, \dfrac{1 - 2d}{2d - 3} \leqslant a \leqslant -\dfrac{1 + 3d}{3 + 5d}$	$\kappa_T^2 \leqslant \langle \kappa^2 \rangle \leqslant \kappa_H^2$

4.3.2 Case 2: known x

When the depolarization factor $x = d_T^X$ is known, there are four extrema, two of which are related to the function $\kappa^2(\varphi)$, defined as

$$
\begin{aligned}
\kappa^2(\varphi) = {}&\tfrac{2}{3} + \tfrac{1}{6}a + \tfrac{1}{6}d - ad + \tfrac{1}{12}ad(1 + 2x) + \tfrac{9}{4}ad \cos^2 2\varphi \\
&+ \tfrac{1}{2}(a + d + ad)\cos 2\varphi\sqrt{\tfrac{1}{3} + \tfrac{2}{3}x} + \tfrac{1}{2}(a - d)\sin 2\varphi\sqrt{\tfrac{2}{3} - \tfrac{2}{3}x}.
\end{aligned}
\tag{4.35}
$$

The four extrema are as follows:

$$
\kappa^2_{CB} = \text{maximum of} \kappa^2(\varphi) \text{with respect to variations in } \varphi, \tag{4.36}
$$

$$
\kappa^2_{CS} = \text{minimum of} \kappa^2(\varphi) \text{with respect to variations in } \varphi, \tag{4.37}
$$

$$
\kappa^2_{HP} = \tfrac{2}{3} - \tfrac{1}{3}a - \tfrac{1}{3}d + \tfrac{1}{3}ad(1 + 2x), \tag{4.38}
$$

$$
\kappa^2_{NN} = \tfrac{2}{3} - \tfrac{1}{3}a - \tfrac{1}{3}d - \tfrac{1}{9}(1 - a)(1 - d) + \tfrac{2}{9}\sqrt{3ad(1 - a)(1 - d)(1 + 2x)}. \tag{4.39}
$$

The function $\kappa^2(\varphi)$ represents the value for $\langle \kappa^2 \rangle$ in the case that the donor axis, the acceptor axis and the separation vector are coplanar. Therefore, κ^2_{CB} and κ^2_{CS} are the largest and smallest $\langle \kappa^2 \rangle$-values, respectively, for this case. These extrema can be obtained graphically by plotting $\kappa^2(\varphi)$ versus φ, and they can also be found analytically [22]. The d–a plane is divided into different regions, where κ^2_{CB} or κ^2_{HP} is the maximum $\langle \kappa^2 \rangle$ and $\kappa^2_{CS}, \kappa^2_{HP}$ or κ^2_{NN} is the minimum. These regions are shown in Table 4.8. Some of the borders depend on $H(a, d)$, defined by [22]

$$
\begin{aligned}
&H(a, d) \\
&= \frac{\left(\sqrt{12(7ad + a + d)(1 - a)(1 - d) + (2 + a + d - 4ad)^2} - 2 - a - d + 4ad\right)^2}{24ad(1 - a)(1 - d)} - \frac{1}{2}.
\end{aligned}
\tag{4.40}
$$

Table 4.8 Regions in the (d, a)-plane with different maxima and minima for the case that the transfer depolarization is known

For $-\tfrac{1}{2} \leqslant d \leqslant 0, -\tfrac{1}{2} \leqslant a \leqslant 0, -\tfrac{1}{2} \leqslant x \leqslant 1$	$\kappa^2_{CS} \leqslant \langle \kappa^2 \rangle \leqslant \kappa^2_{HP}$
For $-\tfrac{1}{2} \leqslant d \leqslant 0, 0 \leqslant a \leqslant 1, -\tfrac{1}{2} \leqslant x \leqslant 1$	$\kappa^2_{CS} \leqslant \langle \kappa^2 \rangle \leqslant \kappa^2_{CB}$
For $0 \leqslant d \leqslant 1, -\tfrac{1}{2} \leqslant a \leqslant 0, -\tfrac{1}{2} \leqslant x \leqslant 1$	$\kappa^2_{CS} \leqslant \langle \kappa^2 \rangle \leqslant \kappa^2_{CB}$
For $0 \leqslant d \leqslant 1, 0 \leqslant a \leqslant 1, -\tfrac{1}{2} \leqslant x \leqslant (1 - a - d - 2ad)/(6ad)$	$\kappa^2_{HP} \leqslant \langle \kappa^2 \rangle \leqslant \kappa^2_{CB}$
For $0 \leqslant d \leqslant 1, 0 \leqslant a \leqslant 1, (1 - a - d - 2ad)/(6ad) \leqslant x \leqslant H(a, d)$	$\kappa^2_{NN} \leqslant \langle \kappa^2 \rangle \leqslant \kappa^2_{CB}$
For $0 \leqslant d \leqslant 1, 0 \leqslant a \leqslant 1, H(a, d) \leqslant x \leqslant 1$	$\kappa^2_{CS} \leqslant \langle \kappa^2 \rangle \leqslant \kappa^2_{CB}$

4.4   NOTES ON THE EFFECTS OF ORDER AND MOTION

The first question in many RET studies is: "What is the donor–acceptor distance?" The next is: "If we assume that $\kappa^2 = \frac{2}{3}$, what is the error in the calculated distance if that assumption is not accurate?" Related to these questions is the problem: "Can we distinguish two different donor–acceptor configurations with an identical rate of transfer but different κ^2 values?" This problem is illustrated in Fig. 4.9. The main difficulty in solving this problem is that the kinetics is only straightforward in the four special cases illustrated. In general, when neither the static nor the dynamic averaging regime applies, the effects of motion and orientational order complicate the description of the

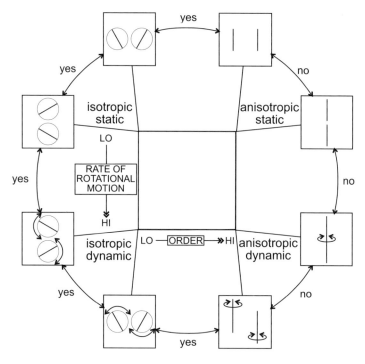

Figure 4.9   A schematic representation of the question: Are two related configurations distinguishable by RET? In the upper right-hand corner are shown two situations (completely static and with perfect orientational order) with identical rates of transfer, one with $\kappa^2 = 4$ and the other $\kappa^2 = 1$. In the $\kappa^2 = 4$ configuration the distance is a factor $4^{1/6}$ larger than that in the $\kappa^2 = 1$ situation. In an isotropic sample it is impossible to distinguish these two using RET and/or fluorescence depolarization. Rapid rotation around the transition moments cannot help (except rotations around an axis at a significant angle to the transition moment). If disorder is introduced without changing the distances, the two configurations become distinguishable

system. Matrix techniques may then be helpful to describe the time-resolved intensity and anisotropy of emission from the donor or the acceptor [23, 24]. In principle, combining intensity and anisotropy RET data allows one to study subtle orientational changes. The full potential of this combination has not yet been completely explored. The accurately measurable quotient κ^2/r^6 allows identification of structural changes that are much more subtle than just changes in the distance r. However, if distance changes and orientational effects are happening at the same time, models are needed to interpret the data [23, 24].

Acknowledgments

I would like to thank Drs. Joe Beechem, Scott Blackman, Dave Piston (Vanderbilt, TN, USA) and Bob Dale (King's College London, UK) for helpful and stimulating discussions. Some of the work discussed in this chapter is supported by NASA NCCW60.

References

1. Abusaleh, A., and C. F. Meares 1984. Excitation and de-excitation processes in lanthanide chelates bearing aromatic sidechains. *Photochem. Photobiol.* 39: 763–769.
2. Bürkli, A., and R. J. Cherry 1981. Rotational motion and flexibility of Ca^{2+}, Mg^{2+}— dependent adenosine 5'-triphosphatase in *Sarcoplasmic-reticulum* membranes. *Biochemistry* 20: 138–145.
3. Dale, R. E. 1978. Fluorescence depolarization and orientation factors for excitation energy transfer between isolated donor and acceptor fluorophore pairs at fixed intermolecular separations. *Acta Phys. Polon.* A54: 743–756.
4. Dale, R. E. 1988. Circulated unpublished lecture notes of contribution to NATO Advanced Study Institute: Excited-state probes in biochemistry and biology, Sicily, 1984.
5. Dale, R. E. 1993. Personal communication.
6. Dale, R. E., and J. Eisinger 1974. Intramolecular distances determined by energy transfer. Dependence on orientational freedom of donor and acceptor. *Biopolymers* 13: 1573–1605.
7. Dale, R. E., and J. Eisinger 1975. Polarized excitation energy transfer. In *Biochemical Fluorescence: Concepts*. R. F. Chen and H. Edelhoch, editors. Marcel Dekker, New York, Vol. 1, pp. 115–284.
8. Dale, R. E., J. Eisinger, and W. E. Blumberg 1979. The orientational freedom of molecular probes. *Biophys. J.* 26: 161–194; and correction in *Biophys. J.* (1980) 30: 365.
9. Eisinger, J., W. E. Blumberg, and R. E. Dale 1981. Orientational effects in intramolecular and intermolecular long range excitation energy transfer. *Ann. N. Y. Acad. Sci.* 366: 155–175.
10. Haas, E., E. Katchalski-Katzir, and I. Z. Steinberg 1975. Brownian motion of the ends of oligopeptide chains in solution as estimated by energy transfer between the chain ends. *Biopolymers* 17: 11–31.

11. Haas, E., E. Katchalski-Katzir, and I. Z. Steinberg 1978. Effect of the orientation of donor and acceptor on the probability of energy transfer involving electronic transitions of mixed polarization. *Biochemistry* 17: 5064–5070.
12. Horrocks, W. DeW., Jr., B. Holmquist, and B. L. Vallee 1975. Energy transfer between terbium (III) and cobalt (II) in thermolysin: a new class of metal – metal distance probes. *Proc. Natl Acad. Sci., USA* 72: 4764–4768.
13. Horrocks, W. DeW., Jr., and W. E. Collier 1981. Lanthanide ion luminescence probes measurement of distance between intrinsic protein fluorophores and bound metal-ions quantitation of energy-transfer between tryptophan and terbium (III) or europium (III) in the calcium-binding protein parvalbumin. *J. Am. Chem. Soc.* 103: 2856–2862.
14. Krejcarek, G. E., and K. L. Tucker 1977. Covalent attachment of chelating groups to macromolecules. *Biochem. Biophys. Res. Commun.* 77: 581–585.
15. Luk, C. K. 1971. Study of the nature of the metal-binding sites and estimate of the distance between the metal-binding sites in transferrin using trivalent lanthanide ions as fluorescent probes. *Biochemistry* 10: 2838–2843.
16. O'Hara, P., S. M. Yeh, C. F. Meares, and R. Bersohn 1981. Distance between metal binding sites in transferrin energy transfer from bound terbium(III) to iron(III) or manganese(III). *Biochemistry* 20: 4704–4708.
17. Rhee, M.-J., W. DeW., Jr. Horrocks, and D. P. Kosow 1984. Laser induced lanthanide luminescence as a probe of metal ion binding sites of human factor xa. *J. Biol. Chem.* 259: 7404–7408.
18. Selvin, P. R. 1995. Fluorescence resonance energy transfer. In *Biochemical Spectroscopy*. K. Sauer, editor. *Meth. Enzymol.* 246: 300–334.
19. Steinberg, I. Z., E. Haas, and E. Katchalski-Katzir 1983. Long-range nonradiative transfer of electronic excitation energy. In *Time-resolved Fluorescence Spectroscopy in Biochemistry and Biology*. R. B. Cundall and R. E. Dale, editors. Plenum Press, New York, pp. 411–450.
20. Tompa, H., and A. Englert 1979. The frequency distribution of the orientation factor of dipole–dipole interaction. *Biophys. Chem.* 9: 211–214.
21. Thomas, D. D., W. F. Carlsen, and L. Stryer 1978. Fluorescence energy transfer in the rapid-diffusion limit. *Proc. Natl Acad. Sci., USA* 75: 5746–5750.
22. van der Meer, B. W., to be published.
23. van der Meer, B. W., G., III Coker, and S.-Y. S. Chen 1994. *Resonance Energy Transfer: Theory and Data*. VCH Publishers, New York.
24. van der Meer, B. W., M. A. Raymer, S. L. Wagoner, R. L. Hackney, J. M. Beechem, and E. Gratton 1993. Designing matrix models for fluorescence energy transfer between moving donors and acceptors. *Biophys. J.* 64: 1243–1263.
25. Wang, C. -L. A., T. Tao, and J. Gergely 1982. The distance between the high affinity sites of troponin-c measured by interlanthanide ion energy transfer. *J. Biol. Chem.* 257: 8372–8375.

Polarization in molecular complexes with incoherent energy transfer

Andrey A. Demidov[1] and David L. Andrews[2]
[1]*Northeastern University, USA*
[2]*University of East Anglia, UK*

5.1   INTRODUCTION

Polarization spectroscopy is one of the key scientific tools for elucidating the structural and kinetic properties of complex molecular systems [21, 26, 41]. Combined with absorption and fluorescence spectroscopy in the time and frequency domains, polarization analysis can afford a source of important additional information relating to both macroscopic and microscopic structure. Where conventional linear (single-photon) spectroscopy is concerned, the process of photoabsorption is not intrinsically sensitive to the input beam polarization unless the system is anisotropic; that is, if it exhibits at least mesoscopic order. However, in any system, whether intrinsically anisotropic or isotropic, the absorption of photons from a polarized beam of light itself confers a degree of anisotropy, which persists only until it is lost through molecular motion or energy randomization. Thus, over a timescale determined by such dynamical effects, any complex system excited by photoabsorption can exhibit its conferred anisotropy through secondary optical processes such as fluorescence. The polarization aspects of such fluorescence emission reflect the effects of local order and structure, with a temporal signature that reveals kinetic features.

   In this chapter, we shall be concerned with the theory of polarization measurements as they relate to condensed phase systems which lack bulk order but

Resonance Energy Transfer, Edited by David L. Andrews and Andrey A. Demidov.
© 1999 John Wiley & Sons Ltd

have microscopic order, as is the case with a wide range of biological and other chemically complex heterogeneous materials. In such media, a key process of excitation randomization is incoherent energy transfer, typically associated with picosecond kinetics. Each elementary process of energy migration in what can often be a multi-step sequence results in a substantial, but not complete, loss of directionality. If fluorescence emission were only to be detected from any such system after a number of such processes, it would be unpolarized and carry only routine spectroscopic information. However, the application of ultrafast laser techniques offers scope for detecting short-lived polarization signatures which carry significantly more structural information.

Some of the most challenging problems concern the spectroscopic properties of individual chromophores embedded in a protein matrix. The light-harvesting systems of photosynthetic organisms represent perhaps the best-known and certainly one of the most important classes of such multichromophore ensembles. Since the protein environment often significantly alters the spectroscopic properties of such chromophores, knowledge of their solution spectra may give little help in understanding the same chromophores in a protein environment. Moreover, the situation is commonly complicated by the fact that the molecular ensemble contains chromophores of more than one spectral form, and these chromophores have strongly overlapped spectra. By developing the general theory and illustrating its application to a number of systems, we hope to demonstrate how polarization spectroscopy can address some of these problems.

5.2 INTERACTION OF LIGHT WITH SINGLE MOLECULES OR CHROMOPHORES

Let us first consider the simple case of light interacting with individual molecules or chromophores, to provide a backdrop for the more intricate cases to be considered later. Since we shall be concerned with the processes of absorption, fluorescence, and energy transfer between any kind of linear absorber with distinct optical and electronic properties, we shall use the terms *molecule* and *chromophore* interchangeably, in the following. To begin, Fig. 5.1 shows the relative orientations of the transition dipole moments for absorption ($\mathbf{i}_{ab}$) and fluorescence ($\mathbf{i}_{fl}$) in one such species, with the (linear) polarization of the incident light represented by a unit vector $\mathbf{e}$. In general, the transition dipole moments are not collinear; the states connected by the downward transition differ from those of the upward excitation (as reflected in the usual Stokes shift of the fluorescence wavelength), and the geometry of the electronic excited state may also to some extent differ from that of the initial ground state. Relative to the Cartesian axes as shown, fixed in the molecular frame, the polarization vector has coordinates $\mathbf{e} = \{\sin \theta_e \cos \varphi_e,\ \sin \theta_e \sin \varphi_e,\ \cos \theta_e\}$,

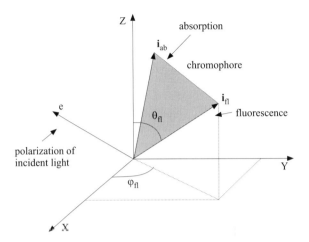

Figure 5.1 The scheme of reference for defining orientations of the polarization of incident light **e**, and the absorption and fluorescence transition dipole moments of chromophore *i*. The angles θ_{fl} and φ_{fl} define the orientation of the fluorescence transition dipole moment $\mathbf{i}_{fl}$. Other angles are omitted for clarity

while $\mathbf{i}_{ab} = \{\sin\theta_{ab}\cos\varphi_{ab},\ \sin\theta_{ab}\sin\varphi_{ab},\ \cos\theta_{ab}\}$ and $\mathbf{i}_{fl} = \{\sin\theta_{fl}\cos\varphi_{fl},\ \sin\theta_{fl}\sin\varphi_{fl},\ \cos\theta_{fl}\}$; θ is the angle between the transition dipole moments: $(\mathbf{i}_{ab}\cdot\mathbf{i}_{fl})^2 = \cos^2\theta$. The probability for the molecule to absorb light is expressible through the absorbance as $A = C_1\sigma(\lambda_{ab})(\mathbf{e}\cdot\mathbf{i}_{ab})^2$, where $\sigma(\lambda_{ab})$ is the absorption cross-section at the wavelength of excitation λ_{ab} and C_1 is a constant determined by instrumental parameters – the latter will not feature in the final expressions for polarization and thus is not of any further concern. The molecular fluorescence has a rate given by the expression $\Phi = C_2\eta\tau^{-1}f(\lambda_{fl})$ and a polarization defined by a direction vector $\mathbf{i}_{fl}$. Here η is the fluorescence quantum yield, τ is the fluorescence lifetime and $f(\lambda_{fl})$ is the normalized spectrum of fluorescence ($\int f(\lambda)\mathrm{d}\lambda = 1$).

We will consider a system of rigid molecules, the mutual orientation of $\mathbf{i}_{ab}$ and $\mathbf{i}_{fl}$ being fixed within each molecular frame, that are randomly distributed and oriented in space. Our initial aim is to calculate the fluorescence components parallel and perpendicular in polarization to the incident light **e**, the latter being a vector fixed in the laboratory frame. Thus we need to conduct an angular average to reflect the random distribution of the sample molecules. The conventional way to do that [2] is to average over all possible orientations of the molecular frame relative to a fixed laboratory frame. To achieve the same result, we find it much more straightforward to consider the molecular frame as fixed, and to randomize the laboratory coordinates – the formal theory of isotropic averaging readily establishes the equivalence of the two methods [5]. In this approach, the component of fluorescence parallel to the polarization of incident light is defined as $\Phi_{\parallel} = C_2 \eta \tau^{-1} f(\lambda_{fl})(\mathbf{e} \cdot \mathbf{i}_{fl})^2$, and to calculate the averaged value $\langle \Phi_{\parallel} \rangle$ we rotationally average all possible orientations of the vector $\mathbf{e}$ with respect to the molecular frame. The following formulae define the measurable parameters:

$$\langle \Phi \rangle = C_1 C_2 \sigma(\lambda_{ab}) \eta \tau^{-1} f(\lambda_{fl}) \langle (\mathbf{e} \cdot \mathbf{i}_{ab})^2 \rangle,$$
$$\langle \Phi_{\parallel} \rangle = C_1 C_2 \sigma(\lambda_{ab}) \eta \tau^{-1} f(\lambda_{fl}) \langle (\mathbf{e} \cdot \mathbf{i}_{ab})^2 (\mathbf{e} \cdot \mathbf{i}_{fl})^2 \rangle,$$
$$\langle \Phi \rangle = \langle \Phi_{\parallel} \rangle + 2 \langle \Phi_{\perp} \rangle, \tag{5.1}$$

where angle brackets denote rotational averaging. The parameter $\langle \Phi_{\perp} \rangle$ is the fluorescence component perpendicular in polarization to the incident light, and $\langle \Phi \rangle$ is the total (unresolved) fluorescence. To quantify the directionality of the fluorescence emission, two parameters are commonly employed. One is the anisotropy of fluorescence, r, and the other the fluorescence polarization, P, the two being directly related as follows:

$$r = \frac{\langle \Phi_{\parallel} \rangle - \langle \Phi_{\perp} \rangle}{\langle \Phi \rangle}, \qquad P = \frac{\langle \Phi_{\parallel} \rangle - \langle \Phi_{\perp} \rangle}{\langle \Phi_{\parallel} \rangle + \langle \Phi_{\perp} \rangle}, \qquad r = \frac{2P}{3 - P}, \tag{5.2}$$

or

$$r = \frac{3\xi - 1}{2}, \qquad P = \frac{3\xi - 1}{\xi + 1}, \qquad \text{where } \xi = \frac{\langle \Phi_{\parallel} \rangle}{\langle \Phi \rangle}. \tag{5.3}$$

On performing the detailed calculations, it is relatively straightforward to show that the results can be expressed in terms of the angle between the transition dipole moments in the final form (see Appendix A):

$$r = \frac{3 \cos^2 \theta - 1}{5}, \qquad P = \frac{3 \cos^2 \theta - 1}{3 + \cos^2 \theta} \tag{5.4}$$

This result was first obtained in the 1920s by Levshin [31, 32] and Perrin [36], and we shall refer to Eqn. 5.4 as the Levshin–Perrin formula.

The physical reason for polarization of fluorescence is the process of molecular *photoselection*, which simply means that it is mainly those molecules whose absorption transitions are more or less parallel to the polarization of the incident light that are excited – and these molecules will make the major contribution to the fluorescence, thus defining its polarization. The dependence on the angle θ in Eqn. 5.4. yields the following special results: (i) transition dipole moments parallel $(\theta = 0), r = 0.4$: (ii) transitions perpendicular $(\theta = 90°)$, $r = -0.2$; and (iii) at $\theta = 54.7°$ (the "magic angle") the fluorescence is completely depolarized, with $r = 0$.

In pump–probe absorption experiments one typically measures a change in absorption ΔA of a probe beam at a wavelength λ_{pr} induced by a pump beam of wavelength λ_p. In the simple case, in which a diminution of the probe beam absorption (*bleaching*) reflects ground-state depletion, the change of absorption at λ_{pr} is given by

$$-\langle \Delta A \rangle = C_1 C_3 \sigma(\lambda_p) \sigma(\lambda_{pr}) \langle (\mathbf{e} \cdot \mathbf{i}_p)^2 \rangle,$$
$$-\langle \Delta A_{||} \rangle = C_1 C_3 \sigma(\lambda_p) \sigma(\lambda_{pr}) \langle (\mathbf{e} \cdot \mathbf{i}_p)^2 (\mathbf{e} \cdot \mathbf{i}_{pr})^2 \rangle,$$
$$\langle \Delta A \rangle = \langle \Delta A_{||} \rangle + 2 \langle \Delta A_{\perp} \rangle, \tag{5.5}$$

similar in form to Eqns 5.2–5.4. The situation changes if in addition to ground-state depletion the molecules exhibit stimulated emission ($\sigma_{st}(\lambda_{pr})$) or absorption ($\sigma_{ex}(\lambda_{pr})$) from the excited state at the wavelength of the probe beam λ_{pr}. The anisotropy in this case is defined by

$$r = \frac{(\sigma(\lambda_{pr}) + \sigma_{st}(\lambda_{pr}))(3\cos^2 \theta_1 - 1) - \sigma_{ex}(\lambda_{pr})(3\cos^2 \theta_2 - 1)}{5(\sigma(\lambda_{pr}) + \sigma_{st}(\lambda_{pr}) - \sigma_{ex}(\lambda_{pr}))}. \tag{5.6}$$

Here θ_1 and θ_2 are, respectively, the angles between the transition dipole moment for absorption from the ground state at wavelength λ_p and (i) that for absorption at λ_{pr}, and (ii) that for absorption from the excited state. The above result is an extension of the similar formula derived in [37] for $\theta_1 = 0$. One interesting result immediately follows from Eqn 5.6: with certain values for the absorption cross-sections, in particular where $[\sigma(\lambda_{pr}) + \sigma_{st}(\lambda_{pr}) - \sigma_{ex}(\lambda_{pr})] \to 0$, then for some angles the anisotropy might significantly exceed one or other of the limiting values $r = 0.4$ or $r = -0.2$.

5.2.1   Nonlinear excitation

5.2.1.1   *Saturation of absorption*

The theory presented thus far is based on the assumption of a linear dependence between the absorption and the intensity of incident light. Under certain experimental conditions – such as when the input radiation very strongly drives an upward transition in the sample – this may not apply, and the optical response develops a nonlinearity due to *saturation* of absorption, i.e. significant bleaching of the ground state. To model the general case, we shall consider fluorescence detected when molecules exhibit absorption from the ground state ($\sigma_{01}(\lambda_{ab})$) and stimulated emission from the excited state ($\sigma_{10}(\lambda_{ab})$). Under steady state excitation, the probability of finding a molecule excited is given by

$$p = \frac{\tau \sigma_{01} F(\mathbf{e} \cdot \mathbf{i})^2}{1 + \tau(\sigma_{01} + \sigma_{10})F(\mathbf{e} \cdot \mathbf{i})^2},$$ (5.7)

and Eqn. 5.1 is modified as follows:

$$\langle \Phi \rangle = C_1 C_2 \eta \tau^{-1} f(\lambda_{fl}) \left\langle \frac{\tau \sigma_{01} F(\mathbf{e} \cdot \mathbf{i})^2}{1 + \tau(\sigma_{01} + \sigma_{10})F(\mathbf{e} \cdot \mathbf{i})^2} \right\rangle,$$

$$\langle \Phi_{\parallel} \rangle = C_1 C_2 \eta \tau^{-1} f(\lambda_{fl}) \left\langle \frac{\tau \sigma_{01} F(\mathbf{e} \cdot \mathbf{i})^2}{1 + \tau(\sigma_{01} + \sigma_{10})F(\mathbf{e} \cdot \mathbf{i})^2} (\mathbf{e} \cdot \mathbf{i}_{fl})^2 \right\rangle,$$

$$\langle \Phi \rangle = \langle \Phi_{\parallel} \rangle + 2\langle \Phi_{\perp} \rangle.$$ (5.8)

After performing all of the necessary averages, we arrive at the following expression for the anisotropy:

$$r(\mu) = \frac{1}{4}(3 \cos^2 \theta - 1) \left[\frac{1}{1 - \frac{a \tan(\sqrt{\mu})}{\sqrt{\mu}}} - \frac{3}{\mu} - 1 \right], \qquad \text{where } \mu = \tau(\sigma_{01} + \sigma_{10})F.$$

(5.9)

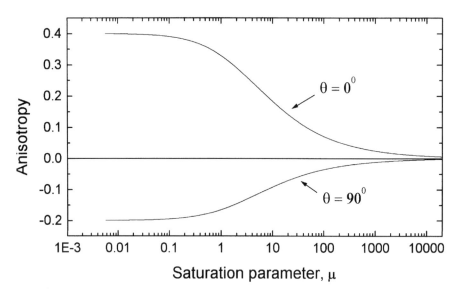

Figure 5.2   The dependence of the fluorescence anisotropy, for angles $\theta = 0°$ and $90°$, on the intensity of excitation, as expressed through the saturation parameter $\mu = \tau \sigma F$

We call μ the saturation parameter; F is the photon flux (photons per unit area per unit time). Figure 5.2 shows the dependence of the anisotropy on the saturation parameter. When there is no significant saturation and the absorption is linear, i.e. when $\mu \ll 1$, the anisotropy has the proper limit $\lim_{\mu \to 0}(r(\mu)) = r_0 = \frac{1}{5}(3\cos^2\theta - 1)$. The opposite extreme, $\lim_{\mu \to \infty}(r(\mu)) = 0$, yields depolarized fluorescence. The former case is obvious; the latter is readily understood from the following. When the level of excitation is sufficiently high, photoselection is no longer a significant feature: essentially, all molecules are amenable to excitation save for the very few whose transition moments are precisely orthogonal to the electric field of the excitation beam. For example, when $\sigma_{10} = 0$ (no stimulated emission), practically all molecules are excited and photoselection no longer features. It is also evident that for any angle θ, $\lim_{\mu \to \infty}(r(\mu)) = 0$. Thus, it is very important, when making polarization measurements, to be able to have confidence that the excitation is "linear," i.e. that the saturation parameter $\mu \ll 1$. An expression analogous to Eqn 5.9 (expressed in terms of the degree of polarization P and for $\theta = 0$) was also derived in [21].

5.2.1.2 Two-photon excitation of fluorescence

Now we will consider another example of nonlinear excitation, where conventional single-photon fluorescence is excited by a process of two-photon absorption. Such processes entail direct transition to the excited state in a concerted manner, as distinct from two-step (i.e. two successive single-photon) processes involving transition through a lower excited state [3]. The rate of concerted two-photon excitation depends quadratically on the incident laser intensity and, as such, cannot be characterized by a normal cross-section; however, we can invoke an effective cross-section $\sigma'(\lambda_{ab})$ which is itself in linear dependence on the input [34]. A further complication is the fact that the efficiency of absorption is determined not by a single transition dipole vector but by a two-photon tensor of intricate structure. However, relatively simple results can be generated if it can be assumed that the excitation is dominated by consecutive absorptions through a single virtual state. Such an assumption proves acceptable for many $S_0 \to S_1$ transitions – and it is almost always valid for transitions to the more highly excited states commonly populated by two-photon absorption.

Given an excited state connected to the ground state by two parallel transition moments, against which the fluorescence transition dipole makes an angle θ, then the following set of equations applies:

$$\langle \Phi \rangle = C_1 C_2 \sigma'(\lambda_{ab})\eta\tau^{-1}f(\lambda_{fl})\langle(\mathbf{e} \cdot \mathbf{i}_{ab})^4\rangle,$$
$$\langle \Phi_{\parallel} \rangle = C_1 C_2 \sigma'(\lambda_{ab})\eta\tau^{-1}f(\lambda_{fl})\langle(\mathbf{e} \cdot \mathbf{i}_{ab})^4(\mathbf{e} \cdot \mathbf{i}_{fl})^2\rangle,$$
$$\langle \Phi \rangle = \langle \Phi_{\parallel} \rangle + 2\langle \Phi_{\perp} \rangle. \tag{5.10}$$

These equations yield an anisotropy ratio:

$$r = \frac{2}{7}(3 \cos^2 \theta - 1). \tag{5.11}$$

One can see that the values for extreme cases ($\theta = 0$ and $\theta = 90°$) cover a larger range than for single-photon absorption, but the depolarized case ($r = 0$) arises under the same "magic angle" condition, where $\theta = 54.7°$. Similar results have been reported elsewhere [1, 7, 9, 29, 30].

Equation 5.11 can be easily extended further to a general case of n-photon absorption. Then, subject again to the assumption of absorption mediated by consecutive transitions with parallel monents, one finds that the anisotropy is

$$r = \frac{n}{2n + 3}(3 \cos^2 \theta - 1) \tag{5.12}$$

In particular, this yields $r(\theta = 0) = 2/3$ for the three-photon-induced fluorescence anisotropy, consistent with [25]. The trend for r to increase with n reflects the higher degree of photoselection which generally accompanies higher order multiphoton processes.

5.3   BICHROMOPHORE MOLECULAR COMPLEXES

In this section, we consider the polarization features of bichromophore molecular complexes, accommodating the effects of energy transfer between the constituent chromophore groups. Without such effects, the response from each type of chromophore would be simply as described in the previous section; however, there are a great many systems in which energy transfer plays a highly significant role, and one which intricately modifies the polarization effects. As will be evident, photobiological systems are very much of this latter kind. In formulating the theory, the precise mechanism and distance dependence of the elementary pair transfer process as described in detail in other chapters will not concern us, although it is principally the Förster mechanism that we have in mind. For present purposes, we can cast the theory directly in terms of the appropriate pair rate constants.

When two chromophores exhibit energy transfer and each acceptor is allowed to be randomly oriented relative to each donor, a single act of energy "hopping" causes a significant drop in polarization: $r = r_0/25$, where r_0 is the anisotropy of the isolated donor or acceptor species [2]. Perhaps surprisingly, the drop is less dramatic ($r = 7r_0/25$) over long distances; that is, distances substantially in excess of the wavelength which would be associated with the energy transferred [4]. However, the efficiency of transfer over such distances is in any case very small; either way, it is clear that it takes very few acts of energy transfer to totally destroy polarization. The situation dramatically changes when the donor and acceptor have a mutual orientation which is fixed in a molecular frame. This condition widely applies in pigment–protein complexes; for example, C-phycocyanin, the light-harvesting antenna of photosynthetic organisms, the photosynthetic reaction center, etc. [42]. Here, there is substantial retention of polarization, and its measurement carries important structural and dynamical information.

Let us consider a simple molecular complex (Fig. 5.3) that comprises two chromophores separated by distance R, for generality assuming non parallel absorption and fluorescence transition dipole moments [16]. We also assume that they interact only weakly, so that we can apply the conditions of incoherent energy transfer, i.e. we consider excitation to be instantaneously localized on either one or other of them. Both chromophores can absorb light and fluoresce, and we allow for energy exchange between them in both directions. Figure 5.4 shows the angular parameters of the considered chromophores with reference to a common molecular frame.

In the case of steady state excitation, the probability n_i of finding chromophore i in its excited state is represented by the pair of equations

$$n_1 = g_{11}A_1 + g_{12}A_2,$$
$$n_2 = g_{21}A_1 + g_{22}A_2, \tag{5.13}$$

where $A_1 = C\sigma_1(\lambda_{ab})(\mathbf{e} \cdot \mathbf{1}_{ab})^2$ and $A_2 = C\sigma_2(\lambda_{ab})(\mathbf{e} \cdot \mathbf{2}_{ab})^2$. The coefficients g_{ij} are elements of the inverse of the excitation balance matrix:

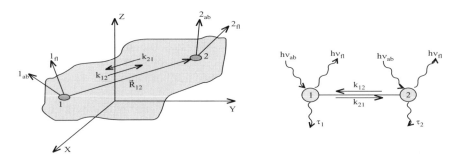

Figure 5.3 The geometry of a bichromophore molecular complex with chromophores 1 and 2 separated in space by the displacement vector $\mathbf{R}_{12}$. Both chromophores can absorb light (transition moments $\mathbf{1}_{ab}$ and $\mathbf{2}_{ab}$) and fluoresce ($\mathbf{1}_{fl}$ and $\mathbf{2}_{fl}$). There is bidirectional energy transfer between the chromophores (k_{12}, k_{21}); $\tau_{1,2}$ are the intrinsic fluorescence lifetimes

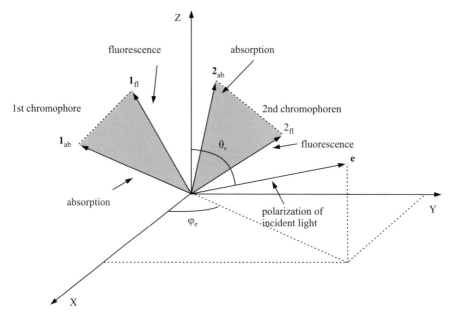

Figure 5.4 Orientations of chromophore transitions and the incident optical polariza-tion vector in a molecular system of axes. The angles between the transition dipole moments are as follows: θ_{11} is the angle between vectors $\mathbf{1}_{ab}$ and $\mathbf{1}_{fl}$; θ_{12}, between $\mathbf{1}_{ab}$ and $\mathbf{2}_{fl}$; θ_{21}, between $\mathbf{2}_{ab}$ and $\mathbf{1}_{fl}$; and θ_{22}, between $\mathbf{2}_{ab}$ and $\mathbf{2}_{fl}$. The angles θ_e and φ_e define the orientation of the polarization vector of the incident light

$$[\bar{g}] = D^{-1} \begin{bmatrix} \tau_2^{-1} + k_{21} & k_{21} \\ k_{12} & \tau_1^{-1} + k_{12} \end{bmatrix}, \tag{5.14}$$

taken with the minus sign: $g_{ij} = -\bar{g}_{ij}$. Here D is the determinant given by $D = (\tau_2^{-1} + k_{21})(\tau_1^{-1} + k_{12}) - k_{12}k_{21}$ and $\tau_{1,2}$ are the intrinsic fluorescence lifetimes, with k_{12} and k_{21} being the total rates of energy transfer, by any mechanism, from the first to the second chromophore and vice versa (see also Appendix B).

In general, fluorescence at any wavelength λ_{fl} is emitted by both chromo-phores in proportion to n_1 and n_2, in accordance with Eqn 5.1. After conduct-ing the orientational averaging for $\langle \Phi \rangle$ and $\langle \Phi_\| \rangle$ with regard to a randomly oriented vector $\mathbf{e}$, we arrive at the formula

$$r = \frac{(3\cos^2\theta_{11} - 1) + q_1(3\cos^2\theta_{21} - 1) + q_2(3\cos^2\theta_{12} - 1) + q_3(3\cos^2\theta_{22} - 1)}{5(1 + q_1 + q_2 + q_3)},$$

$$\tag{5.15}$$

where

$$q_1 = \alpha \frac{g_{12}}{g_{11}} = \alpha \frac{k_{21}}{k_{21} + \tau_2^{-1}}, \qquad q_2 = \gamma \frac{g_{21}}{g_{11}} = \gamma \frac{k_{12}}{k_{21} + \tau_2^{-1}},$$
$$q_3 = \alpha\gamma \frac{g_{22}}{g_{11}} = \alpha\gamma \frac{k_{12} + \tau_1^{-1}}{k_{21} + \tau_2^{-1}}. \tag{5.16}$$

Here the parameters α and γ are, respectively, defined as the relative efficiencies of photoabsorption, $\alpha = \sigma_2(\lambda_{ab})/\sigma_1(\lambda_{ab})$, and of fluorescence, $\gamma = (\eta_2\tau_1 f_2(\lambda_{fl}))/(\eta_1\tau_2 f_1(\lambda_{fl}))$, by the two chromophores, and the angles are defined as $\theta_{11} = \angle(\mathbf{1}_{ab}, \mathbf{1}_{fl}), \theta_{22} = \angle(\mathbf{2}_{ab}, \mathbf{2}_{fl}), \theta_{12} = \angle(\mathbf{1}_{ab}, \mathbf{2}_{fl})$ and $\theta_{21} = \angle(\mathbf{2}_{ab}, \mathbf{1}_{fl})$.[†] Equations 5.15 and 5.16 relate to fluorescence, but the same formulate can be applied in the determination of absorption anisotropy in steady state pump–probe measurements: in this case, we have $\alpha = \sigma_2(\lambda_p)/\sigma_1(\lambda_p)$ and $\gamma = \sigma_2(\lambda_{pr})/\sigma_1(\lambda_{pr})$.

In the "degenerate" case where $\theta_{11} = \theta_{22} = 0$ and $\theta_{12} = \theta_{21} = \theta$, i.e. where each chromophore has parallel absorption and emission transition dipole moments, the anisotropy of fluorescence polarization has the following form:

$$r = \frac{3\cos^2\theta - 1 + 2B}{5(1 + B)}, \qquad B = \frac{\tau_2^{-1} + k_{21} + \alpha\gamma(\tau_1^{-1} + k_{12})}{\alpha k_{21} + \gamma k_{12}}. \tag{5.17}$$

This special case of the more general theory was treated in [13]. When both chromophores are identical ($\alpha = \gamma = 1, k_{12} = k_{21}$) and $1/\tau_{1,2} \ll k_{12}, k_{21}$ then, more simply still, we have $B = 1$ and

$$r = \frac{3\cos^2\theta + 1}{10}. \tag{5.18}$$

In such a system, the anisotropy may vary from $r = 0.4(\theta = 0)$ to $r = 0.1(\theta = 90°)$.

5.3.1 Alternative approach

Before proceeding further, we consider an alternative approach to calculating the anisotropy of polarization in molecular complexes featuring energy transfer. This is an approach directly based on the Levshin–Perrin formula 5.4, and widely used elsewhere:

[†] The angles θ_{12} and θ_{21} also play a part in determining the values of k_{12} and k_{21}, respectively. For simplicity, this intricate dependence is not presented explicitly, since the angles between the transition moments and the interchromophore vector **R** are also involved.

$$r(\lambda_{ab}, \lambda_{fl}, \tau) = \frac{\sum_j \sum_i \sigma_i(\lambda_{ab}) F_{ij}(\lambda_{fl}, \tau) r_{ij}}{\sum_j \sum_i \sigma_i(\lambda_{ab}) F_{ij}(\lambda_{fl}, \tau)}, \qquad r_{ij} = \frac{(3 \cos^2 \theta_{ij} - 1)}{5}. \qquad (5.19)$$

The anisotropy is here calculated as the weighted sum of the anisotropies r_{ij} of the individual chromophores $i = j$ and chromophore pairs $i \neq j$, using the weighting parameters $\sigma_i(\lambda_{ab}) F_{ij}(\lambda_{fl}, \tau)$. The parameters F_{ij} signify the probability that chromophore j fluoresces, given initial excitation (photon absorption) at chromophore i. To be precise, the "diagonal" elements F_{ii} represent the fluorescence quantum yields of the individual chromophores and the "off-diagonal" parameters F_{ij} represent solutions to the system of balance equations (master equations) for energy migration in the molecular complex (see above). Calculation by this strategy comprises the following stages:

1. Calculation of the anisotropies r_{ij} for single chromophores and chromophore pairs after rotational averaging (Eqn 5.19) – taking no account of the interplay between these "single chromophores" and "chromophore pairs."
2. Pumping energy into the molecular complex through absorption by individual chromophores, as characterized by $\sigma_i(\lambda_{ab})$, and solution of the system of master equations to obtain $F_{ij}(\lambda_{fl}, \tau)$.
3. Calculation of the final anisotropy by combining the r_{ij} with the weighting parameter $\sigma_i(\lambda_{ab}) F_{ij}(\lambda_{fl}, \tau)$.
4. Spectral sorting of the emitted fluorescence.

The strategy of our calculations is different:

1. Pumping energy into the molecular complex through the absorption transitions of individual chromophores.
2. Energy equilibration between the chromophores – solving the master equation (Eqns 5.13 and 5.14).
3. Emission of light from all chromophores via radiative transitions.
4. Spectral sorting of the emitted fluorescence.
5. Rotational averaging the components of fluorescence emitted by the whole complex.

One can clearly see that the principal difference concerns the stage of rotational averaging: When should one conduct such an averaging – before or after solving the master equation? In the "linear" case, where the system of balance equations contains only linear components, both methods are equivalent and lead to the same result – and below we give an example that verifies this statement. We nonetheless prefer the latter strategy because, in particular, it works in both linear and nonlinear cases. The former method may fail in nonlinear cases, because the ensemble rotational average is not properly effected on the experimental observable.

Now, let us consider an example of the steady state excitation of a bichromophore complex, where each pair consists of mutually perpendicular but otherwise identical chromophores. Following Eqn 5.17, the anisotropy of fluorescence is determined through our preferred strategy as

$$r = \frac{1}{5}\left(\frac{2+k\tau}{1+2k\tau}\right). \tag{5.20}$$

The alternative approach yields first

$$r = \frac{\sigma(2F_{11} + F_{12}(3\cos^2\theta_{12} - 1) + F_{21}(3\cos^2\theta_{21} - 1) + 2F_{22})}{5\sigma(F_{11} + F_{12} + F_{21} + F_{22})}, \qquad \theta_{12} = \theta_{21} = 90° \tag{5.21}$$

and, further, for the considered case

$$r = \frac{2F_1 - F_2}{5(F_1 + F_2)}, \tag{5.22}$$

where $F_1 = F_{11} = F_{22}$ and $F_2 = F_{12} = F_{21}$. To estimate F_1 and F_2 one has to solve the master equation system with excitation of only one chromophore. For the sake of brevity, we omit these calculations and present the final result: $F_1/F_2 = (1 + k\tau)/k\tau$. After substitution in the above formula, one thereby obtains the identical result:

$$r = \frac{1}{5}\left(\frac{2+k\tau}{1+2k\tau}\right). \tag{5.23}$$

5.3.2   Kinetics of depolarization

Thus far, we have been concerned with the case of continuous excitation, within which the fluorescence lifetimes of each chromophore simply play the role of determining the relative efficiencies of excited-state decay channels. Now we consider the theoretical prototype for the case of pulsed excitation, where the input is represented by a δ-pulse. Although simplistic, the δ-pulse model has the advantage that the ensuing rate equations are not cluttered by information relating only to the source. Moreover, the results are directly applicable if the chromophore decay kinetics are associated with lifetimes that are appreciably longer than the input pulse duration, as with many ultrafast laser studies. When the lifetimes and the pulse duration are of a comparable timescale, the δ-pulse results can readily be adapted for application to a specific form of pulse shape and duration.

In the case of δ-pulse excitation [15, 16], the starting point for calculations is the system of balance equations:

$$dn_1/dt = -(\tau_1^{-1} + k_{12})n_1 + k_{21}n_2,$$
$$dn_2/dt = k_{12}n_1 - (\tau_2^{-1} + k_{21})n_2, \tag{5.24}$$

for which the general solution has the form

$$n_1(t) = g_{11}(t)A_1 + g_{12}(t)A_2,$$
$$n_2(t) = g_{21}(t)A_1 + g_{22}(t)A_2. \tag{5.25}$$

with the initial conditions of excitation $n_1(0) = A_1$ and $n_2(0) = A_2$ associated with the absorption of incident light with polarization $\mathbf{e}$ (A_1 and A_2 being determined as in the case of steady state excitation). The $g_{ij}(t)$ entail the same parameters as the former g_{ij}, as well as exponential functions of the characteristic times for the processes of energy equilibration, t_1, and overall decay, t_2:

$$t_1^{-1} = 0.5(\tau_1^{-1} + \tau_2^{-1} + k_{12} + k_{21} + \sqrt{\xi^2 + 4k_{12}k_{21}}),$$
$$t_2^{-1} = 0.5(\tau_1^{-1} + \tau_2^{-1} + k_{12} + k_{21} - \sqrt{\xi^2 + 4k_{12}k_{21}},$$
$$\xi = \tau_2^{-1} - \tau_1^{-1} + k_{21} - k_{12}. \tag{5.26}$$

Since the system of Eqns 5.25 is analogous to the steady state system (Eqns 5.13) with elements g_{ij} substituted by $g_{ij}(t)$, the resultant anisotropy will have the same form as Eqns 5.15 but with time-dependence included in the q parameters. Specifically, dynamical features of the fluorescence depolarization kinetics are manifest through the following expressions:

$$r(t) = \frac{(3\cos^2\theta_{11} - 1) + q_1(t)(3\cos^2\theta_{21} - 1) + q_2(t)(3\cos^2\theta_{12} - 1) + q_3(t)(3\cos^2\theta_{22} - 1)}{5(1 + q_1(t) + q_2(t) + q_3(t))}, \tag{5.27}$$

where

$$q_1(t) = \alpha\frac{g_{12}(t)}{g_{11}(t)} = \alpha\frac{k_{21}(1 - \exp(-\Omega t))}{(k_{12} + \tau_1^{-1} - t_2^{-1})\exp(-\Omega t) - (k_{12} + \tau_1^{-1} - t_1^{-1})},$$
$$q_2(t) = \gamma\frac{g_{21}(t)}{g_{11}(t)} = \gamma\frac{k_{12}(1 - \exp(-\Omega t))}{(k_{12} + \tau_1^{-1} - t_2^{-1})\exp(-\Omega t) - (k_{12} + \tau_1^{-1} - t_1^{-1})},$$
$$q_3(t) = \alpha\gamma\frac{g_{22}(t)}{g_{11}(t)} = \alpha\gamma\frac{(k_{12} + \tau_1^{-1} - t_2^{-1}) - (k_{12} + \tau_1^{-1} - t_1^{-1})\exp(-\Omega t)}{(k_{12} + \tau_1^{-1} - t_2^{-1})\exp(-\Omega t) - (k_{12} + \tau_1^{-1} - t_1^{-1})},$$
$$\Omega = t_1^{-1} - t_2^{-1}.$$

$$\tag{5.28}$$

Again, the parameters α and γ are as described above in connection with steady state excitation. In the special case in which $\theta_{11} = \theta_{22} = 0$ and $\theta_{12} = \theta_{21} = \theta$, we obtain the following simplified expressions:

$$r(t) = \frac{3\cos^2\theta - 1 + 2B(t)}{5(1 + B(t))}, \quad B(t) = \frac{\xi(1 - \alpha\gamma) + (t_2^{-1} - t_1^{-1})(1 + \alpha\gamma)\coth\frac{1}{2}\Omega t}{2(\alpha k_{21} + \gamma k_{12})}.$$

$$(5.29)$$

Here, one finds that $\lim_{t \to 0}(r(t)) = 0.4$ and $\lim_{t \to \infty}(r(t)) = r$ (*steady state*) for any values of the parameters involved. Indeed, at $t \to 0$, where energy equilibrium has not yet occurred, the contributions of the two chromophores are not correlated and each yields an independent contribution to the fluorescence with $r(t = 0) = 0.4$ (under the condition of incoherent energy transfer).

5.3.3 Some applications to steady state excitation

5.3.3.1 C-Phycocyanin

Demidov and Mimuro [19] have demonstrated (with an example of the β-sub-units of C-phycocyanin, C-PC) that Eqn 5.17 can be used to extract the individual spectra of absorption and fluorescence for each chromophore contained in a pigment–protein complex, without its denaturation, i.e. *in vivo*. The approach is purely spectroscopic, and involves measurements of absorption and fluorescence spectra as well as the polarization spectra of fluorescence excitation and emission. For example, the absorption spectrum of a bichromophoric complex can be expressed as $\sigma(\lambda) = \sigma_1(\lambda) + \sigma_2(\lambda)$ (see Fig. 5.5), and the ratio $\alpha = \sigma_2(\lambda)/\sigma_1(\lambda)$ can be obtained from the polarization spectrum $r(\lambda)$ or $P(\lambda)$. Thus the individual absorptions $\sigma_1(\lambda)$ and $\sigma_2(\lambda)$ can be determined; for details, see [19]. Figure 5.6 shows the spectra of the $\beta - 84$ and $\beta - 155$ chromophores as determined by this procedure. The spectra of the $\alpha - 84$ chromophores, contained in the α-sub-unit of C-PC, are included on the same graphs to complete the picture of C-PC chromophores. Figure 5.7 presents the fluorescence spectra of these chromophores [19].

5.3.3.2 P680 of photosystem II reaction center

In a recently introduced model [20], P680 is considered as a pair of weakly interacting chlorophylls (Fig. 5.8) that form a structure similar to the analogous special pair in a bacterial reaction center, but without excitonic interaction. Only weak interaction is achieved, because of a specific orientation of the chlorophylls that makes the orientation parameter (presented in the Förster

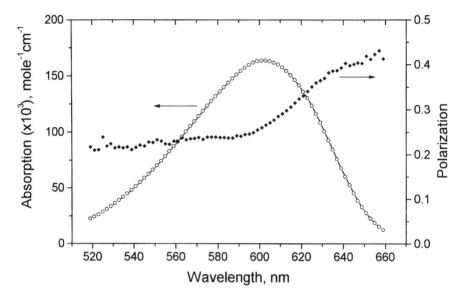

Figure 5.5   C-PC β-sub-unit absorption and fluorescence polarization excitation spectra measured by Mimuro *et al.* [17, 35]. The wavelength of fluorescence detection is $\lambda_{fl} = 670$ nm

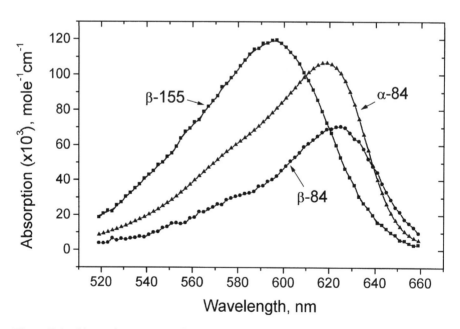

Figure 5.6   Absorption spectra of the $\beta - 84$, $\beta - 155$, and $\alpha - 84$ chromophores [19] (the spectra of the β chromophores are deconvoluted)

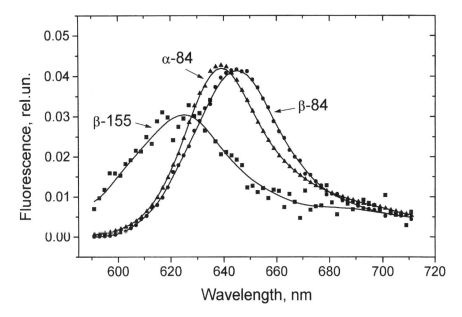

Figure 5.7 Normalized fluorescence spectra of the $\beta - 84$, $\beta - 155$, and $\alpha - 84$ chromophores [19]: $\int f(\lambda)d\lambda = 1$ (the spectra of the β chromophores are deconvoluted). Points are the results of calculation, and lines are from a spline fit

formula [22]) small. According to this model, P680 is a quasi-dimer having C_2 symmetry (Fig. 5.8), where chlorophylls exhibit monomeric behavior. Under these conditions we can apply

$$r = \frac{3 \cos\theta \cos(\theta - 2\varphi) + 1}{10}, \tag{5.30}$$

where φ is the angle between absorption and fluorescence transitions in the individual chlorophyll molecule and θ is the angle between absorption transitions of two chlorophylls. In recent triplet-minus-singlet experiments at 4 K, Kwa *et al.* [27] measured an anisotropy of $r \approx 0.28$ at $\lambda_{\text{exc}} = 682$–$683$ nm, i.e. in the 0–0 band, with an anisotropy of $r = 0.22$ [28] resulting for excitation in the 0–0 band and fluorescence measured in a vibrational band. These data yield $\theta = 141°$ and $\varphi = 10°$ (see Fig. 5.8). The latter angle is in good agreement with polarization data obtained by van Gurp *et al.* [43] for chlorophyll a *in vitro*. The calculated angle θ perfectly matches the analogous angle in the reaction center of purple bacteria, which is $\sim 141°$ [11, 12, 38]. This result can be considered as one more argument to support a direct analogy between these systems.

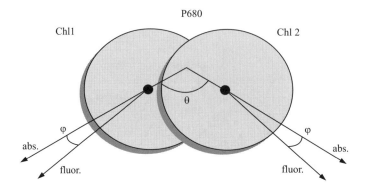

Figure 5.8 A schematic structure of P680 as a quasi-dimer consisting of two monomeric chlorophylls *a*. The scheme accommodates a general case in which the transition dipole moments for absorption and fluorescence are not necessarily parallel, but have an angle φ between them

5.4 TRICHROMOPHORE COMPLEXES

We now turn our attention to molecular complexes that have three chromophores (Fig. 5.9). Again, all three chromophores can absorb light and fluoresce, and there is energy transfer between them. For simplicity, we will consider the case in which the absorption and fluorescence transitions of individual chromophores are parallel; the C-PC monomer containing α-84, β-84, and β-155 chromophores is one such example [14, 23, 39]. The distribution of excitation between the chromophores under steady state excitation is described by formulae similar to Eqn 5.13:

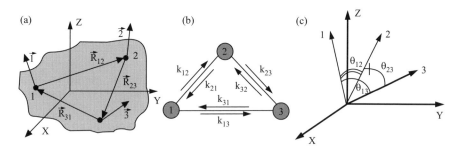

Figure 5.9 A scheme for a trichromophore complex, where $i = 1, 2, 3$ is the chromophore number: (a) shows the molecular structure, (b) the routes for energy transfer, and (c) the relative orientations of the molecular transition dipole moments

$$n_1 = g_{11}A_1 + g_{12}A_2 + g_{13}A_3,$$
$$n_2 = g_{21}A_1 + g_{22}A_2 + g_{23}A_3,$$
$$n_3 = g_{31}A_1 + g_{32}A_2 + g_{33}A_3. \tag{5.31}$$

Explicit expressions for the g_{ij} are given in Appendix B. The ensuing calculations follow the procedure described above and yield [14]:

$$r = \frac{\Phi_\parallel - \Phi_\perp}{\Phi} = \frac{3(q_{12}\cos^2\theta_{12} + q_{13}\cos^2\theta_{13} + q_{23}\cos^2\theta_{23}) - 1 + 2B}{5(1 + B)},$$
$$q_{12} + q_{13} + q_{23} = 1. \tag{5.32}$$

This formula is the three-chromophore counterpart to Eqn 5.17, the relative efficiencies of excitation absorption and fluorescence expressed through the parameters B and q_{ij}, here as given in Appendix C. It is easy to see that in a particular case – for example, when $\sigma_3 = 0$ and $k_{13} = k_{23} = 0$ – the trichromophore complex behaves as a bichromophore complex, with $q_{12} = 1, q_{13} = q_{23} = 0$ and $B = (g_{11} + g_{22}\alpha_{21}\gamma_{21})/(g_{12}\alpha_{21} + g_{21}\gamma_{21}) = [\tau_2^{-1} + k_{21} + (\tau_1^{-1} + k_{12}) \alpha_{21}\gamma_{21}]/[\alpha_{21}k_{21} + \gamma_{21}k_{12}]$; hence we revert to Eqn 5.17.

Figures 5.10 and 5.11 illustrate the application of Eqn 5.32 to the monomers of C-PC [19]: the calculations are based on the degree of polarization, P, rather than the anisotropy, r. Structural information is taken from the work by Shirmer *et al.* [39], energy transfer rates from [18, 19], and the absorption and fluorescence spectra of β-84, β-155, and α-84 are as determined earlier (see Figs 5.6 and 5.7). It should be made clear that *no fitting parameters* are used to match the calculated [19] and experimentally measured polarization [35] data – which serves to emphasize the significance of the good agreement between them.

5.4.1   Kinetics of depolarization

The depolarization kinetics can be tackled in much the same way as for bichromophore complexes. The complete calculations reported in [15] lead us to the expected result:

$$r(t) = \frac{3(q_{12}(t)\cos^2\theta_{12} + q_{13}(t)\cos^2\theta_{13} + q_{23}(t)\cos^2\theta_{23}) - 1 + 2B(t)}{5(1 + B(t))},$$
$$q_{12}(t) + q_{13}(t) + q_{23}(t) = 1. \tag{5.33}$$

Here the parameters B and q_{ij} are time-dependent, and the characteristic lifetimes t_1, t_2 and t_3 represent three eigenvalues for excitation equilibration and decay; all are detailed in Appendix D. The number of time constants involved in

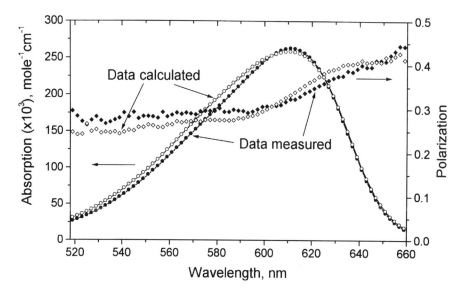

Figure 5.10   Absorption and fluorescence polarization excitation spectra of C-PC monomers: open symbols are experimentally determined; closed symbols denote calculated data. The wavelength of fluorescence detection is $\lambda_{fl} = 670$ nm

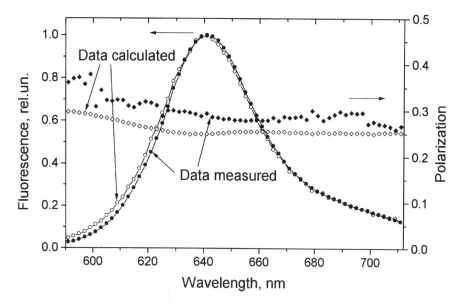

Figure 5.11   Fluorescence and fluorescence polarization spectra of C-PC monomers: open symbols are experimentally determined; closed symbols denote calculated data. The wavelength of fluorescence excitation is $\lambda_{exc} = 550$ nm

the kinetics of anisotropy is two, and they are directly related to the time constants that can be observed in fluorescence decay kinetics:

$$\Phi(t) = c_1 e^{-t/t_1} + c_2 e^{-t/t_2} + c_3 e^{-t/t_3}, \tag{5.34}$$

where the three terms relate to the presence of three chromophore types. Clearly, the anisotropy of depolarization, obtained by rationing, can contain only two independent exponential factors, which can be chosen as $\exp(-t/\Delta t_1)$ and $\exp(-t/\Delta t_2)$, where $\Delta t_1^{-1} = t_1^{-1} - t_3^{-1}$ and $\Delta t_2^{-1} = t_2^{-1} - t_3^{-1}$.

Analysis of the extreme values for the anisotropy at $t \to 0$ and $t \to \infty$ lead to results similar to those obtained for bichromophore complexes: $\lim_{t\to 0}(r(t)) = 0.4$ and $\lim_{t\to\infty}(r(t)) = r$ (*steady state*). While the latter is obvious, the result $r(0) = 0.4$ is in fact valid for molecular complexes of any structure if the energy transfer is incoherent, for the reasons presented above.

5.5 MULTICHROMOPHORE COMPLEXES WITH C_3 SYMMETRY

Trimer structures are remarkably common in biomolecular complexes. Such structures generally exhibit C_3 rotational symmetry, which means that there is the axis of symmetry about which the system "repeats" itself on rotation by $120°$. The hexamers and rod structures which also arise provide other examples of complexes with C_3 rotational symmetry; C-PC aggregates [23, 39] are of this kind. C-PC contains chromophores of only three spectral types α-84, β-84, and β-155; Fig. 5.12 shows the structure of the C-PC trimer [17, 44]. Allophycocyanin (APC) is another example that exhibits threefold symmetry [6]; the structure of APC is very similar to that of C-PC, but contains chromophores of only two spectral types: α-84 and β-84. For generality, we shall consider C_3 molecular complexes with different kinds of aggregation and with the number of chromophore types varying from one to three [17].

5.5.1 Monochromophoric trimer

5.5.1.1 Steady state excitation

Figure 5.13 shows a schematic structure for a monochromophoric trimer. All chromophores in this trimer are identical, and if the polar coordinates of the "first" chromophore are $\{\theta_1, \varphi_1\} = \{\theta, \varphi\}$, the coordinates of other two are $\{\theta_2, \varphi_2\} = \{\theta, \varphi + 2\pi/3\}$ and $\{\theta_3, \varphi_3\} = \{\theta, \varphi - 2\pi/3\}$. Here θ is the inclination or azimuthal angle (relative to the axis of symmetry Z), and φ is the "planar" angle. Application of our calculation strategy described above first gives us the total amount of energy absorbed by an individual trimer:

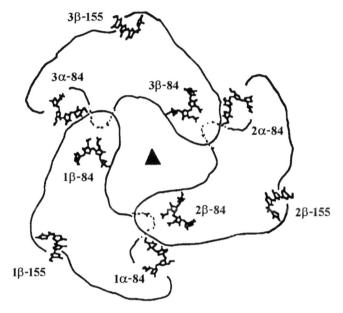

Figure 5.12   A scheme of the C-PC trimer [17] (redrawn from Xie *et al.* [44]). The trimer consists of three monomers, each with three chromophores: $\beta - 84$, $\beta - 155$, and $\alpha - 84$. The distance between $\beta - 84$ and $\alpha - 84$ in adjoining monomers is about 20 Å [39]

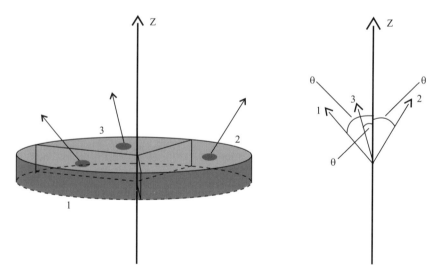

Figure 5.13   A schematic structure of a monochromophoric trimer having C_3 rotational symmetry. Indices 1, 2, and 3 denote the chromophore (monomer) number; arrows show the orientation of their transition dipole moments. All chromophores have an equal angle of inclination θ to the axis of symmetry Z

$$A = \sum_{i=1}^{3} A^i = C\sigma(\lambda_{ab}) \sum_{i=1}^{3} [\sin \theta_i \sin \theta_e \cos(\varphi_i - \varphi_e) + \cos \theta_i \cos \theta_e]^2$$

$$= 3C\sigma(\lambda_{ab})(\cos^2 \theta \cos^2 \theta_e + 0.5 \sin^2 \theta \sin^2 \theta_e), \tag{5.35}$$

where $\{\theta_e, \varphi_e\}$ are the corresponding polar coordinates of the polarization vector **e**. From Eqn 5.35, it is clear that the total amount of absorbed energy is not dependent on the angle φ. After performing other stages of the calculation [17], including averaging over the relative orientations of the incident light, and applying the condition $\tau_i^{-1} \ll k_{ij}$, we obtain the anisotropy:

$$r = \frac{1}{10}(9 \cos^4 \theta - 6 \cos^2 \theta + 1). \tag{5.36}$$

This formula gives a positive r for any θ and, in particular, $r(0) = 0.4$, $r(90°) = 0.1$ and $r(\theta_m) = 0$, where θ_m is the magic angle, $54.7°$. The condition $\tau_i^{-1} \ll k_{ij}$ is very important in our calculations, because it stipulates that the rate of energy equilibration between chromophores is much faster than decay, so that we have a uniform distribution of the absorbed energy between chromophores of the same spectral type.

It is worth noting that Eqn 5.36 could be derived in an alternative way from Eqn 5.32. Our trimer is a regular trichromophore complex with additional symmetry and with all three chromophores of the same spectral type. Thus, $\theta_{12} = \theta_{13} = \theta_{23} = \theta'$, $k_{12} = k_{13} = k_{21} = k_{23} = k_{31} = k$, $\tau_1^{-1} = \tau_2^{-1} = \tau_3^{-1} \ll k$ and $\alpha_{ij} = \gamma_{ij} = 1$. This leads to $q_{ij} = 1/3$ and $B = 1/2$, so that direct application of Eqn 5.32 yields

$$r = \frac{2}{5} \cos^2 \theta' = \frac{1}{10}(9 \cos^4 \theta - 6 \cos^2 \theta + 1), \tag{5.37}$$

since from simple geometry we have

$$\cos \theta' = \frac{1}{2}(3 \cos^2 \theta - 1). \tag{5.38}$$

5.5.1.2 Kinetics of depolarization

The case of a monochromophoric trimer excited by a δ-pulse is the simplest to consider, and performing the necessary calculations [17] yields

$$r(t) = \left[\frac{1}{10}(9 \cos^4 \theta - 6 \cos^2 \theta + 1)\right] + \left[\frac{1}{10}(3 + 6 \cos^2 \theta - 9 \cos^4 \theta)\right]e^{-3kt}. \tag{5.39}$$

The analogous dependence $r(t) = a + be^{-3kt}$ was previously obtained by Lyle and Struve [33]. In both steady state and pulsed-excitation cases, the anisotropy depends only on the inclination angle θ; this is the result of C_3 symmetry.

Further aggregation (as hexamers, rods of hexamers, or trimers) does not influence the anisotropy at a steady state excitation, i.e. Eqn 5.36 remains valid. The depolarization kinetics can be different, but with the same long-time limit $r(t \to \infty)$ independent of aggregation. As far as the depolarization kinetics of aggregates larger than a trimer are concerned, the depolarization kinetics cannot be described by a formula as simple as Eqn 5.39, and more intricate and general calculations are then required (see below).

5.5.2 Bichromophoric trimers

Bichromophore complexes contain chromophores of two distinct spectral types. Macroscopically, energy exchange between them is described by Eqns 5.13 and 5.14; one cannot spectrally distinguish photons emitted by different chromophores of the same spectral type. Thus, on the macroscopic level, we have two pools of chromophores and energy exchange between them. The way to determine the averaged rates of energy transfer k_{12} and k_{21} between these pools is described in [18]. Typical examples of bichromophoric complexes with C_3 symmetry are APC [6, 40] and C-PC of the mutant strain PR6235 (cpcB/C155S) [10].

Following [17] we present the following expression for $\xi = \langle \Phi_\parallel \rangle / \langle \Phi \rangle$:

$$\xi = \tfrac{1}{5}\{g_{11}(3\cos^4\theta_1 + 2 - 2\cos^2\theta_1) + \gamma_{21}\alpha_{21}g_{22}(3\cos^4\theta_2 + 2 - 2\cos^2\theta_2)$$
$$+ (\alpha_{21}g_{12} + \gamma_{21}g_{21})(3\cos^2\theta_2\cos^2\theta_1 + 2 - \cos^2\theta_2 - \cos^2\theta_1)\}/$$
$$\{g_{11} + \gamma_{21}g_{21} + \alpha_{21}(g_{12} + \gamma_{21}g_{22})\}$$

$$(5.40)$$

The anisotropy r equals $(3\xi - 1)/2$:

$$r = \{g_{11}(9\cos^4\theta_1 - 6\cos^2\theta_1 + 1) + \gamma_{21}\alpha_{21}g_{22}(9\cos^4\theta_2 - 6\cos^2\theta_2 + 1)$$
$$+ (\alpha_{21}g_{12} + \gamma_{21}g_{21})(9\cos^2\theta_2\cos^2\theta_1 - 3\cos^2\theta_2 - 3\cos^2\theta_1 + 1)\}/$$
$$10\{g_{11} + \gamma_{21}g_{21} + \alpha_{21}(g_{12} + \gamma_{21}g_{22})\},$$

$$(5.41)$$

where θ_1 and θ_2 are the inclination angles of chromophores of the first and the second spectral type. Similar to the monochromophoric case, there is no dependence on the planar angles φ. The other parameters in the above equations have the same meanings as before.

If only "donor" molecules can absorb light ($\alpha_{21} = 0$) and only "acceptors" fluoresce ($\gamma_{21} = \infty$), then Eqn 5.41 is simplified further:

$$r = \frac{1}{10}(9 \cos^2 \theta_2 \cos^2 \theta_1 - 3 \cos^2 \theta_2 - 3 \cos^2 \theta_1 + 1). \tag{5.42}$$

The other extreme case of interest is when both types of chromophores absorb and fluoresce, but there is *unidirectional* energy transfer from the donor to acceptor chromophores $(k_{12} \gg k_{21} \gg \tau^{-1})$, as may result from the relative position of absorption and fluorescence bands of these chromophores. Here, we obtain

$$r = \frac{1}{10}\{\alpha_{21}(9 \cos^4 \theta_2 - 6 \cos^2 \theta_2 + 1) + 9 \cos^2 \theta_2 \cos^2 \theta_1$$
$$- 3 \cos^2 \theta_2 - 3 \cos^2 \theta_1 + 1\}/\{\alpha_{21} + 1\}. \tag{5.43}$$

Equation 5.43 does not contain the parameter γ_{21}, and thus the excitation spectrum of polarized fluorescence does not depend on the wavelength of fluorescence detection. The physical explanation for this is that the substantially more efficient flow of energy from chromophores of the first type to the second produces an excitation predominantly located on chromophores of the second type. The latter thus provide the major contribution to the emitted fluorescence, while chromophores of the first type give only a minor contribution.

5.5.3 Trichromophoric trimers

The case of trichromophore complexes is the last to be considered. C-PC trimers, hexamers and rods provide examples of such complexes – not that this represents the ultimate degree of complexity in biological systems; one can find examples of molecular complexes with four or more distinct chromophores, as for example phycoerethryn [8] with chromophores of five spectral types or FMO trimer (see Chapter 10) that contains 21 chlorophylls. The procedure of calculating the polarization anisotropy is nonetheless similar, and one could follow it through explicitly in the same way as before. In particular, our calculations [17] for trichromophore complexes with C_3 symmetry produce the following result for the anisotropy:

$$\begin{aligned}
r = \frac{1}{10}\{&g_{11}(9 \cos^4 \theta_1 - 6 \cos^2 \theta_1 + 1) + \gamma_{21}\alpha_{21}g_{22}(9 \cos^4 \theta_2 - 6 \cos^2 \theta_2 + 1) \\
&+ \gamma_{31}\alpha_{31}g_{33}(9 \cos^4 \theta_3 - 6 \cos^2 \theta_3 + 1) \\
&+ (\alpha_{21}g_{12} + \gamma_{21}g_{21})(9 \cos^2 \theta_2 \cos^2 \theta_1 - 3 \cos^2 \theta_2 - 3 \cos^2 \theta_1 + 1) \\
&+ (\alpha_{31}g_{13} + \gamma_{31}g_{31})(9 \cos^2 \theta_3 \cos^2 \theta_1 - 3 \cos^2 \theta_3 - 3 \cos^2 \theta_1 + 1) \\
&+ (\alpha_{31}\gamma_{21}g_{23} + \gamma_{31}\alpha_{21}g_{32})(9 \cos^2 \theta_3 \cos^2 \theta_2 - 3 \cos^2 \theta_3 - 3 \cos^2 \theta_2 + 1)\}/ \\
&\{g_{11} + \gamma_{21}g_{21} + \gamma_{31}g_{31} + \alpha_{21}(g_{12} + \gamma_{21}g_{22} + \gamma_{31}g_{32}) \\
&+ \alpha_{31}(g_{13} + \gamma_{21}g_{23} + \gamma_{31}g_{33})\}, \tag{5.44}
\end{aligned}$$

necessarily a function of three different angles of inclination for each chromophore type.

5.5.3.1   Kinetics of polarization excited by a δ-pulse

In previous sections, we have considered a few particular cases of the kinetics associated with polarization anisotropy. They are examples of our calculational procedure for molecular complexes with a relatively small number of chromophores, where an analytic solution is reasonably simple. When the number of chromophores is large, such analytic expressions become unduly cumbersome. At this stage one can consider an alternative approach, based on numerical solution of the complete system of master equations:

$$\frac{dn_1}{dt} = -\frac{n_1}{\tau_1} - \sum_{i=2}^{N} k_{1i} n_1 + \sum_{i=2}^{N} k_{i1} n_i,$$

$$\frac{dn_j}{dt} = -\frac{n_j}{\tau_j} - \sum_{\substack{i=1 \\ i \neq j}}^{N} k_{ji} n_j + \sum_{\substack{i=1 \\ i \neq j}}^{N} k_{ij} n_i,$$

$$\frac{dn_N}{dt} = -\frac{n_N}{\tau_N} - \sum_{i=1}^{N-1} k_{Ni} n_N + \sum_{i=1}^{N-1} k_{iN} n_i. \tag{5.45}$$

The condition of initial excitation of any chromophore j is defined by

$$n_j(t=0) = C\sigma(\lambda_{ab}, j)(\mathbf{j} \cdot \mathbf{e})^2 = C\sigma(\lambda_{ab}, j)$$

$$[\sin\theta_j \sin\theta_e \cos(\varphi_j - \varphi_e) + \cos\theta_j \cos\theta_e]^2, \qquad j = 1, \cdots, N, \tag{5.46}$$

where

$$(\lambda_{ab}, j) = \sigma_{s(j)}(\lambda_{ab}), \quad \text{if chromophore } j \text{ is of type } s(j), \quad s(j) = 1, 2, 3. \tag{5.47}$$

Here $s(j)$ denotes the spectral type. In this system of equations, n_j is the probability of finding chromophore j in its excited state; τ_j is the intrinsic decay lifetime (by any processes other than energy transfer) and k_{ji} is the rate of energy transfer from a chromophore j to a chromophore i.

For the C-PC trimer the number of equations $N = 9$; for its hexamer, 18; and for the three-hexamer rod, 54. The resulting function $n_j(t), j = 1, \ldots, N$, describes the kinetics of de-excitation of chromophore j. The intensity of fluorescence emitted from each such chromophore at a wavelength λ_{fl} is given by $F_j(\lambda_{fl}, t) = \eta_j \tau_j^{-1} f(\lambda_{fl}, j) n_j(t)$, where the fluorescence spectrum function $f(\lambda_{fl}, j)$ is defined by

$$f(\lambda_{fl},j) = f_{s(j)}(\lambda_{fl}); \eta_j = \eta_{s(j)}; \tau_j = \tau_{s(j)}\}, \quad \text{if chromophore } j \text{ is of type } s(j).$$

$$(5.48)$$

The polarization of the fluorescence emitted by chromophore j is parallel to its transition dipole moment $\mathbf{j}$: the component parallel to the polarization of incident light $\mathbf{e}$ is given by $F_j^{\parallel}(\lambda_{fl}, t) = F_j(\lambda_{fl}, t)(\mathbf{j} \cdot \mathbf{e})^2$. Thus, after solving Eqns 5.45–5.48, we have data on the individual contribution of any chosen chromophore j, as determined by the appropriate inclination and planar angles. The next step is numerical averaging over all orientations of the input polarization and spectral sorting.

Let us make it clear just what is meant by that numerical averaging. The numerical calculation described above gives us just one solution for one chosen set of angles, both for the excitation light and the chromophores. To effect rotational averaging, we have to conduct the same procedure for a meaningful number of different sets of angles, i.e. solve Eqns 5.45–5.48 many times, and to average the calculated solutions for total fluorescence and its component parallel to the polarization of incident light. Following our strategy, we perform variations of the $\{\theta_e, \varphi_e\}$ angles of the incident light, whilst the chromophore angles are fixed. For example, if one has 30 points of angle φ_e (varying from 0 to 2π) and 30 points of θ_e (varying from 0 to π), there will be 900 solutions to calculate and involve in averaging. (Do not forget the weighting factor $\sin\theta_e$ in averaging over θ_e.)

In the case of δ-pulse excitation, the dynamics of excitation equilibration between all of the chromophores is highly important, and its incorporation must include equilibration amongst chromophores of identical spectral type. Polarization spectroscopy, in contrast to conventional fluorometry, allows one to trace such inter-chromophore energy migration among spectroscopically identical chromophores. Thus, in general, we cannot conduct an average over chromophores of the same spectral type as for the case of steady state excitation, and all chromophores must be taken into account at the stage of equilibration, i.e. with regard to their orientations in a molecular frame. Later, we shall see what impact this has on the kinetics of anisotropy in the particular case of C-PC. The spectroscopic and structural parameters used in these calculations are taken from [18, 19, 39].

The first graph (Fig. 5.14) represents the calculated anisotropies for mono-chromophore C_3 complexes comprising trimer, hexamer and three-hexamer rods of C-PC without the $\alpha - 84$ and $\beta - 155$ chromophores (hypothetic model). As expected, each trace starts at $r(t = 0) = 0.4$ and has different behavior over a short timescale, until full energy equilibration occurs ($t < 600$ ps). Then the anisotropy is the same for all aggregates and equal to $r(steady\ state)$. At this stage we have an angular dependence only on the inclination angle θ. The numerically calculated kinetics of anisotropy in the trimer is monoexponential and precisely matches the analytic result given in, Eqn 5.39.

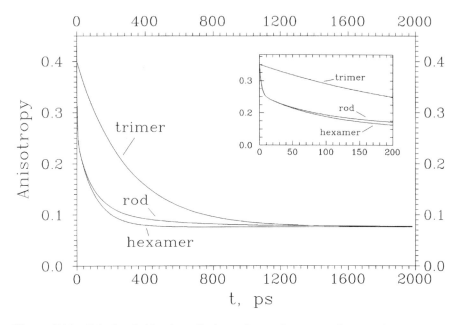

Figure 5.14 Calculated kinetics of absorption anisotropy of monochromophoric ensembles: trimer, hexamer, and three-hexamer rod. Calculations are performed for excitation with a δ-pulse having wavelength $\lambda_{ab} = 615$ nm. The inclination angle of the chromophores equals that of the $\alpha - 84$ chromophore determined by Schirmer *et al.* [39], while its spectroscopic parameters are as defined elsewhere [17, 19]; see also Figs 5.6 and 5.7. The inset is a zoom view of the major graph in a range 0–200 ps

Figure 5.15 shows the kinetics calculated for a trichromophore complex; in this case C-PC aggregates with all (α-84, β-84, and β-155) chromophores present. On this graph, one can observe the behavior of individual anisotropies analogous to that outlined above. The numerical calculations confirm that in the initial stages of energy equilibration both θ and φ angles strongly affect the kinetics of depolarization, whereas at longer times we see a dependence on only the inclination angles. Here, then, there is a potential utility for structure elucidation, such as the determination of chromophore orientations.

To further demonstrate the sensitivity of polarization data on the inclination and planar angles of the chromophores and the dynamics of anisotropy, we consider a bichromophore complex that comprises the C-PC trimer with only α-84 and β-84 chromophores. The angles θ and φ for the first β-84 chromophore vary; the other two β-84 chromophores have the same angle θ and planar angles $\varphi_{2,3} = \varphi \pm 2\pi/3$. Figure 5.16 represents the angular dependence of the anisotropy calculated at (a) 5 ps, (b) 30 ps, (c) 200 ps, and (d) 2 ns after excitation. One can see a strong dependence of the anisotropy on both θ and φ in the early stages, that later flattens out. After energy equilibration

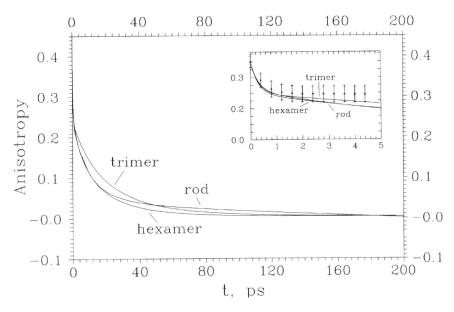

Figure 5.15 Calculated kinetics of absorption anisotropy of trichromophore ensembles: trimer, hexamer, and three-hexamer rod of C-PC. Calculations are performed for excitation with a δ-pulse having wavelength $\lambda_{ab} = 615$ nm. The inclination angles of the chromophores are equal to those of the C-PC chromophores determined by Schirmer *et al.* [39] for cyanobacteria *M. laminosus*, whereas their spectroscopic parameters are defined elsewhere [17, 19]; see also Figs 5.6 and 5.7. The inset is a zoom view of the major graph, also showing experimental data measured by Gillbro *et al.* [24], as asterisks with the appropriate error bars

comes fully into play, the anisotropy becomes dependent on the inclination angle θ only.

In our numerical calculations, the pairwise rates of energy transfer, k_{ij}, are determined for each pair of chromophores and thus depend on their relative orientation; in particular, on angles θ and φ, and more precisely on the orientation factor [22]. So why is the behavior of $r(\infty) \simeq r(2\text{ns})$ as given by Eqn 5.41 (see Fig. 5.16) and not more complicated? The answer is that, under the considered conditions, (i) a balance of energy distribution has been achieved which is inversely proportional to the ratio of energy-transfer rates; and (ii) the forward and backward energy-transfer rates are identically dependent on the orientation factor, and thus their ratio is free from this parameter. This situation is also reflected in the analytic result under steady state excitation (Eqn 5.41): one can there find that, in fact, the anisotropy of polarization is dependent on the ratios $g_{22}/g_{11} \simeq k_{12}/k_{21}, g_{12}/g_{11} \simeq 1$ and $g_{21}/g_{11} \simeq k_{12}/k_{21}$, when $k_{12}, k_{21} \gg \tau_1^{-1}, \tau_2^{-1}$.

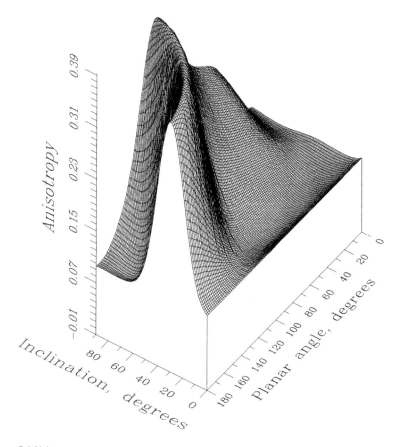

Figure 5.16(a)

5.6   CONCLUSION

In this chapter, we have analysed the phenomena of fluorescence and absorption polarization, and described their application in the polarization spectroscopy of complex molecular ensembles that have chromophores of up to three spectral pools. Although a great many biological systems fit into this category, the approach is amenable to further extension for the study of complexes with chromophores of four or more spectral types. Two time regimes have been considered: these relate to (i) steady state and (ii) δ-pulse excitation. Most experimental measurements can be related to one or other of these cases, and we have shown through several illustrations the close agreement that exists between observed and predicted behavior. This gives confidence that the results

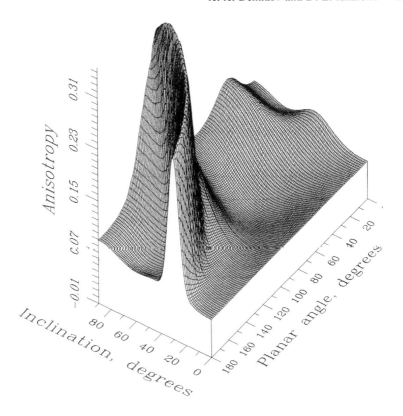

Figue 5.16(b)

can meaningfully be employed to derive valuable new forms of structural information from experimental results on various chemical and biological multichromophore ensembles.

Appendix A

Equations 5.1 contain two averages, $\langle(\mathbf{e} \cdot \mathbf{i}_{ab})^2\rangle$ and $\langle(\mathbf{e} \cdot \mathbf{i}_{ab})^2(\mathbf{e} \cdot \mathbf{i}_{fl})^2\rangle$. One method to significantly simplify their calculation is to exploit the freedom of choosing the orientation of the molecular Cartesian frame. Let us choose the axes such that $\mathbf{i}_{ab} = \{0, 0, 1\}$ and $\mathbf{e} = \{\sin\theta_e \cos\varphi_e, \sin\theta_e \sin\varphi_e, \cos\theta_e\}$. Then,

$$\langle(\mathbf{e} \cdot \mathbf{i}_{ab})^2\rangle = \frac{\displaystyle\int\int \cos^2\theta_e \sin\theta_e \, d\theta_e \, d\varphi_e}{\displaystyle\int\int \sin\theta_e \, d\theta_e \, d\varphi_e} = \frac{\displaystyle\int_{-1}^{1} x^2 \, dx}{\displaystyle\int_{-1}^{1} dx} = \frac{1}{3}, \tag{A.1}$$

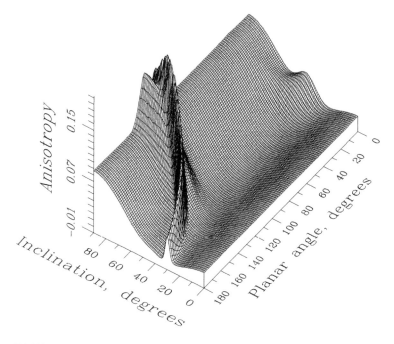

Figure 5.16(c)

where $x = \cos \theta_e$. Similarly, for the second average we have

$$
\begin{aligned}
&\langle (\mathbf{e} \cdot \mathbf{i}_{ab})^2 (\mathbf{e} \cdot \mathbf{i}_{fl})^2 \rangle \\
&= \frac{\displaystyle\int\int (\sin \theta_e \sin \theta_{fl} \cos(\varphi_e - \varphi_{fl}) + \cos \theta_e \cos \theta_{fl})^2 \cos^2 \theta_e \sin \theta_e \, \mathrm{d}\theta_e \, \mathrm{d}\varphi_e}{\displaystyle\int\int \sin \theta_e \, \mathrm{d}\theta_e \, \mathrm{d}\varphi_e} \\
&= \frac{\displaystyle\int_{-1}^{1} [\tfrac{1}{2}(1 - x^2)\sin^2 \theta + x^2 \cos^2 \theta] x^2 \, \mathrm{d}x}{\displaystyle\int_{-1}^{1} \mathrm{d}x} \\
&= \tfrac{1}{2}\left(\tfrac{1}{2} \sin^2 \theta \int_{-1}^{1} x^2 \, \mathrm{d}x + (\cos^2 \theta - \tfrac{1}{2} \sin^2 \theta) \int_{-1}^{1} x^4 \, \mathrm{d}x \right) \\
&= \frac{2 + \cos^2 \theta}{15}, \hspace{4cm} \text{(A.2)}
\end{aligned}
$$

where the angle $\theta = \theta_{fl}$ is that which lies between the absorption and fluorescence transition dipole moments.

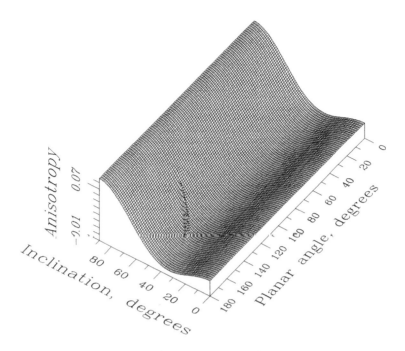

Figure 5.16(d) The dependence of the absorption anisotropy of bichromophoric tri-mers on the angular parameters (inclination θ and planar angle φ) of the "$\beta - 84$" chromophore in the first monomer, following excitation by a δ-pulse with wavelength $\lambda_{ab} = 615$ nm: (a) 5 ps; (b) 30 ps; (c) 200 ps; (d) 2000 ps after excitation. The spectro-scopic parameters of the chromophores are equal to those for the $\alpha - 84$ and $\beta - 84$ chromophores of C-phycocyanin (see text). The planar angles of other "$\beta - 84$" chro-mophores in the trimer are $\varphi \pm 120°$, and all have the same inclination angle θ, whereas the angles of the "$\alpha - 84$" chromophores are those determined by Schirmer *et al.* [39] for C-phycocyanin

An alternative method which provides a short route to the most intricate averages is to refer all parameters to a common laboratory-fixed frame, denoted by regular indices $i, j, \ldots$, and then transform the molecular compon-ents to a molecule-fixed frame, denoted by indices $\lambda, \mu, \ldots$, and finally to average over the relative orientations of the two frames. Using the rule that repeated Cartesian indices signify summation (e.g. $(\mathbf{ab}) = a_i b_i$), and writing the nth order average as $I^{(n)}_{ij\ldots;\lambda\mu\ldots}$, then for $\langle (\mathbf{e} \cdot \mathbf{i}_{ab})^2 \rangle$ we thus have

$$\langle (\mathbf{e} \cdot \mathbf{i}_{ab})(\mathbf{e} \cdot \mathbf{i}_{ab}) \rangle = \langle e_i i_{(ab)i} e_j i_{(ab)j} \rangle = e_i e_j i_{(ab)\lambda} i_{(ab)\mu} I^{(2)}_{ij;\lambda\mu}$$

$$= \tfrac{1}{3} e_i e_j i_{(ab)\lambda} i_{(ab)\mu} \delta_{ij} \delta_{\lambda\mu} = \tfrac{1}{3} (\mathbf{e} \cdot \mathbf{e})(\mathbf{i}_{ab} \cdot \mathbf{i}_{ab}) = \tfrac{1}{3}, \quad \text{(A.3)}$$

using the result for $I^{(2)}$ from [5] and, similarly,

$$\langle (\mathbf{e} \cdot \mathbf{i}_{ab})^2 (\mathbf{e} \cdot \mathbf{i}_{fl})^2 \rangle = \langle e_i i_{(ab)i} e_j i_{(ab)j} e_k i_{(fl)k} e_l i_{(fl)l} \rangle$$

$$= e_i e_j e_k e_l i_{(ab)\lambda} i_{(ab)\mu} i_{(fl)\nu} i_{(fl)o} I^{(4)}_{ijkl;\lambda\mu\nu o}$$

$$= \tfrac{1}{30} e_i e_j e_k e_l i_{(ab)\lambda} i_{(ab)\mu} i_{(fl)\nu} i_{(fl)o}$$

$$\begin{bmatrix} \delta_{ij} & \delta_{kl} \\ \delta_{ik} & \delta_{jl} \\ \delta_{il} & \delta_{jk} \end{bmatrix}^{\mathrm{T}} \begin{bmatrix} 4 & -1 & -1 \\ -1 & 4 & -1 \\ -1 & -1 & 4 \end{bmatrix} \begin{bmatrix} \delta_{\lambda\mu} & \delta_{vo} \\ \delta_{\lambda v} & \delta_{\mu o} \\ \delta_{\lambda o} & \delta_{\mu v} \end{bmatrix}$$

$$= \tfrac{1}{30} \left[2(\mathbf{i}_{ab} \cdot \mathbf{i}_{ab})(\mathbf{i}_{fl} \cdot \mathbf{i}_{fl}) + 4(\mathbf{i}_{ab} \cdot \mathbf{i}_{fl})^2 \right] = \tfrac{1}{15}(1 + 2\cos^2\theta) \quad \text{(A.4)}$$

By either calculational route, we then have the parameter ξ (see Eqn 5.3) given by

$$\xi = \frac{\langle \Phi_{\parallel} \rangle}{\langle \Phi \rangle} = \frac{\langle (\mathbf{e} \cdot \mathbf{i}_{ab})^2 (\mathbf{e} \cdot \mathbf{i}_{fl})^2 \rangle}{\langle (\mathbf{e} \cdot \mathbf{i}_{ab})^2 \rangle} = \frac{1 + 2\cos^2\theta}{5} \quad \text{(A.5)}$$

and finally the anisotropy:

$$r = \frac{3\xi - 1}{2} = \frac{3\cos^2\theta - 1}{5}. \quad \text{(A.6)}$$

Appendix B

The excitation population of chromophores of different spectral types $n_i (i = 1, 2, 3)$ is described by a system of balance equations $\mathbf{Wn} + \mathbf{A} = \mathbf{0}$, where

$$\mathbf{W} = \begin{bmatrix} -\tau_1^{-1} - k_{12} - k_{13} & k_{21} & k_{31} \\ k_{12} & -\tau_2^{-1} - k_{21} - k_{23} & k_{32} \\ k_{13} & k_{23} & -\tau_3^{-1} - k_{31} - k_{32} \end{bmatrix}. \quad \text{(B.1)}$$

The inverse matrix $\bar{\mathbf{G}} = \mathbf{W}^{-1}$, with elements $\bar{g}_{ij}$, gives the solution for the $n_i : \mathbf{n} = -\mathbf{GA}$, i.e. $n_i = -\sum_j \bar{g}_{ij} A_j$, where the parameters A_j represent the excitation (absorption) of the chromophores. In our calculations (for historical reasons), we rather prefer to use parameters $g_{ij} = -\bar{g}_{ij}$; see, for example, Eqns 5.13 and 5.14. The difference is only in the sign, which is canceled anyway when the ratio of the fluorescence components is calculated. Hence the elements g_{ij} used in calculations of polarization are given by

$$g_{11} = \tilde{D}^{-1}(w_{22}w_{33} - w_{23}w_{32}) = \tilde{D}^{-1}[(\tau_2^{-1} + k_{21} + k_{23})(\tau_3^{-1} + k_{31} + k_{32}) - k_{23}k_{32}],$$

$$g_{12} = \tilde{D}^{-1}(w_{13}w_{32} - w_{12}w_{33}) = \tilde{D}^{-1}[k_{31}k_{23} + k_{21}(\tau_3^{-1} + k_{31} + k_{32})],$$

$$g_{13} = \tilde{D}^{-1}(w_{12}w_{23} - w_{13}w_{22}) = \tilde{D}^{-1}[k_{21}k_{32} + k_{31}(\tau_2^{-1} + k_{21} + k_{23})],$$

$$g_{21} = \tilde{D}^{-1}(w_{31}w_{23} - w_{21}w_{33}) = \tilde{D}^{-1}[k_{13}k_{32} + k_{12}(\tau_3^{-1} + k_{31} + k_{32})],$$

$$g_{22} = \tilde{D}^{-1}(w_{11}w_{33} - w_{31}w_{13}) = \tilde{D}^{-1}[(\tau_1^{-1} + k_{12} + k_{13})(\tau_3^{-1} + k_{31} + k_{32}) - k_{13}k_{31}],$$

$$g_{23} = \tilde{D}^{-1}(w_{21}w_{13} - w_{11}w_{23}) = \tilde{D}^{-1}[k_{12}k_{31} + k_{32}(\tau_1^{-1} + k_{12} + k_{13})],$$

$$g_{31} = \tilde{D}^{-1}(w_{21}w_{32} - w_{22}w_{31}) = \tilde{D}^{-1}[k_{12}k_{23} + k_{13}(\tau_2^{-1} + k_{21} + k_{23})],$$

$$g_{32} = \tilde{D}^{-1}(w_{12}w_{31} - w_{11}w_{32}) = \tilde{D}^{-1}[k_{21}k_{13} + k_{23}(\tau_1^{-1} + k_{12} + k_{13})],$$

$$g_{33} = \tilde{D}^{-1}(w_{11}w_{22} - w_{21}w_{12}) = \tilde{D}^{-1}[(\tau_1^{-1} + k_{12} + k_{13})(\tau_2^{-1} + k_{21} + k_{23}) - k_{12}k_{21}],$$

$$\tilde{D} = -[w_{11}(w_{22}w_{33} - w_{23}w_{32}) - w_{12}(w_{21}w_{33} - w_{31}w_{23}) + w_{13}(w_{21}w_{32} - w_{31}w_{22})].$$

$$(\text{B.2})$$

Appendix C

The parameters B and q_{ij} can be defined as follows: $B = E/D$ and $q_{ij} = H_{ij}/D$, where

$$E = g_{11} + g_{22}\alpha_{21}\gamma_{21} + g_{33}\alpha_{31}\gamma_{31},$$

$$H_{12} = g_{12}\alpha_{21} + g_{21}\gamma_{21},$$

$$H_{13} = g_{13}\alpha_{31} + g_{31}\gamma_{31}, \qquad\qquad (\text{C.1})$$

$$H_{23} = g_{23}\gamma_{21}\alpha_{31} + g_{32}\gamma_{31}\alpha_{21},$$

$$D = g_{12}\alpha_{21} + g_{13}\alpha_{31} + g_{21}\gamma_{21} + g_{31}\gamma_{31} + g_{23}\gamma_{21}\alpha_{31} + g_{32}\gamma_{31}\alpha_{21}.$$

The matrix elements g_{ij} are as defined in Appendix B. For fluorescence polarization, the parameters α_{ij} and γ_{ij} are given by

$$\alpha_{ij} = \frac{\sigma_i(\lambda_{ab})}{\sigma_j(\lambda_{ab})}, \quad \gamma_{ij} = \frac{\eta_i\tau_i f_i(\lambda_{fl})}{\eta_j\tau_j f_j(\lambda_{fl})}, \qquad (\text{C.2})$$

and for the "bleach" signal in pump–probe experiments,

$$\alpha_{ij} = \frac{\sigma_i(\lambda_p)}{\sigma_j(\lambda_p)}, \qquad \gamma_{ij} = \frac{\sigma_i(\lambda_{pr})}{\sigma_j(\lambda_{pr})}. \qquad (\text{C.3})$$

Appendix D

The characteristic lifetimes t_1, t_2, and t_3 of energy equilibration and overall decay can be found by calculating the eigenvalues ν_i ($t_i = 1/\nu_i$) of the determinant equation det $(\mathbf{W} + \nu\mathbf{E}) = 0$, where $\mathbf{E}$ is the unit matrix. The elements w_{ij} are given in Appendix B. To find $B(t)$ and $q_{ij}(t)$, it is first necessary to calculate a set of intermediate parameters:

$$y_{0i} = \frac{w_{13}w_{21} - w_{23}(w_{11} + \nu_i)}{w_{12}w_{23} - w_{13}(w_{22} + \nu_i)},$$

$$z_{0i} = \frac{\nu_i^2 + (w_{11} + w_{22})\nu_i + (w_{11}w_{22} - w_{12}w_{21})}{w_{12}w_{23} - w_{13}(w_{22} + \nu_i)}, \quad i = 1, 2, 3,$$

$$\Delta = (y_{01} - y_{03})(z_{02} - z_{03}) - (y_{02} - y_{03})(z_{01} - z_{03}),$$

$$a_1 = \frac{y_{02}z_{03} - y_{03}z_{02}}{\Delta}, \quad a_2 = \frac{z_{02} - z_{03}}{\Delta}, \quad a_3 = \frac{y_{03} - y_{02}}{\Delta},$$

$$b_1 = \frac{y_{03}z_{01} - y_{01}z_{03}}{\Delta}, \quad b_2 = \frac{z_{03} - z_{01}}{\Delta}, \quad b_3 = \frac{y_{01} - y_{03}}{\Delta},$$

$$d_1 = \frac{y_{01}z_{02} - y_{02}z_{01}}{\Delta}, \quad d_2 = \frac{z_{01} - z_{02}}{\Delta}, \quad d_3 = \frac{y_{02} - y_{01}}{\Delta}, \quad \text{(D.1)}$$

which feature in the following calculations for the time-dependent elements $g_{ij}(t)$:

$$g_{11}(t) = a_1 e^{-t/t_1} + b_1 e^{-t/t_2} + d_1 e^{-t/t_3},$$

$$g_{12}(t) = a_2 e^{-t/t_1} + b_2 e^{-t/t_2} + d_2 e^{-t/t_3},$$

$$g_{13}(t) = a_3 e^{-t/t_1} + b_3 e^{-t/t_2} + d_3 e^{-t/t_3},$$

$$g_{21}(t) = a_1 y_{01} e^{-t/t_1} + b_1 y_{02} e^{-t/t_2} + d_1 y_{03} e^{-t/t_3},$$

$$g_{22}(t) = a_2 y_{01} e^{-t/t_1} + b_2 y_{02} e^{-t/t_2} + d_2 y_{03} e^{-t/t_3},$$

$$g_{23}(t) = a_3 y_{01} e^{-t/t_1} + b_3 y_{02} e^{-t/t_2} + d_3 y_{03} e^{-t/t_3},$$

$$g_{31}(t) = a_1 z_{01} e^{-t/t_1} + b_1 z_{02} e^{-t/t_2} + d_1 z_{03} e^{-t/t_3},$$

$$g_{32}(t) = a_2 z_{01} e^{-t/t_1} + b_2 z_{02} e^{-t/t_2} + d_2 z_{03} e^{-t/t_3},$$

$$g_{33}(t) = a_3 z_{01} e^{-t/t_1} + b_3 z_{02} e^{-t/t_2} + d_3 z_{03} e^{-t/t_3}.$$

Now we can write the explicit expressions for $B(t)$ and $q_{ij}(t)$:

$$q_{12}(t) = D^{-1}(t)[g_{12}(t)\alpha_{21} + g_{21}(t)\gamma_{21}],$$

$$q_{13}(t) = D^{-1}(t)[g_{13}(t)\alpha_{31} + g_{31}(t)\gamma_{31}],$$

$$q_{23}(t) = D^{-1}(t)[g_{23}(t)\alpha_{31}\gamma_{21} + g_{32}(t)\gamma_{31}\alpha_{21}],$$

$$B(t) = D^{-1}(t)[g_{11}(t) + g_{22}(t)\alpha_{21}\gamma_{21} + g_{33}(t)\gamma_{31}\alpha_{31}],$$

$$D(t) = g_{12}(t)\alpha_{21} + g_{21}(t)\gamma_{21} + g_{13}(t)\alpha_{31} + g_{31}(t)\gamma_{31} + g_{23}(t)\alpha_{31}\gamma_{21} + g_{32}(t)\gamma_{31}\alpha_{21}.$$

$$\text{(D.2)}$$

Obviously, $q_{12}(t) + q_{13}(t) + q_{23}(t) = 1$ at any time t. The limiting value $g_{ij}(t) \to 0$ as $t \to 0$ for $i \neq j$, and that yields $r(t = 0) = 0.4$ for any values of the given parameters.

References

1. Allock, P., and D. L. Andrews 1997. Two-photon fluorescence: resonance energy transfer. *J. Chem. Phys.* 108: 3089–3095.
2. Agranovich, V. M., and M. D. Galanin 1982. *Electron Excitation Energy Transfer in Condensed Matter.* North-Holland, Amsterdam.
3. Andrews, D. L. 1997. *Lasers in Chemistry*, third edition. Springer-Verlag, Berlin.
4. Andrews, D. L., and G. Juzeliūnas 1991. The range dependence of fluorescence anisotropy in molecular energy transfer. *J. Chem. Phys.* 95: 5513–5518.
5. Andrews, D. L., and T. Thirunamachandran 1977. On three-dimensional rotational averages. *J. Chem. Phys.* 67: 5026 5033.
6. Brejc, K., R. Ficner, R. Huber, and S. Steinbacher 1995. Isolation, crystallization, crystal structure analysis and refinement of allophycocyanin from the cyanobacterium *Spirulinaplatensis* at 2.3 angström resolution. *J. Mol. Biol.* 249: 424–440.
7. Callis, P. R. 1993. On the theory of two-photon induced fluorescence anisotropy with application to indoles. *J. Chem. Phys.* 99: 27–37.
8. Chang, W. R., T. Jiang, Z. L. Wan, J. P. Zhang, Z. X. Yang, and D. C. Liang 1996. Crystal structure of R-phycoerythrin from *Polysiphonia-urceolata* at 2.8 angström resolution. *J. Mol. Biol.* 262: 721–731.
9. Chen, S. Y., and B. W. van der Meer 1993. Theory of two-photon induced fluorescence anisotropy decay in membranes. *Biophys. J.* 64: 1567–1575.
10. Debreczeny, M. P., K. Sauer, J. H. Zhou, and D. A. Bryant 1993. Monomeric C-phycocyanin at room temperature and 77-K resolution of the absorption and fluorescence spectra of the individual chromophores and the energy transfer rate constants. *J. Phys. Chem.* 97: 9852–9862.
11. Deisenhofer, J., O. Epp, K. Miki, R. Huber, and H. Michel 1985. Structure of the protein subunits in the photosynthetic reaction center of *Rhodopseudomonas viridis* at 3 Å resolution. *Nature* 318: 618–624.
12. Deisenhofer, J., O. Epp, K. Miki, R. Huber, and H. Michel. *Protein Databank.* Brookhaven National Laboratory. http://www.pdb.bnl.gov.
13. Demidov, A. A. 1994. Determination of fluorescence polarization of double chromophore complexes. *J. Theoret. Biol.* 170: 355–358.
14. Demidov, A. A. 1994. Fluorescence polarization of triple-chromophore complexes with energy transfer. *Appl. Opt.* 33: 6303–6306.
15. Demidov, A. A. 1994. Quantitative calculations of fluorescence polarization and absorption anisotropy kinetics of double-chromophore and triple-chromophore complexes with energy transfer. *Biophys. J.* 67: 2184–2190.
16. Demidov, A. A., and D. L. Andrews 1995. Theory of polarized fluorescence and absorption in molecular complexes comprising two chromophores with nonparallel absorption and emission transition dipole moments. *Chem. Phys. Lett.* 235: 327–333.
17. Demidov, A. A., and D. L. Andrews 1996. Determination of fluorescence polarization and absorption anisotropy in molecular complexes having threefold rotational symmetry. *Photochem. Photobiol.* 63: 39–52.

18. Demidov, A. A., and A. Y. Borisov 1993. Computer simulation of energy migration in the C-phycocyanin of the blue–green algae *Agmenellum quadruplicatum*. *Biophys. J.* 64: 1375–1384.
19. Demidov, A. A., and M. Mimuro 1995. Deconvolution of C-phycocyanin β-84 and β-155 chromophore absorption and fluorescence spectra of cyanobacterium *Mastigocladus laminosus*. *Biophys. J.* 68: 1500–1506.
20. Demidov, A., R. Sension, B. Donovan, L. Walker, and Ch. Yokum 1996. The re-evaluation of PS-II structure using polarization spectroscopy. In *24th Annual Meeting of American Society for Photobiology, Works-in-Progress*, June 15–20, Atlanta, Georgia.
21. Feofilov, P. P. 1961. *The Physical Basis of Polarized Emission*. New York: Consultants Bureau. Authorized translation from the Russian.
22. Förster, T. 1948. Zwischenmolekulare Energiewanderung und Fluoreszenz. *Ann. Phys.* 2: 55–75.
23. Gantt, E. 1981. Phycobilisomes. *Ann. Rev. Plant Physiol.* 32: 327–347.
24. Gillbro, T., A. V. Sharkov, I. V. Kryukov, E. V. Khoroshilov, P. G. Kryukov, R. Fischer, and H. Scheer 1993. Förster energy transfer between neighbouring chromophores in C-phycocyanin trimers. *Biochim. Biophys. Acta* 1140: 321–326.
25. Gryczynski, I., H. Malak, and J. R. Lakowicz 1995. Three-photon induced fluorescence of 2,5-diphenyloxazole with a femtosecond Ti–sapphire laser. *Chem. Phys. Lett.* 245: 30–35.
26. Kliger, D. S., J. W. Lewis, and C. E. Randall 1990. *Polarized Light in Optics and Spectroscopy*. Academic Press, Boston.
27. Kwa, S. L. S., C. Eijckelhoff, R. van Grondelle, and J. P. Dekker 1994. Site-selection spectroscopy of the reaction center complex of photosystem-II. 1. Triplet-minus-singlet absorption difference search for a 2nd exciton band of P-680. *J. Phys. Chem.* 98: 7702–7711.
28. Kwa, S. L. S., W. R. Newell, R. van Grondelle, and J. P. Dekker 1992. The reaction center of photosystem-II studied with polarized fluorescence spectroscopy. *Biochim. Biophys. Acta* 1099: 193–202.
29. Lakowicz, J. R., I. Gryczynski, and E. Danielsen 1992. Anomalous differential polarized phase angles for two-photon excitation with isotropic depolarizing rotations. *Chem. Phys. Lett.* 191: 47–53.
30. Lakowicz, J. R., I. Gryczynski, Z. Gryczynski, E. Danielsen, and M. J. Wirth 1992. Time resolved fluorescence intensity and anisotropy decays of 2, 5-diphenyloxazole by two-photon excitation and frequency-domain fluorometry. *J. Phys. Chem.* 96: 3000–3006.
31. Levshin, V. L. 1925. Poliarizovannaia fluorescenciia i phosphorescenciia rastvorov krasok. *Zhurnal Russkogo Physiko-Chimicheskogo Obschestva, Fizika (Russian)* 57: 283–300.
32. Lewshin, W. L. 1925. Polarisierte Fluoreszenz und Phosphoreszenz der Farbstoflosungen. IV *Z. Phys.* 32: 307–326.
33. Lyle, P. A., and W. S. Struve 1991. Dynamic linear dichroism in chromoproteins. *Photochem. Photobiol.* 53: 359–365.
34. McClain, W. M., and R. A. Harris 1977. Two-photon molecular spectroscopy in liquids and gases. In *Excited States*. E. C. Lim, editor. Academic Press, New York, Vol. 3, pp. 1–56.
35. Mimuro, M., P. Füglistaller, R. Rumbeli, and H. Zuber 1986. Functional assignment of chromophores and energy transfer in C-phycocyanin isolated from the thermophilic cyanobacterium *Mastigocladus laminosus*. *Biochim. Biophys. Acta* 848: 155–166.

36. Perrin, F. 1929. La fluorescence des solutions. *Ann. Phys.* 12: 169–275.
37. Savikhin, S., and W. S. Struve 1994. Femtosecond pump–probe spectroscopy of bacteriochlorophyll a monomers in solution. *Biophys. J.* 67: 2002–2007.
38. Schiffer, M., and J. R. Norris 1993. Structure and function of the photosynthetic reaction center of *Rhodobacter sphaeroides*. In *The Photosynthetic Reaction Center*. J. Deisenhofer and J. R. Norris, editors. Academic Press, San Diego.
39. Schirmer, T., W. Bode, and R. Huber 1987. Refined three-dimensional structures of two cyanobacterial C-phycocyanins at 2.1 and 2.5 Å resolution. A common principle and phycobilin–protein interaction. *J. Mol. Biol.* 196: 677–695.
40. Sharkov, A. V., I. V. Kryukov, E. V. Khoroshilov, P. G. Kryukov, R. Fischer, H. Scheer, and T. Gillbro 1992. Femtosecond energy-transfer between chromophores in allophycocyanin trimers. *Chem. Phys. Lett.* 191: 633–638.
41. Thulstrup, E. W., and J. Michl 1989. *Elementary Polarization Spectroscopy*. VCH Publishers, New York.
42. van Grondelle, R., J. P. Dekker, T. Gillbro, and V. Sundstrom 1994. Energy transfer and trapping in photosynthesis. *Biochim. Biophys. Acta* 1187: 1–65.
43. van Gurp, M., G. van Ginkel, and Y. K. Levine 1989. Fluorescence anisotropy of chlorophyll-a and chlorophyll-b in castor-oil. *Biochim. Biophys. Acta* 973: 405–413.
44. Xie, X., M. Du, L. Mets, and G. R. Fleming 1992. Femtosecond fluorescence depolarization study of photosynthetic antenna proteins: observation of ultrafast energy transfer in trimeric C-phycocyanin and allphycocyanin. In *Time-resolved Laser Spectroscopy in Biochemistry III*. SPIE, pp. 690–706.

6

Theory of coupling in multichromophoric systems

Gregory D. Scholes
University of California, USA

6.1   INTRODUCTION

In the condensed phase, the time-dependent transition probability for energy transfer involves an entanglement between the electronic factors and the dynamics of chromophore–bath interactions [1, 47, 54, 62] (i.e. interactions of the type which effectuate homogenous line broadening). Here we use "chromophore" to denote any atom, molecule or supramolecular component upon which electronic excitation may be localized. For a (weakly coupled) bichromophoric system, in the nonadiabatic limit, the rate of energy transfer from the derivation of Förster [21, 22] is given by $w = 1/c\hbar^2 |T_{RP}|^2 \bar{J}$, written here in a form such that $\bar{J}$ is the spectral overlap integral between the area-normalized donor fluorescence, $f_R(\nu)$, and acceptor absorption, $a_P(\nu)$, spectra on a wavenumber scale: $\bar{J} = \int_0^\infty d\nu^{-3} f_R(\nu) a_P(\nu)$; T_{RP} is the electronic coupling between reactant (R) and product (P) states. The spectral overlap factor accounts for nuclear effects such as Franck–Condon factors [34, 53]. The electronic coupling determines the mechanism of the energy transfer and contributes a major influence to the rate of energy transfer through separation and orientation factors. Its origin, nature, and determination will be discussed here.

In the absence of direct interchromophore orbital overlap, two chromophores (one in an excited electronic state) interact via the long-range *coulombic* mechanism. When electron correlation is neglected, this interaction is specified by a two-electron integral which describes a simultaneous de-excitation and excitation of the two chromophores, mediated by an interaction between the

Resonance Energy Transfer. Edited by David L. Andrews and Andrey A. Demidov.
© 1999 John Wiley & Sons Ltd

transition moments of the two molecules (and in many ways analogous to the classical electrostatic interaction between two real dipoles). The coulombic potential may be represented as a multipolar expansion, and in the usual treatment of molecular exciton interactions or energy transfer just the dipole–dipole term is considered. The coulombic interaction dominates the coupling in many cases of singlet–singlet energy transfer, and is discussed in Section 6.5.

When the electron density distributions (i.e. orbitals) on the two chromophores overlap, additional effects arise which can dramatically increase the electronic coupling [52, 56]. For example, the Dexter exchange interaction [18], which is directly related to the coulombic integral by the antisymmetrizing operator (as a consequence of the Pauli principle), is often considered to be significant for energy transfer associated with forbidden transitions. The recent work which is summarized in this chapter demonstrates that by defining appropriate reactant and product states the origins of the short-range coupling may be properly elucidated for various cases [25, 58]. It is shown that coupling at close separations (or for forbidden transitions) is promoted primarily by terms other than the Dexter exchange integral. Although the exponential distance-dependence for the complete interaction is analogous to that introduced by Dexter (because the orbital-overlap-dependence is the same), the physical picture and hence some important physical consequences and interpretations differ.

A typical variation of the total singlet–singlet coupling is illustrated in Fig. 6.1. The coupling was calculated only approximately from an *ab initio* CI-singles calculation using a minimum (STO-3G) basis set [52]. Typically, the coulombic interaction is overestimated (owing to the neglect of electron correlation), and the short-range coupling is underestimated owing to the small size of the basis set. Provided that the coupling is weak, then if T_{RP} is a dipole–dipole (i.e. large separation) interaction we have the Förster limit (i.e. R^{-6} distance-dependence for the rate); for triplet–triplet energy transfer we have the Dexter limit (i.e. exponential distance-dependence of the rate); otherwise, we have some combination of the two (i.e. $R^{-6}+$ exponential + interference term) [11].

An historical synopsis of the evolution and application of theories for electronic coupling is given in Table 6.1. Only a selection from the literature is cited, since I have tried to provide key references only. It is evident that investigations of overlap-dependent interactions have focused on the Dexter exchange interaction, although inspection of the work dealing with the stabilization of excimers [5, 41] suggests that this is not a good approximation. It was only recently that Harcourt *et al.* [25, 58] formulated a detailed theory for singlet–singlet and triplet–triplet coupling which revealed the origin of overlap-dependent coupling.

We begin here by presenting a fairly detailed discussion of the development of the theory and how it may be applied to calculation of the coupling. Despite

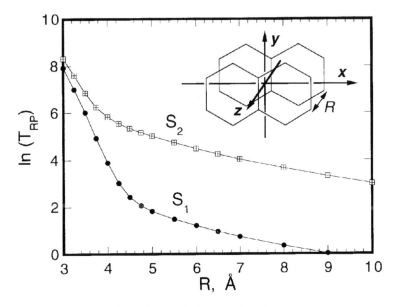

Figure 6.1 The distance-dependence of the coupling for the S_1 and S_2 states of a naphthalene sandwich dimer (obtained from approximate model calculations, *ab initio* CIS/STO-3G [52]). T_{RP} is in units of cm^{-1}. Note the marked influence of the short-range (orbital overlap-dependent interactions)

Table 6.1   A history of coupling models in resonance energy transfer

References	Application	Comments
[16, 17, 30, 35]	Exciton states (Davydov, 1948)	Application of the dipole–dipole coupling ($\kappa\|\mu\|^2/[4\pi\varepsilon_0 R^3]$) to define the electronic states of molecular aggregates. This dipole–dipole coupling arises from the coulombic integral: $2J = (a'a\|bb')$
[21, 22]	Electronic energy transfer (Förster, 1948)	Development of a theory for the rate of energy transfer, through the dipole–dipole coupling.
[8]	The interaction of ground- and excited-state He atoms (Buckingham and Dalgarno, 1952)	The form of the interaction between a ground-state He atom and one excited to the first triplet or singlet metastable state, following the Heitler–London method

References	Application	Comments
[18, 39]	Triplet–triplet energy transfer (Dexter, 1953)	Orbital overlap effects considered to arise via an exchange integral obtained from the coulombic integral by permutation of two orbitals: $Z = (a'b'|ba)$
[31, 32]	Transannular interactions (Koutecky and Paldus, 1962)	Calculation of the interactions between close molecules (e.g. cyclophanes) and perturbations of their absorption spectra
[3, 4, 15, 37, 50]	Very long range coupling (McLone and Power, 1964)	Quantum-electrodynamical theories for the form of the dipole–dipole coupling over very large distances; the R^{-3} distance dependence becomes R^{-1}
[5, 41]	Excimers (Murrell and Tanaka, 1964)	Calculation of spectra based on the LMO prescription involving locally excited and charge-transfer configurations
[42, 43]	Exchange-induced resonance energy transfer (Naqvi and Steel, 1970)	More detailed comments on the exchange interaction in singlet/triplet–singlet/doublet energy transfer
[25, 54, 55, 58]	LMO coupling model (Harcourt, Scholes, and Ghiggino, 1994)	Orbital-overlap-dependent coupling (LMO model), revealing that the significant overlap dependent coupling is mediated via charge-transfer configurations
[59]	Special cases: photosynthesis (1997)	A study of couplings involving the carotenoid S_1 state

a focus on the case of 'normal' singlet–singlet and triplet–triplet energy transfer (i.e. π–π^*-type transitions involving single-electron excitations), sufficient detail has been included so that it may be adapted for any case, as Harcourt *et al.* [58] have demonstrated for various examples involving n–π^* transitions and Scholes *et al.* [59] have done for energy transfer involving the symmetric carotenoid S_1 state. In Section 6.4 the results are summarized and the effects of extensions to the basic theory, such as inclusion of electron correlation, are discussed qualitatively. Coulombic coupling is discussed in Section 6.5. In Sections 6.6 and 6.7, superexchange-mediated coupling is discussed, along with an experimental example of exciton splitting, the interpretation of the splitting and the significance of through-bond-mediated coupling. In Section 6.8, a protocol for calculating couplings is outlined.

6.2　REACTANT AND PRODUCT STATES: LMO MODEL

A natural way to treat the perturbation which leads to the coupling is to define appropriate reactant and product wavefunctions, Ψ_R and Ψ_P respectively, in terms of a number of configuration wavefunctions, ψ_i. The definition of the ψ_i may be based on a localized molecular orbital (LMO) description, which permits us to formulate a description of the reactant and product to an arbitrary degree of complexity, but in a manner which is well-suited to physical interpretation. The ψ_i are configurations, written in terms of localized molecular orbitals (i.e. in a diabatic, or fragment, representation). Note that the ψ_i are *configuration* wavefunctions, not *state* wavefunctions, and consequently they are not orthogonal. This can make calculation a difficult numerical problem. Recently, Zhang *et al.* [67] have described an efficient approach for calculating electron transfer matrix elements which could be adapted for calculating energy transfer couplings.

I shall present the original model, based upon a model four-electron, four-orbital picture, which contains the essence of the problem. It may readily be generalized to accommodate a larger active space and/or electron correlation effects and superexchange-mediated interactions, as will be indicated. In this picture, various configurations are "interacted"; primarily those for which excitation is localized at a chromophore (which we write as $A^* \ldots B$ or $A \ldots B^*$, labeled ψ_1 and ψ_4 respectively, for chromophores A and B). This basic treatment is improved by adding further configurations (contributing to the coupling with smaller weight). Of particular significance are the charge transfer configurations $A^+ \ldots B^-$ and $A^- \ldots B^+$ (ψ_2 and ψ_3), which help properly to incorporate into the model orbital overlap effects, analogously to the valence bond treatment of bonding [45] and intermolecular interactions [2, 38]. Such a description has been applied previously to supermolecule calculations of excimers and dimers. Harcourt *et al.* [25] have developed an approach which enables the coupling between two (or more) chromophores to be determined and characterized.

The model has been formulated as an interaction between reactant and product wavefunctions (Ψ_R and Ψ_P), each expressed in terms of a linear combination of configuration wavefunctions (ψ_i). The ψ_i, in turn, are written as antisymmeterized spin orbital products using MOs (*sc.* molecular orbitals) localized on donor and acceptor chromophores ($a, a', \ldots$ and $b, b', \ldots$); i.e. localized molecular orbitals (LMOs). We label the donor HOMO/LUMO orbitals as a/a' and corresponding acceptor orbitals as b/b' for molecules A and B respectively. Various configuration wavefunctions which we have considered are collected in Table 6.2. The singly excited configurations are those used for the basic treatment of energy transfer involving $\pi-\pi^*$-type transitions [25]. The

Table 6.2 Descriptions of the various LMO configuration wavefunctions which have been considered. The wavefunctions may be visualized in the four-electron, four-orbital picture, which depicts the orbitals of A on the left and those of B on the right, with the LUMO orbitals above the HOMOs (cf. [25]). The spin-paired electrons are represented as circles of the same shade

Configuration	Four-electron, four-orbital picture	Wavefunction
Singly excited configurations		
$^{1,3}\psi_1$ A*B		$\left(|ab\bar{a}'\bar{b}| \pm |a'b\bar{a}\bar{b}|\right)/\sqrt{2}$
$^{1,3}\psi_2$ A$^+$B$^-$		$\left(|ab\bar{b}'\bar{b}| \pm |b'b\bar{a}\bar{b}|\right)/\sqrt{2}$
$^{1,3}\psi_3$ A$^-$B$^+$		$\left(|ab\bar{a}\bar{a}'| \pm |aa'\bar{a}\bar{b}|\right)/\sqrt{2}$
$^{1,3}\psi_4$ AB*		$\left(|ab\bar{a}\bar{b}'| \pm |ab'\bar{a}\bar{b}|\right)/\sqrt{2}$
Doubly excited configurations		
$^{1}\psi_0$ AB		$|a\bar{a}b\bar{b}|$
$^{1}\psi_5$ A**B		$|a'\bar{a}'b\bar{b}|$
$^{1}\psi_6$ A**B**		$|a'\bar{a}'b'\bar{b}'|$
$^{1}\psi_7$ A*B*		$\left(|a\bar{a}'b\bar{b}'| + |a'\bar{a}b'\bar{b}| + |a\bar{a}'b'\bar{b}| + |a'\bar{a}b\bar{b}'|\right)/2$
$^{1}\psi_8$ AB**		$|a\bar{a}b'\bar{b}'|$
$^{1,3}\psi_{2'}$ A^{+*}B$^-$		$\left(|a'\bar{b}'b\bar{b}| + |b'\bar{a}'b\bar{b}|\right)/\sqrt{2}$
$^{1,3}\psi_{3'}$ A$^-$B^{+*}		$\left(|a'\bar{b}'a\bar{a}| + |b'\bar{a}a\bar{a}|\right)/\sqrt{2}$

Table 6.2 (*contd.*)

Configuration	Four-electron, four-orbital picture	Wavefunction
Correlated configurations		
$^{1,3}\psi_{1^*}$ A^*B^{**}		$\left(|ab'\bar{a}'\bar{b}'| \pm |a'b'\bar{a}\bar{b}'|\right)/\sqrt{2}$
$^{1,3}\psi_{2^*}$ $A^{+*}B^{-*}$		$\left(|a'b'\bar{b}'\bar{b}| \pm |b'b\bar{a}'\bar{b}'|\right)/\sqrt{2}$
$^{1,3}\psi_{3^*}$ $A^{-*}B^{++}$		$\left(|b'a'\bar{a}'\bar{a}| \pm |a'a\bar{b}'\bar{a}'|\right)/\sqrt{2}$
$^{1,3}\psi_{4^*}$ $A^{**}B^*$		$\left(|a'b\bar{a}'\bar{b}'| \pm |a'b'\bar{a}'\bar{b}|\right)/\sqrt{2}$

doubly excited configurations are required to treat properly the energy transfer involving a carotenoid S_1 state [59]. The correlated configurations are used to add consideration of electron correlation (double substitutions) to the basic treatment [55].

In other words, we (i) define our active space of HOMO and LUMO orbitals (i.e. those involved in the electronic transitions) for each of A and B (a, a', ... and b, b', ...); (ii) formulate appropriate configuration wavefunctions for the problem being considered (cf. Table 6.2); and (iii) determine reactant and product wavefunctions as linear combinations of configuration wavefunctions.

Matrix elements are nonzero only when a spin-conserving operator connects states of the same *composite* spin multiplicity. We explicitly project out wavefunctions of certain spin multiplicity from our spin orbital product wavefunctions. (It is not necessary to do this as long as the spin wavefunctions are retained throughout the calculation.) We write the wavefunctions pertaining to, for example, configuration ψ_1 by application of the "antisymmetrizer" $A = (N!)^{-1}\sum_P \zeta_P P$ (where P denotes an orbital permutation and ζ_P accounts for the associated parity: $+1$ for an even number of interchanges, -1 for an odd number) to a single spin orbital product to obtain the antisymmetrized Slater determinants [38],

$$^{1,3}\psi_1 = \frac{\left(|ab\bar{a}'\bar{b}| \pm |a'b\bar{a}\bar{b}|\right)}{\sqrt{2\left(1 - S_{ab}^2 - S_{a'b}^2 + S_{ab}^2 S_{a'b}^2 - 2S_{ab}S_{a'b}\right)}}$$
$$\approx \left(|ab\bar{a}'\bar{b}| \pm |a'b\bar{a}\bar{b}|\right)/\sqrt{2}, \tag{6.1}$$

where the spin of the electron associated with each orbital is α or β when a bar is absent or present, respectively, above the orbital symbol, and the normalization constant is obtained using $N_i^2 \langle \psi_i | \psi_i \rangle = 1$. The spin wavefunctions are constructed from one-electron spin eigenfunctions (corresponding here to $^{1,3}A^*$ and ^1B) and we have used the singlet and triplet spin states corresponding to the spin eigenfunctions $S = M = 0$, $2^{-1/2}(\alpha\beta - \beta\alpha)$, and $S = 1$, $M = 0$, $2^{-1/2}(\alpha\beta + \beta\alpha)$, respectively [46], as implied by the superscripts 1 and 3.

6.2.1   Relationship to canonical MO configurations

Canonical molecular orbitals (CMOs) may be related to the LMOs discussed above. It is always possible to transform canonical MOs to localized MOs because expectation values and the Fock operator are invariant to a unitary transformation of the orbital basis set. A general practical method may be to use the natural bond orbital localization method [48]. Hence, the LMO description discussed above can be shown to be isomorphic to a CIS MO calculation. The advantages of a full MO treatment is that it is easily undertaken, using a standard *ab initio* package. Since each of the a and b is orthogonal to a' and b', the CMOs for the dimer are approximately as follows:

$$\varphi_1 = (a+b)/\sqrt{2+2S_{ab}}, \qquad \varphi_2 = (a-b)/\sqrt{2-2S_{ab}}, \qquad (6.2a, b)$$

$$\varphi_3 = (a'+b')/\sqrt{2+2S_{a'b'}}, \qquad \varphi_4 = (a'-b')/\sqrt{2-2S_{a'b'}}, \qquad (6.2c, d)$$

where S_{ab} is the overlap integral between LMOs a and b. These MOs may be used to construct the CMO configurations of Eqns 6.3, as discussed in [59]. Using identities of the type $|\varphi_1\varphi_2\ldots| = -|ab\ldots|/\sqrt{1-S_{ab}^2}$, the Φ_i may be expressed as linear combinations of the ψ_i:

$$^{1,3}\Phi_1 = (|\varphi_1\bar{\varphi}_1\varphi_2\bar{\varphi}_3| \pm |\varphi_1\bar{\varphi}_1\varphi_3\bar{\varphi}_2|)/\sqrt{2}$$
$$= K_1(\psi_1 + \psi_2 - \psi_3 - \psi_4), \qquad (6.3a)$$

$$^{1,3}\Phi_2 = (|\varphi_1\bar{\varphi}_1\varphi_2\bar{\varphi}_4| \pm |\varphi_1\bar{\varphi}_1\varphi_4\bar{\varphi}_2|)/\sqrt{2}$$
$$= K_2(\psi_1 - \psi_2 - \psi_3 + \psi_4), \qquad (6.3b)$$

$$^{1,3}\Phi_3 = (|\varphi_2\bar{\varphi}_2\varphi_1\bar{\varphi}_3| \pm |\varphi_2\bar{\varphi}_2\varphi_3\bar{\varphi}_1|)/\sqrt{2}$$
$$= K_3(\psi_1 + \psi_2 + \psi_3 - \psi_4), \qquad (6.3c)$$

$$^{1,3}\Phi_4 = (|\varphi_2\bar{\varphi}_2\varphi_1\bar{\varphi}_4| \pm |\varphi_2\bar{\varphi}_2\varphi_4\bar{\varphi}_1|)/\sqrt{2}$$
$$= K_4(\psi_1 - \psi_2 + \psi_3 - \psi_4), \qquad (6.3d)$$

where $K_1^{-1} = -2\sqrt{(1 + S_{ab})(1 + S_{a'b'})}$, $K_2^{-1} = -2\sqrt{(1 + S_{ab})(1 - S_{a'b'})}$, $K_3^{-1} = -2\sqrt{(1 - S_{ab})(1 + S_{a'b'})}$ and $K_4^{-1} = -2\sqrt{(1 - S_{ab})(1 - S_{a'b'})}$. Hence, we adduce Eqns 2.4, which represent the LMO configuration wavefunctions employed in the elucidation of T_{RP} in terms of the canonical molecular orbitals of the dimer:

$$\psi_1 = (\Phi_1/K_1 + \Phi_2/K_2 + \Phi_3/K_3 + \Phi_4/K_4)/4, \tag{6.4a}$$

$$\psi_2 = (\Phi_1/K_1 - \Phi_2/K_2 + \Phi_3/K_3 - \Phi_4/K_4)/4, \tag{6.4b}$$

$$\psi_3 = (-\Phi_1/K_1 - \Phi_2/K_2 + \Phi_3/K_3 + \Phi_4/K_4)/4, \tag{6.4c}$$

$$\psi_4 = (-\Phi_1/K_1 + \Phi_2/K_2 + \Phi_3/K_3 - \Phi_4/K_4)/4. \tag{6.4d}$$

6.3   THE ORIGIN OF COUPLING MATRIX ELEMENTS

In this section the electronic coupling matrix element that promotes the transition is defined. This amounts to a separation of the Hamiltonian H into a part which is unperturbed by the transition and a perturbation which instigates the energy transfer: $H = H_0 + V$. This may be achieved by appropriately defining reactant and product wavefunctions, Ψ_R and Ψ_P, such that $T_{RP} \equiv \langle \Psi_P | V | \Psi_R \rangle$, where $V = H - H_{11}$ is the coupling operator, then solving the secular equations,

$$\sum_j (T_{ij} - \varepsilon S_{ij})C_j = 0, \qquad \mathbf{C}_i^T(\mathbf{H}_{ij} - \varepsilon \mathbf{S}_{ij})\mathbf{C}_j = 0, \tag{6.5a, b}$$

with

$$\Psi = \sum_j C_j \psi_j \tag{6.6}$$

and with $T_{ij} = H_{ij} - S_{ij}H_{11}$ and $\varepsilon = E - H_{11}$, where $H_{ij} \equiv \langle \psi_j | H | \psi_i \rangle$ and $S_{ij} \equiv \langle \psi_j | \psi_i \rangle$ are the Hamiltonian and overlap matrix elements between reactant and product configurations or states. In Eqn 6.6b, the $\mathbf{C}_i$ are vector matrices of the C_i coefficients and T indicates a matrix transpose.

In the theory developed by Harcourt *et al.* [25, 58], the reactant and product wavefunctions of Eqns 6.7 below were considered. These provide the essential overlap-dependent contributions to the coupling (primarily via the configurations ψ_2 and ψ_3). An approach based on these reactant and product states is directly comparable to the CI-singles method involving CMOs:

$$\Psi_R = C_1\psi_1 + C_2\psi_2 + C_3\psi_3$$
$$= N_R(\psi_1 + \lambda_R\psi_2 + \mu_R\psi_3),$$

$$\Psi_P = D_4\psi_4 = D_2\psi_2 + D_3\psi_3$$
$$= N_P(\psi_4 + \mu_p\psi_2 + \lambda_p\psi_3). \tag{6.7}$$

The coefficients here relate to the λ and μ defined earlier ($\lambda_R \equiv C_2/C_1$, $\mu_R \equiv C_3/C_1$, etc.) and N_R and N_P are normalizing coefficients. For small overlap between ψ_i and ψ_j, the λ and μ are given approximately by $-T_{12}/A_{12}$ and $-T_{13}/A_{13}$ respectively (from second-order perturbation theory). Once the Hamiltonian matrix elements and configuration overlap integrals have been evaluated, the mixing coefficients are obtained by solving the secular determinant of Eqn 6.5 for the reactant and product wavefunctions (cf. Eqn 6.5b, matrix notation), i.e. for each of the Ψ_R and Ψ_P, where the locally excited final and initial states, respectively, are omitted from the secular equations. Note that the configurations being mixed are *not* orthogonal. The general method is similar to a valence bond approach [23, 38].

The Hamiltonian and overlap matrix elements required for the Eqns 6.5 may be obtained explicitly from the configuration wavefunctions of Table 6.2 by writing

$$
\begin{aligned}
^{1,3}H_{pq} &= \langle {}^{1,3}\psi_p|H|{}^{1,3}\psi_q\rangle \\
&= \langle 2^{-1/2}(|ij\bar{k}l| \pm |kl\bar{i}j|)|H|2^{-1/2}(|uv\bar{w}\bar{x}| \pm |wx\bar{u}\bar{v}|)\rangle \\
&= \langle ij\bar{k}l|H|(uv\bar{w}\bar{x} - vu\bar{w}\bar{x} - uv\bar{x}\bar{w} + vu\bar{x}\bar{w}) \\
&\quad \pm (wx\bar{u}\bar{v} - xw\bar{u}\bar{v} - wx\bar{v}\bar{u} + xw\bar{v}\bar{u})\rangle,
\end{aligned}
\tag{6.8}
$$

where $H = 1$ for an overlap matrix element or $H = \sum_\mu h(\mu) + \sum_{\mu<\nu} r_{\mu\nu}^{-1}$ for a Hamiltonian matrix element (in atomic units), with $h(\mu) \equiv (1/2)\nabla^2(\mu) - \sum_n Z_n/r_{n\mu}$ denoting the usual one-electron Hamiltonian operator, for electrons μ and ν and atomic centres n with charge Z_n. Hence we obtain

$$
\begin{aligned}
^{1,3}H_{pq} &= h_{iu}S_{jv}S_{hw}S_{lx} + S_{iu}h_{jv}S_{hw}S_{lx} + S_{iu}S_{jv}h_{hw}S_{lx} + S_{iu}S_{jv}S_{hw}h_{lx} \\
&\quad + (iu|jv)S_{kw}S_{lx} + (iu|kw)S_{jv}S_{lx} + (iu|lx)S_{jv}S_{hw} \\
&\quad + (jv|kw)S_{ij}S_{lx} + (jv|lx)S_{iu}S_{hw} + (kw|lx)S_{iu}S_{jv} - \ldots \pm \ldots,
\end{aligned}
\tag{6.9}
$$

$$
^{1,3}S_{pq} = S_{iu}S_{jv}S_{hw}S_{lx} - S_{iv}S_{jju}S_{hw}S_{lx} - S_{iu}S_{jv}S_{kx}S_{lw} + S_{iv}S_{ju}S_{kw}S_{lx} \pm \ldots,
\tag{6.10}
$$

where orbital overlap integrals and one-electron integrals are denoted by S and h respectively, and two-electron integrals are written using the notation $(iu|jv) \equiv \langle i(\mu)j(\nu)|r_{\mu\nu}^{-1}|u(\mu)v(\nu)\rangle$. In the LMO picture, some simplification is possible using the orthogonality between orbitals on each molecule such that $S_{aa'} = S_{bb'} = 0$, and for normalized wavefunctions $S_{aa} = S_{a'a'} = S_{bb} = S_{b'b'} = 1$.

The one-electron integrals considered here involve explicitly only active space orbitals (i.e. the *a, b,* etc.). Implicitly, however, they account for the partial screening which the core orbitals effect upon the one-electron integrals involving active-space orbitals. If we consider the set of doubly occupied core orbitals c_n and the active space orbitals i and j, then the one-electron integrals which are discussed in the present work are represented as

$$h_{ij} \approx h_{ij}^{\text{unscreened}} + \sum_n 2(ij|c_n c_n) - (ic_n|c_n j). \tag{6.11}$$

In this manner, it can be shown that h_{ij} may be obtained without explicit consideration of the core orbitals by defining an effective core charge for each atomic centre. In other words, inner-shell electrons are treated as part of the core, whose charge is equal to the nuclear charge minus the number of inner-shell electrons. A further approximation is to include only some of the valence electrons in the Slater determinants, assigning the rest to the core (Hamilton's approximation) [24].

The Mulliken approximation is a simple but reasonably accurate procedure for simplifying three- and four-centre integrals when necessary:

$$(iu|jv) \approx \tfrac{1}{4} S_{iu} S_{jv} [(ii|jj) + (ii|vv) + (uu|jj) + (uu|vv)]. \tag{6.12}$$

A more drastic approach (which is implemented in some semi-empirical MO schemes) involves the zero-differential overlap approximation wherein it is assumed that $\langle i|u \rangle = \delta_{iu}$, whence

$$(iu|jv) \approx \delta_{iu} \delta_{jv} (ii|jj). \tag{6.13}$$

Note that using either approximation, the coulombic (Förster-type) interaction is found to be zero in the four-electron–four-orbital model.

Consideration of the secular Eqns 6.5 leads us to define the coupling matrix elements between configurations as

$$T_{pq} = H_{pq} - S_{pq} H_{11}. \tag{6.14}$$

These more transparent coupling expressions are better suited for interpretation and for use in approximate expressions. Moreover, they provide quantitative couplings which reflect the partitioning of the total Hamiltonian (H) into an unperturbed contribution (H_0) and a perturbation which induces the transition (V). If there were no interaction between A and B, then the system would be described by ψ_1 (i.e. $A^* \ldots B$); V extracts the interaction energy due to the coupling of ψ_1 with other configurations. This is important, because H_{pq} includes large contributions from the core (spectator) orbitals which remain unchanged between configurations ψ_p and ψ_q. This is illustrated in Table 6.3, where it is seen clearly that the H_{ij} are core-dependent, whereas the T_{ij} are not.

Table 6.3 Comparison of the Hamiltonian, overlap and coupling matrix elements obtained from model LMO calculations of an ethene dimer. The difference between the overlap-dependent couplings is due to the approximate treatment of nuclear screening employed in the four-electron calculations. The ethene molecules are arranged in a sandwich configuration with a separation of 4 Å (see [58])

	Four-electron calculation	All-electron calculation
$H_{11}(= H_{44})$	26.957	−6913.90
$H_{12}(= H_{34})$	−0.0897	−33.011
$H_{13}(= H_{24})$	0.0509	66.200
H_{14}	0.251	0.564
S_{12}	0.00488	0.00476
S_{13}	−0.00979	−0.00955
S_{14}	-4.77×10^{-5}	-4.55×10^{-5}
T_{12}	−0.221	−0.111
T_{13}	0.315	0.142
T_{14}	0.252	0.250

6.4 PARADIGMATIC RESULTS

In this section the form of the coupling is examined in detail for π–π^*-type singlet–singlet and triplet–triplet energy transfer. The primary contributions to the coupling are ascertained and their relative magnitudes and orientation/distance-dependencies are discussed.

We begin by providing general expressions for the T_{ij} needed for Eqns 6.5. These expressions enable us to interpret the primary mechanism for energy transfer, and to elicit how the coupling may be estimated. The T_{12}, T_{13}, and T_{14} matrix elements are given by Eqns 13a, 14a, and 15a of [60]. They may be formulated as follows [55]:

$$^{1,3}T_{12}[A^*B \to A^+B^-] = B' + Q' \pm \left\{ -S_{ab}J - S_{a'b'}J_0^{(A)} \right\},$$

$$^{1,3}T_{34}[A^-B^+ \to AB^*] = B' + \bar{Q}' \pm \left\{ -S_{ab}J - S_{a'b'}J_0^{(A)} \right\}, \tag{6.15a}$$

$$^{1,3}T_{24}[A^+B^- \to AB^*] = -B - \bar{Q},$$

$$^{1,3}T_{13}[A^*B \to A^-B^+] = -B - Q, \tag{6.15b}$$

$$^{1,3}T_{14}[A^*B \to AB^*] = J - S_{ab}B' - S_{a'b'}B - Q'' \pm \left\{ J + S_{ab}S_{a'b'}J_0^{(A)} \right\}, \tag{6.15c}$$

in which some of the integrals have been grouped into the penetration (B and B') terms of Eqn 6.16 or the Q, Q', and Q'' terms of Eqn 6.17 below. The (small) ionic integrals identified in [55, 58] have been omitted. We have retained terms up to second order in overlap. Note that Q'' contains the Dexter "exchange integral," $Z = (a'b'|ab)$ and that the "barred" symbols ($\bar{Q}$ and $\bar{Q}'$) are included so as to account for nonidentical donor and acceptor chromophores. The notation suggested here is useful for preserving the complete electron and hole transfer terms, B and B', rather than reducing them with the Mulliken approximation to the bond integrals β_{ab} and $\beta_{a'b'}$ respectively. This may be important, since it is these terms which dominate the overlap-dependent part of the coupling,

$$B = h_{ab} + (ab|bb) + (ab|aa) - S_{ab}\{h_{bb} + (bb|bb) + (aa|bb)\}$$
$$\approx \beta_{ab},$$
$$B' = h_{a'b'} + (a'b'|bb) + (a'b'|aa) - S_{a'b'}\{h_{a'a'} + (a'a'|aa) + (a'a'|bb)\}$$
$$\approx \beta_{a'b'}, \tag{6.16}$$

$$Q = (a'a'|ab) - S_{ab}(a'a'|bb)$$
$$\approx \frac{1}{2}S_{ab}^3 A^{(A)},$$
$$\bar{Q} = (ab|b'b') - S_{ab}\{(aa|b'b') - 2(a'a'|bb) + \bar{D}\},$$

$$Q' = (a'b'|bb) - S_{a'b'}(a'a'|bb),$$
$$\approx \frac{1}{2}S_{a'b'}{}^3 A^{(B)},$$
$$\bar{Q}' = (aa|a'b') - S_{a'b'}\{(a'a'|bb) + \bar{D}'\},$$

$$Q'' = (a'b'|ab) - S_{ab}S_{a'b'}(a'a'|bb),$$
$$\approx \frac{1}{4}S_{ab}S_{a'b'}\{{}^3A^{(A)} + {}^3 A^{(B)}\}. \tag{6.17}$$

In Eqn 6.17, ${}^3A^{(B)} = {}^3H_{22} - {}^3H_{11}$ for molecule B (i.e. the energy gap between the charge transfer state and the localized triplet state), and we have used the abbreviated notations

$$\bar{D} = EX(B) - IP(B) - EX(A) + IP(A)$$
$$\approx h_{b'b'} + (bb|b'b') - h_{a'a'} - (aa|a'a') \pm \left(J_0^{(B)} - J_0^{(A)}\right),$$
$$\bar{D}' = IP(B) - IP(A)$$
$$\approx h_{aa} - (aa|aa) - h_{bb} + (bb|bb),$$

where EX and IP denote "excitation energy" and "ionization potential" respectively.

The simplifications indicated in Eqns 6.16 and 6.17 arise through an analysis of two-electron integrals using the Mulliken approximation. The β_{ab} and $\beta_{a'b'}$

are the bond integrals $\beta_{ij} \equiv h_{ij} - S_{ij}h_{ii}$, and the superscript on the energy gaps signifies that they refer to either molecule A or B, as indicated. It is noted that the penetration terms are approximately proportional to the constructive interference energies, $\beta_{ab}/(1 + S_{ab})$ and $\beta_{a'b'}/(1 + S_{a'b'})$, which arise when degenerate donor and acceptor orbitals overlap to form bonding (dimer) MOs. The constructive interference energies also correspond to resonance stabilization energies for one-electron transfer processes.

The coupling may then be examined, as follows:

$$T_{RP} = N_R N_P (T_{14} + \lambda_R T_{24} + \mu_R T_{34} + \lambda_P T_{12} + \mu_P T_{13} + \lambda_R \mu_P T_{22} + \mu_R \lambda_P T_{33})$$
$$\approx T_{14} - T_{12}T_{24}/A_{12} - T_{13}T_{34}/A_{13}.$$

$$(6.18)$$

The overlap-dependence of the couplings provides useful clues as to the relative magnitudes of each contribution. The configuration overlaps are collected in Table 6.4. It is evident that the overlap-dependence of Eqn 6.18 therefore goes as $S_{ab}S_{a'b'}$. A number of approximations to the couplings which comprise Eqn 6.18 have been investigated, revealing that the simplest expression for the coupling based on Eqn 6.18 is that which would be obtained in the zero differential overlap approximation:

$$^1T_{RP} \approx 2J + 2\beta_{ab}\beta_{a'b'}/^1A, \quad ^3T_{RP} \approx 2\beta_{ab}\beta_{a'b'}/^3A. \qquad (6.19)$$

The distance-dependence for various terms which contribute to Eqns 6.18 and 6.19 are illustrated in Fig. 6.2 for a model system. These calculations were undertaken for a model ethene sandwich dimer using the LMO method with an STO-6G basis set (and hence with the orbital overlap-dependence considerably underestimated) as described in [58]; T^{Coul} was obtained by taking a simple correlation correction to $2J$ ($T^{Coul} \approx 0.65 \times 2J$) as described in [55]. It is seen that Eqn 6.19 is a good approximation to the orbital-overlap-dependent contribution to the coupling (i.e. T^{short}) and that the Dexter coupling, $-Z$, is very small in comparison.

Table 6.4 Overlap integrals between configurations 1 to 4 of Table 2

	1	2	3	4
1	1	$-S_{a'b'}$	S_{ab}	$-S_{ab}S_{a'b'}$
2		1	$-S_{ab}S_{a'b'}$	S_{ab}
3			1	$-S_{a'b'}$
4				1

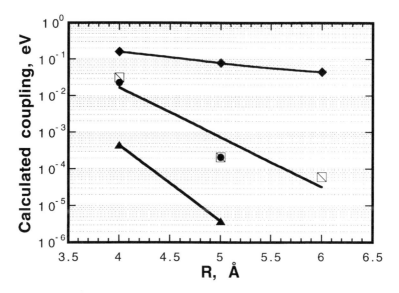

Figure 6.2 Contributions to the coupling between ethene molecules (sandwich geometry) at various separations from the LMO/STO-6G method [60]. ◆, T^{Coul}; ◻, T^{short}; ●, T^{short} from Eqn 6.19; ▲ Z (the Dexter exchange integral)

However, it is not always the case that the coulombic and the overlap-dependent couplings have the same sign; it is possible that for certain orientations they interfere with each other. In Fig. 6.3, the CIS/STO-3G excitation energies for the lowest two pairs of exciton states of the naphthalene dimer (D_2 symmetry) are plotted. The calculations are the same as those depicted in Fig. 6.1 for the corresponding sandwich (D_{2h} symmetry) dimer and are therefore very approximate, but will not illustrate anything unphysical. An interference between coulombic and overlap-dependent couplings is clearly evident at certain geometries.

For approximate numerical evaluation, the resonance integrals may be approximated by $\beta_{ij} \approx -I_{2p}S_{ij}$, where $I_{2p} = 10.5$ eV for carbon 2p orbitals [56, 66]. The As are the energy differences between the initially excited configuration and the charge transfer configuration, and may sometimes be estimated from experiment. The orbital-overlap integral may be estimated by summing the contributing atomic orbital overlaps $S_{\mu\nu}$, weighted by their MO coefficients c_μ and c_ν (obtained by SCF calculation). That is, $S_{ij} = \sum_{\mu,\nu} c_\mu c_\nu S_{\mu\nu}$, when μ represents an atomic orbital on A and ν on B. The atomic orbital overlaps, in turn, may be estimated by resolving the 2p–2p overlap (contributing to the π orbitals) into σ- and π-type overlaps (using simple trigonometry). Then using the analytic expressions of Mulliken *et al.* [40] with an exponent $\zeta = 1.2$ Bohr^{-1} and $p = \zeta R$ (where R is the orbital separation in Bohr units):

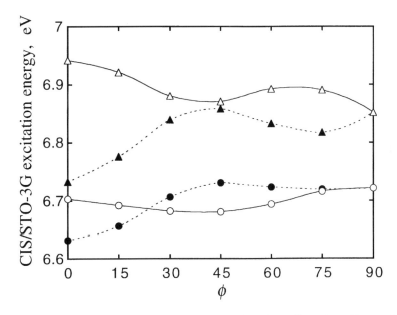

Figure 6.3 Excitation energies for the model naphthalene dimer in a D_2 sandwich geometry, the intermolecular separation (between the planes) fixed at 3.5 Å. The angle ϕ is the angle between the long-axes of the two molecules. The split S_1 state is represented by circles and the S_2 by triangles. Filled symbols correspond to the symmetric (g) states, open symbols denote antisymmetric (u) states. The coupling at any point equals half the splitting. The interference caused by T^{Coul} and T^{short} having opposite signs is evident at some orientations

$$
\begin{aligned}
S(2\mathrm{p}\sigma, 2\mathrm{p}\sigma) &= \mathrm{e}^{-p}\left(-1 - p - \frac{1}{5}p^2 + \frac{2}{15}p^3 + \frac{1}{15}p^4\right), \\
S(2\mathrm{p}\pi, 2\mathrm{p}\pi) &= \mathrm{e}^{-p}\left(1 + p + \frac{2}{5}p^2 + \frac{1}{15}p^3\right).
\end{aligned}
\tag{6.20}
$$

The key conclusions that may be drawn from Eqns 6.19 are as follows.

1. The Dexter exchange integral itself has negligible magnitude and is not present in Eqns 6.19 at all.
2. To treat properly the overlap effects which provide coupling for energy transfer, we need to consider the charge transfer configurations if we are employing a model based on LMOs (i.e. a perturbation-type model as opposed to a supermolecule model).
3. The overlap-dependence of Eqns 6.19 goes as $S_{ab} S_{a'b'}$.
4. $2J$ is the coulombic integral. In a more general treatment (i.e. including configuration interaction) it involves a more complex expression, which we write as T^{Coul} and discuss in the following section.

5. There is an isomorphism between the mechanisms of singlet–singlet and triplet–triplet energy transfer, but the overlap-dependent terms have different magnitudes owing to the difference between ^1A and ^3A. Generally, the singlet–singlet overlap-dependent coupling is larger, because the singlet locally excited state is higher in energy than the corresponding triplet excited state by $2J_0$.

6. $\beta_{a'b'}$ and β_{ab} are the couplings involved in photoinduced electron transfer and photoinduced hole transfer respectively. Hence there exists a firm relationship between the mechanisms of energy and electron transfer. This provides another route for evaluation of these terms.

7. The coulombic and the orbital overlap-dependent couplings do not always have the same sign, and may therefore sometimes interfere destructively. This is illustrated in Fig. 6.3 for a set of model calculations.

6.5   COULOMBIC COUPLING

At donor–acceptor separations beyond their van der Waals radii, the coupling between allowed transitions is described primarily by the coulombic interaction, which takes the form of a two-electron integral, $T^{\text{Coul}} = 2(a'a|bb')$, in the four-electron, four-orbital model. It is an interaction between *transition densities*. Account of electron correlation modifies significantly the explicit form of this interaction; however, in the general theory discussed below such effects are implicitly accommodated by dealing with transition densities for wavefunctions of any accuracy. For electronically allowed emission and absorption transitions, an interaction between *ideal transition dipoles* (i.e. the point-dipole approximation) is generally assumed to be a reasonable approximation. At this level of approximation, the coupling is proportional to $\mu^A \mu^B / R^3$, where μ^N is the dipole transition moment of molecule N and R is the center-to-center separation between donor and acceptor. If one associates a length, l, with each transition moment, $\mu = ql$, then the *ideal dipole approximation* (IDA) is valid when $R \gg l$.

Deviation from the IDA arises because details of the spatial extent of the transition densities (mainly determined by the shape of the molecule's HOMO and LUMO orbitals) become important at close separations. Especially in biological studies, this coupling has been under scrutiny because the sizes of the chromophores involved are not small relative to their separation.

6.5.1   The transition density

The transition density determines the transition probability (dipole strength) between two electronic states and is calculated from the wavefunctions of each state and is defined by

$$M_{eg}(\mathbf{r}; s) = |\Psi_g\rangle\langle\Psi_e| d\mathbf{r}, \tag{6.21}$$

where the spin variables may be integrated out. It may be written as a charge distribution in space (such as a cube-shaped grid), such that the integrated charge vanishes, or $\sum_i q_i = 0$, by

$$\mathbf{M}_{eg}(x, y, z) = \int_z^{z+\delta_z} \int_y^{y+\delta_y} \int_x^{x+\delta_x} \int \Psi_g \Psi_e^* \, ds \, dx \, dy \, dz, \tag{6.22}$$

with s representing the spin variables and the δ_α define the grid size of the cube in direction α.

6.5.2 Multipole moments of the transition density

The spatial extent of the transition density may be described (exactly) by a sum of multipole moments, the following Cartesian tensors:

$$\mu_\alpha = \sum_i q_i (\mathbf{r}_i - \mathbf{R}_N)_\alpha,$$

$$\Theta_{\alpha\beta} = \frac{1}{2} \sum_i q_i (\mathbf{r}_i - \mathbf{R}_N)_\alpha (\mathbf{r}_i - \mathbf{R}_N)_\beta,$$

$$\Omega_{\alpha\beta\gamma} = \frac{1}{6} \sum_i q_i (\mathbf{r}_i - \mathbf{R}_N)_\alpha (\mathbf{r}_i - \mathbf{R}_N)_\beta (\mathbf{r}_i - \mathbf{R}_N)_\gamma,$$

$$\vdots$$

$$Q^{(l)}_{\alpha_1\alpha_2\ldots\alpha_l} = \sum_i \frac{q_i}{l!} (\mathbf{r}_i - \mathbf{R}_N)_{\alpha_1} (\mathbf{r}_i - \mathbf{R}_N)_{\alpha_2} \cdots (\mathbf{r}_i - \mathbf{R}_N)_{\alpha_1}, \tag{6.23}$$

where a sum over Greek indices (which represent the Cartesian direction unit vectors) is implicit: $\mathbf{r}_i$ is the position vector of charge q_i and $\mathbf{R}_N$ specifies the position vector of the centre of molecule N: μ_α is the dipole transition moment vector (tensor of rank 1), and $\Theta_{\alpha\beta}$ and $\Omega_{\alpha\beta\gamma}$ are the quadrupole and octopole transition moment tensors respectively, corresponding to the transition density $\mathbf{M}_{eg}$, and $Q^{(l)}_{\alpha_1\alpha_2\ldots\alpha_l}$ is a multipole moment of rank l. In other words, a set of such multipole moments may be defined for each multipolar transition density.

It is also useful to define here a set of quantities Q_n^m which are related to the multipole moments of Eqn 6.23, defined in terms of the transition density cube and the associated Legendre polynomials, $P_n^m(\cos\theta)$, such that

$$Q_n^m = \sum_i q_i r_i^n P_n^m(\cos\theta_i) \exp(im\phi_i). \tag{6.24}$$

Here n determines the order of the multipole moment (e.g. $n = 1$ are components of the dipole moment, $n = 2$ the quadrupole moment, etc.). The relationship between the Cartesian components and those defined by m is described elsewhere [27, 64].

Hence, for the dipole transition moment (of an E1 transition), we have

$$\mu_\alpha^{eg} = \langle \Psi_e | r_\alpha | \Psi_g \rangle$$
$$\cong \sum_z \sum_y \sum_x r_\alpha M_{eg}(x, y, z), \tag{6.25}$$

where g and e denote the electronic ground and excited states, respectively, and the transition density cube is given by Eqn 6.22.

If, however, the electric dipole transition moment for the transition is zero owing to the symmetries of the states, the leading interactions which promote the transition arise from electric quadrupole (E2) and magnetic dipole (M1) terms (coupling the molecule to the radiation field). This is interpreted as arising from variations in the vector potential (of the "emitted" or "absorbed" photon) over the spatial extent of the molecule. Magnetic dipole and electric quadrupole transitions are generally considered to be 10^{-3} to 10^{-2} times as probable as an electric dipole-allowed transition [15]. The higher multipole transition moments may be determined in an analogous manner to the dipole transition moment, for example for E2 transitions we must consider the transition density associated with $\Theta_{\alpha\beta}^{eg} = \langle \Psi_e | r_\alpha r_\beta | \Psi_g \rangle$, or for M1 transitions $m_\alpha^{eg} = \langle \Psi_e | (\mathbf{r} \times \mathbf{p})_\alpha | \Psi_g \rangle$ [65].

6.5.3   Approximations to the coupling

In terms of the depiction of arbitrary donor and acceptor transition densities in Fig. 6.4, we may write the coulombic coupling between two transition densities as

$$T^{\text{Coul}} = e^2 \int \frac{M_{aa'}(\mathbf{r}_1) M_{b'b}(\mathbf{r}_2)}{r_{12}} \, d\mathbf{r}_1 d\mathbf{r}_2$$
$$\cong \sum_{i,j} q_i q_j / (4\pi\varepsilon_0 r_{ij}). \tag{6.26}$$

The only approximation involved here is in the graining of space over which the summation is carried out (cf. Section 6.8).

To express Eqn 6.26 as an interaction between multipolar contributions to these transition moments one begins by taking a "two-center" expansion of the interaction potential between the transition densities, such that the separation r_{ij} is implicitly accounted for indirectly via the vectors $\mathbf{r}_i, \mathbf{r}_j$, and $\mathbf{R}$,

$$\frac{1}{r_{ij}} = \sum_{n_A=0}^{\infty} \sum_{n_B=0}^{\infty} \sum_{m=-n_<}^{+n_<} B_{n_A,n_B}^{|m|}(r_i, r_j; R) P_{n_A}^m(\cos\theta_i) P_{n_B}^m(\cos\theta_j)\exp(im\varphi), \quad (6.27)$$

where $\varphi = \phi_j - \phi_i$ and $n_<$ indicates the smaller of n_A and n_B. The coefficients $B_{n_A,n_B}^{|m|}$ are complicated functions and have been evaluated by Buehler and Hirschfelder for various cases of overlapping charge [9, 10, 27]. Equation 6.27 is valid for any two interacting charge distributions. When $R > r_i + r_j$, it is found that $B_{n_A,n_B}^{|m|}(r_i, r_j; R) \propto r_i^{n_A} r_j^{n_B}/R^{n_A+n_B+1}$; hence the coupling is given by

$$T^{\text{Coul}} = \sum_{n_A=0}^{\infty} \sum_{n_B=0}^{\infty} \sum_{m=-n_<}^{+n_<} \frac{(-1)^{n_B+m}(n_A+n_B)!}{(n_A+|m|)!(n_B+|m|)!} \frac{r_i^{n_A} r_j^{n_B}}{R^{n_A+n_B+1}} Q_{n_A}^{m*} Q_{n_B}^{m}, \quad (6.28)$$

where the asterisk indicates a complex conjugate.

Similar to expressions that have been developed previously for intermolecular forces [7, 27, 64], if the charge distributions of the donor and acceptor transition densities do not overlap, then we may write

$$\begin{aligned}
T^{\text{Coul}} = \Big[&\Big\{\mu_\alpha^{0n}\mu_\beta^{m0} + m_\alpha^{0n}m_\beta^{m0}\Big\}V_{\alpha\beta} + \Big\{\mu_\alpha^{0n}\Theta_{\beta\gamma}^{m0} - \Theta_{\beta\gamma}^{0n}\mu_\alpha^{m0}\Big\}V_{\alpha\beta\gamma} \\
&+ \Big\{\Theta_{\alpha\gamma}^{0n}\Theta_{\beta\delta}^{m0} + \mu_\alpha^{0n}\Omega_{\beta\gamma\delta}^{0n} + \Omega_{\beta\gamma\delta}^{0n}\mu_\alpha^{m0}\Big\}V_{\alpha\beta\gamma\delta} + \cdots \\
&+ \text{Im}\Big\{\mu_\alpha^{0n}m_\beta^{m0} + m_\beta^{0n}\mu_\alpha^{m0}\Big\}V'_{\alpha\beta} + \text{Im}\Big\{m_\alpha^{0n}\Theta_{\beta\delta}^{m0} - \Theta_{\alpha\delta}^{0n}m_\beta^{m0}\Big\}V'_{\alpha\beta\gamma} \\
&+ \text{Im}\Big\{\mu_\alpha^{0n}m_{\beta\gamma}^{m0} + m_{\beta\gamma}^{0n}\mu_\alpha^{m0}\Big\}V_{\alpha\beta\gamma} + \cdots \Big].
\end{aligned} \quad (6.29)$$

Here μ_α, $\Theta_{\alpha\beta}$, and $\Omega_{\alpha\beta\gamma}$ are components of the electric dipole (E1), quadrupole (E2), and octopole (E3) transition moment tensors, respectively, while m_α is the magnetic dipole (M1) and $m_{\alpha\beta}$ is the magnetic quadrupole (M2) transition moment for transitions $0 \leftarrow n$ (molecule A) or $m \leftarrow 0$ (molecule B), as indicated by superscripts. The V_j are the transition moment coupling tensors discussed below and described in detail in [50], obtained via $V_{i_1 i_2 \ldots i_n} = \nabla_{i_1} \ldots \nabla_{i_{n-2}} (\nabla^2 \delta_{i_{n-1} i_n} - \nabla_{i_{n-1}} \nabla_{i_n})(e^{i\kappa R}/R)$ and $V'_{i_1 i_2 \ldots i_n} = (4\pi\epsilon_0)^{-1}\{\nabla_{i_1}\nabla_{i_2} \ldots \nabla_{i_{n-3}}(\varepsilon_{i_{n-2} i_{n-1} i_n}\nabla_{i_n}) (\kappa e^{i\kappa R}/R)\}$. They provide dipole–dipole, dipole–quadrupole, etc. coupling contributions to the total coupling. Note that those associated with E–E or M–M interactions differ from those for E–M coupling (highlighted by the primes), essentially owing to their different time-reversal signature.

A general quantum-electrodynamical treatment of higher multipole contributions to the coupling between transition moments was derived in [50]. A new formulation was presented wherein expressions for the multipolar coupling tensors were obtained in terms of spherical Bessel functions, providing a clear, compact representation of the retarded coupling interaction and its distance-dependence for multipolar couplings of arbitrary order. The irreducible tensor formulation of the coupling was discussed, highlighting features concerning the exact form of the orientation factors that have often in the past escaped notice, and the detailed method of implementing a rotational averaging of the resultant interaction tensors was demonstrated. For separations significantly less than the wavelength of the transferred photon, the coupling tensors were shown to be given by

$$V_{\alpha\beta} \approx \frac{1}{4\pi\varepsilon_0} \left(\delta_{\alpha\beta} - 3\hat{R}_\alpha\hat{R}_\beta\right)/R^3, \tag{6.30a}$$

$$V_{\alpha\beta\gamma} \approx \frac{1}{4\pi\varepsilon_0} \left[15\hat{R}_\alpha\hat{R}_\beta\hat{R}_\gamma - 3\left(\delta_{\alpha\beta}\hat{R}_\gamma + \delta_{\alpha\gamma}\hat{R}_\beta + \delta_{\beta\gamma}\hat{R}_\alpha\right)\right]/R^4, \tag{6.30b}$$

$$\begin{aligned}V_{\alpha\beta\gamma\delta} \approx \frac{1}{4\pi\varepsilon_0}[&-105\hat{R}_\alpha\hat{R}_\beta\hat{R}_\gamma\hat{R}_\delta - 3(\delta_{\alpha\beta}\delta_{\gamma\delta} + \delta_{\alpha\gamma}\delta_{\beta\delta} + \delta_{\alpha\delta} + \delta_{\beta\gamma}) \\ &+ 15(\delta_{\alpha\beta}\hat{R}_\gamma\hat{R}_\delta + \delta_{\alpha\gamma}\hat{R}_\beta\hat{R}_\delta + \delta_{\alpha\delta}\hat{R}_\beta\hat{R}_\gamma + \delta_{\beta\gamma}\hat{R}_\alpha\hat{R}_\delta + \delta_{\beta\delta}\hat{R}_\alpha\hat{R}_\gamma \\ &+ \delta_{\gamma\delta}\hat{R}_\alpha\hat{R}_\beta)]/R^5,\end{aligned} \tag{6.30c}$$

$$V'_{\alpha\beta} \approx \frac{1}{4\pi\varepsilon_0}\varepsilon_{\alpha\beta\gamma}\hat{R}_\gamma\kappa/R^2, \tag{6.30a}$$

$$V'_{\alpha\beta} \approx \frac{1}{4\pi\varepsilon_0}\varepsilon_{\alpha\beta\delta}\left(\delta_{\gamma\delta} - 3\hat{R}_\gamma\hat{R}_\delta\right)\kappa/R^3, \tag{6.30b}$$

where $\kappa = (E_n - E_0)/\hbar c$, $\delta_{\alpha\beta} \equiv \mathbf{r}_\alpha \cdot \mathbf{r}_\beta$, and $R_\alpha \equiv \mathbf{r}_\alpha \cdot \hat{\mathbf{R}}$. Generally, it is considered a good approximation to truncate this expansion at the dipole–dipole term (except for dipole–forbidden or weakly allowed transitions); hence

$$T^{\text{Coul}} \approx T^{d-d} = \frac{1}{4\pi\varepsilon_0} \frac{\kappa\mu^A\mu^B}{R^3}, \tag{6.31}$$

where here μ^N is the magnitude of the transition moment of molecule N and the orientation factor is $\kappa = \boldsymbol{\mu}^A \cdot \boldsymbol{\mu}^B - 3(\boldsymbol{\mu}^A \cdot \mathbf{R})(\boldsymbol{\mu}^B \cdot \mathbf{R}) = 2\cos\theta_A\cos\theta_B + \sin\theta_A\sin\theta_B\cos\phi$ (see Fig. 6.4).

Model calculations have been used to examine the usefulness of the dipole approximation for large molecules [51, 57]. Some typical results for bacteriochlorophyll *a* are given in Fig. 6.5. The coupling between the ^1L_a transitions of two anthracene molecules, on the other hand, was found to be well described by the T^{d-d} term at separations greater than 5 Å. This is attributed to the combination of strongly allowed, short-axis-polarized transition and high molecular symmetry. A significant separation and orientation dependence was observed for the couplings calculated between bacteriochlorophyl *a* molecules, which suggests that the dipole approximation may either overestimate or underestimate the coupling to a significant extent for many biological

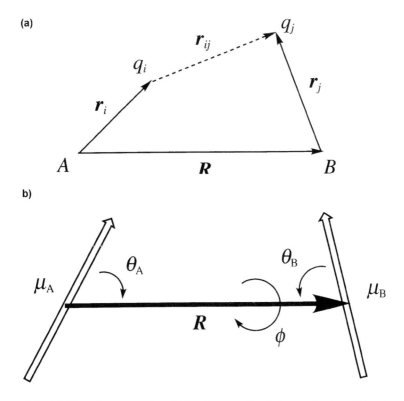

Figure 6.4 (a) Coordinates used to define the coupling between the transition densities of molecules A and B. (b) Angles between the dipole transition moments and interchromophore separation of A and B used in the definition of the orientation factor

arrangements. Indeed, this was found to be the case for couplings in the bacterial light harvesting complex LH2 studied recently [51, 57]. However, the multipole expansion (Eqns 6.30) does not appear to be ideal for accurate calculation of these couplings, since the higher multipole terms become particularly significant at separations where the condition $r_i + r_j < R$ no longer obtains. It is simpler to use Eqn 6.26.

Our studies of electronic couplings in photosynthetic light-harvesting complexes [51, 57] have suggested that a useful method for the estimation of the coulombic interaction is to determine first the transition densities, e.g. $M_{aa'}(\mathbf{r}_1)$, for each chromophore. This may be accomplished by first calculating the electronic excited states using an *ab initio* CI-singles calculation, and then obtaining the transition density between the ground and first excited state and writing this to a matrix, $\mathbf{M}(\mathbf{r})$, with elements q_i (cf. Eqn 6.26). The two matrices are then interacted by coupling elements, q_i and q_j, via $1/r_{ij}$, just as indicated in Eqn 6.26.

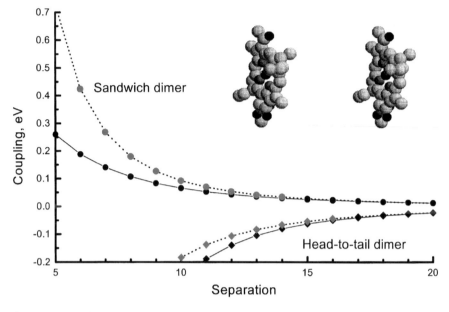

Figure 6.5 Model results showing the comparison between the dipole–dipole coupling
(.....) and the Coulombic coupling determined from Eqn. 6.26 (——) involving the Q_y
transition of bacteriochlorophyll *a* for a sandwich arrangement (●) and a head-to-tail
arrangement (◆) [51, 57]

6.6   SUPEREXCHANGE

The first evidence that a superexchange mechanism could significantly enhance
rates and the range of energy transfer was reported by Robinson and co-workers [14, 19, 63], in studies of mixed aromatic crystals consisting of host doped
with a small concentration of isotopic guest molecules. They postulated that the
most important mechanism for promoting long-range triplet–triplet energy
migration amongst the traps (isotopic guests) at liquid helium temperature
involved indirect trap-to-trap interactions through the virtual states of the host.

The involvement of σ-bond "spacer" groups in promoting intramolecular
triplet–triplet energy transfer was first demonstrated by Closs *et al.* [12, 13]. It is
now well established that such indirect electronic couplings effect long-range
intramolecular electron transfer [29, 44], even in complex biological systems
[6, 33]. Some recent studies reported by Ghiggino and co-workers [60, 61]
suggest that a "through-bond" superexchange mechanism is capable of mediating significant overlap-dependent coupling over reasonable separations (e.g. six
σ bonds), resulting in a substantial discrepancy between the measured rate of
energy transfer and that predicted by the Förster model.

A superexchange interaction mediates the coupling between donor and acceptor via relay chromophores, solvent or connecting bonds. The main consequence, as far as energy transfer is concerned, is that the short-range, orbital-overlap-dependent, interactions may be promoted over large through-space separations. This occurs by the delocalization (or mixing) of the donor and acceptor wavefunctions into the bridge orbitals, such that donor and acceptor wavefunctions may overlap appreciably, even at separations far exceeding their van der Waals radii. In the context of the four-configuration theory presented in this chapter, the eigenvalue problem for a donor + acceptor + m bridge unit system $2^m \times 2^m$ Hamiltonian may be reduced to an eigenvalue problem for the 4×4 Hamiltonian involving the configurations of Eqn 6.7 by using the Löwdin partitioning technique whereby the "pre-mixed" configuration–bridge wavefunctions are interacted,

$$\mathbf{H'} = \begin{pmatrix} \varepsilon_1 & \tilde{V}_{12} & \tilde{V}_{13} & \tilde{V}_{14} \\ \tilde{V}_{21} & \tilde{\varepsilon}_2 & \tilde{V}_{23} & \tilde{V}_{24} \\ \tilde{V}_{31} & \tilde{V}_{32} & \tilde{\varepsilon}_3 & \tilde{V}_{34} \\ \tilde{V}_{41} & \tilde{V}_{42} & \tilde{V}_{43} & \tilde{\varepsilon}_4 \end{pmatrix}. \tag{6.32}$$

The $\tilde{\varepsilon}_i$ are thus the *effective* configuration energies, and the *effective* electronic couplings, as described by Evenson and Karplus [20], are indicated by the tilde notation.

If one employs a tight-binding Hamiltonian and there is a large energy gap between bridge states and donor/acceptor states as well as weak coupling between bridge units, the McConnell formula [36] is obtained for identical donor and acceptor chromophores. Then, for m bridge units, the through-bond mediated electronic coupling is given by

$$\tilde{T}_{\text{RP}} \approx \left(\frac{T^2}{-D}\right)\left(\frac{t}{-D}\right)^{m-1} \tag{6.33}$$

where T is the chromophore–bridge coupling, t is the bridge–bridge coupling, and D is the energy gap between chromophore and bridge. For long bridges, such a model predicts an exponential attenuation of the magnitude of electronic coupling with increasing bridge length. Some simple, approximate, relationships between the distance dependence of (orbital-overlap-dependent) couplings for through-bond electron and energy transfer may thus be derived [11]. Given the McConnell relationship (Eqn 6.33), we may write

$$\tilde{T}_{\text{RP}} \approx T^{\text{Coul}} + T_0 \exp[-\beta(m - m_0)/2], \tag{6.34}$$

where β describes the attenuation of the interaction with distance and T_0 is the interaction with m_0 bridge units. The through-bond attenuation exponents are related by Eqn 6.35 and the pre-exponential factor via Eqn 6.36:

$$\beta^{EET} = \beta^{ET} + \beta^{HT}, \tag{6.35}$$

$$T_0^{EET} \approx \frac{T_0^{ET} T_0^{HT}}{A}. \tag{6.36}$$

The abbreviations ET and HT refer to electron transfer and hole transfer, respectively, and A is the energy gap between the locally excited state and the charge-separated configurations (cf. Eqn 6.19).

Such a relationship between electron transfer and triplet–triplet energy transfer was suggested by the experimental studies of Closs *et al.* [12, 13]. The effect of the through-bond interaction is effectively to increase the range of donor–acceptor orbital overlap; thus, by employing the effective two-state model, the theory for the two-chromophore system is preserved. The exponential attenuation of the overlap-dependent contribution to the coupling does not apply if the mediating chromophores (i.e. bridge units) are close in energy to that of the donor and acceptor. In that case, the superexchange-mediated coupling should often be large, and a more sophisticated description is required [28].

Superexchange-mediated coulombic interactions should generally be significant only if the intermediate molecules (i.e. those promoting the superexchange) have fairly low-lying, allowed singlet–singlet transitions. Because of the closure rule, spectral overlap terms involving these intermediate states do not enter the equation.

6.7   INTERPRETATION OF STEADY STATE SPECTRA

The absorption spectrum of a dimer consisting of two strongly interacting chromophores exhibits a splitting of the monomer bands [16, 17, 30, 35]. This is because the stationary excited (exciton) states of the system consisting of two identical molecules are given as linear combinations of the reactant and product wavefunctions:

$$\Psi_+ = (\Psi_R + \Psi_P)/(2 + 2S_{RP})^{1/2}$$
$$\Psi_- = (\Psi_R - \Psi_P)/(2 - 2S_{RP})^{1/2} \tag{6.37}$$

From consideration of the energies of these states, it follows that half the energy difference between these states (i.e. half the "splitting") is given as

$$(E_- E_+)/2 \equiv \Delta = \left[-H_{RP} + \frac{1}{2}(H_{RR} + H_{PP})S_{RP} \right] / \left[1 - S_{RP}^2 \right] \tag{6.38a}$$

$$\approx -H_{RP} + \tfrac{1}{2}(H_{11} + H_{44})S_{RP}$$
$$= -T_{RP}, \tag{6.39b}$$

where Eqn. 6.39 is accurate to second order in interchromophore orbital overlap (see Eqn 6 of [25]). Similarly, we may consider the sum of these energies, and therefore elucidate the origin of the stabilization of the dimer states relative to the monomer state. It is found that, in the simple four-electron, four-orbital model, this stabilization is due primarily to interaction between the locally excited and charge-transfer states, and is quite significant. These interactions act to red-shift (or blue-shift) the dimer electronic absorption bands and to alter the transition oscillator strength [49].

The results of a recent study [61] emphasize the importance of through-bond interactions in enhancing energy delocalization in rigidly linked bichromophoric systems, and for the first time the efficiency of a through-bond interaction in mediating a strongly allowed molecular exciton interaction was demonstrated. Substantially larger exciton resonance splittings were observed for the naphthalene $^1A_g \rightarrow 2^1B_{3u}$ electronic transition in the series of rigidly linked dinaphthyl molecules, DN-2, DN-4, and DN-6 (DN-n, n being the number of connecting σ bonds), than predicted by the usual electric dipole–dipole interaction model (Fig. 6.6).

Two contrasting cases were encountered. For DN-2, direct through-space orbital overlap is considered to dominate the spectroscopy, although a

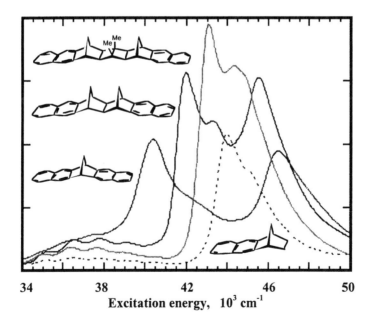

Figure 6.6 Absorption spectra of the DN-n rigidly-linked bichromophores compared to the model chromophore (- - - -), showing the strong splittings

contribution from a through-bond interaction may also be present. For DN-4 and DN-6, however, direct through-space interchromophore overlap is very small. It was concluded that the strong delocalization of electronic excitation through both chromophores is dominated by a through-bond coupling mechanism involving the intervening σ and σ^* orbitals of the bridge, while the through-space coulombic mechanism plays a less significant role.

These results indicate that there is a substantial contribution to the exciton interaction by a through-bond mechanism, although this may not necessarily be the case for lower energy transitions or for less favorable bridge conformations. It is likely that one of the factors acting to increase the chromophore–bridge electronic coupling is the high energy of the electronic transition being studied, which decreases the π^*–σ^* energy gap. This can be understood within the context of the simple McConnell model, from which it can be seen that the effectiveness of a bridge in propagating interactions depends both on the intrabridge coupling matrix elements, t, and on the energy gap, A, separating the chromophore $\pi(\pi^*)$ orbitals with the localized bridge $\sigma(\sigma^*)$ orbitals. As the π^*–σ^* energy gap, A, diminishes, the distance dependence becomes weaker. Through-bond interactions should therefore be more significant for this high-energy electronic transition than for, say, the S_1 transition.

The salient conclusion is that chromophores do not necessarily have to be very close for orbital-overlap-dependent coupling to become significant.

6.8   CALCULATION OF COUPLINGS

The chapter is concluded with a précis of the consequences of the theory presented here for the calculation of couplings for energy transfer and molecular exciton interactions. Generally, MO calculations will underestimate the short-range couplings (because of atomic orbital basis set limitations, especially for large systems) and will overestimate the coulombic interaction (primarily because of neglect of electron correlation). Hence it is desirable to obtain separately each contribution to the coupling. With this in mind, the following protocol is suggested:

1. Choose a set of diabatic configuration wavefunctions appropriate to the problem (Table 6.2).
2. Calculate couplings and overlaps.
3. Calculate transition densities using the same basis set and method (Eqn 6.21 *et seq.*) and use them to determine T^{Coul} (Eqn 6.26) and subtract this from H_{14}, then $T^{\text{short}} = T_{14} - T^{\text{Coul}}$.
4. Scale this coupling to experimentally determined transition moments such that T^{Coul} (scaled) $= T^{\text{Coul}}$ (calc) $\mu^A(\exp)\mu^B(\exp)/\mu^A(\text{calc})\mu^B(\text{calc})$.

5. Put the scaled T^{Coul} back into H_{14}: H_{14} (new) $= H_{14}(\text{calc}) - T^{Coul}(\text{calc}) + T^{Coul}$ (scaled).
6. Solve Eqns 6.5 as described in Section 6.4 to determine the mixing coefficients C_i, and therefore the coupling T_{RP}.

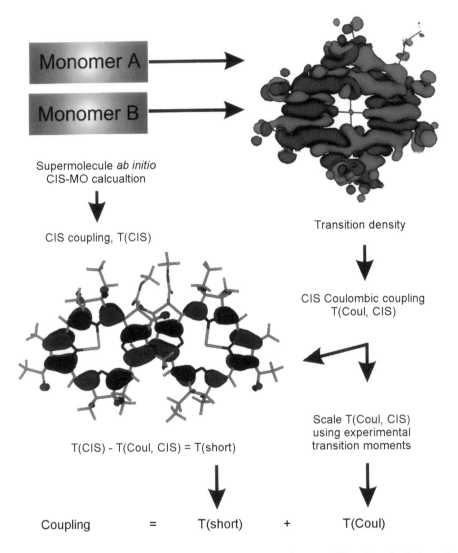

Figure. 6.7 A schematic illustration of the scheme we have used for the calculation of electronic couplings using very large CI-singles MO calculations. The coulombic coupling is calculated by interacting transition densities which are calculated for each monomer. The molecules illustrated here are bacteriochlorophyll *a*

For extremely large calculations, an approach based on diabatic wavefunctions is not practical. We have been able to overcome this by using the CI-singles (CIS) method to calculate excited states of quite large aggregate systems and their monomeric components, and then extract and scale the coulombic coupling *post hoc*. We have used this approach for the study of some photosynthetic light-harvesting antenna systems [51, 57]. Such calculations involve 800–2000 basis functions, and a schematic outline of the approach is given in Fig. 6.7.

Acknowledgments

I gratefully thank the following co-workers with whom the work reported in this chapter has been developed: Professor David Andrews, Professor Graham Fleming, Dr. Ken Ghiggino, Dr. Dick Harcourt and Professor Mike Paddon-Row. In particular, Dr. Dick Harcourt's role in the development of the four-electron, four-orbital theory cannot be overemphasized. I also thank the Ramsay Memorial Fellowship Trust.

References

1. Agranovich, V. M., and M. D. Galanin 1982. *Electronic Excitation Energy Transfer in Condensed Matter*. North-Holland, Amsterdam.
2. Amovilli, C., and R. McWeeny 1995. Molecular interactions: a study of charge transfer effects. *Chem. Phys.* 198: 71–77.
3. Andrews, D. L. 1989. A unified theory of radiative and radiationless molecular energy transfer. *Chem. Phys.* 135: 195–201.
4. Avery, J. 1984. Use of the S-matrix in the relativistic treatment of resonance energy transfer. *Int. J. Quantum Chem.* 25: 79–96.
5. Azumi, T., and S. P. McGlynn 1965. Energy of excimer luminescence. III. Group theoretical considerations of molecular exciton and charge resonance states. *J. Chem. Phys.* 42: 1675–1680.
6. Bertrand, P. 1991. Application of electron transfer theories to biological systems. *Structure and Bonding* 75: 3–47.
7. Buckingham, A. D. 1978. In *Intermolecular Forces – from Diatomics to Biopolymers*. B. Pullman, editor. John Wiley, New York.
8. Buckingham, R. A., and A. Dalgarno 1952. The interaction of normal and metastable helium atoms. *Proc. R. Soc. London* A 213: 327–349.
9. Buehler, R. J., and J. O. Hirschfelder 1951. Bipolar expansion of coulombic potentials. *Phys. Rev.* 83: 628–633.
10. Buehler, R. J., and J. O. Hirschfelder 1952. Errata. *Phys. Rev.* 85: 149–149.
11. Clayton, A. H. A., G. D. Scholes, K. P. Ghiggino, and M. N. Paddon-Row 1996. Through-bond and through-space coupling in photoinduced electron and energy transfer: an *ab initio* and semiempirical study. *J. Phys. Chem.* 100: 10 912– 10 918.
12. Closs, G. L., M. D. Johnson, J. R. Miller, and P. Piotrowiak 1989. A connection between intramolecular long-range electron, hole and triplet energy transfers. *J. Am. Chem. Soc.* 111: 3751–3753.

13. Closs, G. L., P. Piotrowiak, J. M. MacInnis, and G. R. Fleming 1988. Determination of long distance intramolecular triplet energy transfer rates. A quantitative comparison with electron transfer. *J. Am. Chem. Soc.* 110: 2652–2653.
14. Colson, S. D., and G. W. Robinson 1968. Trap–trap triplet energy transfer in isotopic mixed benzene crystals. *J. Chem. Phys.* 48: 2550–2556.
15. Craig, D. P., and T. Thirunamachandran. 1984. *Molecular Quantum Electrodynamics*. Academic Press, New York.
16. Craig, D. P., and S. H. Walmsley 1968. *Excitons in Molecular Crystals*. Benjamin, New York.
17. Davydov, A. S. 1948. Theory of absorption spectra of molecular crystals. *J. Exptl Theor et. Phys. (USSR)* 18: 210–218.
18. Dexter, D. L. 1953. A theory of sensitised luminescence in solids. *J. Chem. Phys.* 21: 836–850.
19. El-Sayed, M. A., M. T. Wauk, and G. W. Robinson 1962. Retardation of singlet and triplet excitation migration in organic crystals by isotopic dilution. *Mol. Phys.* 5: 205–208.
20. Evenson, J. W., and M. Karplus 1992. Effective coupling in bridged electron transfer molecules: computational formulation and examples. *J. Chem. Phys.* 96: 5272–5278.
21. Förster, T. 1948. Intermolecular energy transference and fluorescence. *Ann. Phys.* 2: 55–75.
22. Förster, T. 1965. Delocalized excitation and excitation transfer. In *Modern quantum chemistry*, Vol. III. O. Sinanoglu, editor. Academic Press, New York.
23. Gerratt, J. 1974. Valence bond theory. *Chem. Soc. Specialist Periodical Reports. Theoretical Chemistry* 1: 60–80.
24. Harcourt, R. D., and W. Roso 1978. Valence-bond studies of 4-electron 3-centre bonding units. I. The π-electrons of O_3, NO_2, and CH_2N_2. *Can. J. Chem.* 56: 1093–1101.
25. Harcourt, R. D., G. D. Scholes, and K. P. Ghiggino 1994. Rate expressions for excitation transfer: II. Electronic considerations of direct and through-configuration exciton resonance interactions. *J. Chem. Phys.* 101: 10 521–10 525.
26. Harcourt, R. D., K. P. Ghiggino, G. D. Scholes, and S. Speiser 1996. On the origin of matrix elements for electronic excitation (energy) transfer. *J. Chem. Phys.* 105: 1897–1901.
27. Hirschfelder, J. O., C. F. Curtis, and R. B. Bird 1954. *Molecular Theory of Gases and Liquids*. John Wiley, New York, pp. 26–27, 843–846.
28. Hu, Y., and S. Mukamel 1989. Tunneling versus sequential long-range electron transfer: analogy with pump–probe spectroscopy. *J. Chem. Phys.* 91: 6973–6988.
29. Jordan, K. D., and M. N. Paddon-Row 1992. Analysis of the interactions responsible for long-range through-bond-mediated electronic coupling between remote chromophores attached to rigid polynorbornyl bridges. *Chem. Rev.* 92: 395–410.
30. Kasha, M., H. R. Rawls, and M. A. El-Bayoumi 1966. Excited states of nitrogen base pairs and polynucleotides. *Pure Appl. Chem.* 11: 371–392.
31. Koutecky, J., and J. Paldus 1962. Quantum-chemical study of transannular interaction I. Model of (n,n)paracyclophanes, not considering the benzene ring distortion. *Coll. Czech. Chem. Commun.* 27: 599–617.
32. Koutecky, J., and J. Paldus 1963. A study of the interaction between two ethylene molecules by semi-empirical complete configuration interaction method in the π-electron approximation. *Theor. Chim. Acta (Berl.)* 1: 268–281.
33. Kuki, A. 1991. Electronic tunneling paths in proteins. *Structure and Bonding* 75: 49–81.

34. Lin, S. H. 1971. Isotope effect, energy gap law and temperature effect in resonance energy transfer. *Mol. Phys.* 21: 853–863.
35. McClure, D. S. 1959. *Electronic Spectra of Molecules and Ions in Crystals.* Academic Press, New York.
36. McConnell, H. M. 1961. Intramolecular charge transfer in aromatic free radicals. *J. Chem. Phys.* 35: 508–515.
37. McLone, R. R., and E. A. Power 1964. On the interaction between two identical neutral dipole systems, one in an excited state and the other in the ground state. *Mathematika* 11: 91–94.
38. McWeeny, R. 1992. *Methods of Molecular Quantum Mechanics,* second edition. Academic Press, London.
39. Merrifield, R. E. 1955. Exciton multiplicities. *J. Chem. Phys.* 23: 402–402.
40. Mulliken, R. S., C. A. Rieke, and H. Orloff 1949. Formulas and numerical tables for overlap integrals. *J. Chem. Phys.* 17: 1248–1267.
41. Murrell, J. N., and J. Tanaka 1964. The theory of the electronic spectra of aromatic hydrocarbon dimers. *Mol. Phys.* 7: 363–380.
42. Naqvi, K. R. 1981. Spin selection rules concerning intermolecular energy transfer. Comments on "Energy-transfer studies using doublet-state acceptors." *J. Phys. Chem.* 85: 2303–2304.
43. Naqvi, K. R., and C. Steel 1970. Exchange-induced resonance energy transfer. *Chem. Phys. Lett.* 6: 29–32.
44. Newton, M. D. 1991. Quantum chemical probes of electron transfer kinetics: the nature of donor–acceptor interactions. *Chem. Rev.* 91: 767–792.
45. Pauling, L. 1960. *The Nature of the Chemical Bond,* third edition. Cornell University Press, Ithaca, New York.
46. Paunz, R. 1979. Spin eigenfunctions. Construction and use. Plenum, New York.
47. Pullerits, T., and V. Sundström 1996. Photosynthetic light-harvesting pigment–protein complexes: toward understanding how and why. *Acc. Chem. Res.* 29: 381–389.
48. Reed, A. E., L. A. Curtiss, and F. Weinhold 1988. Intermolecular interactions from a natural bond orbital, donor–acceptor viewpoint. *Chem. Rev.* 88: 899–926.
49. Scholes, G. D. 1996. Energy transfer and spectroscopic characterization of multichromophoric assemblies. *J. Phys. Chem.* 100: 18 731–18 739.
50. Scholes, G. D., and D. L. Andrews 1997. Damping and higher multipole effects in the quantum electrodynamical model for electronic energy transfer in the condensed phase. *J. Chem. Phys.* 107: 5374–5384.
51. Scholes, G. D., and G. R. Fleming 1997. The coupling between transition moments of large molecules. Unpublished results.
52. Scholes, G. D., and K. P. Ghiggino 1994. Electronic interactions and interchromophore excitation transfer. *J. Phys. Chem.* 98: 4580–4590.
53. Scholes, G. D., and K. P. Ghiggino. 1994. Rate expressions for excitation transfer: I. radiationless transition theory perspective. *J. Chem. Phys.* 101: 1251–1261.
54. Scholes, G. D., and K. P. Ghiggino 1996. Electronic interactions and interchromophore energy transfer. In *Advances in Multiphoton Processes and spectroscopy.* S. H. Lin, A. A. Villaeys, and Y. Fujimura, editors. World Scientific, Singapore, Vol. 10, pp. 95–331.
55. Scholes, G. D., and R. D. Harcourt 1996. Configuration interaction and the theory of coulombic interactions in energy transfer and molecular exciton interactions. *J. Chem. Phys.* 104: 5054–5061.
56. Scholes, G. D., K. P. Ghiggino, and G. J. Wilson 1991. Excimer geometries in bichromophoric molecules: a theoretical and experimental study of 1,2-(bis-9-anthryl)-ethane. *Chem. Phys.* 155: 127–141.

57. Krueger, B. P., G. D. Scholes and G. R. Fleming 1998. Calculation of couplings and energy transfer pathways between the pigments of LH2 by the *ab initio* transition density cube method. *J. Phys. Chem.* B102: 5378–5386.
58. Scholes, G. D., R. D. Harcourt, and K. P. Ghiggino 1995. Rate expressions for excitation transfer: III. an *ab initio* study of electronic factors in excitation transfer and excitation resonance interactions. *J. Chem. Phys.* 102: 9574–9581.
59. Scholes, G. D., R. D. Harcourt, and G. R. Fleming 1997. Electronic interactions in photosynthetic light-harvesting complexes: the role of carotenoids. *J. Phys. Chem.* 101: 7302–7312.
60. Scholes, G. D., K. P. Ghiggino, A. M. Oliver, and M. N. Paddon-Row 1993. Intramolecular electronic energy transfer between rigidly linked naphthalene and anthracene chromophores. *J. Phys. Chem.* 97: 11 871–11 876.
61. Scholes, G. D., K. P. Ghiggino, A. M. Oliver, and M. N. Paddon-Row 1993. Through-space and through-bond effects on exciton interactions in rigidly linked dinaphthyl molecules. *J. Am. Chem. Soc.* 115: 4345–4349.
62. Speiser, S. 1996. Photophysics and mechanisms of intramolecular electronic energy transfer in bichromophoric molecular systems: solution and supersonic jet studies. *Chem. Rev.* 96: 1953–1976.
63. Sternlicht, H., G. C. Nieman, and G. W. Robinson 1963. Triplet–triplet annihilation and delayed fluorescence in molecular aggregates. *J. Chem. Phys.* 38: 1326–1335.
64. Stone, A. J. 1991. In *Theoretical Models of Chemical Bonding*. Z. B. Maksic, editor. Springer-Verlag, Berlin, Part 4.
65. Struve, W. S. 1989. *Fundamentals of Molecular Spectroscopy*. John Wiley, New York.
66. Warshel, A., and E. Huler 1974. Theoretical evaluation of potential surfaces, equilibrium geometries and vibronic transition intensities of excimers: the pyrene crystal excimer. *Chem. Phys.* 6: 463–468.
67. Zhang, L. Y., R. A. Friesner, and R. B. Murphy 1997. *Ab initio* quantum chemical calculation of electron transfer matrix elements for large molecules. *J. Chem. Phys.* 107: 450–459.

7

Exciton annihilation in molecular aggregates

Leonas Valkunas, Gediminas Trinkunas, and Vladas Liuolia

Institute of Physics, Lithuania

7.1 INTRODUCTION

Excitation energy transfer from one molecule to another is responsible for a wide range of phenomena observed in such diverse systems as molecular solutions, crystals, films, and polymers [2, 14, 32, 73, 78], as well as in photosynthetic membranes [21]. Experimentally, this transfer manifests itself as an increase in fluorescence quenching, as acceleration of the excitation decay due to impurities or other excitation quenchers present in aggregates, and in the depolarization kinetics [80]. There are a few experimental requirements, however, if the energy-transfer rates are to be determined.

First of all, the system under consideration should be characterized by an intrinsic spatial scale parameter; for example, the mean distance between the quenching centers, or the scaling of the excitation spatial inhomogeneity in transient grating experiments [20]. In systems with impurities, the excitation quenching has to be (close to) migration-limited, while in transient grating experiments the spatial scaling has to be considerably smaller than the physical size of the aggregate under consideration. Moreover, these experiments have to be carried out under low-excitation conditions to avoid the nonlinear processes which are the result of the exciton–exciton interaction. Both demands are not so easily fulfilled. The excitation trapping by impurities is in most cases trap-limited, while the grating experiments are difficult to apply to measurements of the energy migration in a small molecular aggregate.

Resonance Energy Transfer. Edited by David L. Andrews and Andrey A. Demidov.
© 1999 John Wiley & Sons Ltd

Besides impurities and quenchers caused by structural imperfections or defects and by chemically modified molecules present in the system, the molecules in excited singlet (S) or triplet (T) states can also act as mobile quenching centers for other S excitations. Such phenomena, called singlet–singlet (S–S) or singlet–triplet (S–T) annihilation, are dependent on the excitation pulse intensity and may become apparent at high excitation intensities. These nonlinear processes do not manifest themselves experimentally in a trivial exponential decay fashion. Both the amplitude and the rate of such processes are intensity dependent.

Many examples of annihilation nowadays come from studies of photosynthetic pigment–protein complexes and molecular aggregates. This is due to renewed interest in the primary processes of photosynthesis following advances in determining the detailed structure, and the highly developed biotechnology of pigment–protein complex isolation and preparation [6, 42, 49, 53, 60]. Pigment–protein complexes containing aggregates of chlorophylls (Chl) and carotenoids, involved in light energy harvesting and charge separation, show a huge variety in size and, consequently, in the number of pigments within these aggregates, spanning from a few to thousands. It is not surprising that, in order to get a better response in time-resolved spectroscopic studies of pigment–protein complexes, high excitation intensities are applied, and therefore these experiments are susceptible to excitation annihilation processes [25, 40, 96, 97]. For example, in a complex containing just six to eight closely spaced porphyrin chromophores [64], a recent analysis under variable excitation conditions revealed that most of the earlier transient absorption spectroscopy studies were performed under conditions that were far from annihilation-free.

The processes of S–S and S–T annihilation in molecular aggregates are generally migration-limited, and investigations of these processes therefore provide possibilities for the determination of excitation migration parameters. The S–S and S–T annihilation processes are similar to the diffusion-limited chemical reactions $A + A \rightarrow A$ and $A + B \rightarrow B$, respectively, which, however, until recently [44] were studied mainly in infinitely large systems [48, 52, 68]. It is evident that the annihilation processes under consideration have two limiting cases when comparing the excitation diffusion radius with the natural size of the aggregate. For large molecular aggregates, where size is commensurate with the excitation diffusion radius, the nonlinear annihilation is sensitive to the rate of the excitation migration. For such systems, the multiparticle distribution functions can be used to determine the statistics of the excitation relative distribution within an individual aggregate and in an ensemble of aggregates [81, 94]. Further extension to infinitely large aggregates (the approach widely used for molecular crystals) is straightforward from this multiparticle distribution function description [68, 81]. In the opposite case, for small aggregates – when their sizes are much smaller than the excitation diffusion radius – the nonlinear annihilation process is no longer diffusion-limited. Excitations in such systems equilibrate on a very fast timescale, and the annihilation process itself is determined by the static rate of annihilation between two excitations already equilibrated in the system. The main statistical effect then is due to the distribution of the excitations in the ensemble of such aggregates [59, 70]. Both of these approaches will be discussed in this chapter.

It is noteworthy that the excess energy produced by S–S and S–T annihilation is accumulated in the molecular aggregate. The dissipated excitation energy redistributed among the vibrational modes can manifest itself, in some special cases, as local heating of the pigment molecule and/or its immediate surroundings. This can be characterized by a single parameter T, the nonequilibrium temperature. The rise in temperature T evidently leads to significant changes in the electronic transient absorption spectrum. For molecular aggregates, the absorption changes caused by local heating can be comparable to, or even larger than, the absorption changes caused by electronic excitation [33, 34, 86]. Energy redistribution over the multiple vibrational modes, and subsequent thermalization of these modes in the vicinity of the initially excited pigment molecule, have been considered in the context of phase relaxation [29, 54] or energy transfer in solutions [3]. Molecular dynamics simulations of local heating and heat diffusion in proteins [39] reveal fast thermalization kinetics, i.e. the local temperature is established within tens of picoseconds. Similar rates are obtained in other molecular structures [29, 54]. These investigations consider the direct relaxation from highly excited electronic levels of the pigment molecules [39] or the distribution of vibrational modes in solution [3]. Nonlinear relaxation processes, which are pronounced in molecular aggregates, essentially enhance the radiationless relaxation rate at high excitation densities [33, 86, 94]. On the other hand, the slower rate of vibrational energy exchange with the surrounding protein or solution results in energy accumulation within the aggregate on the pigment molecules or in a limited volume of their surrounding. Here we will focus on the local heating effects caused by high excitation intensities, when nonlinear relaxation of excitations via S–S annihilation dominates.

The chapter is organized in three parts: theory, applications, and discussion. The theory part is divided into sections where we present in detail the S–S annihilation process descriptions for large (Section 7.2.1) and small (Section 7.2.2) aggregates, the distinctive features of S–T annihilation (Section 7.2.3); we then briefly describe processes which could be stimulated by local molecular heating or cooling (Section 7.2.4) and the manifestation of these processes in pump–probe absorption spectroscopy (Section 7.2.5). The applications part includes analysis of spectroscopic measurements in photosynthetic molecular aggregates and phthalocyanine films at high excitation intensity conditions. Our main aim is to show how the annihilation process can provide us with energy transfer parameters, as well as with structural information inherent to the molecular aggregates under study.

7.2   THEORY

7.2.1   Large aggregates

The process of excitation annihilation occurring in finite-size molecular aggregates is very dependent on the aggregate dimensions. The critical parameter which enables us to distinguish the annihilation conditions is the excitation diffusion radius:

$$R_{\text{dif}} = \sqrt{Za^2\tau/\tau_h},\tag{7.1}$$

where a is the average intermolecular spacing, Z is the molecular coordination number, and τ and τ_h are the excitation mean lifetime and hopping time, respectively. When R_{dif} does not exceed the characteristic aggregate dimension $L(R_{\text{dif}} < L)$ the aggregate is termed "extended" or "large". For such aggregates, the excitation diffusion time is commensurate with the excitation lifetime and, as will be shown, the annihilation process is then diffusion-limited.

7.2.1.1   *Mathematical formulation of the kinetic equations*

The starting point for our analysis [94] is an aggregate of M pigment molecules located on the sites of a lattice. Each molecule is characterized by a set of singlet states $S_0, S_1, \ldots, S_n$. Here we will consider the high-temperature case and neglect the inhomogeneous distribution of molecular states in the aggregate. The effect of inhomogeneous broadening further complicates the description of the annihilation process, and will be discussed below. Upon single pulse excitation, the molecular transition $S_0 \to S_1$ is determined by an excitation rate $J(t)$ per molecule. The excitation of already excited molecules to a higher excited state occurs at the rate $\alpha J(t)$ per molecule, or the $S_1 \to S_0$ stimulated transition at the rate $\bar{\alpha} J(t)(\alpha = \sigma_1/\sigma_0, \bar{\alpha} = \bar{\sigma}_1/\sigma_0$; where σ_0 and σ_1 are the absorption cross-sections of molecules in the S_0 and S_1 states, respectively, and $\bar{\sigma}_1$ is the cross-section for stimulated emission from S_1). Owing to diffusion, the excitations move on the lattice and upon approaching one another may annihilate according to the scheme shown in Fig. 7.1a.

Due to the process depicted in Fig. 7.1a, or by direct excitation of already excited molecules, at high excitation intensities a considerable population of higher excited states can be created. Assuming that among the higher excited states, S_2 has the longest lifetime, then by a similar annihilation mechanism to that shown in Fig. 7.1a, molecules in the S_2 state may serve as mobile quenchers for S_1 excitations (Fig. 7.1b) and vice versa (Fig. 7.1c).

To describe the ensemble of excitations, we introduce the distribution function $P^M_{N=N_1+N_2}(\mathbf{r}_1, \mathbf{r}_2, \cdots, \mathbf{r}_{N_1}; \mathbf{r}'_1, \mathbf{r}'_2, \cdots \mathbf{r}'_{N_2})$, which represents the probability of

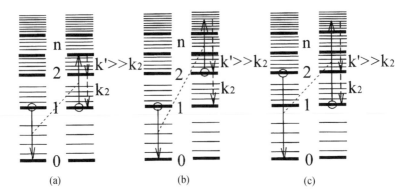

Figure 7.1   Singlet electronic and vibronic levels of aggregate molecules, singlet–singlet annihilation scheme, and excitation relaxation paths involving two sorts of excitations: (a) S_1–S_1 annihilation; (b) and (c) S_1–S_2 annihilation

finding the aggregate of M molecules containing N_1 molecules with the coordinates $\mathbf{r}_i, i = 1, \ldots, N_1$, in the excited state S_1 and N_2 molecules with the coordinates $r_j, j = 1, \ldots, N_2$, in the excited state S_2. The function P_N^M is normalized as follows:

$$\sum_{N=0}^{M}\sum_{N_1=0}^{N}\frac{1}{N_1!\,N_2!}\sum_{\{\mathbf{r}\}^{N_1}\cap\{\mathbf{r}'\}^{N_2}=\varnothing} P_N^M(\{\mathbf{r}\}^{N_1};\{\mathbf{r}'\}^{N_2}) = 1, \tag{7.2}$$

where $\mathbf{r}_1, \mathbf{r}_2, \ldots, \mathbf{r}_N \in N$. For compactness of the presentation, time is omitted in the list of arguments of the function P_N^M. Assuming detailed balance, an expression for the probability distribution is given by

$$\frac{\partial}{\partial t} P_N^M(\{\mathbf{r}\}^{N_1};\{\mathbf{r}'\}^{N_2}) = \left(D_1\sum_{i=1}^{N_1}\nabla_{\mathbf{r}_i}^2 + D_2\sum_{j=1}^{N_2}\nabla_{\mathbf{r}_j'}^2 \right) P_N^M(\{\mathbf{r}\}^{N_1};\{\mathbf{r}'\}^{N_2})$$

$$+ J(t)\left[\sum_{\mathbf{r}_{N_1+1}}\left(\alpha\sum_{i=1}^{N_2} P_N^M(\{\mathbf{r}\}^{N_1+1};\{\mathbf{r}'\}_i^{N_2})\delta(\mathbf{r}_{N_1+1},\mathbf{r}_i') + \bar{\alpha}P_{N+1}^M(\{\mathbf{r}\}^{N_1+1};\{\mathbf{r}'\}^{N_2}) \right) \right.$$

$$+ \sum_{i=1}^{N_1} P_{N-1}^M(\{\mathbf{r}\}_i^{N_1};\{\mathbf{r}'\}^{N_2}) - (1+\alpha+\bar{\alpha})N_1 P_N^M(\{\mathbf{r}\}^{N_1};\{\mathbf{r}'\}^{N_2}) \bigg]$$

$$+ k_1(t)\left(\sum_{\mathbf{r}_{N_1+1}} P_{N+1}^M(\{\mathbf{r}\}^{N_1+1};\{\mathbf{r}'\}^{N_2}) - N_1 P_N^M(\{\mathbf{r}\}^{N_1};\{\mathbf{r}'\}^{N_2}) \right)$$

$$
+ k_2 \left(\sum_{\mathbf{r}'_{N_2+1}} \sum_{i=1}^{N_1} P_N^M(\{\mathbf{r}\}_i^{N_1}; \{\mathbf{r}'\}^{N_2+1}) \delta(\mathbf{r}_i, \mathbf{r}'_{N_2+1}) - N_2 P_N^M(\{\mathbf{r}\}^{N_1}; \{\mathbf{r}'\}^{N_2}) \right)
$$

$$
+ \frac{1}{2} \sum_{\mathbf{r}_{N_1+1}, \mathbf{r}_{N_1+2}} \sum_{i=1}^{N_2} P_{N+1}^M(\{\mathbf{r}\}^{N_1+2}; \{\mathbf{r}'\}_i^{N_2})[\delta(\mathbf{r}_{N_1+1}, \mathbf{r}'_i) + \delta(\mathbf{r}_{N_1+2}, \mathbf{r}'_i)]\lambda(\mathbf{r}_{N_1+1} - \mathbf{r}_{N_1+2})
$$

$$
- 2 \sum_{j>i=1}^{N_1} P_N^M(\{\mathbf{r}\}^{N_1}; \{\mathbf{r}'\}^{N_2})\lambda(\mathbf{r}_i - \mathbf{r}_j) + \sum_{\mathbf{r}_{N_1+1}} \sum_{i=1}^{N_2} P_{N+1}^M(\{\mathbf{r}\}^{N_1+1}; \{\mathbf{r}'\}^{N_2})\mu(\mathbf{r}_{N_1+1} - \mathbf{r}'_i)
$$

$$
- \sum_{i=1}^{N_1} \sum_{j>i}^{N_2} P_N^M(\{\mathbf{r}\}^{N_1}; \{\mathbf{r}'\}^{N_2})\mu(\mathbf{r}_i - \mathbf{r}_j).
$$

$$(7.3)$$

The first term on the right-hand side of Eqn 7.3 determines the increase in the probability due to excitation diffusion. The next term (within square brackets) reflects the generation of excitations. The terms multiplied by rate constants k_1 and k_2 arise from excitation monomolecular decay and trapping for S_1 and S_2 excitations, respectively. The next terms, multiplied by the annihilation probabilities λ and μ, represent the loss of excitations due to annihilation. D_1 and D_2 define the diffusion coefficients of S_1 and S_2 excitations, respectively: D_1 can be connected with the excitation mean hopping time, τ_h, between the nearest pigments, via the relation $D_1 = a^2/\tau_h$, which can be considered as the definition of the latter. Here it is assumed that due to pairwise excitation collision a doubly excited state is produced (see Fig. 7.1), which in a very short time relaxes to the state S_1 resulting in the loss of one excitation in the pair.

To account for the lowest-order excitation correlations, the excitation probability distribution functions have to be averaged as follows:

$$
f_{n=i+j}(\mathbf{r}_1, \cdots, \mathbf{r}_i; \mathbf{r}'_1, \cdots \mathbf{r}'_j)
$$

$$
= \sum_{N=n}^{M} \sum_{N_1=i}^{N-j} \frac{1}{(N_1 - i)!(N_2 - j)!} \sum_{\{\mathbf{r}\}_{1,\dots,i}^{N_1} \cap \{\mathbf{r}'\}_{1,\dots,j}^{N_2} = \emptyset} P_N^M(\{\mathbf{r}\}^{N_1}; \{\mathbf{r}'\}^{N_2}).
\qquad (7.4)
$$

The semicolon in the parenthesis of the function argument separates the coordinates of the molecules in the excited states S_1 (on the left) and S_2 (on the right). The kinetic equations for the n-particle distribution functions f_n are then obtained by a direct summation of Eqn 7.3 using Eqn 7.4. The expressions that one then obtains for the one-particle as well as two-particle excitation distribution functions are presented below:

$$\frac{\partial}{\partial t} f_1(\mathbf{r}_1;) = [D_1 \nabla_r^2 - k_1(t) - (\alpha + \bar{\alpha})J(t)] f_1(\mathbf{r}_1;)$$

$$+ k_2 f_1(;\mathbf{r}_1) + J(t) f_1^0 - 2 \sum_{\mathbf{r}_2} \lambda(\mathbf{r}_1 - \mathbf{r}_2) f_2(\mathbf{r}_1, \mathbf{r}_2;)$$

$$- \sum_{\mathbf{r}_1'} \mu(\mathbf{r}_1 - \mathbf{r}_1') f_2(\mathbf{r}_1;\mathbf{r}_1'), \tag{7.5a}$$

$$\frac{\partial}{\partial t} f_1(;\mathbf{r}_1') = [D_2 \nabla_{\mathbf{r}'1}^2 - k_2] f_1(;\mathbf{r}_1') + \alpha J(t) f_1(\mathbf{r}_1';)$$

$$+ \sum_{\mathbf{r}_1} \lambda(\mathbf{r}_1 - \mathbf{r}_1') f_2(\mathbf{r}_1, \mathbf{r}_1';), \tag{7.5b}$$

$$\frac{\partial}{\partial t} f_2(\mathbf{r}_1, \mathbf{r}_2;) = [D_1(\nabla_{\mathbf{r}_1}^2 + \nabla_{\mathbf{r}_2}^2) - 2(\alpha + \bar{\alpha})J(t) - 2k_1(t) - 2\lambda(\mathbf{r}_1 - \mathbf{r}_2)] f_2(\mathbf{r}_1, \mathbf{r}_2;)$$

$$+ J(t)[f_1(\mathbf{r}_1;) + f_1(\mathbf{r}_2;)] + k_2[f_2(\mathbf{r}_1;\mathbf{r}_2) + f_2(\mathbf{r}_2;\mathbf{r}_1)]$$

$$- 2 \sum_{\mathbf{r}_3} [\lambda(\mathbf{r}_1 - \mathbf{r}_2) + \lambda(\mathbf{r}_2 - \mathbf{r}_3)] f_3(\mathbf{r}_1, \mathbf{r}_2, \mathbf{r}_3;)$$

$$+ \sum_{\mathbf{r}_1'} [\mu(\mathbf{r}_1 - \mathbf{r}_1') + \mu(\mathbf{r}_2 - \mathbf{r}_1')] f_3(\mathbf{r}_1, \mathbf{r}_2;\mathbf{r}_1'), \tag{7.5c}$$

$$\frac{\partial}{\partial t} f_2(\mathbf{r}_1;\mathbf{r}_1') = \left[D_1 \nabla_{\mathbf{r}_1}^2 + D_2 \nabla_{\mathbf{r}_1'}^2 - \mu(\mathbf{r}_1 - \mathbf{r}_1') - k_1(t) - k_2 - (\alpha + \bar{\alpha})J(t) \right]$$

$$f_2(\mathbf{r}_1;\mathbf{r}_1') + J(t)[\alpha f_2(\mathbf{r}_1, \mathbf{r}_1';) + f_1(;\mathbf{r}_1')] + k_2 f_2(;\mathbf{r}_1, \mathbf{r}_1')$$

$$+ \sum_{\mathbf{r}_2} \lambda(\mathbf{r}_1 - \mathbf{r}_2)[f_3(\mathbf{r}_1, \mathbf{r}_2, \mathbf{r}_1';) - 2 f_3(\mathbf{r}_1, \mathbf{r}_2;\mathbf{r}_1')]$$

$$- \sum_{\mathbf{r}_2'} \mu(\mathbf{r}_1 - \mathbf{r}_2') f_3(\mathbf{r}_1;\mathbf{r}_1', \mathbf{r}_2'). \tag{7.5d}$$

The function f_1^0 in Eqn 7.5a determines the probability for a single molecule to be in the ground state S_0, as follows from Eqn 7.3 if the sum over the excited molecules is taken to $M - 1$.

In the case of homogeneous excitation conditions, and taking into account the local character of the excitation trapping, the excitation distribution is assumed to be homogeneous:

$$f_1^0 \equiv n_0, \qquad f_1(\mathbf{r}_1;) \equiv n_1, \qquad f_1(;\mathbf{r}_1') \equiv n_2, \tag{7.6}$$

n_i being the fractional population of molecules in the states S_i ($n_0 + n_1 + n_2 = 1$). By introducing the relative distribution functions

$$g(\mathbf{r}_1 - \mathbf{r}_2) = f_2(\mathbf{r}_1, \mathbf{r}_2;)/n_1^2, \tag{7.7a}$$

$$h(\mathbf{r}_1 - \mathbf{r}_1') = f_2(\mathbf{r}_1; \mathbf{r}_1')/n_1 n_2, \tag{7.7b}$$

$$l(\mathbf{r}_1' - \mathbf{r}_2') = f_2(; \mathbf{r}_1', \mathbf{r}_2')/n_2^2, \tag{7.7c}$$

$$g(\mathbf{r}_1 - \mathbf{r}_2, \mathbf{r}_1 - \mathbf{r}_3) = f_3(\mathbf{r}_1, \mathbf{r}_2, \mathbf{r}_3;)/n_1^3, \tag{7.7d}$$

$$q(\mathbf{r}_1 - \mathbf{r}_2, \mathbf{r}_1 - \mathbf{r}_1') = f_3(\mathbf{r}_1, \mathbf{r}_2; \mathbf{r}_1')/n_1^2 n_2, \tag{7.7e}$$

$$p(\mathbf{r}_1 - \mathbf{r}_1', \mathbf{r}_1 - \mathbf{r}_2') = f_3(\mathbf{r}_1; \mathbf{r}_1', \mathbf{r}_2')/n_1 n_2^2, \tag{7.7f}$$

the following equations for the excitation densities are obtained:

$$\frac{dn_1}{dt} = -[k_1(t) + (1 + \alpha + \bar{\alpha})J(t)]n_1 - 2\gamma(t)n_1^2$$
$$- \beta(t)n_1 n_2 + [k_2 - J(t)]n_2 + J(t), \tag{7.8a}$$

$$\frac{dn_2}{dt} = -k_2 n_2 + \gamma(t)n_1^2 + \alpha n_1 J(t), \tag{7.8b}$$

where $1/k_2$ is the S_2 excitation lifetime, and $k_1(t)$ denotes the generalized rate of (quasi-)linear excitation quenching processes due to the internal conversion, intersystem crossing and the presence of excitation traps (see Eqn 7.34 below). The time-dependence of the latter is included to cover the transient trap saturation phenomena. The parameters $\gamma(t)$ and $\beta(t)$ define the rates of $S_1 + S_1 \rightarrow S_2$ and $S_1 + S_2 \rightarrow S_2$ annihilation processes, respectively (see Eqns 7.5 and 7.7), and are given by

$$\gamma(t) = \sum_{\mathbf{r}} \lambda(\mathbf{r})g(\mathbf{r}, t), \qquad \beta(t) = \sum_{\mathbf{r}} \mu(\mathbf{r})h(\mathbf{r}, t), \tag{7.9}$$

where $g(\mathbf{r}, t)$ and $h(\mathbf{r}, t)$ are two-particle correlation functions, already determined by Eqns 7.7, that describe the relative distribution of a pair of S_1 excitations and a pair of S_1 and S_2 excitations, respectively. The $\lambda(\mathbf{r})$ and $\mu(\mathbf{r})$ define the respective pairwise annihilation rates, which in the case of Förster-type interaction are given by

$$\lambda(\mathbf{r}) = k_f (R_1^0/r)^6; \qquad \mu(\mathbf{r}) = k_f (R_2^0/r)^6, \tag{7.10}$$

where R_1^0 and R_2^0 are the Förster radii of S_1–S_1 and S_1–S_2 annihilation, respectively, and k_f is the S_1 radiative rate. Values for the S_2 state cross-sections σ_2 and $\bar{\sigma}_2$ are not known. Here, we assume that both of these values are smaller than $\sigma_0 \simeq \sigma_1$ and $\bar{\sigma}_1$, respectively, in the spectral region of the ground-state absorption. Taking further into account the fact that the S_2 lifetime is much smaller than that of S_1, the process of S_1–S_1 annihilation will strongly dominate over S_1–S_2 annihilation. To obtain the two-particle correlation functions $g(\mathbf{r}, t)$ and $h(\mathbf{r}, t)$ in Eqns 7.9, the following kinetic equations have to be solved:

$$\frac{\partial g(\mathbf{r}, t)}{\partial t} = 2D_1 \nabla_{\mathbf{r}}^2 g(\mathbf{r}, t) + H(g(\mathbf{r}, t)), \tag{7.11a}$$

$$\frac{\partial h(\mathbf{r}, t)}{\partial t} = (D_1 + D_2) \nabla_{\mathbf{r}}^2 h(\mathbf{r}, t) + F(h(\mathbf{r}, t)), \tag{7.11b}$$

where

$$\begin{aligned}
H(g(\mathbf{r}, t)) = & -2\left[\lambda(\mathbf{r}) + \frac{J(t) + n_2[k_2 - J(t)]}{n_1} - J(t)\right] g(\mathbf{r}, t) \\
& + 2\frac{J(t) + k_2 n_2 h(\mathbf{r}, t)}{n_1} \\
& + 2n_1\left[2\gamma(t) - \sum_{\mathbf{r}'}\left[\lambda(\mathbf{r}') + \lambda(\mathbf{r} - \mathbf{r}')\right]\frac{g(\mathbf{r}, \mathbf{r}', t)}{g(\mathbf{r}, t)}\right] g(\mathbf{r}, t) \\
& + n_2\left[2\beta(t) - \sum_{\mathbf{r}'}\left[\mu(\mathbf{r}') + \mu(\mathbf{r} - \mathbf{r}')\frac{q(\mathbf{r}, \mathbf{r}', t)}{g(\mathbf{r}, t)}\right]\right] g(\mathbf{r}, t),
\end{aligned} \tag{7.12}$$

and

$$\begin{aligned}
F(h(\mathbf{r}, t)) = & -\left[\mu(\mathbf{r}) + \frac{J(t) + n_2(k_2 - J(t))}{n_1} + J(t)\left(\alpha\frac{n_1}{n_2} - 1\right)\right] h(\mathbf{r}, t) \\
& + \frac{J(t) + k_2 n_2 l(\mathbf{r}, t)}{n_1} + \frac{\alpha J(t) n_1 g(\mathbf{r}, t)}{n_2} \\
& - \frac{n_1^2}{n_2}\left[\gamma(t) - \sum_{\mathbf{r}'}\lambda(\mathbf{r} - \mathbf{r}')\frac{g(\mathbf{r}, \mathbf{r}', t)}{h(\mathbf{r}, t)}\right] h(\mathbf{r}, t) \\
& + n_1\left[2\gamma(t) - 2\sum_{\mathbf{r}'}\lambda(\mathbf{r}')\frac{q(\mathbf{r}, \mathbf{r}', t)}{h(\mathbf{r}, t)}\right] h(\mathbf{r}, t) \\
& + n_2\left[\beta(t) - \sum_{\mathbf{r}'}\mu(\mathbf{r}')\frac{p(\mathbf{r}, \mathbf{r}', t)}{h(\mathbf{r}, t)}\right] h(\mathbf{r}, t).
\end{aligned} \tag{7.13}$$

The solution of Eqns 7.11 through Eqns 7.12 and 7.13 implies that we have to know the three-particle correlation functions $g(\mathbf{r}, \mathbf{r}', t)$, $q(\mathbf{r}, \mathbf{r}', t)$ and $p(\mathbf{r}, \mathbf{r}', t)$, where $g(\mathbf{r}, \mathbf{r}', t)$ involves three S_1 excitations, $q(\mathbf{r}, \mathbf{r}', t)$ two S_1 and one S_2 excitations, and $p(\mathbf{r}, \mathbf{r}', t)$ two S_2 and one S_1 excitations; $l(\mathbf{r}, t)$ in Eqn 7.13 defines the distribution function of a pair of S_2 excitations. The solution of Eqns 7.11 in terms of a series of kinetic equations for the multiple particle correlation functions is truncated after Eqns 7.12 and 7.13 and implies that we will approximate the three-particle correlation functions as products of two-particle correlation functions (see below).

The initial conditions are given by

$$n_1(0) = n_2(0) = 0, \qquad g(\mathbf{r}, 0) = h(\mathbf{r}, 0) = l(\mathbf{r}, 0) = 1, \qquad (7.14)$$

and the following boundary conditions apply, which are obtained from the corresponding kinetic equations [24, 94]:

$$\left.\frac{dg(\mathbf{r}, t)}{dt}\right|_{r=R_1} = \left.\frac{2\Omega D_1}{R_1^2}\frac{\partial g(\mathbf{r}, t)}{\partial r}\right|_{r=R_1} + H(g(\mathbf{r}, t))|_{r=R_1}$$

$$\left.\frac{dh(\mathbf{r}, t)}{dt}\right|_{r=R_2} = \left.\frac{2\Omega(D_1 + D_2)}{2R_2^2}\frac{\partial h(\mathbf{r}, t)}{\partial r}\right|_{r=R_2} + F(h(\mathbf{r}, t))|_{r=R_2}, \qquad (7.15)$$

where Ω is the steric angle (for 3-D aggregates, $\Omega = 4\pi$; for 2-D, $\Omega = 2\pi$; for 1-D, $\Omega = 2$), with R_1 and R_2 representing reaction radii to be defined later. These boundary conditions are obtained from Eqns 7.11 by rewriting them for two corresponding excitations separated by a distance equal to the radius of the reaction spheres, i.e. R_1 and R_2 respectively (see Eqns 7.29 for definitions).

7.2.1.2   *Analysis of kinetic equations*

Let us now qualitatively discuss the time evolution of Eqns 7.8. The stationary solution of these equations is given by

$$n_2 = \frac{J + n_1[J(\alpha - \bar{\alpha} - 1) - k_1]}{k_2 + J + \beta n_1}, \qquad (7.16)$$

which is independent of the singlet annihilation rate, $\gamma(t)$. From Eqn 7.16, it follows that at low excitation intensities, where the nonlinearities can be neglected, $n_2 = 0$. At high intensities, where $n_2 \gg n_1$ but assuming that conditions $J \ll k_2; \beta n_1 \ll k_2$ are fulfilled, it follows that

$$n_2 \simeq J/k_2. \qquad (7.17)$$

The stationary solution of Eqn 7.8b then becomes

$$\gamma n_1^2 + J\alpha n_1 - J = 0, \qquad (7.18)$$

and Eqns 7.17 and 7.18 are then valid at the time when the kinetic traces reflecting the populations n_1 and n_2 reach their maximum amplitudes. For a description of the kinetics after termination of the excitation pulse, on a time-scale slower than k_2^{-1} (implying that $dn_2/dt = 0$), it follows from Eqn 7.8b that

$$n_2 = \gamma(t)\frac{n_1^2}{k_2}, \tag{7.19}$$

and according to Eqn 7.8a the excitation decay kinetics follows the simplified equation

$$dn_1/dt = -k_1(t)n_1 - \gamma(t)n_1^2. \tag{7.20}$$

Thus, the effect due to population of the S_2 state is only important during the action of the excitation pulse.

The time-dependence of $\gamma(t)$ and $\beta(t)$ is determined by the correlation functions $g(\mathbf{r}, t)$ and $h(\mathbf{r}, t)$, respectively. The corresponding equations that describe the time evolution of these functions are in general very cumbersome and contain contributions from the three-particle correlation functions; see Eqns 7.12 and 7.13. However, in the following we will show that these terms can be substantially simplified for application to the rising part of the pulse [1, 94]. If the following inequality is satisfied,

$$2n_1\gamma(t) + n_2\beta(t) < \lambda(\mathbf{r}) + \frac{J(t) + n_2[k_2 - J(t)]}{n_1} - J(t), \tag{7.21}$$

then the nonlinear terms and terms containing the three-particle correlation function in Eqn 7.12 can be neglected. By using Eqn 7.8a, the inequality 7.21 can be rewritten as

$$-\frac{1}{n_1}\frac{dn_1}{dt} < \lambda(\mathbf{r}) + k_1(t) + (\alpha + \bar{\alpha})J(t). \tag{7.22}$$

An analogous inequality can be obtained for the correlation function $h(\mathbf{r}, t)$:

$$-\left(\frac{1}{n_1}\frac{dn_1}{dt} + \frac{1}{n_2}\frac{dn_2}{dt}\right) < \mu(\mathbf{r}) + k_1(t) + (\alpha + \bar{\alpha})J(t). \tag{7.23}$$

Now it is evident that, during the initial period when the generation term is dominant, where

$$\frac{1}{n_1}\frac{dn_1}{dt} > 0 \quad \text{and} \quad \frac{1}{n_2}\frac{dn_2}{dt} > 0,$$

the inequalities 7.22 and 7.23 are valid and, as a consequence, Eqns 7.11 can be simplified in the following way:

$$\frac{dg(\mathbf{r}, t)}{dt} = 2D_1\Delta_r^2 g(\mathbf{r}, t) - 2[\lambda \mathbf{r} + 2\frac{J(t)}{n_1}]g(\mathbf{r}, t) + 4\frac{J(t)}{n_1}, \tag{7.24a}$$

$$\frac{dh(\mathbf{r}, t)}{dt} = (D_1 + D_2)\nabla_r^2 h(\mathbf{r}, t) + 2\frac{J(t)}{n_1} + \alpha J(t)\frac{n_1}{n_2}g(\mathbf{r}, t)$$

$$- [\mu(\mathbf{r}) + 2\frac{J(t)}{n_1} + J(t)(\alpha\frac{n_1}{n_2} - 1)]h(\mathbf{r}, t). \tag{7.24b}$$

During the decay of the generation term,

$$\frac{1}{n_1}\frac{dn_1}{dt} \quad \text{and} \quad \frac{1}{n_2}\frac{dn_2}{dt}$$

change their signs, and the inequalities 7.22 and 7.23 may be violated. The physical explanation for the simplification shown in Eqns 7.24 is that during the random generation of the excitations within the domain, correlations are continuously destroyed, while this is not the case when the generation is switched off.

When the excitation pulse is over, the excitation kinetics is determined by Eqn 7.20 and, therefore, we must consider the correlation function $g(\mathbf{r}, t)$, which then satisfies the following equation:

$$\frac{dg(\mathbf{r}, t)}{dt} = 2D_1\Delta_r^2 g(\mathbf{r}, t) - 2\lambda(\mathbf{r})g(\mathbf{r}, t)$$

$$+ 2n_1\left\{\gamma(t) - \sum_{\mathbf{r}'}[\lambda(\mathbf{r}') + \lambda(\mathbf{r} - \mathbf{r}')\frac{g(\mathbf{r}, \mathbf{r}', \mathbf{t})}{g(\mathbf{r}, t)}\right\}g(\mathbf{r}, t). \tag{7.25}$$

The three-particle correlation function $g(\mathbf{r}, \mathbf{r}', t)$ can be estimated by means of the Kirkwood approximation:

$$g(\mathbf{r}, \mathbf{r}', t) = g(\mathbf{r}, t)g(\mathbf{r}', t)g(\mathbf{r} - \mathbf{r}', t). \tag{7.26}$$

After substitution of Eqn 7.26 into Eqn 7.25, and developing $g(\mathbf{r} - \mathbf{r}')$ in a Taylor series at the point $\mathbf{r}$, Eqn 7.25 takes the following form:

$$\frac{dg(\mathbf{r}, t)}{dt} = 2D_1\nabla_r^2 g(\mathbf{r}, t) - 2\lambda(\mathbf{r})g(\mathbf{r}, t) + 2\gamma(t)n_1 g(\mathbf{r}, t)[1 - g(\mathbf{r}, t)]. \tag{7.27}$$

Thus, a comprehensive set of equations is obtained for complete analysis of the excitation decay kinetics under high excitation intensities, both during the generating pulse and the following decay.

7.2.1.3   Nonlinear excitation quenching

For a diffusion-limited annihilation process, i.e. when the excitation diffusion radius from Eqn 7.1 is smaller than the domain size, the annihilation rates of

Eqn 7.9 are determined by diffusion of the excitations towards the so-called "black sphere" of the reaction, defined by the radii R_1 and R_2 for S_1–S_1 and S_1–S_2 annihilation, respectively [68]:

$$\gamma_{\text{dif}}(t) = \pi \frac{D_1}{(2R_1)^{d-1}} \frac{\partial}{\partial r} g(\mathbf{r}, t) \bigg|_{r=R_1}, \tag{7.28a}$$

$$\beta_{\text{dif}}(t) = \frac{\pi}{2} \frac{D_1 + D_2}{(2R_2)^{d-1}} \frac{\partial}{\partial r} h(\mathbf{r}, t) \bigg|_{r=R_2}, \tag{7.28b}$$

where d is the dimension of the aggregate, and the reaction radii R_1 and R_2 are defined via the following relations [2]:

$$\frac{R_1^2}{D_1} = \frac{1}{\lambda(R_1)}, \qquad 2\frac{R_2^2}{D_1 + D_2} = \frac{1}{\mu(R_2)}. \tag{7.29}$$

By taking into account Eqn. 7.10 and the Förster-type expression for D_1, i.e. $D_1 = k_{fl}(R_0/a)^6 a^2$, this results in the following expression for the reaction radius R_1:

$$R_1 = \left(R_1^0/R_0\right)^{3/2} a. \tag{7.30}$$

For diffusion-limited annihilation, R_1 must be close to a, implying – according to Eqn 7.30 – that $R_1^0 \sim R_0$. Neglecting the quasi-linear quenching ($k_1(t) \equiv k$ in Eqn 7.20), this case provides us with simple asymptotic expressions for the annihilation rate $\gamma(t)$ as $t \to \infty$ [48, 72], depending on the dimension, d, of the aggregate:

$$\gamma(t) \sim \begin{cases} 4\pi D_1 R_1, & d = 3, \\ (\ln t)^{-1}, & d = 2, \\ t^{-d/2}, & 2 > d \geqslant 1. \end{cases} \tag{7.31}$$

Note that only in the three-dimensional aggregate case the annihilation rate limit is constant; in lower dimensions this is not the case. Therefore for aggregates where the linear decay timescale greatly exceeds the characteristic annihilation timescale, the $\gamma(t)$ asymptotes can give additional information about the aggregate dimension d.

In the opposite case when $R_1 \gg a$, static annihilation must also be taken into account. In that case, the following approximation can be used:

$$\gamma(t) = \gamma_{\text{dif}}(t) + \gamma_{\text{stat}}(t). \tag{7.32}$$

Here, $\gamma_{\text{dif}}(t)$ is defined according to Eqns 7.28a and $\gamma_{\text{stat}}(t)$ is given by

$$\gamma_{\text{stat}}(t) = \int_0^{R_1} g_{\text{stat}}(\mathbf{r}, t)\lambda(\mathbf{r})d\mathbf{r}, \tag{7.33}$$

where $g_{\text{stat}}(\mathbf{r}, t)$ is the solution of Eqn 7.11a in the static approximation within the sphere of reaction radius R_1, i.e.

$$\frac{\partial g_{\text{stat}}(\mathbf{r}, t)}{\partial t} = H(g_{\text{stat}}(\mathbf{r}, t)). \tag{7.34}$$

However, as will become evident from analysis of the difference absorption spectra of photosynthetic membranes (see Section 7.3), $R_1^0 \sim R_0$ and thus, for aggregates with a size of the same order as the excitation diffusion radius (see Eqn 7.1), the annihilation rate is determined by excitation diffusion.

7.2.1.4   Quasi-linear excitation quenching

For large aggregates, the generalized rate of (quasi-)linear excitation quenching processes due to the excitation traps introduced above can be expressed as follows [16, 26, 50, 69]:

$$k_1(t) = k + k_0[1 - \eta(t)] + k_c\eta(t), \tag{7.35}$$

where k is the S_1 relaxation rate in the aggregate without traps; rates k_0 and k_c take into account a change of the trapping rate due to the changes of the state of the trap. Here we assume two state traps, i.e. k_0 corresponds to the initially nonactivated (open) state, k_c to the activated (closed) state of the trap. This kind of trapping can be applied to analysing excitation kinetics in photosynthetic light-harvesting antenna (LHA), where the so-called reaction centers (RCs) are traps for excitation, transforming the neutral excitation into a pair of separated charges (see [96] for a review). In Eqn 7.35, $\eta(t)$ is the fraction of closed traps. For the large aggregates under consideration – or, in the terminology used by the photosynthetic community, in the "lake model" [26, 69], which describes the common LHA with large RC concentrations – the time-dependence of $\eta(t)$ is given by the integral equation:

$$\eta(t) = N_{\text{tr}} \, k_0 \int_0^t [1 - \eta(t')]n_1(t')d't + \eta(0), \tag{7.36}$$

where N_{tr} is the number of pigments per a single trap. This integral equation gives the following solution:

$$\eta(t) = 1 - [1 - \eta(0)]\exp[-N_{\text{tr}} \, k_0 \int_0^t n_1(y)dy], \tag{7.37}$$

where the initial fraction of closed traps $\eta(0)$ reflects the initial conditions. Thus, in terms of the "lake model" the temporal evolution of the fraction of closed traps due to excitation trapping is determined by the excitation flux from the LHA to the RC.

7.2.1.5 Fractal model for spectrally inhomogeneous aggregates

The theory presented above is applicable to the case of spectrally homogeneous aggregates. However, for most molecular aggregates there is a characteristic disorder in the molecular excitation energies. When the characteristic energy distribution width matches the thermal energy ($\delta_{inh} \approx k_B T$), the theory is no longer valid and requires fundamental extension. This task is rather problematic, although it is expected to be accomplished in the future. Nevertheless, a simplified description of annihilation in spectrally inhomogeneous aggregates is possible, by applying the concept of fractal structure [37, 92].

The energetic disorder manifests itself mainly at lower temperatures, when excitations migrate mainly among the low-energy molecules and the shorter-wavelength pigments can be considered to be like energy barriers for the excitations. The map of "visited" molecule sites has a fractal structure [37, 47]. By investigating the time window starting a few picoseconds after excitation (relaxation to the randomly distributed "reddest" molecules in, e.g., photosynthetic LHA takes a few hundred femtoseconds [97]) Equation 7.8a is still valid. However, N and $n \equiv n_1$ now refer to only those longest-wavelength, or "reddest," molecules in the aggregate. Here, the solution of Eqn 7.8a for times longer than the pulse duration (i.e. for Eqn 7.20) and for $k_1(t) = k$ can be written as

$$n(t) = \frac{n_0 \exp(-kt)}{1 + n_0 \int_0^t \gamma(t') \exp(-kt') dt'}, \tag{7.38}$$

where n_0 is the initial concentration of excited pigments.

For three-dimensional large aggregates, γ is time-independent (see Eqn 7.31). In such a case, Eqn 7.38 can be rewritten as

$$n(t) = n_0 \exp(-kt)/\{1 + \frac{n_0\gamma}{k}[1 - \exp(-kt)]\}. \tag{7.39}$$

In the general case, $\gamma(t)$ is determined by the pair correlation function of the excitations (see Eqn 7.9). For diffusion-limited annihilation the asymptotic time-dependence of $\gamma(t)$ corresponds to a power law [37], i.e.

$$\gamma(t) = \gamma_0/t^\alpha, \tag{7.40}$$

where

$$\alpha = 1 - d_s/2, \qquad \text{if } d_s < 2,$$
$$\alpha = 0, \qquad \text{if } d_s \geqslant 2, \tag{7.41}$$

and where d_S is the spectral dimension of the fractal structure, or d for Euclidian structures. For the short-time kinetics where $kt < 1$, substituting Eqn 7.40 into Eqn 7.38 gives the following analytic solution:

$$n(t) = \frac{(1 - \alpha)n_0}{1 - \alpha + n_0\gamma_0 t^{1-\alpha}}. \tag{7.42}$$

7.2.2 Small aggregates

When the molecular aggregate consists of relatively few molecules and the size of the aggregate is much smaller than the excitation diffusion radius, it can be viewed as a supermolecule being characterized mainly by energy levels of single and multiple excitation.

7.2.2.1 Distribution function approach

To illustrate the use of the distribution function approach (Section 7.2.1) we consider below the problem of excitation annihilation in an ensemble of three-monomer aggregates which serves as a model for excitation relaxation in the trimeric Fenna–Matthews–Olson (FMO) complexes [35] (see also Chapter 10 in this book). For noninteracting trimers, it follows that

$$f_M(\mathbf{r}_1, \mathbf{r}_2, \cdots, \mathbf{r}_N, t) = \prod_{\nu}^{M/3} f_3(\mathbf{r}_1^{\nu}, \mathbf{r}_2^{\nu}, \mathbf{r}_3^{\nu}), \tag{7.43}$$

where the index ν enumerates the trimers. The naturally truncated chain of kinetic equations with the immobile three-particle distribution function $f_3(1, 2, 3)$ then reads

$$\frac{\partial}{\partial t} f_1(i) = J(t) - kf_1(i) + \sum_{j \neq i} [w_{ji} f_1(j) - w_{ij} f_1(i)]$$
$$- \sum_{j \neq i} [\lambda_{ij} f_2(i, j) + \lambda_{ji} f_2(i, j)], \tag{7.44a}$$

$$\frac{\partial}{\partial t} f_2(i, j) = J(t)(f_1(i) + f_1(j)) - 2k f_2(i, j)$$
$$+ w_{li} f_2(i, l) + w_{lj} f_2(l, j) - f_2(i, j)(w_{il} + w_{jl})$$
$$- (\lambda_{ij} + \lambda_{ji}) f_2(i, j) - (\lambda_{il} + \lambda_{jl} + \lambda_{li} + \lambda_{lj}) f_3(1, 2, 3), \tag{7.44b}$$

$$\frac{\partial}{\partial t} f_3(1,2,3) = J(t)[f_2(1,2) + f_2(1,3) + f_2(2,3)] - 3kf_3(1,2,3)$$
$$- \sum_{\substack{i,j \\ i \neq j}} w_{ij} f(1,2,3), \tag{7.44c}$$

where $l \neq i,j; i,j = 1 \cdots 3$; $f_1(i)$ is a single-excitation function, $f_2(i,j)$ is the two-excitation function (i and j enumerating the excited monomers within the trimer), w_{ij} is the excitation energy transfer rate between the ith and jth monomers within the trimer, and λ_{ij} is the rate of excitation energy transfer from the excited monomer to another excited monomer in the trimer (see Eqn 7.10); enumeration of trimers ν and the time-dependence of the functions are omitted for clarity. We now introduce normalized one-, two-, and three-particle distribution functions for the entire trimer:

$$n(t) = \frac{1}{3} \sum_i f_1(i), \tag{7.45a}$$

$$g_2(a,t) = \frac{1}{6n^2(t)} \sum_{\substack{i,j \\ i \neq j}} f_2(i,j), \qquad g_3(a,t) = \frac{1}{n^3(t)} f_3(1,2,3), \tag{7.45b, c}$$

where a is the distance between monomers within the trimer. Thus, for a symmetric trimer, and defining $\lambda(a) = \lambda_{ij}$, we obtain the following:

$$\frac{\partial}{\partial t} n(t) = 3J(t) - kn(t) - 2\gamma(t)n^2(t), \tag{7.46a}$$

$$\gamma(t) = \lambda(a)g(a,t), \tag{7.46b}$$

and

$$\frac{\partial}{\partial t} g(a,t) = 2\frac{J(t)}{n(t)}(1 - g(a,t)) - \lambda(a)g(a,t) + 4\gamma(t)n(t)g(a,t) - 3\lambda(a)n(t)g_3(t),$$
$$\tag{7.47}$$

where $n(t)$ is the average number of excitations per monomer (the concentration of excitation normalized to the monomer) and $g(a,t)$ is the pair-correlation function for two excitations situated in the same trimer; $g_3(t)$ is the three-particle correlation function that satisfies the following evolution equation:

$$\frac{\partial}{\partial t} g_3(t) = \frac{J(t)}{n(t)} g(a,t) - 9\frac{J(t)}{n(t)} g_3(t) - 3\lambda(a)g_3(t) + 2\gamma(t)n(t)g_3(t). \tag{7.48}$$

Equations 7.46–7.48 can be solved numerically. However, some functional time-dependence is evident directly from the analysis of Eqn 7.47 at moderate excitation intensities.

After the excitation pulse action, i.e. when $J(t) = 0$, by neglecting the last two terms on the right-hand side of Eqn 7.47 (which describe negligible higher correlation aspects) the following analytic function for $g(a, t)$ can be obtained:

$$g(a, t) = g^0(a)e^{-\lambda(a)t}. \tag{7.49}$$

The initial value $g^0(a)$ can be determined from the solution of Eqn 7.47 during the excitation pulse action. As already shown (see the discussion following Eqn 7.20), in the course of the excitation pulse the higher-order correlations are weakened (see Eqn 7.21) and $g^0(a)$ can be evaluated by assuming a rectangular pulse shape:

$$g^0(a) = \frac{J}{n} \Big/ \left[\frac{J}{n} + \lambda(a)\right]. \tag{7.50}$$

Thus, the annihilation rate $\gamma(t)$ due to Eqn 7.46b is as follows:

$$\gamma(t) = \gamma_0(a)e^{-\lambda(a)t}, \tag{7.51}$$

giving an exponential decay ($\gamma_0(a) = \lambda(a)g^0(a)$). Taking into account the fact that the annihilation process is much faster than the rates of linear decay, when the excitation pulse is over, Eqn 2.4a gives

$$\frac{\partial}{\partial t}n(t) = -2\gamma_0(a)e^{-\lambda(a)t}n^2(t), \tag{7.52}$$

the solution of which reads

$$n(t) = n_0 \Big/ \left\{1 + n_0 \frac{\gamma_0(a)}{\lambda(a)} \left[1 - e^{-\lambda(a)t}\right]\right\}. \tag{7.53}$$

When $n_0 \ll \lambda(a)/\gamma_0(a)$, i.e. at low initial excitation concentrations, it reads

$$n(t) = n_0 \left[1 - n_0 \frac{\gamma_0(a)}{\lambda(a)}\right] + n_0^2 \frac{\gamma_0(a)}{\lambda(a)} e^{-\lambda(a)t}. \tag{7.54}$$

Thus, it is worth noting that at moderate excitation intensities the annihilation kinetics in small aggregates can be exponential, which is in contrast to the conventional hyperbolic law for extended aggregates (see Eqn 7.39).

7.2.2.2 *Statistical approach*

In the supermolecular approach, the time-dependent spatial distribution of the excitations within the domain does not play any essential role in the solution of Eqns 7.11ab, and the annihilation processes are then limited to static annihilation between nearest neighbors according to the probabilities given in Eqn. 7.10. A good approximation is then to assume that the annihilation rate is time-independent, and that the only statistical effects that are taken into account are those due to the initial distribution of the excitations over all the domains. This is the basic approach commonly applied for small systems [59, 70]. The process of annihilation thus can be viewed as state relaxation in a multi-excitation state scheme (see Fig. 7.2), described by the following Pauli master equation:

$$\frac{\partial p_i(n, t)}{\partial t} = -[ki + \gamma i(i - 1)]p_i(n, t) + [k(i + 1) + \gamma i(i + 1)]p_{i+1}(n, t), \quad (7.55)$$

where $p_i(n, t)$ is the probability that at time t there are i excitations present, given n excitations at $t = 0 (p_i(n, 0) = \delta_{n,i}); k$ is the overall rate of decay due to loss or trapping for a single excitation in the domain, and γ is the overall rate of

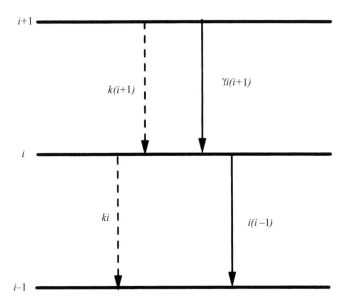

Figure 7.2 A scheme of excitation relaxation in a small aggregate. The horizontal bars represent aggregate states characterized by the excitation numbers (indicated on the left) in the domain. Solid and dashed vertical lines indicate the monomolecular decay and bimolecular excitation annihilation processes, respectively. The symbols by the vertical lines are the corresponding process rates

loss of single excitation from the pair of excitations due to their annihilation upon mutual interaction (note that the annihilation rate as defined above is twice as large as defined in the original work [70]; the convenience of this redefinition becomes clear when performing the comparative analysis of the fluorescence quantum yield expressions for small and large systems). Equation 7.55 can be modified trivially to the case of annihilation of both excitations by including the extra probability gain $((i+1)(i+2)p_{i+2}(n,t))$ and loss $(i(i-1)p_i(n,t))$ terms. To avoid cumbersome expressions, we do not present here how Eqn 7.55 could be extended to also include the excitation traps and S_2 state excitations, instead referring to [16] and [36], respectively.

For n excitations in the domain at $t = 0$, the average number of excitations at time t reads

$$\langle i \rangle_n = \sum_{i=1}^{n} i p_i(n, t).$$ (7.56)

The above summation applied to each term in Eqn 7.55 gives

$$\frac{d\langle i \rangle_n}{dt} = -k\langle i \rangle_n - \gamma \langle i(i-1) \rangle_n.$$ (7.57)

In an actual experiment, the initial number of excitations in the domain obeys the Poisson distribution. In that case after averaging by Poissonian distribution $\langle i(i-1) \rangle = \langle i \rangle^2$; thus Eqn 7.57 is analogous to Eqn 7.20.

The normalized intensity of fluorescence for the fixed initial number of excitations in the domain (n) can be determined as

$$F_n(t) = \frac{\langle i \rangle_n}{n}.$$ (7.58)

Therefore, in the case of a mean number y of excitations created per domain, after averaging of Eqn 7.58 on Poissonian distribution for the fluorescence intensity, it follows that

$$F(t) = \sum_{n=1}^{\infty} \frac{y^{n-1}e^{-y}}{(n-1)!} F_n(t).$$ (7.59)

By applying the generating function method, the solution of Eqn 7.57 can be obtained and, thus, $F(t)$ is then described by a sum of exponentials:

$$F(t) = \sum_{p=0}^{\infty} (-1)^p \exp[-(p+1)(p+r)\tau] A_p,$$ (7.60)

where

$$A_p = \sum_{q=p}^{\infty} \frac{(-1)^q q! y^q (r + 1 + 2p)}{p!(q-p)!(r+p+1)\cdots(r+p+1+q)}, \qquad (7.61)$$

with $\tau = \gamma t$ and $r = k/\gamma$. The fluorescence quantum yield is then expressed as follows:

$$\phi = \phi_0 r \sum_{q=0}^{\infty} \frac{(-1)^q y^q}{r(r+1)\cdots(r+q)} \frac{1}{q+1}. \qquad (7.62)$$

When the annihilation rate constant γ is much larger than the monomolecular decay rate k, $r \ll 1$,

$$F(t) = \frac{1 - e^{-y}}{y} e^{-kt}, \qquad (7.63)$$

and Eqn 7.62 results in

$$\phi = \phi_0 \frac{1 - e^{-y}}{y}. \qquad (7.64)$$

Equation 7.64 was first obtained while analysing fluorescence quenching in photosynthetic membranes [59]. In the opposite limit, when $r \gg 1$,

$$F(t) = 1 \Big/ \left[e^{kt} \left(1 + \frac{y}{r} \right) - \frac{y}{r} \right] \qquad (7.65)$$

and

$$\phi = \phi_0 \frac{r}{y} \ln\left(1 + \frac{y}{r} \right). \qquad (7.66)$$

It is important to note that the latter expression is identical to that which would be obtained by using the solution of Eqn 7.39, thus approaching the case of diffusion-limited annihilation in a large aggregate [82].

7.2.2.3   *Coherent excitons*

For small aggregates which might be well defined by coherent excitons, the supermolecule approach of coherently linked molecules has to be used, and the annihilation process in general corresponds to relaxation. Schematically, these $1-, 2-, \cdots, n$-exciton energy bands are arranged similarly to the multi-excitation states shown in Fig. 7.2. Thus, in this scheme during the excitation pulse, sequential optical transitions into one-exciton, two-exciton, etc. states are stimulated [45, 63, 95]. It is evident that the annihilation process in this scheme corresponds to relaxation from the highest exciton band to the next lower band. In the case in which the relaxation rates are not dependent on the number of exciton bands, the process under consideration might also be well determined by the Pauli master equation (Eqn 7.55). It is evident that the initial statistical distribution of excitations among the exciton bands is now determined by the excitation pulse shape and its intensity, and corresponds in general to the problem of multi-exciton optical processes [63]. However, energy relaxation processes as well as the increase in local temperature generated during the annihilation (for the effects of local heating, see below) also contribute to exciton dephasing. Therefore, coherence might be expected to be lost in a single or few inter-band relaxation steps, and the view of coherent multi-exciton processes in the aggregate could be relevant only over a very short initial timescale. Considering the annihilation kinetics at much longer timescales than the actinic pulse duration, the approach presented in Section 7.2.2.2 seems to be a good approximation.

7.2.3   Singlet–triplet annihilation

For long excitation pulses or at high repetition rates of excitation, some singlet-excited molecules can be converted to triplet states via inter-system crossing.

7.2.3.1   Singlet excitation trapping by triplets

In the presence of triplet molecules, S–T annihilation creates an additional channel for singlet excitation decay. Assuming a random distribution of triplet molecules, the process of S–T annihilation in homogeneous aggregates is independent of the spatial coordinates and is a function of the triplet concentration n_T, singlet migration and the S–T annihilation rate only. The corresponding kinetic equations which determine this process are as follows:

$$\frac{\mathrm{d}}{\mathrm{d}t}n = J(t) - \frac{n}{\tau_S} - \gamma n^2 - \gamma_{ST} n_T n, \qquad (7.67)$$

$$\frac{\mathrm{d}}{\mathrm{d}t}n_T = k_{IC}n - \frac{n_T}{\tau_T}, \qquad (7.68)$$

where γ and γ_{ST} are the rates of S–S and S–T annihilation, respectively, k_{IC} is the rate of inter-system crossing, and τ_T is the triplet lifetime, typical values of the latter being in the microsecond to millisecond range, $\tau_T \gg \tau_S$. Thus, in considering the kinetics of singlet excitation at high repetition rates of the excitation pulses, triplets could be regarded as being under quasi-stationary conditions, i.e. $n_T \approx \tau_T k_{IC} n$, if the quasi-stationary value of the singlet concentration could be defined. Such an approach may be also correct for long excitation pulses, when the pulse duration is of the order of τ_T or longer, while in the case of a high repetition rate of excitation, triplets accumulate over the whole sequence of pulses [90]:

$$n_T = \Phi_{IC} \omega n_0, \tag{7.69}$$

where $\Phi_{IC} = k_{IC} \tau_S$ is the quantum yield of inter-system crossing, ω is the pulse repetition frequency and n_0 is the total concentration of singlets generated by a single pulse. At high pulse repetition rates, it is evident that the S–T effect is more pronounced [90]. Therefore, to avoid this process low repetition rates should be used. The kinetics of singlet excitation, neglecting S–S processes, run as follows:

$$\frac{\mathrm{d}}{\mathrm{d}t} n = -K(t)n, \tag{7.70}$$

where

$$K(t) = \frac{1}{\tau_S} + \gamma_{ST} n_T. \tag{7.71}$$

The time dependence of $K(t)$ is due to the correlative effects of singlets and triplets via the time dependence of γ_{ST}. A general approach to the S–T annihilation problem is similar to the case of S–S annihilation, i.e. the process can be considered via multiparticle correlation functions (see Section 7.2.1) by taking into account the relative distribution of both sorts of particles, i.e. singlets and triplets. This problem is also analogous to the diffusion-limited reactions of two different reactants. As has been shown by the analysis of $A + B \rightarrow B$ reactions [52, 68], inhomogeneous distribution of the reactants is generated in the course of time. Thus, the asymptotic (i.e. long-time) kinetics, is mainly determined by the decay kinetics of singlet excitations which reside in areas that are free of triplets. Such kinetics must evidently relate to the most probable size of such a volume free of triplets, and therefore becomes sensitive to the dimension of the structure under consideration [37],

$$n(t) \propto \exp(-At^{d/(d+2)}), \tag{7.72}$$

where A is a constant that is dependent on n_T, i.e. as determined by the excitation intensity J. Equation 7.72 also holds for fractal structures by replacing the Euclidean dimension d by d_s.

The inhomogeneity in the relative distribution of singlets and triplets which emerge is a result of the S–T annihilation itself. This is because, in the vicinity of any triplet, which already exists in the aggregate, the lifetime of singlets decreases and the probability of new triplet generation reduces. This aspect of S–T annihilation influences the next order of the annihilation process, i.e. it promotes the so-called anti-Smoluchowski triplet–triplet (T–T) annihilation, because of the inhomogeneous distribution of triplets determined by S–T annihilation. This kind of kinetics has been observed in the delayed fluorescence from anthracene in viscous solutions [65].

7.2.3.2   *Fluorescence induction via singlet–triplet annihilation*

The sigmoidal shape of the fluorescence induction curves on the long timescale (from milliseconds to seconds) in chloroplasts [13, 55, 102] (as well as the S–S and S–T annihilation data [15, 62, 96]) support the view that the excitation migrates within the LHA common to a few connected photosynthetic units (PSU, the LHA unit related to a single RC). However, this is not the case for the short (microsecond and less) timescale. By means of two-pulse techniques [22] it has been shown that both the shape and amplitude of the fluorescence induction depends on the duration of the actinic light pulse, i.e. for short excitation pulses ($< 2\mu s$) the shape of fluorescence induction is exponential and the ratio of the maximal fluorescence yield (F_{max}; all of the RCs are closed) to the minimal one (F_0; all of the RCs are open) $R = F_{max}/F_0 < 3.0$, while for long excitation pulses ($> 50\mu s$) the induction behavior is sigmoidal and $R > 3.0$ [22, 88, 93]. Thus, the time interval for the actinic pulse duration related to the change in fluorescence induction spans from 2 to 50 μs. This effect of the decrease in connectivity of the PSUs for shorter actinic pulses is in line with the absence of a change in the absorption cross-section of photosystem II (PS II) in chloroplasts upon closing the RCs [59]. This suggests that the fluorescence induction could be influenced by factors other than connectivity alone. Quenching of singlets by triplets is generally much more efficient than by open RCs, both in photosynthetic bacteria [62, 100] and in chloroplasts [9, 46], and is therefore close to migration-limited. When modeling fluorescence induction and S–T annihilation simultaneously, LHA size has to be taken into account.

In the case of N_{RC} RCs related to the common LHA, the state of the aggregate can be characterized by the number of closed and/or open RCs which are present at a given moment in the course of the excitation lifetime. Let us define i as a number of closed RCs in the domain ($0 \leqslant i \leqslant N_{RC}$). This state of the aggregate changes when the excitation is trapped by the open RC, stimulating at the same time its transition into a closed state. Thus, the process itself can be represented as a transition between quantum states of the domain determined by different numbers of i, as shown in Fig. 7.3. The vertical

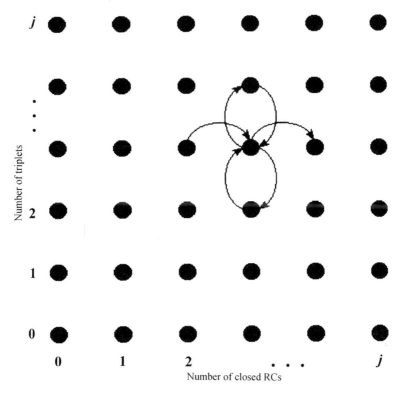

Figure 7.3   A scheme of the domain states characterized by the numbers of open and closed RCs, and the number of triplet states, with corresponding transitions between these states. Within the time window of the experiment and simulations, the closing of RCs is irreversible

transitions in Fig. 7.3 correspond to the formation/decay of a triplet state; the horizontal transitions correspond to a change in the number of closed RCs. Here, the fact that the RC can turn again into an open state – which is correct when this transition time is much longer than the mean time of the process under consideration – is not taken into account. We will restrict ourselves to the case of two RC states, i.e. open and closed.

Let us assume that the aggregate is dark-adapted, i.e. that initially all RCs are in the open state and no triplets are present in the domain. Then the singlet excitations created in the LHA under the action of the excitation pulse stimulate the transition of the RCs into closed states as well as generate triplets. At this point, two timescales can be distinguished: one of them is related to the lifetime of the singlet excitations, as determined by the trapping rate k_0 for open RCs and by k_c for closed RCs (typical values are $10^{-2} - 10^{-3}\mu s^{-1}$), while another is associated with the triplet lifetime ($\tau_T \sim 10\mu$s). Below, the rates of excitation quenching by the RCs and the triplets are defined in units k_0 and $1/\tau_T$, respectively.

In order to determine the kinetic equations for the extent of singlet excitations $n(i,j)$ in the domain characterized by the i closed RCs and j triplets, the corresponding rate parameters and the probability $P(i,j)$ of finding such a domain under the selected conditions have to be defined. It is noteworthy that for $N_{RC} \to \infty$, i.e. for the "lake model" $i/N_{RC} \to \eta(\infty)$ already defined by Eqn 7.37. The overall excitation quenching rates by the open and closed RCs read [69, 87]

$$k_0^i = 1 - \frac{i}{N_{RC}}, \tag{7.73}$$

and

$$k_c^i = (1 - p)\frac{i}{N_{RC}}, \tag{7.74}$$

respectively, where

$$p = 1 - \frac{k_c}{k_0} \tag{7.75}$$

is the PSU connectivity parameter [69, 88, 93]. The relative rates for intersystem crossing $I_T = k_{IC}/k$, the singlet–triplet quenching $\Gamma = \gamma_{ST}/k_0$ and the triplet lifetime τ_T determine the interplay between the singlet and triplet subsystems. Below, the pulse lengths and the delay time between the actinic pump and probe pulses will be expressed in units of α, where $\alpha = \tau_T k_0$ is the parameter which determines the ratio of two timescales mentioned above. The amount of singlets is thus defined as follows [87]:

$$\frac{dn(i,j)}{dt} = J - \left(1 - p\frac{i}{N_{RC}}\right)n(i,j) - I_T n(i,j) - \frac{\Gamma}{N_{RC}}jn(i,j), \tag{7.76}$$

where J is the excitation intensity. Here the rate of S–T annihilation Γ is divided by the domain size to account for different domain sizes in the simulations. This approach is correct for domains smaller than the diffusion radius of the singlet excitations.

The fluorescence induction is considered on a microsecond timescale: thus, in Eqn 7.76 the kinetics determined on the timescale of singlet lifetime can be considered under steady state conditions, giving $n = n^{SS}$,

$$n^{SS}(i,j) = J\left(1 - p\frac{i}{N_{RC}} + I_T + \frac{\Gamma}{N_{RC}}j\right)^{-1}. \tag{7.77}$$

The probability $P(i,j)$ is then given by the following balance equation [69, 87]:

$$\begin{aligned}
\frac{dP(i,j)}{dt} &= \mu\left[k_0^{i-1}n^{SS}(i-1,j)P(i-1,j) - k_0^i n^{SS}(i,j)P(i,j)\right] \\
&+ I_T\left[n^{SS}(i,j-1)P(i,j-1) - n^{SS}(i,j)P(i,j)\right] \\
&+ \frac{1}{\alpha}[(j+1)P(i,j+1) - jP(i,j)],
\end{aligned} \tag{7.78}$$

where μ is the quantum yield of the excitation trapping by the RC. Equations 7.77 and 7.78 can be used for either calculating single-pulse fluorescence induction curves or for two-pulse (pump–probe) fluorescence induction curves. In the latter case, calculations of $n\,(i,j)$ and $P\,(i,j)$ have to be carried out in three steps: during the pump pulse action (the excitation fluence J_{pump}); between pulses; and during the probe pulse action (probe fluence $J_{probe} \ll J_{pump}$). The fluorescence induction in the i-th state can be expressed as follows:

$$F(i,j) = \frac{k_f}{J\tau}\int_{t_0}^{t_0+\tau} n(i,j,t')dt', \tag{7.79}$$

where k_f is the radiative decay rate. In the one-pulse experiment, τ corresponds to the pulse duration. In the two-pulse case, J corresponds to J_{probe}, and τ is the duration of the probe pulse (t_0 is the start time of the probe pulse). Under steady state conditions (see Eqn 7.77)

$$F(i,j) = \frac{k_f n^{SS}(i,j)}{J}. \tag{7.80}$$

Thus the observed fluorescence induction F is given by

$$F = \sum_{i=0}^{N_{RC}}\sum_j P(i,j)F(i,j). \tag{7.81}$$

Equations 7.77 and 7.78 can be solved numerically for each i-th state of the aggregate, which contains j triplets. From Eqn 7.81 the fluorescence quantum yields are then obtained as a function of the intensity J.

Numerical calculations have shown that the sigmoidal shape of the fluorescence induction curve can be distinguished only when the connectivity parameter $p > 0.5$ [13, 69, 88]. Moreover, it has also been shown [69] that the difference between domains containing different numbers of RCs becomes indistinguishable when $R < 4$. Thus, due to the relation between these values,

$$R = \frac{1}{1-p}, \tag{7.82}$$

which is correct for the lake model. Let us assume that $p = 0.8$. For this value of the connectivity parameter, the calculated fluorescence induction curve is sigmoidal in the case in which S–T annihilation is not taken into account (see Fig. 7.4 for $\Gamma = 0$). The triplet effects are defined by the following parameters: Γ, which is the S–T annihilation rate, and I_T, the rate of inter-system crossing. According to Eqns 7.77 and 7.82, the effect of the last parameter determines a renormalization of the excitation intensity, i.e. $J/(1 + I_T)$, as well as the connectivity parameter, $p/(1 + I_T)$, and the S–T annihilation rate, $\Gamma/(1 + I_T)$. Therefore, the conditions for the presence of a sigmoidal induction curve can now be reformulated as $p/(1 + I_T) > 0.5$, and the effective value of p becomes smaller when the loss due to inter-system crossing is taken into account.

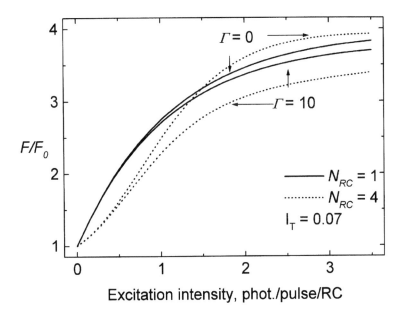

Figure 7.4   Fluorescence induction versus actinic pulse duration for two values of the S–T annihilation rate ($\Gamma = 0; 10$), and two numbers of connected RCs ($N_{RC} = 1; 4$). The time delay between the actinic and the probe pulses is 10α

7.2.4   Local heating during annihilation

In this section, we consider the local heating/cooling processes initiated by the dissipation of excess excitation energy which is expected to follow excitation annihilation. These processes are not yet well understood. We discuss these processes, mainly considering two aspects of their manifestation. First, we consider the local temperature change due to different timescales of the excitation energy dissipation to be distinguished. The possibility of stimulating the generalized nuclear coordinate motion during the relaxation is also briefly discussed. These phenomena are thought to be responsible for the specific absorption changes of highly excited aggregates on a picosecond timescale, as observed in molecular aggregates [31, 33, 34, 74] and pigment–protein complexes [35, 86].

7.2.4.1   *Dissipation of vibrational mode energy*

In the case of the annihilation process under consideration (Fig. 7.1), it is supposed that the energy of the higher excited state is transferred into vibrational modes of the molecule. Thus, the energy conservation law requires the following balance relation to be fulfilled:

$$\frac{dW}{dt} + \frac{dW_0}{dt} = \varepsilon(kn + \gamma n^2), \tag{7.83}$$

where dW/dt expresses the rate of energy transformation into vibrational modes of the relaxing molecule and its nearest surroundings, while dW_0/dt determines the further energy exchange with the thermal bath; ε determines the energy converted into vibrations during relaxation. For simplicity here, we assume δ-pulse excitation. The two terms on the left-hand side of Eqn 7.83 correspond to a well expressed hierarchy of timescales for energy dissipation: relaxation from the highly excited states occurs within picoseconds or less (see Section 7.3.1 for an estimation of the S_2 state relaxation rate for bacteriochlorophyll [94]), while the relaxation of the local excited vibrational modes is much slower and covers hundreds of picoseconds [3, 10, 39]. After fast relaxation the non equilibrium distribution of excited vibrational modes starts to interact with other modes of the surrounding medium, producing heat diffusion, with a typical constant for various proteins, water and other organic aggregates being of the order of $10^{-3}\,\mathrm{cm^2\,s^{-1}}$ [56]. Therefore, we can suppose that after tens of picoseconds the excited local modes become thermalized within nonequilibrium heated surroundings covering distances less than 50 Å.

In the case of a molecular aggregate with dimension L of the order of tens (to hundreds) of angstroms, we can suppose that

$$L > \frac{K}{H}. \tag{7.84}$$

Here, K and H are the coefficients of thermal conductivity and heat exchange, respectively; they describe the heat exchange with thermal bath, entering into the Newton boundary condition for the temperature $K\partial T/\partial r = H(T - T_0)$, T_0 being the temperature of the thermal bath. Thus we characterize each aggregate by a nonequilibrium temperature T and internal energy W ($W = C_V T$; C_V being the heat capacity of the molecular aggregate). The second term on the left-hand side of Eqn 7.83 can then be expressed as follows:

$$\frac{dW_0}{dt} = \beta(T - T_0), \tag{7.85}$$

where $\beta = HS$, S being the total contact surface of excited molecules with the thermal bath (solution, protein, etc.). In application to experiments with more than a few picoseconds, resolution, this confers on Eqn 7.83 the following form:

$$\frac{d}{dt} C_V T = \varepsilon(kn + \gamma n^2) - \beta(T - T_0). \tag{7.86}$$

At high temperatures, assuming $C_V = $ const., the formal solution of Eqn 7.86 then gives

$$\Delta T(t) = \frac{\varepsilon}{C_V} \int \exp\left(-\beta \frac{t - t'}{C_V}\right) \left[kn(t') + \gamma(t')n^2(t')\right] dt', \tag{7.87}$$

where $\Delta T = T - T_0$.

The temporal dependence of T is highly influenced by the temperature change rate β. Only in the case of very large β values does the time-dependence of ΔT replicate the excitation relaxation time-dependence, and taking into account Eqn 7.20 it becomes evident that $\Delta T(t) \propto dn/dt$. When β in contrast is small, the T relaxation kinetics strongly differs from n relaxation kinetics. Initially, the local temperature grows in proportion to the relaxation flux. Such a departure of the temperature time-dependence from the excitation decay kinetics could reveal new nonlinear effects in transient spectroscopy measurements.

7.2.4.2   *Modulation of the nuclear motion*

The local dissipation of excess energy can modify the coupling of the relaxed molecule to its surroundings. For instance, the change of coupling can influence the arrangement of molecules, giving the so-called polaron effect [78]. A similar situation is relevant in proteins where excitation or electron transport is accompanied by conformational changes (see, for instance, [28]). In order to describe the molecular rearrangement effect discussed above, the generalized coordinate Q for the collective nuclear motion of the molecules can be introduced. Assuming linear coupling between the electron and the collective nuclear motion, the excited molecular energy E as well as that for the hole in the ground state,

$$E = E_1 + \chi Q, \tag{7.88}$$

where E_1 describes the energy term of the excited molecule before rearrangement of the surrounding molecules, and χ is a coefficient that describes the strength of coupling of the excited molecule to the coordinate Q. The simplest way to describe the temporal evolution of the coordinate Q is through use of the stochastic damped oscillator equation:

$$\frac{d^2Q}{dt^2} + \eta\frac{dQ}{dt} + \omega_0^2 Q = -\chi n + y(t). \tag{7.89}$$

Here, ω_0 and η are the oscillation frequency and damping coefficient, respectively; n is the occupation number of the excited molecule; and $y(t)$ is a stochastic force, the correlator of which is related to the local temperature:

$$\langle y(t)y(t')\rangle = k_B T \eta \delta(t - t'). \tag{7.90}$$

Angle brackets denote a statistical average and k_B is the Boltzmann constant.

In the case in which the coordinate Q kinetics is fast in comparison with temperature and n evolution, thermal fluctuations can be neglected ($y(t) = 0$) and $n \approx$ const. Assuming the initial condition $Q = dQ/dt = 0$, the solution of Eqn. 7.89 then reads

$$Q = Q_0\left[1 - \exp\left(-\omega_0^2 t/\eta\right)\right] \tag{7.91}$$

for $\eta \gg \omega_0$ and

$$Q = Q_0[1 - \exp(-\eta t/2)\cos(\omega_0 t)] \tag{7.92}$$

for $\eta \ll \omega_0$, when $Q_0 = -\chi n/\omega_0^2$. In the opposite case, when the kinetics of coordinate Q is slow in comparison with T evolution, then $n \approx 0$ and the solution of Eqn. 7.89 gives the same time-dependence for the correlator $\langle Q(t)Q(t')\rangle$ as for $Q(t)$ in Eqns 7.91 and 7.92 while substituting Q_0 with $\langle Q_0^2\rangle = kT/\omega_0^2$.

The absorption spectrum σ_1^T of the excited molecule heated to temperature T is a function of the wavelength, i.e. of E. Therefore, the difference absorption spectrum (see the next section) could involve the kinetics of the "slow" evolution of the molecular rearrangement stimulated by the excitation relaxation. From the above considerations, albeit simplified, it becomes evident that, in some special case ($\eta \ll \omega_0$), this evolution could resemble the behavior of a damped oscillator.

7.2.5 Pump–probe (difference absorption) spectroscopy of excitation annihilation

When the change in optical density of the sample $\Delta A \ll 1$, the difference absorption observed by a weak probe pulse can be estimated from the following expression:

$$\Delta A(\lambda, t) = \frac{Cx}{\ln 10} \int J_{\text{probe}}(t - t') \left[\sum_i n_i(t')\sigma_i(\lambda) - \sigma_0(\lambda) + \Delta A_{\text{corr}}(\lambda, t') \right] d't,$$

(7.93)

where λ and $J_{\text{probe}}(t)$ are the wavelength and the probing pulse shape function, respectively, C is the concentration of molecules in the sample with thickness x and $\Delta A_{\text{corr}}(\lambda, t)$ is the change in the difference spectrum due to correlations between excitations, which is important at high excitation intensities. σ_i here includes the cross-sections of the excited state i absorption and stimulated emission. At low excitation intensities, and if $\Delta A_{\text{corr}}(\lambda, t)$ can be neglected, Eqn 7.93 leads to the well-known result that ΔA is proportional to the cross-correlation of the pump and probe laser pulses. At high excitation intensities, when exciton–exciton annihilation starts to contribute, the evolution of the difference spectra becomes much more complicated, and additional effects due to dispersion-like interactions become possible. If the molecules next to the excited molecule are perturbed, this effect is also included in the term $\Delta A_{\text{corr}}(\lambda, t)$; i.e.

$$\Delta A_{\text{corr}}(\lambda, t) = n_1 \sum_r \left[\sigma_0^r(\lambda) - \sigma_0(\lambda) \right]$$
$$+ n_1^2 \sum_{r, r'} g(r, t) \left[\left(\sigma_0^{r', r}(\lambda) - \sigma_0(\lambda) \right) + \left(\sigma_1^r(\lambda) - \sigma_1(\lambda) \right) \right],$$

(7.94)

where the sum over r goes over all molecules at distance r, $\sigma_i^r(\lambda)$ for $i = 0$ is the ground-state absorption, and for $i = 1$ is the sum of the excited-state absorption and stimulated-emission cross-sections of a molecule situated at distance r from the excited molecule, and $\sigma_0^{r', r}(\lambda)$ is the ground-state absorption of a molecule situated close to two excited molecules.

7.2.5.1 Higher excited state relaxation

If the correlative effect on the ground and excited state absorption is ignored, i.e. $\sigma_i^r(\lambda) = \sigma_i(\lambda)$, the time-dependence of the difference spectra displays excitation annihilation kinetics only. The stationary solution of Eqns 7.8a,b that leads to Eqn 7.16 in the case of very high intensities, when $n_2 \gg n_1$, can be rewritten as Eqn 7.17. In that case, and for "broad" pulses, i.e. when $J_0\tau_{\text{pump}} > n_1 + n_2(\tau_{\text{pump}}$ is the pump pulse duration), Eqn 7.17 is valid during the pump pulse action and thus

$$\sum_i n_i(t)\sigma_i(\lambda) - \sigma_0(\lambda) = n_2(t)[\sigma_2(\lambda) - \sigma_0(\lambda)]. \tag{7.95}$$

Assuming the pump and probe pulses to be Gaussian, and substituting Eqn 7.17 into Eqn. 7.95, the maximal value of the spectral changes is given by

$$\Delta A^{\max}(\lambda) = \frac{J_0 A(\lambda)[\sigma_2(\lambda) - \sigma_0(\lambda)]}{\sqrt{2}k_2\tau_{\text{pump}}\sigma_0(\lambda)}, \tag{7.96}$$

where J_0 is the maximal value of the excitation pulse and $A(\lambda) = Cx\sigma_0(\lambda)\ln 10$ is the absorption of the sample. One can see from Eqn. 7.96 that the excitation intensity dependence of the difference absorption spectrum can be used to determine the higher excited state relaxation rate k_2.

At low excitation intensities, i.e. when $n_1 \gg n_2$, $\Delta A(\lambda, t)$ is given by

$$\Delta A(t, \lambda) = \frac{A(\lambda)}{\sigma_0(\lambda)} n_1(t)[\sigma_1(\lambda) - \sigma_0(\lambda)]. \tag{7.97}$$

In order to obtain the maximal value of ΔA, the stationary value of n_1 obtained from Eqn 7.8a has to be substituted into Eqn 7.97.

7.2.5.2 Excitation correlation effects

The spectral changes associated with excitation correlations are described in Eqn 7.94. The absorption cross-section of molecules situated in the vicinity of the excited molecule, $\sigma_i^r(\lambda) = \sigma_i^r(E)$ and $\sigma_0^{r',r}(\lambda) = \sigma_0^{r',r}(E)$, can be estimated as follows:

$$\sigma_i^r(E) = \sigma_i[E - E_i - V_i(r)], \tag{7.98a}$$

$$\sigma_0^{r',r}(E) = \sigma_0[E - E_0 - V_0(r') - V_0(r - r')], \tag{7.98b}$$

where $\sigma_i(E - E_i) = \sigma_i(\lambda)$ is the homogeneously broadened absorption of pigments in the ith state with absorption at E_i, and $V_i(r)$ determines the correlative interaction between two excited molecules at a distance r from each other, i.e.

$$V_i(r) = V_i(a)\left(\frac{a}{r}\right)^\zeta, \tag{7.99}$$

where $V_i(a)$ is the strength of such an interaction between "nearest neighbors," and the power-law exponent ζ is determined by the order of multipole intermolecular interaction. For instance, in the case of dipole–dipole type interaction, $\zeta = 3$ [14]. Thus, it is evident that a significant deviation of $\sigma_i^r(\lambda)$ from $\sigma_i(\lambda)$ occurs only at small r, when $r < r_i$, the latter value being determined by the following equality:

$$\sigma_i^{r_i}(\lambda) = \sigma_i(\lambda). \tag{7.100}$$

Therefore, due to the definition of the reaction radius R_1 (see Eqn 7.30), the correlation function can be considered in the static approximation (Eqn 7.34). During the action of the excitation pulse, Eqn 7.22 is fulfilled and the terms due to three-particle correlations in Eqn 7.34 can be neglected, yielding the following stationary solution:

$$g_{\text{stat}}(r) = \frac{J}{n_1} \frac{1}{\lambda(r) + J/n_1}. \tag{7.101}$$

From Eqn 7.18, the ratio J/n_1 increases with increasing excitation intensity, and thus at the highest intensities g_{stat} approaches unity. The kinetics of the static correlation function after the pulse action (neglecting higher correlations) gives the following analytic expression:

$$g_{\text{stat}}(r, t) = e^{-2\lambda(r)t}. \tag{7.102}$$

Thus, due to correlation effects induced during the action of the excitation pulse, a fast kinetic component may be observed, as determined by Eqn 7.102, faster than the decay of n_1, which occurs on the timescale of the annihilation rate γ. Therefore, the spectral changes at high excitation intensities which arise either from the population of higher excited states or from the correlation effects (in the static approximation) are indistinguishable. Both introduce fast kinetics, and according to Eqns 7.17 and 7.18 both are proportional to J. The spectral changes that remain after the excitation pulse are determined mainly by the population of the first excited states of pigment molecules (Eqn 7.97), the spectral difference of $\sigma_0^r(\lambda)$ and $\sigma_1(\lambda)$ being negligible. This implies that we can ignore the contribution of ΔA_{corr}. The unknown cross-section $\sigma_1(\lambda)$ of Eqn

7.97 can then be determined by analysis of the difference spectra at low excitation conditions:

$$\sigma_1(\lambda) = \frac{\Delta A(\lambda)\sigma_0(\lambda)}{n_1 A(\lambda)} + \sigma_0(\lambda). \qquad (7.103)$$

7.2.5.3   Local temperature effect

At high excitation intensity, when nonlinear excitation decay dominates even during the course of the excitation pulse, local heating starts to change the spectra of absorption and stimulated emission. That is because these depend parametrically on temperature T, i.e. the spectral bands can change position and width. Despite the fact that the influence of such temperature effects is difficult to distinguish from the kinetic dependence, local heating can reveal itself in the measurements of the difference spectra.

The unperturbed molecular absorption at a thermal bath temperature T_0 reads

$$A_0 = \frac{Cx}{\ln 10}\sigma_0^{T_0}. \qquad (7.104)$$

After excitation (assuming, for simplicity, first singlet excited molecules only) the sample absorption changes due to (i) the optical transition of some fraction n of molecules to the excited state and (ii) the change of temperature T at the excited and relaxed molecules:

$$A_1 = \frac{Cx}{\ln 10}\left[n\sigma_1^T + (1-n)\sigma_0^T\right]. \qquad (7.105)$$

The difference absorption spectrum then reads

$$\Delta A = A_1 - A_0 = \frac{Cx}{\ln 10}\left[n(\sigma_1^T - \sigma_0^T) + (\sigma_0^T - \sigma_0^{T_0})\right]. \qquad (7.106)$$

The ΔA kinetics depends on $n(t)$ as well as on the temporal dependence of T. The spectral dependence of ΔA is determined by the differences between cross-sections. In the case in which there is no local heating, i.e. $T = T_0$, Eqn 7.106 reverts to the well-known Eqn 7.97. In the case of slow temperature exchange with the thermal bath, relative to $n(t)$, the temporal dependence of ΔA reflects two timescales, the first corresponding to $n(t)$ and the second following the temperature relaxation.

7.3   APPLICATIONS

7.3.1   Photosynthetic bacteria chromophores

One of the experimental strategies to study excitation transfer in the LHA is to measure the decay of excitation density due to nonlinear annihilation. As has been demonstrated in the previous section, the kinetics of the excitations is determined by their mutual interaction and, therefore, this process is diffusion-limited, at least at medium excitation intensities. The energy migration parameters in the LHA can be obtained from the analysis of S–S or S–T annihilation.

7.3.1.1   *Room temperature*

The spectrum of $\sigma_1(\lambda)$ can be used for estimation of the values of the Förster radii of S_1–S_1 annihilation and that of the diffusion R_0 (see [94]). Inspection of the S_1 state absorption for *Rhodospirillum (Rh.) rubrum* [94] shows that spectral overlaps of the fluorescence spectrum with the absorption spectra $\sigma_0(\lambda)$ and $\sigma_1(\lambda)$ are very similar and indicates directly that the radius ratio R_1^0/R_0 must be close to unity. Thus, according to Eqn 7.30 the "black sphere" radius R_1 equals a, the average intermolecular spacing, and, therefore, the annihilation process is purely diffusion-limited. This is an essential simplification of the theoretic analysis of singlet–singlet annihilation, because only two semi-empirical parameters are required to fit the experimental data; namely, D_1 (or τ_h) and N_{tr}, the amount of pigments per RC (see Eqn 7.36).

The annihilation kinetics has been analysed using the two-dimensional "bubble" model of the domain (chromatophore), i.e. assuming that pigment molecules are situated on the surface of a sphere. The number of RCs in a domain was calculated from an analysis of fluorescence quantum yield as a function of total laser energy (see Fig. 7.5a), with the conclusion that $\tau_h = 0.65$ ps and $N_{tr} = 12$ (or even less). Taking into account the fact that the LHA of *Rh. rubrum* contains 24 Bchls per RC [97], this implies that the main LHA "building blocks" are at least dimers, while by assuming that it has 36 Bchls in the PSU [42], trimers has to be assumed as being coherently coupled. A detailed spectroscopic analysis of the B820 sub-unit of the LHA of *Rh. rubrum* and the intact LHA has also strongly indicated that the basic functional unit is a dimer of Bchl a [97, 98, 101]. It is noteworthy that the fluorescence quantum yield analysis is not sensitive to a variation in the size of the domain. This is in contrast to the annihilation rate time-dependence, $\gamma(t)$ (see Fig. 7.6), which is sensitive to the domain size due to the correlative effects in the excitation dynamics. The smaller the system, the higher is the sensitivity to these correlative effects. It is evident that in large domains, within the lifetime of the excitation, the excitation annihilation can be well approximated by $\gamma = $ const., which is strictly valid only in three-dimensional systems of infinite size according to the general theory (see Section 7.2.1.3). The approximation $\gamma = $ const. can be used to fit the excitation kinetics in RC-free chromophores of *Rh. rubrum* at low excitation intensities ($J_0 < 0.025$ photons/Bchl per RC per pulse; see [94]). Assuming $\gamma = $ const. and analysing the annihilation kinetics at moderate excitation intensities (see Fig. 7.7) yields the same values for N_{tr} and τ_h (12 and 0.65 ps) as obtained from analysis of the fluorescence quantum yield. Note, however, that the resulting kinetics is sensitive to variations in the numerical values of these parameters, as demonstrated in Fig. 7.8. Also, the occupation of the higher excited states, S_2 (or spectral changes due to correlations between excitations), must explicitly be taken into account.

The essential effect due to correlations between excitations, displayed through the time-dependence of γ, can be observed at high excitation intensities. Qualitatively, this effect is evident from the experimental data [8], which indicate that the excitation kinetics after normalization is independent of the excitation intensity. It shows that, after termination of the excitation pulse, the amount of excitation that remains in the domain is the same. There are several possible reasons for this effect, which we will mention briefly: (i) an increase in the temporal dependence of γ; (ii) the occupation of higher excited states S_2; (iii) spectral changes due to the higher correlations between the excitations; and (iv) an increase in the local temperature due to the stimulated relaxation processes

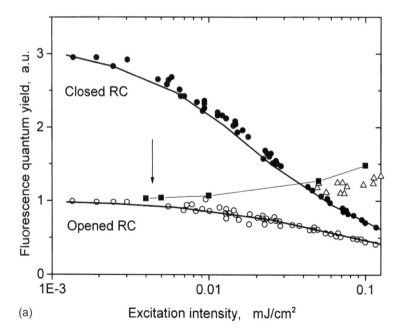

(a)

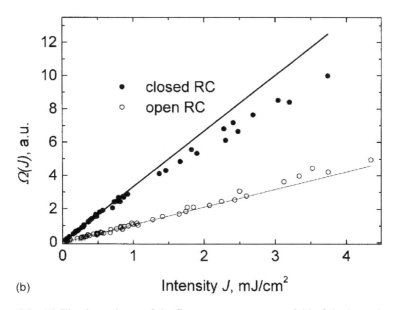

(b)

Figure 7.5 (a) The dependence of the fluorescence quantum yield of the bacterium *Rh. rubrum* on excitation intensity [94]. Experimental points [4] are for open (○) and closed (●) RCs. Theoretic data (solid lines) are obtained for diffusion-limited annihilation ($R_0^1 \simeq R_0$) by numerical solution of Eqns 7.8a,b, 7.9, and 7.27, with initial and boundary conditions according to Eqns 7.14 and 7.15 and by neglecting correlations between excitations S_1 and S_2. The chromatophore is assumed to be a sphere of radius $R = 10a$, a being the mean space between molecules [84]. The only fitting parameters are N_{tr}, the number of Bchl molecules per RC, and R_0, the Förster migration radius. The former insures the correct ratio of the fluorescence quantum yield traces for initially open and closed RC's, ($N_{tr} = 12$) the variation of R_0 is matching the slope of experimental curves ($R_0 = 5.5a$). Other parameters used in the calculations are as follows: $k^{-1} = 3.5$ ns, $k_f^{-1} = 18$ ns, $k_0^{-1} = 60$ ps, $k_c^{-1} = 200$ ps, $k_2^{-1} = 1$ ps, $\tau_{pump} = 35$ ps, $\alpha = \bar{\alpha} = 1$. The arrow indicates the excitation intensity corresponding to one photon per domain, from the asymptotics of the RC charge separation quantum yield (■), obtained from the experimental RC fluorescence yield (△) through $\phi_f = 1/[1 - (k_0 - k_c)\eta/k_0]$ [16]. (b) The dependence of the experimental quantum yield function $\Omega(J) = \exp[\phi_f(J)J\gamma/k] - 1$ on the excitation intensity J for closed (●) and open (○) RCs. Experimental data are calculated according to fluorescence quantum yield measurements [4]. Solid lines represent respective fits by Eqn. A.3, using the corresponding normalization factors between z and J according to the fitting presented in Figure 7.5a

that occur during S–S annihilation (see Section 7.2.4.1). The consequences of most of these effects are demonstrated in Fig. 7.8. The absorption spectrum $\sigma_2(\lambda)$ is a continuous function of λ and here we assume that it does not contain any well-expressed bands in the spectral region of the ground-state absorption. Thus, according to Eqn 7.96, the difference spectrum reflects the bleaching of the ground state absorption $\sigma_0(\lambda)$ of the antenna pigments. This is indeed

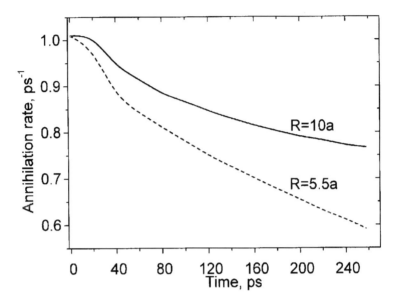

Figure 7.6 The time course of the annihilation rate for small ($R = 5.5a$) and large ($R = 10a$) chromatophores. Excitation intensity $J = 0.65$ photons per RC per pulse; $\eta = 1$; $R_0^1 = 5.5a$; the remaining parameters are as used in Figure 7.5a

observed in the purple photosynthetic bacterium *Rh. rubrum* under very high excitation conditions [7, 8]. Another feature of Eqn 7.96 is the linear dependence on excitation intensity. Moreover, the slope of this linear dependence is directly connected with k_2, the rate of excitation relaxation from the higher excited state S_2 (see Eqn 7.17). The experimental results [7] demonstrate such a linear dependence and the estimated value of k_2 is of the order of $1ps^{-1}$. Measurable excitation of the S_2 state occurs only during the action of the excitation pulse and, as follows from Eqn 7.19, its value becomes small due to the fast relaxation rate k_2. Thus, the difference spectrum (Eqn 7.96) obtained at high excitation intensities is transformed after termination of the excitation pulse into the spectrum (Eqn 7.96) which evolves with the time-dependence of $n_1(t)$ according to Eqn 7.20. A similar interpretation of such experiments [7, 8] has been presented [50, 51] to explain the spectral changes at high excitation intensities.

7.3.1.2   77 K results

To illustrate the analysis of annihilation kinetics in a spectrally inhomogeneous (see Section 7.2.1.5) molecular aggregate, we discuss the annihilation rate time dependence $\gamma(t)$ obtained from the kinetic measurements of *Rh. rubrum* at 77 K,

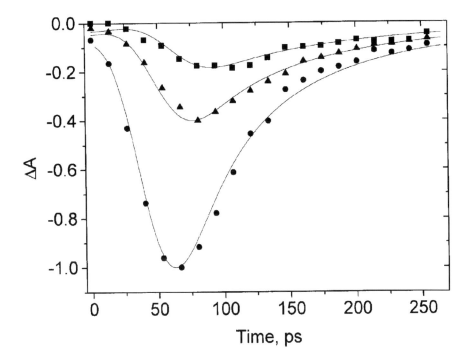

Figure 7.7 A comparison of the model kinetics calculated according to Eqns 7.8a,b, 7.9, and 7.27, with Förster annihilation radius $R_0 = 5.5a$ ($a = 20$ Å; remaining parameters as used in Figure 7.5a) with the experimental kinetic curves [66] at different excitation intensities: (■) 0.031 mJ cm^{-2} (0.12 photons per RC), (▲) 0.081 mJ cm^{-2} (0.32 photons per RC), and (●) 0.34 mJ cm^{-2} (1.4 photons per RC)

using a pulse energy of 2.5 nJ and with 2 ps time resolution [92]. The time-dependence of the annihilation rate can be obtained by transforming Eqn 7.42 and substituting Eqn 7.41, giving

$$\frac{n_0}{n(t)} - 1 = \frac{2n_0\gamma_0}{d_s} t^{\frac{d_s}{2}}. \tag{A.1}$$

Here on the left-hand side we have the reciprocal of the normalized excitation density which can be directly related to the experimental decay curve. In order to avoid unreasonably short decay components, the time interval from 5 to 50 ps was chosen; thus, the linear quenching with $k^{-1} = 210$ ps in this interval is negligible. It is evident that a linear plot of the time dependence versus $t^{d_s/2}$ according to Eqn A.1 is convenient for determining the spectral dimension d_s (see Fig. 7.9). A number of independent experimental curves have been analysed giving the average value $d_s = 1.5^{+0.3}_{-0.2}$. For the other parameter, a

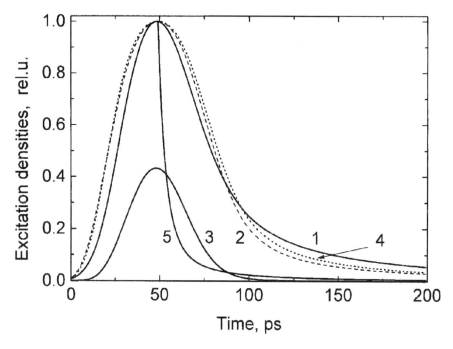

Figure 7.8   Calculated decays of S_1 and S_2 excitation densities at various excitation intensities (Eqns 7.8a,b, 7.9, and 7.27). Kinetic traces 1 and 2 for S_1 excitations show no significant differences despite excitation intensities, which differ by an order of magnitude: 0.5 and 5.0 photons per pulse per molecule, respectively. Trace 3 is for the S_2 excitation density. Trace 4 is calculated with annihilation rate $\gamma = 0.5$ ps^{-1} (compare with $\gamma(t)$ in Figure 7.6 for $R = 10a$) and neglecting the S_2 state occupation. Trace 5 is obtained for $\delta(t)$ pulse excitation with the same conditions as for trace 4. Curves 3–5 are calculated for $J_0 = 5.0$ photons per pulse per pigment. Other parameters are as follows: $k_c^{-1} = 200$ ps, $k_2^{-1} = 1$ps, $\tau_{\text{pump}} = 35$ps, $N_{tr} = 12$

value of $n_0\gamma_0 = 0.04(\pm0.005)ps^{d_s/2}$, was obtained, which by estimating $n_0 = 0.01$ (1 excitation per 100 molecules per pulse) at the given excitation intensity leads to the conclusion that $\gamma_0 = 4ps^{d_s/2}$.

The physical meaning of d_s can be related to the spectral dimension, which is due to fractality of the structure [37, 47]. This parameter contains information about the changes in the symmetry and structure of the paths along which the excitation moves. For an ideal two-dimensional LHA aggregate the most straightforward expectation at room temperature would be $d_s = d = 2$ (non-fractal structure). The value obtained above is somewhat smaller, and at 77 K it is due to the spectral inhomogeneity of the LHA. The inhomogeneous distribution of the site energies forms a potential surface for the excitation, a random landscape of "valleys" and "ridges." The excitation then migrates through

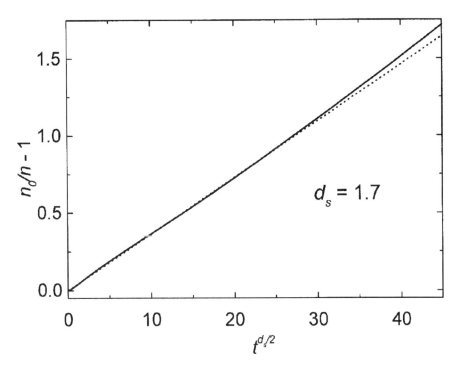

Figure 7.9   The power-law time-dependence of the annihilation rate in LHA of photo-synthetic bacteria *Rh. rubrum* at 77 K. The solid line represents experimental data [92]. The dotted line corresponds to an exact power-law dependence, $t^{0.85}$

"valleys," and consequently the pathway of excitation migration has a fractal-like structure. This interpretation is supported by a much weaker time dependence of $\gamma(t)$ at room temperature (see Fig. 7.6).

7.3.1.3   Fluorescence quantum yield

The process of excitation trapping after an intense laser flash competes with S–S annihilation, as is evident from multiple flash experiments with photosynthetic systems in plants [18, 26]. Similar experiments with bacteria [4] have shown that using picosecond intense laser pulses it is very difficult to convert all the traps into a closed state, demonstrating that even at medium to high excitation intensities, S–S annihilation is more effective than trapping. From Eqn 7.37, the ratio of the quantum yield of excitation trapping $Q(z)$ and the relative fluorescence quantum yield ϕ_f – which ratio equals 1 under annihilation-free conditions – can be calculated [26, 85]:

$$Q(z) = \frac{1}{z}[1 - \eta(0)][1 - \exp(-Q_0\phi_f z)], \tag{A.2}$$

where $Q_0 = k_0/(k_0 + k)$ and $z = \int J(t)\mathrm{d}t/N_{tr}$ is the number of excitations per RC. Equation A.2 is obtained by definition that $Q(z) = [\eta(\infty) - \eta(0)]/z$. It is evident that at low excitation intensities ($z \ll 1$) and in the case of open RCs ($\eta(0) = 0$), we have $Q(z) = Q_0$. We note that the relationship between the fluorescence and trapping quantum yields, given by Eqn A.2, is a direct consequence of the lake model of the LHA [4, 16, 26, 85]. In the puddle model, i.e. for the model of disconnected RCs each with their own LHA, a relation similar to Eqn A.2 can be obtained by replacing ϕ_f by 1 [26].

Eqn A.2 is sensitive to N_{tr}, the number of the LHA pigments per RC, i.e. it is dependent on the normalization of z. For instance, from the experimental data for *Rh. rubrum* [4] illustrated in Fig. 7.5a, the point of the intensity at which $zN_{RC} = 1$ (N_{RC} is the number of RCs per domain) is indicated (see Fig. 7.5a) and by applying Eqn. A.2 where $Q(z)$ becomes close to Q_0, we obtain a value for N_{RC} between 20 and 25 at room temperature. At high excitation intensities, when $z \gg 1$, Eqn 7.37 gives $Q(z) = [1 - \eta(0)]/z$, i.e. the excitation trapping efficiency becomes independent of the structure and parameters of the LHA. This is the main reason why calculations based on the statistics of excitation distribution in the domain alone (and not taking into account the actual dynamics of the processes under consideration) fit the experimental data [61].

Equation 7.35 determines the average rate of excitation trapping in a domain with open and closed RCs. The averaging procedure assumes that the state of a particular RC is independent of the state of other RCs or, in other words, that the RCs are not combined into clusters within the domain. Therefore, the average of effective trapping rate in a domain containing such a mixture of open and closed traps is obtained by taking the sum of the excitation trapping rates for open (k_0) and closed (k_c) RCs, multiplied by the relative probability of finding the RC in each of these states, i.e. by $[1 - \eta(t)]$ and $\eta(t)$, respectively.

A final assumption made in obtaining Eqn 7.35 is that the process of trapping by an individual RC is mono-exponential. This is supported by calculations for two- and three-dimensional arrays of pigments assuming periodic boundary conditions and a homogeneous initial distribution of the excitation [38, 89]. Therefore, this assumption is consistent with the averaging procedure discussed above, and Eqn 7.35 can be used to describe intensity-dependent kinetics for systems containing one sort of LHA pigment and at high temperatures, where spectral inhomogeneity of the LHA is not essential.

If the excitation decay kinetics $n_1(t)$ can be approximated as a single exponential, i.e. $n_1(t) \simeq n_1(0)\exp(-kt)$, it follows from Eqn 7.36 that the excitation trapping rate is equal to $n_1(0)N_{tr}k_0$. Through the initial value $n_1(0)$, the trapping rate is proportional to the excitation intensity. Only when $n_1(0) \ll 1$ does the excitation trapping rate equal k. Therefore, only at low excitation intensities the excitation trapping rate can be determined by analysing $\eta(t)$ kinetics. It is noteworthy that according to Eqn 7.66 and also by using the normalization of the excitation per RC (z), instead of excitation per domain (y), the relative fluorescence quantum yield $\phi_f(z)$ for large aggregates (the lake model) can be determined as follows:

$$\phi_f(z) = \frac{k}{z\gamma}\ln(1 + \frac{z\gamma}{k}), \tag{A.3}$$

The normalization of the annihilation rate (γ) per PSU, correspondingly, also has to be undertaken. For a sufficiently large domain (containing, for instance, 20–25 RCs) as is the case for photosynthetic bacteria *Rh. rubrum*, the initial slope of the fluorescence quantum yield (the low-intensity side in Fig. 7.5a) always can be described well according to Eqn. A.3. Such approximation is a useful definition of the ratio γ/k. The bounds of its applicability can be evidently demonstrated in terms of dependence of value $\Omega(z) = \exp[\phi_f(z)z\gamma/k] - 1$ on the excitation intensity (see Fig. 7.5b). The deviation of the values calculated according to the experimental results from the corresponding linear dependences is evident at high excitation intensities. For the closed RCs, the number of annihilated excitations is larger in comparison with that for the initially open RCs. Moreover, the effect of closing of RCs, described by function $k_1(t)$ (see Eqn 7.35), increases the fluorescence quantum yield at higher excitation intensities in comparison with the prediction of Eqn A.3. Both of these aspects are related to larger departure of the experimental data from the straight line for the closed RCs (see Fig. 7.5b).

Recent analysis of the temperature dependence of excitation trapping by RCs has suggested that two distance scaling parameters determine the organization of the LHA pigments: the pigment–pigment distance within the LHA and the pigment–RC distance [79, 91]. It was concluded that the distance scaling parameter from the LHA to the RC is about 1.7 times larger than the corresponding parameter within the LHA and, thus, the energy trapping rates k_0 and k_c are mainly limited by the slow rate of energy transfer from the nearest surrounding pigments to the RC. Also in that case, the mono-exponential approximation for the kinetics of excitation trapping holds [79, 89]. However, these differences in the distance scaling parameters imply that the excitation equilibrates within the domain of the LHA before being trapped by the RC, and the kinetic processes taking place within the LHA become independent of the trapping processes. This implies that the S–S annihilation and trapping in fact probe different energy-transfer steps.

7.3.2   Fenna–Matthews–Olson complexes

7.3.2.1   Room temperature

Let us now consider the annihilation process with regard to the structural data of Fenna–Matthews–Olson (FMO) complexes and the spectral peculiarities of its pigments. Biochemical assays and spectroscopic analysis [71, 99] show that in solution the FMO complexes are arranged in trimers; thus, it is natural to assume the absence of the energy transfer between different trimers. As already determined, the mean timescale for inter-pigment energy migration within the monomer is of the order of 100–900 fs [77] at room temperature. This conclusion is also in line with the assumption of the presence of excitonic interactions in the monomers at lower temperatures [41, 99]. Thus, taking into account an initial Poissonian distribution of excitations among the trimers, which means that annihilation takes place only in trimers in which more than one excitation is created, the 7 ps relaxation component evidently corresponds to energy transfer between excited monomers within the trimer. According to the S–T annihilation data [99], energy migration between monomers is incoherent even at very low temperatures. Thus, because the migration of energy within a monomer is much faster than its transfer between monomers, at least at room temperature, the whole energy migration process can be separated into two stages: excitation equilibration within each separate monomer and energy transfer between monomers. Such a conclusion is supported by consideration of the different inter-pigment distances within the monomers and between them.

Equilibration of excitation within the monomer leads toward localization of the excitation on Bchl *a* No. 7 [71]. Thus, the approximation that the energy transfer between monomers proceeds mainly between Bchl *a* No. 7 molecules seems to be reasonable. According to the Boltzmann distribution, at room temperature the excitation resides on such molecules for about 40% of its lifetime; therefore, the annihilation between excitations located on different monomers could proceed in two different ways: (i) annihilation as a one-step process due to the interaction of two excited Bchl *a* molecules located on different monomers; or (ii) transfer of the excitation energy to the same monomer by interaction between the excited and unexcited molecules in different monomers, and then annihilation between two excitations within the same monomer. For incoherent (Förster-type) energy transfer between molecules, the ratio of these rates can be estimated by analysing the spectral overlap of the donor molecule fluorescence and the acceptor molecule absorption [94]. Rough deconvolution of the difference spectrum at room temperature [35] evidently leads us to conclude that the emission spectrum overlap with the excited state absorption spectrum is about seven times smaller than that with the ground-state absorption spectrum and, correspondingly, the rate of the first mechanism is much slower. Thus, the annihilation rate should be slower than the excitation energy transfer rate. This finding is in agreement with the inter-monomer energy transfer being of the order of several picoseconds, as obtained from femtosecond excitation kinetic measurements [77]. At intermediate excitation intensities, when no more than two excitations are created within the trimer, the annihilation is a single-exponential process; while at higher intensities, where three excitations per trimer are created, the annihilation rate becomes a complex function of excitation intensity and time (see Eqn 7.53). Thus, the kinetic processes taking place at room temperature are completely explained within the context of the inhomogeneous energy migration.

7.3.2.2   77 K results

At 77 K, the absorption spectrum is more complex, containing three sub-bands within the Q_y molecular transition band that are most probably due to coherent interaction between the Bchl *a* molecules in the monomer [71] (see also Chapter 10). However, according to excitonic calculations [71, 99], the excitation of the longest-wavelength sub-band (825 nm) also corresponds to exciton localization, presumably on Bchl *a* No. 7. This could be the reason why the annihilation rate at 77 K is the same as at room temperature.

Due to the exciton interaction within the monomer, the relaxation process and the excitation localization on Bchl *a* No. 7 has to be fast. Femtosecond spectral equilibration studies show that "downhill" energy transfer at room temperature has several stages, with lifetimes ranging from ~ 100 to ~ 900 fs [35, 75–77]. However, evolution of the transient difference spectra at 77 K suggests a 26 ps equilibration time. On lowering the temperature, the equilibration rate can become slower due to changes in the density of vibrational modes involved in the relaxation process; however, it is very unlikely that these changes could be so strong. Moreover, the annihilation time is found to be 7 ps, implying that the excitation equilibration should be even faster – because the excitations, as discussed, should be localized on Bchl *a* No. 7 prior to annihilation. Thus, within a time resolution of 1 ps the spectral changes have to be related to the thermalized exciton redistribution. According to the Boltzmann distribution, at 77 K about 95% of the excitons should be situated on the lowest excitonic level when the system is thermalized. However, as already shown [98], the lowest excitonic state corresponds mainly to the localization of the excitation on Bchl *a* No. 7. Therefore, the difference spectrum, when the energy redistribution process within the whole trimer is over, has to correspond to the difference spectrum that is mainly determined by bleaching of the longest-wavelength band with a slight ($\sim 5\%$) mixing of the higher exciton states. However, this is not the case for the difference spectrum even at 90 ps delay (see Fig. 7.10). The bleaching bands of the higher excitonic states cannot be related

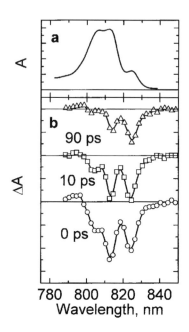

Figure 7.10 The absorption spectrum (a) and transient difference absorption spectra (b) at various delay times after excitation of the FMO pigment–protein complexes at 77 K

to a shift of the exciton bands through the occupation of the lowest exciton state, because in this case an increase in the absorption in some spectral regions would have to appear, and the bleaching spectrum could hardly resemble the steady state absorption spectrum so closely.

Estimating the local temperatures which, according to the Boltzmann distribution, would correspond to the experimentally measured intensities of the bleaching bands at 814 nm and 805 nm compared to the lowest one at 825 nm, gives approximately 260 K at 0 ps, 220 K at 5 and 10 ps, 170 K at 30 ps, and 150 K at 90 ps. It is noteworthy that estimations obtained from the ratio of the bleaching intensities at 825 nm and 814 nm bands, and those at 814 nm and 805 nm, provide very similar temperature values, confirming that some sort of thermalization is really taking place.

The highest possible thermal energy excess supported by the Bchl a molecules in the FMO complex during the relaxation could be estimated by taking into account the heat capacity of the Bchl a molecule. However, since its value at 77 K is not available, for estimations the known heat capacity values for similar organic molecules can be used, and by taking into account the number of atoms in the molecule (the Bchl a) the heat capacity of the molecule is estimated to be equal to 100–150 J mol^{-1} K^{-1}. Thus, for excitation at 760 nm, the increase in local temperature by $\sim$ 10 K for all seven Bchl a molecules can be estimated assuming that the thermal energy excess created during relaxation to the lowest exciton state is concentrated only on the Bchl a molecules, and is equally distributed over all seven of them. However, much higher values of the local temperature can be stimulated by S–S annihilation. The thermal energy created during a single event of the S–S annihilation is enough to raise the temperature of all seven Bchl a molecules by more than 100 K. If we assume that only two excited Bchl a No. 7 molecules are involved in the nonlinear annihilation, then the increase in temperature for the excited molecule can be as high as 300–500 K, even if an increase in the heat capacity with temperature is also taken into account. This result is not very surprising. Similar values of molecular heating were achieved in numerous organic molecules in solution [19] and even in the solid state [33]. The thermal relaxation of the Bchl a molecules should be defined by further redistribution of the vibrational energy initially concentrated in the "heated" molecular modes among the vibrational modes of the surrounding protein. Typical times for energy redistribution between solute molecules and their nearest solvent surrounding range between 15 and 40 ps, while further energy redistribution over the entire solvent is still far slower and can extend to the nanosecond timescale [19]. Thus the nonequilibrium temperature, calculated assuming a Boltzmann distribution between the exciton energy states of the monomer, corresponds reasonably to the value of the temperature obtained from estimates of the thermal energy excess, the 26 ps spectral evolution time also being in good agreement with typical cooling rates for the molecules in solution [3]. Such a slow cooling might be also related to the slow kinetics observed in FMO at low temperatures when exciting at the blue side of absorption band [23, 57].

The model which allows us to understand the redistribution kinetics at 77 K based on the heating/cooling processes of the Bchl a molecules is as follows. Two stages of the heating process can be distinguished. The first concerns heating of the Bchl a molecules during the excitation equilibration through the excitonic states. During this stage, which lasts $\sim$ 1 ps, the temperature of all seven pigment molecules in the monomer increases by 10 K. The second stage involves heating stimulated by S–S annihilation. During this stage, which lasts for about 10 ps, the Bchl a No 7 molecules involved in the nonlinear annihilation are heated up to 300–500 K. The timescale for heating is comparable to the excitation pulse duration, and therefore the largest heat values are achieved during the pulse action. Further redistribution of the thermal energy to the nearest protein surroundings is defined by a mean time of 26 ps when the local temperature of the Bchl a No. 7 molecule and that of its nearest surrounding reaches the same values, which could be estimated as about 150 K. This local thermal relaxation occurs on the nanosecond timescale.

7.3.3 The effect of singlet–triplet annihilation on fluorescence induction in photosystem II

The fluorescence from the common LHA is dependent on the state of the RCs in a nonlinear way, generally ascribed to large-scale connectivity [55, 102]. The dependence of the fluorescence yield on the excitation light intensity has been widely used in one-pulse [13, 55, 102] and two-pulse [22, 88, 93] experiments to investigate the connectivity between PSII on the millisecond to second time-scale, as well on shorter timescales [17, 22]. Here, we will not present an analysis of the fluorescence induction but briefly discuss the possible effect of S–T annihilation on the fluorescence induction [12, 87] by using the theoretical description of the processes presented in Section 7.2.3.2. This approach, described by Eqns 7.77–7.81, can be easily generalized for the two-pulse case, when the first pulse prepares the RC states and generates triplets while the second (probing) pulse creates some further excitation. The fluorescence quantum yield of these excitations is sensitive to the states of the RCs and the amount of triplets; hence the actinic pulse intensity determines the fluorescence induction.

The calculated results for the fluorescence induction just after the actinic pulse, when the triplets can also directly quench the fluorescence induced by the probe pulse, as shown in Fig. 7.11, demonstrate a strong effect of S–T annihilation. However, the most easily measurable parameter is the R value, which in the case of S–T annihilation also becomes dependent on the actinic pulse duration (see Fig. 7.12) [87]. Thus, the given analysis clearly demonstrates that when using microsecond flashes to cause fluorescence induction, S–T annihilation will perturb the induction curve, which might be the possible explanation for the fluorescence induction shape and the R-value dependence on the actinic pulse duration [22].

7.3.4 Phthallocyanines

At high excitation intensities phthalocyanine (Pc) structures of various molecular arrangements, such as thermally evaporated crystalline and amorphous films, or Pc dispersed in polymer matrices, exhibit ultrafast annihilation kinetics which determines the main excitation relaxation timescale [11, 30, 31, 83]. It has also been observed [30, 31] that the annihilation rate is strongly time-dependent. It is evident from the analysis presented above that the additional effect caused by this relaxation and determining the dynamics of absorption spectra of the film on a picosecond timescale is the local heating of the aggregate.

In the visible and near-infrared regions, thermally evaporated polycrystalline vanadyl and lead phthalocyanine (VOPc and PbPc) film absorption spectra

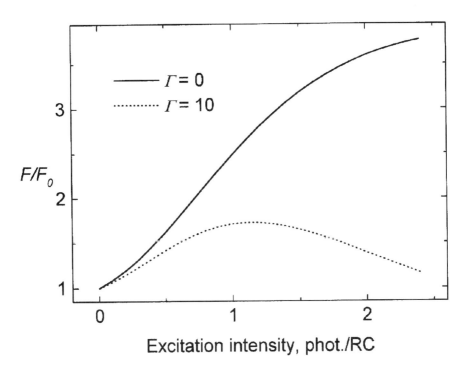

Figure 7.11   Fluorescence induction curves at the end of the actinic pulse, for an actinic pulse length of 0.2α and for two values of the S–T annihilation rate ($\Gamma = 0; 10$). The remaining parameters are $I_T = 0.07; N_{RC} = 4$

contain a wide Q band, i.e. an S_0–S_1 molecular transition around 780 nm with three distinct maxima [34]. Similar spectral features have been observed in numerous studies, and it is evident [103] that the absorption maxima and relative intensities of the bands indicate some variation in quality of the crystalline structure, the size of crystallites, and the ratio of the fractions of the crystalline and amorphous phases, as well as the variation of the central atom. Transient absorption spectra of VOPc films measured at different delays after excitation are shown in Fig. 7.13. At short delay times the spectra correlate well with the normal absorption spectrum, except for the broad induced absorption in the 400–550 nm region [34]. At long delay times (> 50 ps), the difference spectra are quite different in shape from those at short times, indicating a blue shift of the red edge of the induced absorption band, and remaining almost the same for more than 1 ns. These spectral changes are very similar to those induced by heating the samples [11] both with respect to isosbestic points and band shapes.

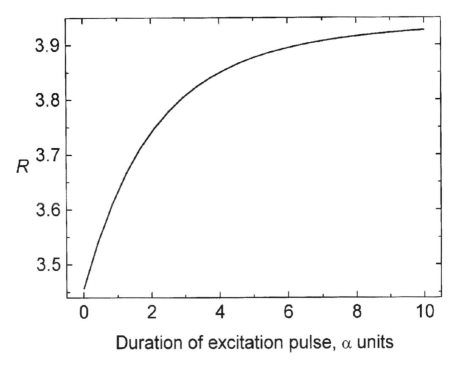

Figure 7.12   The dependence of the fluorescence induction ratio on the actinic pulse duration. $\Gamma = 10; I_T = 0.07; N_{RC} = 4$. The time delay between the actinic and the probe pulses is 10α

7.3.4.1   Local heating

The long-timescale kinetics at the red edge of absorption spectrum of VOPc film clearly indicates a low-amplitude oscillation feature with a period of ~ 100 ps (see Fig. 7.14). Similar kinetic behavior observed in VOPc film [11] with pico-second time resolution has been assumed to be due to local heating of the aggregate and the generation of acoustic waves. Although the density of the excitation used during the femtosecond measurements [34] is 10–100 times lower than it is in picosecond experiments [11], it is nevertheless sufficiently high to induce absorption changes through the essentially local heating effect. With excitation pulses of 1.2 mJ cm^{-2}, the temperature of the VOPc film increases by approximately 15–25 K in the case in which all of the excitation energy is converted into heat, resulting in the difference absorption spectrum observed at 150 ps (Fig. 7.13). If the main exciton relaxation channel is nonradiative, the rate of temperature increase is proportional to the exciton density decay rate. A simple estimate shows that during the first several picoseconds after excitation

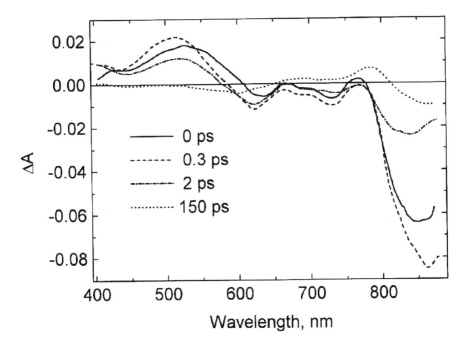

Figure 7.13 Difference absorption spectra of VOPc films excited by 200 fs and 1.2 mJ cm^{-2} energy density pulses at various delays after excitation. Excitation wavelength $\lambda_{ex} = 600$ nm

it exceeds 10^{13} K/s^{-1}. Due to the very fast temperature increase, the structure of the aggregate cannot change as fast as the temperature. In Pc films, the main limiting process is thermal expansion of the aggregate. Therefore, initially after the relaxation, the heated aggregate is in a nonstationary state, with a density higher than it should have at this temperature. The absorption of the aggregate in this state is different when the expansion has already occurred [11]. Due to the stress created in the sample, acoustic waves are generated, causing the oscillations in the absorption. Such oscillations were observed in VOPc film excited by picosecond pulses, as well as in some other materials [11, 33]. An oscillating feature in the transient absorption, although not well expressed, is also observed in the kinetics at 410 nm [34], the period of these oscillations being dependent on the thickness of the film [11]. Alternatively, energy redistribution inside the vibrational manifold can complicate the influence of local heating on the absorption dynamics (see Section 7.2.4.2). The generalized coordinate Q describes the intermolecular distance in the case of VOPc. This is the case when the viscosity is low and the kinetics of coordinate Q slow in comparison with T evolution – thus, damped oscillations can take place according to Eqn 7.12.

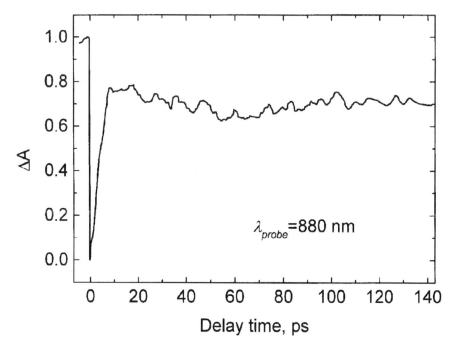

Figure 7.14 Transient absorption kinetics of VOPc film at 880 nm wavelength on a long timescale

7.3.4.2 One-dimensional excitation annihilation dynamics

It is known that intra- and intermolecular vibrational energy redistribution occurs in the femto- and picosecond time domains [19]. Thus, local heating of the aggregate excited by ultrashort light pulses is a very complex process, which is difficult to take into account. The only possibility of obtaining absorption kinetics free of the heating influence is to find spectral points at which this influence is at a minimum [34]. Such a spectral point in the VOPc film could be at ~ 525 nm, close to the blue edge of the S_1 state manifold in the absorption spectrum. At long times the transient spectral features caused by local heating are small in this spectral region. The second advantage of this spectral region is that it corresponds to the maximum of the induced absorption, and the initial spectral evolution observed as a shift of the induced absorption band has a minimum influence on the absorption dynamics at 525 nm. Thus, it can be expected that the transient absorption at 525 nm is proportional to the density of excitations n only, and this spectral region is convenient for analysis of the relaxation dynamics of excitations in the VOPc film. In PbPc film, the point at which the thermally induced spectrum has no effect is at 540 nm, i.e. also close to the absorption minimum.

Excitation relaxation dynamics in the annihilation-dominant time interval, for the first 10–15 ps in both Pc films, are well described by Eqn 7.20, taking $k_1(t) = k$ and using the annihilation rate approximation,

$$\gamma(t) = \gamma_0 t^{-\frac{1}{2}} \tag{A.4}$$

(see Eqn 7.31). This dependence has also been found in polycrystalline thin films of β-hydrogen phthalocyanine [30, 31]. In general, $\gamma(t)$ is determined by excitation correlations via the pair correlation function $g(r, t)$ (see Eqn 7.9a), the latter being determined through Eqn 7.11a. Two limiting cases of annihilation are possible, i.e. static annihilation or a migration-controled process.

In the static annihilation case the first term in Eqn 7.11a can be neglected, i.e. the diffusion constant $D \approx 0$. From Eqn 7.27 higher order correlations – the term $H(g(r, t))$ – can also be neglected for times

$$t < \frac{1}{4n\gamma_{\text{stat}}(t)}. \tag{A.5}$$

These conditions give the solution

$$g(r, t) = e^{-2\lambda(r)t} \tag{A.6}$$

and, according to Eqn 7.9a, it follows that

$$\gamma_{\text{stat}}(t) = \gamma_d^{\text{stat}} t^{-(6-d)/6}, \tag{A.7}$$

where d denotes the dimension of the aggregate. In the time period defined by Eqn A.5, the linear relaxation term is negligible ($kt < 1$) and the solution of Eqn 7.20 reads

$$n(t) = n_0/(1 + \frac{6}{d} n_0 \gamma_d^{\text{stat}} t^{d/6}). \tag{A.8}$$

Here, n_0 is the initial occupation of the excited molecules. At higher intensities, when the inequality

$$\frac{6}{d} n_0 \gamma_d^{\text{stat}} > k^{d/6} \tag{A.9}$$

is fulfilled, Eqn A.5 can be rewritten as $1 < 6/d$. This is fulfilled for all dimensions up to $d = 3$.

In the case of diffusion-limited annihilation, defining $\gamma_{\text{dif}}(t)$ as the flux of the excitations (Eqn 7.28) on the reaction sphere of radius R_1 (Eqn 7.30), it follows from Eqn 7.31 that the observed power law of the annihilation rate time dependence (Eqn A.4) can be expected for dimensions $1 \leqslant d < 2$.

Analysis of the rate time-dependence (see Eqn A.4) shows that both limiting cases could be adequate candidates for explaining the experimental data [34]. However, the reaction radius R_1 here could serve as a critical parameter in distinguishing the relevant dimension of the Pc aggregate. When $R_1 \gg a$ (see Eqn 7.30), static annihilation dominates. This means that the overlap of the fluorescence and excited-state absorption spectra (determining the Förster annihilation radius R_1^0) has to be much larger than that of the fluorescence and the ground-state absorption spectra (determining the Förster diffusion radius R_0). However, comparison of these spectra for VOPc as well as PbPc films does not provide support for this [34].

Thus, the above considerations lead to the conclusion of a one-dimensional diffusion-limited annihilation in both PC films. In this case, γ_0 is related to the excitation hopping time between the neighboring molecules by the following relation [58, 67]:

$$\tau_h = \frac{4}{\pi \gamma_0 M}, \tag{A.10}$$

where M is the number of molecules in the domain, giving hopping times of 0.1–0.4 ps and 0.02–0.08 ps for VOPc and PbPc, respectively.

7.4 DISCUSSION

The intrinsic spatial scaling parameter for the processes under consideration is the mean distance between the excited molecules in singlet (for the S–S annihilation) and triplet (for the S–T annihilation) excited states. The conditions for observation of nonlinear annihilation also generally increase the efficiency for excitations to reach each other by scattering. The latter process is evidently determined by another spatial scaling parameter, i.e. the excitation diffusion radius R_{dif} determined in Eqn 7.1. Thus, the excitation concentration has to be sufficiently high for R_{dif} to exceed the mean distance between excitations.

Considering spatially confined systems, such as molecular aggregates, colloids, films and pigment–protein complexes (but not infinitely extended ones, however), the interplay between the parameters mentioned above and the dimension L of the system under consideration also becomes important. It is evident that L should always exceed the mean distance between excitations in order for nonlinear annihilation to be present – or in other words, on average there have to be at least a few excitations per domain.

The ratio between R_{dif} and L determines the limiting stages of the processes under consideration and, correspondingly, the theoretic approach to be applied. In the case in which Eqn 7.1 is satisfied, we are dealing with large molecular aggregates. Then the excitation diffusion radius R_{dif} determines the domain size, and the annihilation process is diffusion-limited. Conversely, when $R_{dif} \gg L$, the small-aggregate approach can be used, as the domain size is determined by the natural size of the aggregate. In that case, the excitation diffusion is very fast, and excitations are therefore equilibrated over the aggregate before the annihilation takes place. Hence, the annihilation rate is determined by the static interaction of two excitations, while the statistical distribution of the excitations contributes mainly via the initial distribution per domain. In that case, the domain can be considered as a supermolecule consisting of molecules incoherently coupled (via excitation migration), and nonlinear annihilation determines the process of excitation relaxation from the highly excited supermolecule states. However, it is noteworthy that the S–S annihilation of coherent excitons corresponds to the analogous supermolecular approach for coherently coupled molecules. The main difference between the latter and the incoherent supermolecular approach (or small-aggregate approach) is in the initial statistical distribution, which may be affected by the coherence of the light–matter interaction [63]. Moreover, due to the relaxation from the higher excitation states caused by exciton scattering, the excitation also loses its coherence. Therefore, it seems that after a few steps of relaxation the aggregate turns into a supermolecule of incoherently linked molecules, and the excitation diffusion coefficient can also be determined. Hence, on timescales longer than the excitation pulse duration and the multi-excitation relaxation mean time, the excitation kinetics can be well defined by the same approach as incoherent excitation annihilation.

An additional factor stimulating loss of coherence during the annihilation is related to the local heating. It is evident that at higher temperatures the excitation interaction with inter- and intramolecular vibrations increases, also destroying the excitation coherence. However, heating of the local surroundings of the molecules also disturbs the transient spectra, even for incoherently coupled molecules. This effect is distinguishable in the transient absorption spectra but not in the fluorescence kinetics, and the corresponding spectral changes are proportional to $\sigma_0^T(\lambda) - \sigma_0^{T_0}(\lambda)$ (see Eqn 7.106). The relaxation rate of these spectral changes is determined by heat exchange with the thermal bath, i.e. with energy relaxation into other vibrational modes (see Eqn 7.87) or with the diffusion of heat into more distant parts of the surroundings, e.g. protein, solvent, etc. Thus, comparative analysis of the transient absorption spectra with the fluorescence kinetics provides information about the rates of heat exchange. As has been demonstrated [35], the local temperature relaxation kinetics in some molecular aggregates is slower than the nonlinear annihilation of the electronic excitation and, thus, the former determines the long-term changes observed in the transient spectra.

Other aspects which also have to be taken into consideration for nonlinear annihilation are (i) the higher-state occupation during annihilation and (ii) the correlation between excitations. The possibility of populating highly excited molecular states during S–S annihilation has also been demonstrated in anthracene crystals [43]. However, it is important to take these effects into account by analysing the excitation kinetics and not by considering the fluorescence quantum yields. This is because the population of higher excited states occupied in the course of the excitation pulse vanishes by relaxation when the pulse action is over. Spectral effects due to the correlative behavior of excitations can compete with those due to occupation of higher excited states. For example, such an effect disturbs the initial distribution of triplets in the course of S–T annihilation, observed as an anti-Smoluchowski delayed fluorescence in viscous solutions [65]. Both of these effects may be distinguished by comparing fluorescence and transient absorption measurements. The transient absorption spectra exhibit the effects of changes in the ground-state absorption $\sigma_0^{r,r}(\lambda)$, whereas the fluorescence kinetics only reveal correlative spectral changes in the excited state $\sigma_1^r(\lambda)$ (see Section 7.2.5.1).

The correlation of excitations is also an important factor in the overall nonlinear annihilation kinetics. For small aggregates, this determines the initial distribution of excitations in the domain. Then the kinetics corresponds to averaging over the (Poissonian) distribution of domains with various amounts of excitation (see Section 7.2.2.2 for the S–S annihilation and Section 7.2.3.2 for the S–T annihilation). For large aggregates, the correlation has to be taken into the kinetic scheme explicitly, because the excitation during its lifetime visits different regions of the domain containing different amounts of excitations. Thus, the annihilation rate is dependent on the size of the aggregate L as well as on the dimension of the system. The larger the aggregate size and dimensionally, the less the annihilation rate depends on time. All of these aspects have to be taken into account by analysing the nonlinear annihilation kinetics.

The theory presented in this chapter has considered nonlinear annihilation processes that take place in spectrally homogeneous structures. This approach is also valid for high temperatures, when the width of inhomogeneous distribution is much smaller than the thermal energy. However, the theory is also applicable where the opposite holds, and in this case two timescales are present. The first determines excitation transfer to the longest-wavelength pigments (this corresponds to the fast scale determined by a few energy transfer steps). The second timescale defines excitation migration over the longest-wavelength pigments, and the shorter-wavelength pigments can be considered as energy barriers for the excitations. In such a case, the aggregate behaves like a fractal structure, and the excitation kinetics on the second timescale can be considered by relating the spectral inhomogeneity to the spectral dimension d_S of the aggregate (see Section 7.2.1.5).

As has been shown, for large molecular aggregates the nonlinear annihilation processes are diffusion-limited. The possibility of determining the energy transfer parameters for various molecular structures has been demonstrated in Section 7.3. The photosynthetic bacterium *Rh. rubrum* is a good example, since at lower temperatures (77 K), two well-expressed timescales are present: the excitation spectral equilibration is over within 1 ps, while excitation annihilation takes tens of picoseconds. According to the analysis of the S–S annihilation, the mean excitation hopping time (equivalent to the average hopping time determined in Section 7.2.1.1) in this system is 0.5 ps [94]. The energy migration in other photosynthetic pigment–protein complexes is different and slower. For instance, the rate of energy migration in FMO complexes is determined by two timescales: a fast subpicosecond energy equilibration within the monomeric unit and a slower process (over a few picoseconds) between the monomers within the trimer [35]. The S–S annihilation studies for LHA complexes of PS II (LHCII) also indicate a slow (5 ps) excitation hopping time [5, 27].

Analysis of S–S annihilation for metallophthalocyanine films clearly demonstrates that a one-dimensional annihilation is taking place in these aggregates with a very fast (subpicosecond) energy hopping time. It is noteworthy that such fast rates of energy transfer might be determined from analysis of the nonlinear excitation kinetics, obtained even with excitation pulses much longer than the single hopping time itself.

Acknowledgments

The authors wish to acknowledge Dr. J. Connelly for reading the manuscript and for valuable comments. This work was supported in part by the Bundesministerium für Bildung, Wissenschaft, Forschung und Technologie, Bonn, Germany (project no. 0311206).

References

1. Agranovich, V. M., and N. A. Efremov 1980. Luminescence spectra of noncoherent Frenkel excitons at high pumping levels. *Sov. Phys. – Solid State* 22: 583–588.
2. Agranovich, V. M., and M. D. Galanin 1982. *Electron Excitation Energy Transfer in Condensed Matter*. North-Holland, Amsterdam.
3. Bakker, H. J., P. C. M. Planken, and A. Langendijk 1990. Role of solvent on vibrational energy transfer in solution. *Nature* 347: 745–747.
4. Bakker, J. G. C., R. van Grondelle, and W. T. F. den Hollander 1983. Trapping, loss and annihilation of excitations in photosynthetic systems. II. Experiments with the purple bacteria *Rhodospirillum rubrum* and *Rhodopseudomonas capsulata*. *Biochim. Biophys. Acta* 725: 508–518.

5. Barzda, V., G. Garab, V. Gulbinas, and L. Valkunas 1996. Evidence for long-range excitation energy migration in macroaggregates of the chlorophyll *a/b* light-harvesting antenna complexes. *Biochim. Biophys. Acta* 1273: 231–236.
6. Boekema, E. J., B. Hankamer, D. Blad, J. Kruip, J. Nield, F. Boonstra, J. Barber, and M. Rögner 1995. Supramolecular structure of the photosystem II complex from green plants and cyanobacteria. *Proc. Natl Acad. Sci., USA* 92: 175–179.
7. Borisov, A. Y., R. A. Gadonas, R. V. Danielius, A. S. Piskarskas, and A. P. Razjivin 1982. Minor component B-905 of light-harvesting antenna in *Rhodospirillum rubrum* chromatophores and the mechanism of singlet-singlet annihilation as studied by difference selective picosecond spectroscopy. *FEBS Lett.* 138: 25–28.
8. Borisov, A. Y., R. A. Gadonas, R. V. Danielius, A. S. Piskarskas, A. P. Razjivin, and R. Rotomskis 1984. Study of the kinetics of deactivation of excited states in the antenna of the chromatophores of *Rhodospirillum rubrum* by picosecond differential spectrophometry. *Biofizika* 29: 402.
9. Breton, J., N. E. Geacintov, and C. E. Swenberg 1979. Quenching of fluorescence by triplet states in chloroplasts. *Biochim. Biophys. Acta* 548: 616–635.
10. Brown, R., and N. A. Efremov 1991. New Monte Carlo simulations of many-particle stochastic dynamics: growth of correlations and local self-ordering during annihilation of like particles. *Chem. Phys.* 155: 357–368.
11. Butvilas, V., V. Gulbinas, A. Urbas, and A. J. Vachnin 1992. Ultrafast processes in VO-phthalocyanine film caused by intense excitation. *Radiat. Phys. Chem.* 39: 165–169.
12. Cervinskas, V., L. Valkunas, and F. van Mourik 1994. The dependence of the shapes of fluorescence induction curves in chloroplasts on the duration of illumination pulses. The effects of singlet–triplet annihilation. *Lith. J. Phys.* 34: 375–378.
13. Dau, H. 1994. Molecular mechanisms and quantitative models of variable photosystem II fluorescence. *Photochem. Photobiol.* 60: 1–23.
14. Davydov, A. S. 1971. *Theory of Molecular Excitons*. Plenum Press, New York.
15. Deinum, G., T. J. Aartsma, R. van Grondelle, and J. Amesz 1989. Singlet–singlet excitation annihilation measurements on the antenna of *Rhodospirillum rubrum* between 300 and 4 K. *Biochim. Biophys. Acta* 976: 63–69.
16. den Hollander, W. T. F., J. G. C. Bakker, and R. van Grondelle 1983. Trapping, loss and annihilation of excitations in a photosynthetic system. I. Theoretical aspects. *Biochim. Biophys. Acta* 725: 492–507.
17. Deprez, J., A. Dobek, N. E. Geacintov, G. Paillotin, and J. Breton 1983. Probing fluorescence induction in chloroplasts on a nanosecond time scale utilizing picosecond laser pulse pairs. *Biochim. Biophys. Acta* 725: 444–454.
18. Dobek, A., J. Deprez, N. E. Geacintov, G. Paillotin, and J. Breton 1985. Chlorophyll fluorescence in chloroplasts on subnanosecond time-scales probed by picosecond pulse pairs. *Biochim. Biophys. Acta* 806: 81–92.
19. Elsaesser, T., and W. Kaiser 1991. Vibrational and vibronic relaxation of large polyatomic molecules in liquids. *Ann. Rev. Phys. Chem.* 42: 83–107.
20. Fayer, M. D. 1982. Dynamics of molecules in condensed phases: Picosecond holographic grating experiments. *Ann. Rev. Phys. Chem.* 33: 63–87.
21. Fleming, G. R., and R. van Grondelle 1994. The primary steps of photosynthesis. *Phys. Today* 47: 48–55.
22. France, L., N. E. Geacintov, J. Breton, and L. Valkunas 1992. The dependence of the degree of sigmoidicities of fluorescence induction curves in spinach chloroplasts on the duration of actinic pulses in pump–probe experiments. *Biochim. Biophys. Acta* 1101: 105–119.

23. Freiberg, A., S. Lin, K. Timpmann, and R. E. Blankenship 1997. Exciton dynamics in FMO bacteriochlorophyll protein at low temperatures. *J. Phys. Chem.* 101: 7211–7220.

24. Gaididei, Y. B., I. V. Zozulenko, and A. I. Onipko 1985. Diffusion mediated annihilation reactions via short-and long-range interactions: determination of the correct boundary condition. *Chem. Phys. Lett.* 118: 421–426.

25. Geacintov, N. E., and J. Breton 1987. Energy transfer and fluorescence mechanisms in photosynthetic membranes. *CRC Crit. Rev. Plant Sci.* 5: 1–44.

26. Geacintov, N. E., G. Paillotin, J. Deprez, A. Dobek, and J. Breton 1984. Picosecond pulse energy dependence of singlet–singlet annihilation and fluorescence in chloroplasts. In *Advances in Photosynthesis Research*. I. C. Sybesma, editor. Martinus Nijhoff/Dr. W. Junk Publishers, The Hague, The Netherlands, pp. 37–40.

27. Gillbro, T., A. Sandström, M. Spangfort, V. Sundström, and R. van Grondelle 1988. Excitation energy annihilation in aggregates of chlorophyll *a/b* complexes. *Biochim. Biophys. Acta* 934: 369–374.

28. Goushcha, A. O., V. N. Kharkyanen, and A. R. Holzwarth 1997. Nonlinear light-induced properties of photosynthetic reaction centers under low intensity irradiation. *J. Phys. Chem.* 101: 259–265.

29. Graener, H., T. Q. Ye, and A. Laubereau. 1989. Ultrafast dynamics of hydrogen bonds directly observed by time-resolved infrared spectroscopy. *J. Chem. Phys.* 90: 3413–3416.

30. Greene, B. I., and R. R. Millard. 1985. Singlet-exciton fusion in molecular solids: A direct subpicosecond determination of time-dependent annihilation rates. *Phys. Rev. Lett.* 55: 1331–1334.

31. Greene, B. I., and R. R. Millard 1985. Subpicosecond spectroscopic studies of singlet exciton fusion in molecular solids. *J. Phys. (Paris)* 46: 371–377.

32. Guillet, J. 1985. Polymer photophysics and photochemistry. *An Introduction to the Study of Photoprocesses in Macromolecules*. Cambridge University Press, Cambridge.

33. Gulbinas, V., L. Valkunas, and R. Gadonas 1994. Local heating and nonlinear annihilation in molecular aggregates. *Lith. J. Phys.* 34: 348–360.

34. Gulbinas, V., M. Chachisvilis, L. Valkunas, and V. Sundström 1996. Excited state dynamics of phthalocyanine films. *J. Phys. Chem.* 100: 2213–2219.

35. Gulbinas, V., L. Valkunas, D. Kuciauskas, E. Katilius, V. Liuolia, W. Zhou, and R. E. Blankenship 1996. Singlet–singlet annihilation and local heating in FMO complexes. *J. Phys. Chem.* 100: 17 950–17 956.

36. Gülen, D., B. P. Wittmerhaus, and R. S. Knox 1986. Theory of picosecond-laser-induced fluorescence from highly excited complexes with small number of chromophores. *Biophys. J.* 49: 469–477.

37. Havlin, S., and A. Bunde 1991. Percolation II. In *Fractals and Disordered Systems*. A. Bunde and S. Havlin, editors. Springer-Verlag, Berlin, pp. 97–149.

38. Hemenger, R. P., R. M. Pearlstein, and K. Lakatos-Lindenberg 1972. Incoherent exciton quenching on lattices. *J. Math. Phys.* 13: 1056–1063.

39. Henry, E. R., W. A. Eaton, and R. M. Hochstrasser 1986. Molecular dynamic simulations of cooling in laser-excited heme proteins. *Proc. Natl Acad., USA* 83: 8982–8986.

40. Holzwarth, A. R. 1989. Application of ultrafast laser spectroscopy for the study of biological systems. *Q. Rev. Biophys.* 22: 239–295.

41. Johnson, S. G., and G. J. Small 1991. Excited-state structure and energy-transfer dynamics of the bacteriochlorophyll *a* antenna complex from *Prostecochloris aestuarii*. *J. Phys. Chem.* 95: 471–479.

42. Karrasch, S., P. A. Bullough, and R. Ghosh 1995. 8.5Å projection map of the light-harvesting complex I from *Rhodospirillum rubrum* reveals a ring composed of 16 subunits. *EMBO J.* 14: 631–638.

43. Katoh, R., and M. Kotani 1993. Observation of fluorescence from higher excited states in an antracene crystal. *Chem. Phys. Lett.* 201: 141–144.

44. Khairutdinov, R. F., and N. Serpone 1996. Kinetics of chemical reactions in restricted geometries. *Progr. Reaction Kinetics* 21: 1–68.

45. Knoester, J., and F. C. Spano 1996. Theory of pump–probe spectroscopy of molecular J-aggregates. In *J-Aggregates.* T. Kobayashi, editor. World Scientific, Singapore, pp. 111–160.

46. Kolubayev, T., N. E. Geacintov, G. Paillotin, and J. Breton 1985. Domain sizes in chloroplasts and chlorophyll–protein complexes probed by fluorescence yield quenching induced by singlet–triplet exciton annihilation. *Biochim. Biophys. Acta* 808: 66–76.

47. Kopelman, R. 1988. Fractal reaction kinetics. *Science* 241: 1620–1626.

48. Kotomin, E., and V. Kuzovkov 1992. Phenomenological kinetics of Frenkel defects, recombination and accumulation in solids. *Rep. Progr. Phys.* 55: 2079–2188.

49. Krauss, N., W.-D. Schubert, O. Klukas, P. Fromme, H. T. Witt, and W. Saenger 1996. Photosystem I at 4Å resolution represents the first structural model of a joint photosynthetic reaction center and core antenna system. *Nature Struct. Biol.* 3: 965–973.

50. Kudzmauskas, S., V. Liuolia, G. Trinkunas, and L. Valkunas 1985. Nonlinear phenomena in chromatophores of photosynthetic bacteria excited by picosecond laser pulses. *Phys. Lett.* 111A: 378–381.

51. Kudzmauskas, S., V. Liuolia, G. Trinkunas, and L. Valkunas 1986. Minor component of the difference absorption spectra of photosynthetic bacteria chromatophores and nonlinear effect during excitation. *FEBS Lett.* 194: 205–209.

52. Kuzovkov, V., and E. Kotomin 1988. Kinetics of bimolecular reactions in condensed media: critical phenomena and microscopic self-organization. *Rep. Prog. Phys.* 51: 1479–1523.

53. Kühlbrandt, W., D. N. Wang, and Y. Fujiyoshi 1994. Atomic model of plant light-harvesting complex by electron crystallography. *Nature* 367: 614–621.

54. Laubereau, A., and W. Kaiser 1978. Vibrational dynamics of liquids and solids investigated by picosecond light pulses. *Rev. Mod. Phys.* 50: 607–665.

55. Lavorel, J., and P. Joliot 1972. A connected model of the photosynthetic unit. *Biophys. J.* 12: 815–831.

56. Li, P., and P. M. Champion 1994. Investigations of the thermal response of laser-excited biomolecules. *Biophys. J.* 66: 430–436.

57. Louwe, R. J. W., and T. J. Aartsma 1997. On the nature of energy transfer at low temperatures in the Bchl *a* pigment-protein complex of green sulfur bacteria. *J. Phys. Chem.* 101: 7221–7226.

58. Markovitsi, D., I. J. Lecuyer, and J. Simon 1991. One-dimensional triplet energy migration in columnar liquid crystals of octasubstituted phthalocyanine. *J. Phys. Chem.* 95: 3620.

59. Mauzerall, D. 1976. Multiple excitations in photosynthetic systems. *Biophys. J.* 16: 87–91.

60. McDermott, G., S. H. Prince, A. A. Freer, A. M. Hawthornthwaite-Lawless, M. Z. Papiz, R. J. Cogdel, and N. W. Issacs 1995. Crystal structure of an integral membrane light-harvesting complex from photosynthetic bacteria. *Nature* 374: 517–521.

61. Mineev, A. P., and A. P. Razjivin 1987. Statistics of the photon distribution in the set of photosynthetic antenna domains. *FEBS Lett.* 223: 187–190.

62. Monger, T. G., and W. W. Parson 1977. Singlet–triplet fusion in *Rhodopseudomonas sphaeroides* chromatophores. A probe of the organization of the photosynthetic apparatus. *Biochim. Biophys. Acta* 460: 393–407.
63. Mukamel, S. 1995. *Principles of Nonlinear Optical Spectroscopy*. Oxford University Press, New York.
64. Müller, M. G., M. Hucke, M. Reus, and A. R. Holzwarth 1996. Annihilation processess in the isolated D1-D2-cyt-b559 reaction center complex of photosystem II. An intensity-dependence study of femtosecond transient absorption. *J. Phys. Chem.* 100: 9537–9544.
65. Nickel, B., H. E. Wilhelm, and A. A. Ruth. 1994. Anti-Smoluchowski time dependence of the delayed fluorescence from antracene in viscous solution due to triplet–triplet annihilation. Effect of Forster energy transfer $S_1 + T_1 \rightarrow S_0 + T_n$ on the initial spatial distribution of molecules in T_1. *Chem. Phys.* 188: 267–287.
66. Nuijs, A. M., R. van Grondelle, H. L. P. Joppe, A. C. Bochove, and L. N. M. Duysens 1985. Singlet and triplet excited carotenoid and bacteriochlorophyll of the photosynthetic purple bacterium *Rhodospirillum rubrum* as studied by picosecond absorbance difference spectroscopy. *Biochim. Biophys. Acta* 810: 94–105.
67. Onipko, A. I., and I. V. Zozulenko 1989. Kinetics of incoherent exciton annihilation in nonideal one-dimensional structures. *J. Luminescence* 43: 173–184.
68. Ovchinikov, A. A., S. F. Timashev, and A. A. Belyi 1989. *Kinetics of Diffusion-controlled Chemical Processes*. Nova, New York.
69. Paillotin, G., N. E. Geacintov, and J. Breton 1983. A master equation theory of fluorescence induction, photochemical yield, and singlet–triplet exciton quenching in photosynthetic systems. *Biophys. J.* 44: 65–77.
70. Paillotin, G., C. E. Swenberg, J. Breton, and N. E. Geacintov 1979. Analysis of picosecond laser-induced fluorescence phenomena in photosynthetic membranes utilizing a master equation approach. *Biophys. J.* 25: 513–534.
71. Pearlstein, R. M. 1992. Theory of optical spectra of the bacteriochlorophyll *a* antenna protein trimer from *Prostecochloris aestuarii*. *Photosynth. Res.* 31: 213–226.
72. Peliti, L. 1986. Renormalization of fluctuation effects in the A + A → A reaction. *J. Phys. A: Math. Gen.* 19: L365–L367.
73. Pope, M., and C. E. Swenberg 1982. *Electronic Processes in Organic Crystals*. Oxford University Press, New York.
74. Rench, S. K., R. V. Danielius, R. Gadonas, and A. Piskarskas 1982. Picosecond kinetics and transient spectra of pseudocyanine monomers and J-aggregates in aqueous solution. *Chem. Phys. Lett.* 84: 446–449.
75. Savikhin, S., and W. S. Struve 1994. Ultrafast energy transfer in FMO trimers from the green bacterium *Clorobium tepidum*. *Biochemistry* 33: 11 200–11 208.
76. Savikhin, S., and W. S. Struve 1996. Low-temperature energy transfer in FMO trimers from the green photosythetic bacterium *Chlorobium tepidum*. *Photosynth. Res.* 48: 271–276.
77. Savikhin, S., W. Zhou, R. E. Blankenship, and W. S. Struve 1994. Femtosecond energy transfer and spectral equilibration in bacteriochlorophyll *a* protein antenna trimers from the green bacterium *Chlorobium tepidum*. *Biophys. J.* 66: 110–114.
78. Silinsh, E. A., and V. Capek 1994. *Organic Molecular Crystals. Interactions, Localisation and Transport Phenomena*. American Institute of Physics, New York.
79. Somsen, O. J. G., F. van Mourik, R. van Grondelle, and L. Valkunas 1994. Energy migration, and trapping in a spectrally, and spatially inhomogeneous light-harvesting antenna. *Biophys. J.* 66: 1580–1596.

80. Struve, W. S. 1995. Theory of electronic energy transfer. In *Anoxigenic Photosynthetic Bacteria*. R. E. Blankenship, M. T. Madigan, and C. E. Bauer, editors. Kluwer Academic Publishers, Dordrecht, pp. 297–313.
81. Suna, A. 1970. Kinematics of exciton–exciton annihilation in molecular crystals. *Phys. Rev. B* 1: 1716–1739.
82. Swenberg, C. E., N. E. Geacintov, and M. Pope 1976. Bimolecular quenching of excitons and fluorescence in the photosynthetic unit. *Biophys. J.* 16: 93–97.
83. Terasaki, A., T. Wada, H. Tada, A. Koma, A. Yamada, H. Sasabe, A. F. Garito, and T. Kobayashi 1992. Femtosecond spectroscopy of vanadyl phthalocyanines in various molecular arrangements. *J. Phys. Chem.* 96: 10 534–10 542.
84. Trinkunas, G., and L. Valkunas 1989. Exciton–exciton annihilation in picosecond spectroscopy of molecular systems. *Exp. Tech. Phys.* 37: 455–458.
85. Valkunas, L. 1989. Nonlinear process in picosecond spectroscopy of photosynthetic systems. In *Proceedings of the 5th International School on Quantum Electronics*. A. Y. Spasov, editor. World Scientific, Singapore, pp. 541–560.
86. Valkunas, L., and V. Gulbinas 1997. Nonlinear exciton annihilation and local heating effects in photosynthetic antenna systems. *Photochem. Photobiol.* 66: 628–634.
87. Valkunas, L., V. Cervinskas, and F. van Mourik 1997. Energy transfer and connectivity in chloroplasts. The competition between trapping and annihilation in pulsed fluorescence induction. *J. Phys. Chem.* 101: 7327–7331.
88. Valkunas, L., N. E. Geacintov, and L. France 1992. Fluorescence induction in green plants revisited. Origin of variabilities in sigmoidicities on different time scales of irradiation. *J. Luminescence* 51: 67–78.
89. Valkunas, L., S. Kudzmauskas, and V. Liuolia 1986. Noncoherent migration of exciton in impure molecular structure. *Liet. Fiz. Rink. (Sov. Phys. -Coll.)* 26: 1–11.
90. Valkunas, L., V. Liuolia, and A. Freiberg 1991. Picosecond processes in chromatophores at various excitation intensities. *Photosynth. Res.* 27: 83–95.
91. Valkunas, L., F. van Mourik, and R. van Grondelle 1992. On the role of spectral and spatial antenna in process of excitation energy trapping in photosynthesis. *J. Photochem. Photobiol. B*. 15: 159–170.
92. Valkunas, L., E. Akesson, T. Pullerits, and V. Sundström 1996. Energy migration in the light-harvesting antenna of the photosynthetic bacterium *Rhodospirillum rubrum* studied by time-resolved excitation annihilation at 77 K. *Biophys. J.* 70: 2372–2379.
93. Valkunas, L., N. E. Geacintov, L. France, and J. Breton 1991. The dependence of the shapes of fluorescence induction curves in chloroplasts on the duration of illumination pulses. *Biophys. J.* 59: 397–408.
94. Valkunas, L., G. Trinkunas, V. Liuolia, and R. van Grondelle 1995. Nonlinear annihilation of excitations in photosynthetic systems. *Biophys. J.* 69: 1117–1129.
95. van Burgel, M., D. A. Wiersma, and K. Duppen 1995. The dynamics of one-dimensional excitons in liquids. *J. Chem. Phys.* 102: 20–33.
96. van Grondelle, R. 1985. Excitation energy transfer, trapping and annihilation in photosynthetic systems. *Biochim. Biophys. Acta* 811: 147–195.
97. van Grondelle, R., J. P. Dekker, T. Gillbro, and V. Sundström 1994. Energy transfer and trapping in photosynthesis. *Biochim. Biophys. Acta* 1187: 1–65.
98. van Mourik, F., R. W. Visschers, and R. van Grondelle 1992. Energy transfer and aggregate size in the inhomogeneously broadened core light-harvesting complex of *Rhodobacter sphaeroides. Chem. Phys. Lett.* 195: 1–7.
99. van Mourik, F., R. R. Verwijst, J. M. Mulder, and R. van Grondelle 1994. Singlet–triplet spectroscopy of the light-harvesting Bchl *a* complex of *Prostecochloris aestuarii*. The nature of the low-energy 825 nm transition. *J. Phys. Chem.* 98: 10 307–10 312.

100. van Mourik, F., K. J. Visscher, J. M. Mulder, and R. van Grondelle 1993. Spectral inhomogeneity of the light-harvesting antenna of the *Rhodospirillum rubrum* probed by T–S spectroscopy and singlet–triplet annihilation at low temperatures. *Photochem. Photobiol.* 57: 19–23.

101. Visschers, R. W., M. C. Chang, F. van Mourik, P. S. Parkes-Loach, B. A. Heller, P. A. Loach, and R. van Grondelle 1991. Fluorescence polarization and low-temperature absorption spectroscopy of a subunit form of light-harvesting complex I from purple bacteria. *Biochemistry* 30: 5736–5742.

102. Vredenberg, W. J., and L. M. N. Duysens 1963. Transfer of energy from bacterio-chlorophyll to reaction center during bacterial photosynthesis. *Nature* 197: 355–357.

103. Yamashita, A., T. Maruno, and T. Hayashi 1993. Absorption spectra of organic-molecular-beam-deposited vanadyl- and titanylphthalocyanine. *J. Phys. Chem.* 97: 4567–4569.

8

Energy transfer and localization: applications to photosynthetic systems

Sandrasegaram Gnanakaran, Gilad Haran, Ranjit Kumble, and Robin M. Hochstrasser

University of Pennsylvania, USA

8.1 INTRODUCTION

In the field of photobiology, there are many important questions whose answers involve an understanding of the underlying energy transport mechanisms. In this chapter, we will discuss two examples from photosynthesis, where energy transfer is one of the key physical processes involved in the biological function. These are the reaction center, where the initial chemical step occurs, and the light-harvesting units that gather and funnel the solar energy to the useful reactive regions. In each case an assembly of chlorophyll or similar aromatic molecules is maintained in a functionally effective structure by means of a protein which acts as the scaffolding.

These biological assemblies are reminiscent of small molecular aggregates, such as dimers, or even of molecular crystalline materials. It can then be expected that the properties of molecular excitons derived from model systems will help considerably to obtain a complete description of the biological systems. Therefore, this chapter begins by summarizing some of the basic principles of energy transfer obtained from studies of simple systems.

Exciton concepts have proved useful in describing the molecular spectra and relaxation [27, 58, 76, 83, 102, 106, 134, 136, 153] in molecular aggregates and crystals. In such systems, the low-energy excitations resemble those of the

Resonance Energy Transfer. Edited by David L. Andrews and Andrey A. Demidov.
© 1999 John Wiley & Sons Ltd

molecular units inasmuch as they are Frenkel-like, with the excitation consisting of a strongly bound electron–hole pair on a single unit. The delocalized states result from the periodic transfer of these excitations between molecules. An electron might also be excited into an unoccupied state of a different molecule, which is the process that gives rise to the charge transfer exciton states. In reality, the low-energy states are mixtures of these two types of excitations. The properties of molecular excitons can be strongly dependent on the nuclear motions of the molecules themselves, as well as on those of the whole aggregate. As a result, molecular systems also exhibit self-trapping and stochastic or nondeterministic energy transfer [3, 45, 49, 50, 78, 119, 140] from unit to unit when the coupling is sufficiently weak and the number of available states of nuclear motion is large. It is also well known from studies of model systems that the energy transfer in molecular aggregates can be severely influenced by the static disordering of the excitation energies in the system [41, 152]. Since biological assemblies and proteins are intrinsically disordered, we can expect that the concepts learned from disordered molecular systems will have particularly useful applications in biology.

8.2   OPTICAL PROPERTIES OF DIMERS AND AGGREGATES

The purpose of this section is to provide a rather brief introduction to some notation and the principles needed for the discussions of specific examples later in the chapter. This is not intended as a comprehensive review of exciton theory or energy transfer.

8.2.1   Fixed nuclear position excitons

We consider first the case of two atoms, 1 and 2, at a fixed separation, where atom 1 (or 2) can either be excited to its electronic state $|1_f>$ at energy ϵ_f, or be in its ground state $|1_0>$ with energy 0. It is more accurate to regard the energy ϵ_f as incorporating the change in each atomic state due to the presence of the other atom. In the absence of the resonance coupling, the ground state is $|1_0>|2_0>$, the excited states at ϵ_f are $|1_f>|2_0>$ and $|1_0>|2_f>$, and the doubly excited state at $2\epsilon_f$ is $|1_f>|2_f>$. In what follows, we suppress the index f and label these states as $|0>, |1>, |2>$, and $|12>$. In the basis $|0>, |1>, |2>$, corresponding to the ground state and states near ϵ_f, the additional potential from the interaction between the atoms has diagonal elements $D_{11} = D_{22} = D$ and D_{00}, and an off-diagonal element β, so that the Hamiltonian in the so-called site basis, neglecting the coupling between the ground and excited states, is

$$H = (\epsilon_f + D)(|1 >< 1| + |2 >< 2|) + \beta(|1 >< 2| + |2 >< 1|) + D_{00}|0 >< 0|.$$
$$(8.1)$$

When the system consists of N atoms $\{1, 2, \ldots, m, n, \ldots, N\}$ the site basis Hamiltonian is obtained from Eqn 8.1 by replacing 1 with n, 2 with m, and summing over m and n (the second term requires only $m < n$). If the interatomic coupling involves nonnearest neighbors, β should be replaced with β_{nm}. This approach omits the excitation exchange between the state f on one atom and a different state on the other atom, but these terms are readily incorporated if necessary.

In the basis of the symmetric and antisymmetric states $|\pm >= 2^{-1/2}(|1 > \pm|2 >)$, the dimer Hamiltonian is diagonal:

$$H = (\epsilon_f + D \pm \beta)(|\pm >< \pm|).$$
$$(8.2)$$

There are now two transitions with excitation energies $(\epsilon_f + D - D_{00} \pm \beta)$. When the two atoms have different excitation energies with separation $\Delta\varepsilon$, the states become

$$|+ >= \cos \alpha|1 > +\sin \alpha|2 >, \qquad |- >= -\sin \alpha|1 > +\cos \alpha|2 >, \qquad (8.3)$$

where $\tan 2\alpha = 2\beta/\Delta\varepsilon$. In a many-particle assembly, the Hamiltonian is diagonal in the basis of $\{|k >\}$:

$$H = \Sigma(\epsilon_f + D \pm \epsilon_k)(|k >< k|).$$
$$(8.4)$$

The various formulas for ϵ_k depend on the structure of the Hamiltonian in the site basis, which is determined by the topology of the assembly [60, 76]. In the light-harvesting complexes, chlorophyll molecules are arranged in nearly circular arrays. In the nearest-neighbor approximation, the values of ϵ_k and eigenfunctions are given by

$$\epsilon_k = 2\beta \cos[2\pi k/N], \qquad |k\rangle = \frac{1}{\sqrt{N}} \sum_{n=0}^{N-1} e^{2\pi ikn/N}|n\rangle, \qquad (8.5)$$

where k takes N successive integral values, such as from 0 to $N - 1$ or from $-(N/2) + 1$ to $+(N/2)$ for even N. When there are no cyclic boundary conditions, such as for a chain with uncoupled ends, the energies and eigenfunctions are

$$\epsilon_k = 2\beta \cos[\pi k/(N + 1)], \qquad |k\rangle = \sqrt{\frac{2}{N+1}} \sum_{n=1}^{N} \sin(\frac{kn\pi}{N+1})|n\rangle, \qquad (8.6)$$

where k runs from 1 to N.

The site states are nonstationary and when there are many atoms the time dependence is defined by the Green function:

$$G_0(t) = (1/N) \sum_k \exp[-i\epsilon_k t/\hbar], \tag{8.7}$$

which is the Fourier transform of the amplitude of the spectrum of eigenvalues, and $|G_0(t)|^2$ gives the probability that an excitation, initially at a given site, will be found at that site after the time t has elapsed. Since we will use these results later in discussions of the light-harvesting complex, $|G_0(t)|^2$ for an 18-member circle of atoms is shown in Fig. 8.1. For the case of a dimer, this function has the well-known form

$$|G_0(t)|^2 = \cos^2[\beta t/\hbar]. \tag{8.8}$$

This type of energy transfer is periodic, deterministic, and coherent. If the system is initiated with the excitation at any site, say n, the off-diagonal

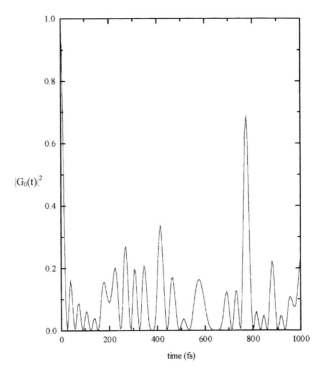

Figure 8.1 The probability for an excitation that is localized at a specific site within an 18-membered circular aggregate at $t = 0$ to recur to the same position at a later time t

elements (the *coherences*) of the site density matrix always exist in the form

$$\rho_{lm}(t) = \sum_{k,k'} \exp[i\epsilon_{kk'}t/\hbar]\langle l|k\rangle \langle k|n\rangle \langle n|k'\rangle \langle k'|m\rangle, \tag{8.9}$$

which, for the 18-atom ring example, takes the simple form

$$\rho_{lm}(t) = \sum_{k,k'} \exp[i\epsilon_{kk'}t/\hbar]\exp[i2\pi(k(l-n) - k'(m-n))/N]. \tag{8.10}$$

The time-dependence of a density matrix element is illustrated in Fig. 8.2, which shows $\rho_{9,11}(t)$, the coherence diametrically across the ring of atoms from the initially excited site $n = 1$. Later, we will consider the effects of static and dynamic disorder on the coherence of exciton states.

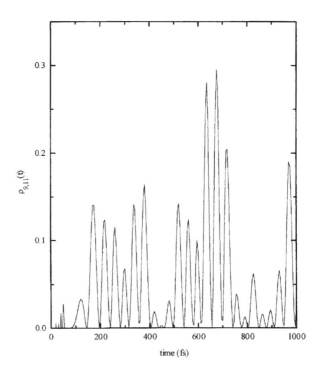

Figure 8.2 The time-dependence of the density matrix element $\rho_{9,11}$ for an 18-membered circular aggregate, given that the initial excitation was at site 1

8.2.2 The exciton coupling

The calculation of the *excitation* exchange interactions β_{nm}, and the D factors has received considerable attention. Both through-space Coulomb interactions and electron exchange interactions, which depend on the overlap of the wavefunctions of the different sites, can contribute to β. The Coulomb potential can be evaluated by calculating the matrix elements of a $1/r$ potential from approximate wavefunctions [110]. Often, it is adequate to compute an approximate point charge distribution for the electronic transition and use this to obtain the Coulomb interaction. The relevant transition charge density is readily defined in terms of the molecular orbitals of the site [24, 139]. When the sites are far apart compared with the size of this transition density, a dipole–dipole interaction becomes a good approximation if the transitions are strongly allowed. Otherwise, higher multipoles are needed to describe the coupling. To calculate the electron exchange contribution, the site wavefunction must be known at typical inter-site distances. It is still a challenging theoretic problem to obtain accurate wavefunctions at distances much larger than normal chemical bonds. A quantitative example of possible electron exchange effects will be given later (Section 8.3.3.4) in regard to photosynthesis.

8.2.3 The effects of initial conditions on energy transfer

A key factor in any nonequilibrium experiment is the initial condition. When the preparation of the system is carried out with light pulses through linear or nonlinear optical interactions, the initial states are well-defined. If the Hamiltonian in the absence of the preparing pulse of radiation $(\varepsilon(t))$ is time-independent, then the state vector at any time t is defined as

$$|\Psi>(t) = |\Psi>(0) - i/\hbar \sum_{j} \exp[-iHt]|j> \int dt \varepsilon(t) \cdot \exp[i(\omega_{j0} - \omega)t]$$

$$<j|\mu|\Psi>(0) + \cdots. \tag{8.11}$$

This expression implies that only the states having transition dipoles to the ground state can be prepared. If the excitation pulse is long compared with the evolution of the free system then, immediately after it, the system might be found in states that do not have the required transition dipole. However, if the pulse is very short compared with the dynamics of the free system, the change in the state vector immediately after the light pulse is proportional to

$$\delta|\Psi>(t \simeq 0) = -i\varepsilon(\omega) \cdot \mu|\Psi>(0), \tag{8.12}$$

where μ is the dipole operator. For a discussion of state preparation by non-linear optical interactions, see [158].

If we are dealing with a distribution of molecules in contact with a bath that can cause energy fluctuations, we should speak of the initially prepared density matrix rather than a state vector, but the essential role of dipole transitions is unchanged. The practical advantage of beginning with the density operator, ρ, rather than the state function, is that medium-induced dynamics such as ensemble dephasing and relaxation can be directly incorporated into the equations of motion. In addition, the observables, B, of such systems involve these ensemble averages and are directly obtained from Trace $[B\rho]$. In terms of density matrix language, the excitation of a system with an electromagnetic field generates coherence between the ground and excited states in the first order. In the second order it generates population changes and coherences within the ground and excited states. Coherences are defined as off-diagonal elements of the density matrix. As an example, we give the very common initial density matrix when an exciton system is excited by a pulse that is long compared with the electronic dephasing time, γ_{el}:

$$\rho_{kk'}(t) = \frac{8\pi(\mu_{k0} \cdot \hat{\varepsilon})(\mu_{0k'} \cdot \hat{\varepsilon})}{\hbar^2 \gamma_{el}} \int_0^\infty d\tau \, I(t - \tau) \exp(i\Omega_{kk'}\tau), \tag{8.13}$$

where $\Omega_{kk'} = \Omega_{kk'} - i\gamma_{kk'}$ is the complex exciton frequency, $I(t)$ is the intensity distribution of the exciting pulse envelope and $\gamma_{kk'}$ is the total inter-exciton dephasing rate, including the scattering rate constant. Thus, the pulse generates exciton coherences ($k \neq k'$) and populations ($k = k'$) governed by the existence of $\hat{\varepsilon} \cdot \mu_{k0}\mu_{0k'} \cdot \hat{\varepsilon}$, and the probability that both $|k >$ and $|k' >$ states are encompassed by the pulse envelope intensity spectrum, i.e. the Fourier transform of $I(t)$.

An important feature of the state preparation is that a light pulse will initiate the system in a nonstationary state; any superposition of eigenstates generated by Eqn 8.12 is an appropriate representation of the initial state. Thus, in an energy-transferring situation, it becomes possible to prepare excited states of individual molecules which later undergo excitation transfer. We will see in Section 8.2.6 that it is also possible to prepare nonstationary distributions of vibrations in an electronic transition. The initial conditions will also prove to be important in anisotropy experiments that involve coherence.

8.2.4   The effects of disorder on energy transfer and exciton motion

There is a large amount of literature on the theory of energy transfer in disordered aggregates, especially the transport of energy from a host material to traps [77, 82, 86, 135]. However, a common situation is that there are many traps within the excitation volume. When pulsed excitation is used, the energy can be distributed evenly over the excitation volume. On the other hand, when the excitation source is narrow-band, the initial excitations can be localized and must diffuse to the vicinity of traps by some type of activated hopping. A principal effect of disorder is to scatter the delocalized excitations. The scattering processes occur on the timescale of motion of the excitations between scattering centers. The scattering can be calculated by replacing the Hamiltonian of the ideal aggregate by an effective Hamiltonian in which there is a contribution V that describes the disorder. Then, the survival probability for a particular delocalized state $|k>$ can be calculated from

$$P_k(t) = | < k | \exp\{-i \int_0^t dt' V(t')\} | k > |^2, \qquad (8.14)$$

where $V(t)$ is the additional potential in the interaction picture of the perfect aggregate. The form of $P_k(t)$, as a matrix element of a time-ordered exponential, renders it suitable to be approximated by cumulant expansions when the perturbation is relatively small [2, 87, 177]. As will be seen later, the quantity in Eqn 8.14 and the associated density matrix can be calculated exactly for certain types of disorder in aggregates that contain a finite number of molecules [88].

In the frequency domain, disorder leads to a redistribution of dipole strength amongst exciton levels and thereby activates transitions that are otherwise forbidden in the perfect aggregate. Site energy disorder introduces an inhomogeneous contribution to the line shape of the exciton transitions, but it is important to note that the degree of inhomogeneous broadening of transitions within the aggregate is *lowered* (relative to the monomer transitions) due to the presence of the inter-site interactions. If the variance σ^2 of the disorder energy distribution is small compared with β, the effective inhomogeneity is reduced to $(\sigma/\beta)\,\sigma$. This effect is referred to as *motional narrowing* [59]. The influence of static disorder upon the exciton spectra of molecular aggregates can be modeled using random sampling procedures [152]. Assuming a specific functional form for the distribution of pigment site energies, the eigenstates of a single aggregate comprised of N pigments can be obtained by constructing and diagonalizing the Hamiltonian matrix, in which the diagonal elements (which correspond to the individual site transition energies) are randomly selected from the specified distribution. Disorder in the inter-site coupling can also be incorporated in a similar fashion by choosing the off-diagonal elements from an appropriate distribution. The spectrum of the single aggregate can be obtained from the eigenstates and transition energies. The ensemble optical spectrum is then obtained by repeatedly averaging the results from many such calculations, in which the pigment site energies within each aggregate are selected independently from the same distribution each time. For a specified extent of disorder, the ensemble-averaged degree of delocalization can be determined by this computational procedure using criteria such as the participation ratio or the quantum connectedness length, which are discussed in detail later in this chapter.

The time evolution of optical excitations within an inhomogeneous aggregate can be followed from the survival probability $P_k(t)$ for the initial state k using the expression given above. Under conditions of impulsive excitation (see Section 8.2.3) the initial state k extends over many pigments within the aggregate, and $P_k(t)$ decreases in time as this state becomes progressively localized due to disorder scattering. If the pigment energies follow a Gaussian distribution and the degree of disorder is sufficiently weak, the Green function used to determine $P_k(t)$ can be simplified by the method of cumulants to yield an analytic expression for the time-dependence of the survival probability. Abram and Hochstrasser [2] have presented the development of a model to describe exciton scattering dynamics in a weakly disordered system, using the cumulant expansion to second order. For finite aggregates, the survival probability can be evaluated exactly for any level of disorder using the random sampling procedures described in the preceding paragraph. Evaluation of $P_k(t)$ is performed for a single disordered aggregate once its exciton energies and eigenfunctions are numerically obtained, and ensemble averaging is then carried out by averaging the results of this procedure for several configurations. An example of the dependence of the scattering timescale upon the degree of disorder (relative to the inter-site interaction energy) is shown in Fig. 8.3 for an 18-membered circular aggregate model of a bacterial light-harvesting assembly.

8.2.5   Two-exciton and charge-resonance states

The full exciton matrix, even for a dimer, should incorporate states that have more than one excitation. One example is the state $| 12 >$ mentioned earlier. The energy of this state is shifted from $2(\varepsilon_f + D)$ by the additional coupling between the excited atoms. Further, the multi-excitation states do not have the simple boson-like character of the single excitation conditions. They are nevertheless very important in pump–probe experiments, when the coupling is not large enough to shift the bi-exciton out of the spectral range of approximately twice the exciton frequency.

To complete the exciton picture in the absence of nuclear motion, the charge resonance states $| CT_\pm \rangle = 1/\sqrt{2}[| 1^+2^- \rangle \pm | 1^-2^+ \rangle]$ must be considered. In a many-atom system, these types of states form a band of levels [136]. Transitions between exciton states are forbidden in the absence of these charge transfer states unless the monomers themselves change dipole moment on excitation. In the presence of an additional potential F from an effective external field E, with $F = -\mu E$ and where μ is the dipole operator, the Hamiltonian for the dimer excitons becomes

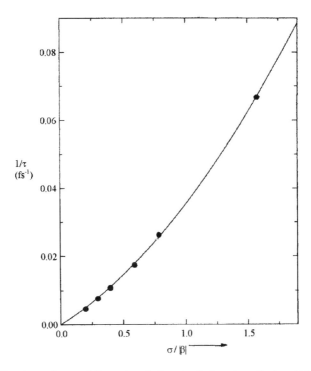

Figure 8.3 The dependence of the rate of disorder-induced scattering $(1/\tau)$ upon the standard deviation (σ) of the Gaussian distribution of pigment transition energies relative to the inter-site interaction β; τ = time taken for $\langle P_k(t) \rangle$ to decrease to $1/e$ times the total decay for an 18-membered circular aggregate of pigments

$$
H = \begin{array}{cccc}
|+\rangle & |CT_+\rangle & |CT_-\rangle & |-\rangle \\
\begin{bmatrix}
\varepsilon_f + D + \beta & V_{++} & 0 & F_{+-} \\
V_{++} & E_{CR}^+ & -\vec{\mu}_{CT} \cdot E & 0 \\
0 & -\vec{\mu}_{CT} \cdot E & E_{CR}^- & V_- \\
F_{-+} & 0 & V_- & \varepsilon_f + D - \beta
\end{bmatrix}
\end{array}, \qquad (8.15)
$$

where V_{++} and V_{--} are the interaction energies between exciton and charge-transfer states (V_{++} is much smaller than V_{--}) [110, 126], $\vec{\mu}_{CT}$ is the charge transfer dipole moment, and $E_{CR}^\pm$ are the degenerate symmetric and antisymmetric charge transfer energies. Figure 8.4 shows a typical distribution of states and the field effect on them. It is readily seen that all levels except $|+\rangle$ become influenced by the external field. The $|+\rangle$ state is decoupled here from the other states, in the absence of any asymmetry. Stark effects on crystalline excitons are well characterized [58] and there have been important applications to biological

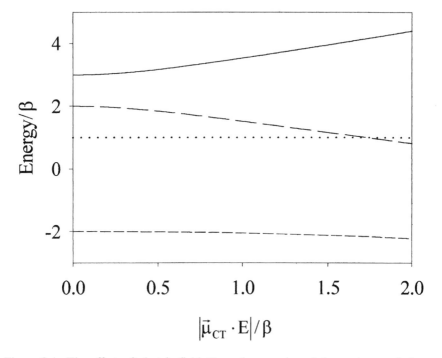

Figure 8.4 The effect of electric field E on the energies of the exciton and charge transfer states of a dimer, obtained by diagonalization of the Hamiltonian of Eqn 8.15. The parameters used are as follows: $\beta = 1$, $E_{CR}^{\pm}/\beta = 2$, $V_{--}/\beta = 2$, and $\varepsilon_f + D = V_{++} = F_{+-} = F_{-+} = 0$

systems in recent years [20]. Note that most experimental work has so far been carried out in the range $|\vec{\mu}_{CT} \cdot E|/\beta \ll 1$.

8.2.6 The effect of nuclear motions on exciton dynamics: weak and strong coupling in aggregates

The strength of the vibrational electronic coupling in a molecule, which determines the moments of its electronic spectrum, is also the key factor determining the energy transfer in an aggregate of the molecules [46, 109, 168, 169]. In an electronic transition, the width of the vibrational progression of a mode with dimensionless shift Δ is $\hbar\omega\Delta$. In the limit $\beta \ll \hbar\omega\Delta$, each vibronic state undergoes energy transfer dynamics with a period on the order of $\hbar/S\beta$, where S is its Franck–Condon factor. The other extreme where $\beta \gg \hbar\omega\Delta$ is usually termed strong coupling, in which case the spectrum manifests well-separated electronic (excitonic) absorption bands. If the coupling, β, becomes sufficiently weak and distributed over a large enough number of vibronic levels (i.e. large electronic–vibrational coupling) that the coherent recurrences of excitation probability at a given site become negligible, the energy transfer is irreversible and there is no longer a periodic motion of excitations between sites. Instead, there is a stochastic hopping of the excitations throughout the aggregate. This coincides with the Golden Rule limit, often equated with Förster energy transfer.

It is of interest in applications of energy transfer to multichromophore systems, such as occur in photosynthesis, to consider how the electronic and internal vibrational excitations are distributed over the monomers. The main points again can be illustrated by the example of a dimer, in this case of diatomic molecules. The nuclear Hamiltonian for the case of two harmonic oscillators is known to be

$$
\begin{aligned}
H(Q) = & (H(Q_1) + H(Q_2) + D(x))[|\ 1 >< 1\ | + |\ 2 >< 2\ |] \\
& + \hbar\omega\Delta(Q_2 - Q_1)[|\ 1 >< 2\ | + |\ 2 >< 1\ |] + \beta(x)[|\ 1 >< 1\ | - |2 >< 2\ |],
\end{aligned}
\tag{8.16}
$$

where $H(Q_1)$ is the nuclear Hamiltonian for molecule 1 with a potential that is the average of its ground and electronically excited potentials, and $D(x)$ is the Hamiltonian for the relative nuclear motion. This problem is exactly solvable in the weak and strong coupling limits mentioned above. A number of other situations are of interest. When the energy transfer is independent of x, the states of relative motion are separable. Therefore displacements along x will give rise to shifted harmonic oscillator dynamics that do not depend on the coupling limits. For example, both of the exciton transitions of a strongly coupled dimer will display side-bands due to the relative motion of the monomers.

Another point of interest is whether the vibrational and electronic excitations move together or separately. To illustrate the importance of this question, the nuclear motion equation 8.16 was solved and used to obtain the state function following coherent excitation of an exciton with a mode having a dimensionless shift of 0.6 and $\omega/\beta = 5$. The ultrafast excitation pulse only generates states (there are four of them) with the vibrational and electronic excitations on the same molecule, as can be readily deduced from Eqn 8.12. As shown in Fig. 8.5, the probability that the vibrational and electronic quanta separate starts at zero but increases significantly with time after excitation. The signal exhibits oscillations, at or near the vibrational period, that are superimposed on a slower component which is at $\hbar/2\beta S_{00}S_{11}$, where S_{00} and S_{11} are the Franck–Condon integrals for the 0–0 and 1–1 vibronic transitions. For this example, the state vector is a superposition of six states, two of which have the vibrational excitation on the molecule that is not electronically excited. Considerations of this sort must be incorporated when impulsive signals from dimers or

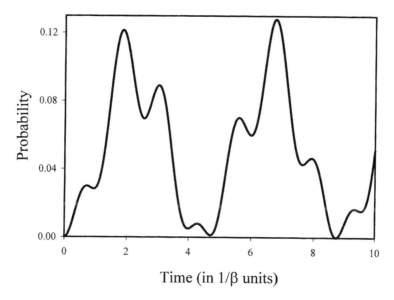

Figure 8.5 The probability that, once optically excited, the vibrational energy will be found on the molecule of the dimer that is not electronically excited (assuming the vibrational frequency does not change on excitation). The fast oscillation is the 00 to 01 frequency modified by the coupling and the slow oscillation is the $2^*\beta^*S_{00}^*S_{11}$. Note that after two vibrational periods the probability peaks

aggregates are being interpreted. There has been considerable study of electronic vibrational coupling applied to the dimers and crystals [46, 104, 105, 109, 133, 168]. The fluorescence signal from a dimer with spectra of nuclear motion incorporated was recently calculated [101].

The usual form for the Hamiltonian of an assembly of identical two-level atoms in the presence of nuclear motion, characterized by the operators x, is

$$H = \sum [\epsilon_n + D_n(x)] \mid n><n \mid + \sum{}' \beta_{nm}(x) \mid m><n \mid + H(x), \qquad (8.17)$$

where $H(x)$ is the Hamiltonian for the nuclear motion of the whole assembly. This Hamiltonian can be written as

$$H = H_0 + H(x) + \Sigma \sigma_{mn}(x) \mid m><n \mid, \qquad (8.18)$$

where H_0 is the rigid exciton Hamiltonian (Eqn 8.1) and the $\sigma_{mn}(x)$ are the nuclear position-dependent Taylor expansion terms of $D_n(x)$ and $\beta_{nm}(x)$. For a dimer, a harmonic bath and linear coupling, this equation is equivalent to the so-called spin-boson Hamiltonian, which can provide a reasonable description even when the atoms are immersed in a liquid [92]. In the linear regime it is the time correlation functions of $\sigma_{mn}(x)$ that define the spectral densities for the solvation effect on energy transfer. In many instances, the greatest influence of nuclear motion on the energy transfer comes from the $D(x)$ term. For excitons composed of chlorophyll molecules, the spectral shifts of the chromophores due to the environment are usually greater than the excitation exchange interactions. Further, relatively small alterations in the local environment of a chromophore can significantly change the transition energy. Therefore it is expected that, in general, dD / dx will significantly exceed $d\beta / dx$, as is the case in molecular crystals [62]. In this case, the correlation functions of $\sigma_{nn}(x)$, the energy fluctuations in the site basis, determine the dynamics.

The effect of nuclear dynamics on the excitonic spectrum has been studied extensively. A clear manifestation of nuclear motion is the temperature-dependent motional narrowing of the excitonic bandwidth, induced by bath fluctuations, which cause transitions between the excitonic levels. The two excitonic bands in a dimer spectrum gradually broaden and coalesce as the temperature is increased, and at a high enough temperature the spectrum shows only one Lorentzian line. Analogous narrowing will occur in multilevel exciton systems. The physics of this behavior can be captured by simple stochastic models, such as the "two-state jump" (TSJ) model of Kubo [87], or by microscopic models (see Sections 8.2.9 and 8.2.10 below).

8.2.7   Localization of excitations

When relaxation occurs during the process of energy transfer, a partial localization of the excitation can result. In molecular systems, this notion was introduced by Davydov [27], and has been frequently discussed since then for molecular aggregates and dimers. Excimer formation by certain electronically excited aromatic molecules is a particularly clear example of this effect [61]. In more general terms, we are dealing here with polaron formation [64, 65]. The interplay between the solvent nuclear dynamics (responsible for the monomer Stokes shift) and exciton motion is illustrated in Fig. 8.6, which schematically depicts the potential surfaces for the states of a molecular dimer. The surfaces are viewed along a solvent coordinate Q_s for which the minima of the localized excitations 1^*2 and 12^* are shifted by equal and opposite amounts from the ground-state equilibrium position ($Q_s = 0$ here). These displacements lead to stabilization of the individual excitations by $\lambda/4$, where λ is the reorganization energy, which is reflected in the emission Stokes shift ($\lambda/2$) seen for the individual monomers. The energy separation gap between 1^*2 and 12^* at one of the minima is λ. The solvent coordinate can be considered to be the energy gap, as is customary in electron transfer. If the monomer excitations are mixed

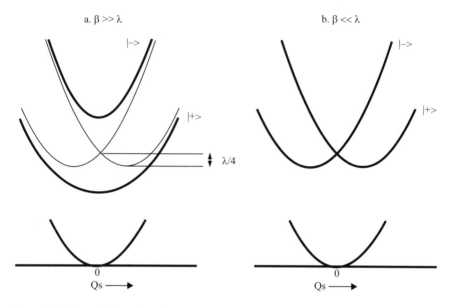

Figure 8.6 Potential surfaces for the ground and excited states of a molecular dimer viewed along a solvent coordinate Q_s. Thick lines represent the surfaces of the dimer eigenstates (the lower surface in each figure represents the ground electronic state of the dimer) while thin lines represent the surfaces of the individual monomer excitations 1^*2 and 12^*. Two cases are shown, corresponding to (a) large and (b) small ratios of the inter-site coupling β to the monomer reorganization energy λ

through the electronic interaction β, the potential surfaces are modified in a fashion that depends upon the magnitude of the interaction relative to the reorganization energy.

Strong inter-site interactions that are significantly greater than the reorganization energy establish surfaces of the form shown in Fig. 8.6a, where the two excited states now correspond to nearly equal admixtures of 1^*2 and 12^* at all points along Q_s. If, however, the reorganization energy λ dominates over the inter-site interaction β, surfaces of the type shown in Fig. 8.6b are established. The potential for the lower excited state now exhibits a maximum at $Q_s = 0$, where the electronic wavefunction is delocalized over both monomers. Minima corresponding to essentially the diabatic site states exist on both sides of the saddle point, and the wavefunction becomes progressively more localized as the dimer structure is distorted along this coordinate from the $Q_s = 0$ geometry in either direction. Exciton delocalization is thus constrained by nuclear motion to an extent determined by the ratio β/λ: the sensitivity of the exciton wavefunctions to nuclear coordinates becomes more pronounced as this ratio decreases.

These considerations are fundamental to electron–nuclear coupling phenomena, and the manifestations are very similar to those encountered in mixed-valence charge-transfer systems, where nuclear motions can lead to self-trapping of an electron (rather than an electron–hole pair, as in the case of a Frenkel exciton). While the dependence of exciton wavefunctions on nuclear coordinates determines whether a localized excitation is created by nuclear motion, the dynamics of this self-trapping process itself is determined by the *timescale* of the Stokes shift, i.e. the relaxation of the solvent structure in response to excitation. Therefore, for the same coupling, a system in a fast responding solvent such as water may localize, whereas it may remain delocalized in a slow solvent such as protein or glass or at lowered temperatures. When the solvent relaxation is fast compared with energy transfer, there will be no coherent recurrence of the excitation on the initially excited molecule – this corresponds to localization. When it is slow, the electronic coherence oscillates at 2β.

8.2.8   Excitonic coherence size

As described in Section 8.2.7, excitons can be fully or partially localized by static or dynamic lattice disorder. A fundamental difference exists in this respect between three-dimensional systems and one- and two-dimensional systems. In a three-dimensional system, weak disorder leads to localization of states on the low- and high-energy sides of the exciton band, while states in the center of the band are delocalized. The delocalized states are Bloch states, and extend over the whole lattice. The "localized" states have exponentially decaying spatial tails, and while they might extend over quite a large range, they do not cover the whole lattice. At a critical level of disorder, a transition occurs (Anderson transition [6]), which leads to localization of all the states. Only in the limit of very strong disorder will all the states fully localize. In one- and two-dimensional systems, disorder localizes all of the states, and there are no Bloch states [1]. Determination of the spatial extension of excitonic states in disordered systems can provide information on the factors that lead to exciton scattering, their dependence on intensive variables such as temperature and their influence on the functional properties of the system under study. We discuss in this section experimental methods and theoretic measures that are used to determine coherence size in excitonic systems.

Direct information on the excitonic coherence size is provided by the measurement of excitation saturation and superradiance. In order to illustrate these factors, we first discuss some simple planar aggregates. Figure 8.7 shows the excitonic levels of four- and six-membered perfect rigid rings (with a negative interaction energy), together with the transition dipole associated with each level for the case of monomer transition dipoles perpendicular to the molecular

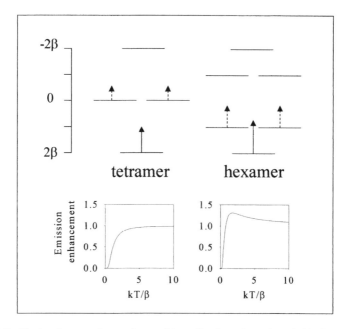

Figure 8.7   Excitonic energies and transition dipoles of perfect rigid rings. Full-line arrows are transition dipoles in the case that the monomer dipoles are perpendicular to the plane of the ring. Dashed-line arrows are for the case of in-plane monomer dipoles. The lower part of the figure shows emission enhancement as a factor of temperature for the in-plane case

plane (full line arrows) and for the case of in-plane transition dipoles (dashed-line arrows). It is seen that for the former case the oscillator strength is concentrated in the lowest-energy state. Transitions between the ground state and this state can therefore be thought of as arising from the coherent oscillation of all of the monomer (transition) dipoles, which add to form a large effective dipole. The oscillator strength and the radiative lifetime become proportional to the number of coupled dipoles. We can think of the *coherence size* of an aggregate as the average number of molecules over which the electronic excitation is delocalized. In the case of the perfect ring, the coherence size is the number of monomers, N. The number of photons needed to saturate the absorption of an ensemble of such aggregates is N times smaller than the corresponding number for an ensemble of monomers. By the same token, the radiative lifetime is N times shorter in the aggregated sample, and the enhanced emission rate is described as *superradiance*. At finite temperatures higher energy, "dark" excitonic states become populated, and the emission enhancement is reduced; at infinite temperature all of the states are equally populated, and the emission rate equals the monomer rate.

The case of in-plane transition dipoles is quite different; the oscillator strength is shared between two degenerate states, which are not at the bottom of the exciton band. At zero temperature the aggregate will show no emission, at least in the limit of negligible electronic–vibrational coupling (see Section 8.2.6). At a higher temperature, where the two degenerate states will be thermally populated, there will be emission. A plot of the emission rate enhancement as a factor of temperature for the in-plane dipole case is given in the lower part of Fig. 8.7. The enhancement factor is never as high as N, and in fact for the tetramer it is actually always less than or equal to unity, with the largest value occurring at infinite temperature.

The extraction of coherence size from spectroscopic measurements is even more complicated when disorder is introduced, and it is useful to compare the experimental result to a theoretical calculation. Several quantitative measures for the coherence size have been developed. The most commonly used one is the participation ratio [35, 108, 129, 151] which, for state k of an excitonic manifold, is defined as

$$\prod_k = \left\langle \sum_n (a_{kn})^4 \right\rangle^{-1}, \qquad (8.19)$$

where a_{kn} is the coefficient of the wavefunction of state k on site n, N is the number of monomers in the aggregate and the brackets denote an ensemble average. If the state k is completely localized ($a_{kn} = \delta_{kn}$), the ratio equals unity. Otherwise, the ratio depends on the identity of the state; in the fully delocalized case, there will be one state, the $k = 0$ state, for which $a_{kn} \approx N^{-1/2}$, and its participation ratio is equal to N. For finite temperatures the participation ratio can be further Boltzman-averaged over all states. It is possible to obtain this ratio from any type of electronic structure calculation on an aggregate, e.g. from Monte Carlo simulations in the presence of disorder and electron–phonon interactions [39, 111, 129].

A participation ratio based on the density matrix [108] can also be defined:

$$\prod_\rho = \left[N \sum_{mn} |\rho_{mn}|^2 \right]^{-1} \left[\left(\sum_{mn} |\rho_{mn}| \right)^2 \right], \qquad (8.20)$$

where ρ_{mn} is the (mn)th element of the density matrix in the site basis. This expression tends to 1 for a localized state (where only diagonal elements of ρ exist) and to N for a fully delocalized state. A direct relationship between $\prod_\rho$ and the degree of superradiant enhancement has been demonstrated for certain specific systems [107, 108] when the participation ratio is computed from the equilibrium density matrix. Another useful measure for coherence size is provided by the "quantum connectedness length" X [137], defined as

$$X^2 = \left(\frac{\sum\limits_{mn} \langle \Xi_{mn} \rangle r_{mn}^2}{\sum\limits_{mn} \langle \Xi_{mn} \rangle} \right), \tag{8.21}$$

where r_{mn} is the distance between sites m and n, and Ξ_{mn} is the quantum connectivity of these sites, given by $\Xi_{mn} = R_{mn}(R_{mm}R_{nn})^{-1/2}$, with $R_{mn} = \sum_k |a_{km}|^2 |a_{kn}|^2$. The sum in the definition of R_{mn} runs over all excitonic states.

The functional dependence of each of these measures on disorder is different. To illustrate this point, we plot in Fig. 8.8 the wavefunction-based participation ratio for two different states and the quantum connectedness length, calculated for an 18-mer (with parameters appropriate for a light-harvesting complex (see Section 8.3.2.2), as a function of the strength of disorder, σ. The value of the participation ratio depends on the state for which it is calculated. Further, it decays much faster than the quantum connectedness length, as the disorder is increased. Thus it is clear that the value of any of these measures does not directly reflect the number of pigments or physical length over which the exciton is delocalized.

Some further description of the experimental methods for determination of coherence size is warranted here. In a saturation experiment, pulsed laser excitation is used to populate an excitonic state, which will typically be the state carrying most of the oscillator strength of the aggregate. The laser pulses used should be much shorter than the spontaneous decay rate of that level. If the laser spectral bandwidth is not smaller than the inter-state energy difference (which for large N scales as $\sim V/N^2$ at the bottom of the exciton band), then more than one excitonic state will be populated. The population in the excited state is probed as a function of pump laser power, I_{pump}. One way to accomplish this is by detecting a signal due to transient absorption to higher-lying states. A saturation curve is obtained, which for a two-level system (i.e. a situation in which only the lowest excitonic state is excited) takes the following form:

$$S = S_{sat} \frac{2cI_{pump}}{1 + 2cI_{pump}}, \tag{8.22}$$

where c is a constant related to molecular parameters (absorption cross-section and relaxation rate) and S_{sat} is the signal extrapolated to saturation, where half of the absorbing entities (each part of an aggregate) are excited. By fitting experimental measurements to such a curve, one obtains the parameters needed to calculate the number of photons, n_{photon}, absorbed by the sample at saturation. The coherence size is then given by half of the ratio of the total monomer number in the solution to n_{photon}.

This approach was used to find the coherence size in a polysilane polymer [150]. It was found that saturation occurred at a level of one excitation per 50

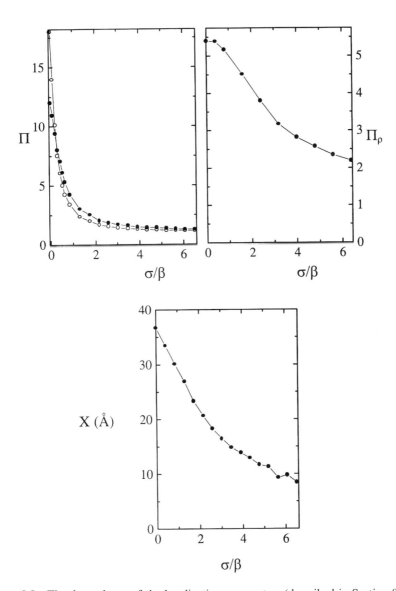

Figure 8.8   The dependence of the localization parameters (described in Section 8.2.8) upon the extent of pigment inhomogeneity (σ) relative to the inter-pigment interaction energy ($\beta = -250\text{cm}^{-1}$) for an 18-membered circular pigment aggregate. Variation of (a) participation ratio $\prod$ (defined in Eqn 8.19) for the lowest and second lowest levels of the aggregate (hollow and solid circles, respectively; (b) equilibrium density matrix participation ratio $\prod_\rho$ (Eqn 8.20) at 300 K; (c) quantum connectedness length X (Eqn 8.21)

silicon atoms, which was interpreted as a coherence size of 25, much smaller than the average polymer length of ~ 2100 silicon atoms. It was concluded that configurational disorder is the cause for the partial localization. Measurements of a polydiacetylne–paratoluene crystal, where the molecular structure dictates formation of 1-D Frenkel excitons [48], were interpreted using a phase-space filling approach. It was found that the Bohr radius of the exciton is 30–40 Å, equivalent to about 5–10 monomers. A general caveat of saturation experiments is that their interpretation becomes very difficult in systems in which nonlinear processes such as bi-exciton absorption or exciton annihilation are important, since at high laser power these might dominate the signal.

In a typical superradiance experiment, time-resolved fluorescence techniques or photon echoes (at low temperatures, where the main dephasing mechanism is radiative) are used to measure the radiative lifetime of a sample of aggregates $(\tau_{\mathrm{rad}}^{\mathrm{ag}})$, which is then compared to the radiative lifetime of a monomer $(\tau_{\mathrm{rad}}^{\mathrm{mon}})$. The ratio of the two is the superradiance enhancement factor, which directly provides the coherence size for the case of the parallel dipole aggregate already mentioned, but should be compared with calculation in other cases. In a superradiance measurement of polysilane, an enhancement factor of ~ 25 was found [80]. A maximum coherence size of 25 was assigned to the polymer by the authors, which is similar to the result obtained by the saturation technique [150]. Another example of excitonic interactions in a linear chain is found in J-aggregates of pseudoisocyanine [28, 29, 38, 40]. For these aggregates, an enhancement factor of 185 was found from an accumulated photon-echo measurement at 1.5 K [28]. Theoretic analysis of J-aggregates [38, 141] led to the conclusion that their actual length is much larger than the number obtained from the superradiance measurement, and the excitons are confined by disorder.

8.2.9   Two-level density matrix description of energy transfer

In condensed phase experiments, relaxation processes influence the energy transfer in profound ways. Some of the essential effects of dephasing and energy relaxation are readily appreciated from the time evolution of the density matrix, incorporating damping parameters which can be considered as phenomenologic yet will be seen later to have a fundamental validity. The contributions to dephasing (T_2) arise from two decay processes: the pure dephasing (T_2^*) caused by the energy fluctuations of the excitonic levels $|+>$ and $|->$, and the equilibration time of the populations (T_1) between the two excitonic levels through exchange of energy with the bath. In the site basis $(|1>, |2>)$, the equations of motion for a system having a Hamiltonian containing an energy transfer term β are given in terms of the population sum, σ, and difference $W(t)$, the imaginary, $-i\Delta$, and real, Σ, parts of the coherence, as follows:

$$dW/dt = 2\beta\Delta - W/T_2^*, d\Sigma/dt = -\Sigma/T_1;$$
$$d\Delta/dt = -2\beta W - \Delta/T_2, d\sigma/dt = 0; \qquad (8.23)$$

where the relationship $1/T_2 = 1/T_1 + 1/T_2^*$ is used.[†]

When the system is prepared by an optical pulse whose spectral bandwidth encompasses the whole spectrum of states of the two-level system, the initial value of Δ can be deduced from Eqn 8.12 to be zero, and the population can be obtained in the form

$$W(t) = W(0)\exp[-(1/T_2 - 1/(2T_1))t]\{(1/T_1)\sinh(\zeta t/2)/\xi$$
$$+ \cosh(\zeta t/2)\} = W(0)f(t), \qquad (8.24)$$

where ζ is defined as $[(1/T_1)^2 - 16\beta^2]^{1/2}$. It follows that in the presence of very fast dephasing, the population difference assumes the simple form

$$W(t) = W(0)\exp[-(4\beta^2 T_2)t]. \qquad (8.25)$$

This is the Golden Rule limit of energy transfer between equivalent molecules, where the population equalizes through the energy transfer involving the term $4\beta^2 T_2$. In effect, the density of final states in the irreversible (stochastic or noncoherent) process is given by T_2. In some cases, $1/T_2$ may represent a significant proportion of the spectrum of the two-level system reached from a ground state [81]. When β becomes large enough, the population $W(t)$ becomes oscillatory and the system displays coherent energy transfer; this amounts to the substitution $\sinh x \to i \sin x, \cosh x \to \cos x$ in the equation for $W(t)$ in the limit $|4\beta| > 1/T_1$. This oscillatory behavior has now been seen in a number of cases [8, 89, 181].

When the system is excited by a short light pulse with polarization E, then $\sigma(0), W(0), \Delta(0)$ and $\Sigma(0)$ are all known from Eqn 8.12, so the anisotropic absorption or gain, $\delta\alpha_{EP}$, of a probing light field with polarization P, can readily be calculated, as

$$\delta\alpha_{EP} = < \sigma(0)(p_1^2 + p_2^2) > /2 + < \Sigma(0)p_1p_2 > \exp[-t/T_1]$$
$$+ < W(0)(p_1^2 - p_2^2) > f(t)/2, \qquad (8.26)$$

---

† We are not considering the temperature dependence of T_1 that arises from the principle of detailed balance; T_1 is the time constant for equilibration of the two-level system, $1/T_1 = K_{+-} + K_{-+}$, where K_{+-} is the rate constant for $|+ >$ to $|- >$ population flow. For levels separated by much less than $k_B T$, we have $K_{+-} \sim K_{-+} = 1/(2T_1)$.

where p_i is the cosine of θ_{pi}, the angle between the transition dipole on the ith molecule and the polarization direction P, and angle brackets represent an orientational average. For these conditions, the limiting pump–probe anisotropy $r(\infty)$ and the zero time anisotropy $r(0)$ depend only on the mean squared cosine of the angle θ_{12} between the transition dipoles on the two sites:

$$r(0) = \{0.7 + 0.1 < \cos^2 \theta_{12} >\}/\{1 + < \cos^2 \theta_{12} >\};$$
$$r(\infty) = 1/5\{1 + < P_2[\cos \theta_{12}] >\}. \tag{8.27}$$

Note that the initial anisotropy contains a coherent term, from the optically prepared off-diagonal density matrix element Σ, that decays with the eigenstate population decay constant.

The results in Eqns 8.26 and 8.27 are very useful; first, because a measurement of the anisotropy determines the angle θ_{12} whether or not the system contains coherence. Of comparable interest is the prediction that, if the anisotropy is found to exceed 0.4, it demonstrates that the system contains coherence.† Figure 8.9 shows the expected variations of $r(0)$ and $r(\infty)$ with θ_{12}, demonstrating that the largest changes due to interference are expected for

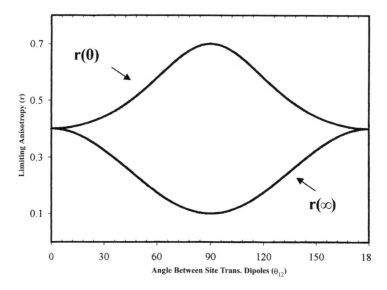

Figure 8.9   The dependence of the initial ($r(0)$) and final ($r(\infty)$) anisotropy values on the angle θ_{12} between the two site transition dipoles

† In pump–probe experiments it is possible to obtain anisotropies in excess of 0.4 without there being any coherence when the probed system is undergoing both loss and gain processes [130].

angles near $\pi/2$. A similar set of results is obtained for a single system in which a doubly degenerate state is excited. In this case, $\theta_{12} = \pi/2$ and $r(0) = 0.7$ [171]. Examples of this can arise with porphyrins [47] and with the circular light-harvesting structures discussed later in Section 8.3.2.

8.2.10   Microscopic picture of energy transfer

The equations of motion for σ, W, Δ, and Σ given above in terms of T_1 and T_2 are equivalent to the quantum-mechanical dynamics results obtained from second order perturbation theories [167, 181] of a two-level system coupled to a bath at high temperature, with the assumption that the fluctuations in the coupling and site energies are statistically independent. However, the equations of motion that can be obtained from a simple (Bloch) two-level system are not in fact equivalent to those given above. The difference arises because the latter model fails to consider the interconversion of coherence between the two levels, which causes Δ and Σ to decay with different time constants. For example, ρ_{12} and ρ_{21} simply interchange roles through the pure dephasing relaxation, so their sum, Σ, decays only with a T_1 process. The phenomenological relaxation parameters T_1 and T_2 are readily given basic physical significance from dynamical theories such as stochastic modeling [49, 50, 78, 119, 134, 140] or by Redfield theory [121]. The pure dephasing (T_2^*) arises from the fluctuations of the resonance coupling in the site basis, but is in fact the pure dephasing of the $|+>, |->$ level pair induced by their energy fluctuations. The T_1 arises from fluctuations in the energies of the sites or, equivalently, the bath-induced coupling between the $|+>$ and $|->$ states that tends to equalize their populations. The parameters that describe the decay of coherence and population in terms of the correlation functions of a fluctuating coupling between the two sites, $\chi(t)$, and fluctuating site energies, $\xi(t)$, are [171]

$$\frac{1}{T_2^*} = 4 \int_0^\infty d\tau < \chi(t)\chi(t-\tau) >,$$

$$\frac{1}{T_1} = \int_0^\infty d\tau < \xi(t)\xi(t-\tau) > \cos(2\beta\tau). \tag{8.28}$$

The above correlation functions that control energy transfer can be estimated from classical simulations of the dynamics [5]. Below, as an illustration, the role of coupling fluctuations on energy transfer of a solvated dimer is examined using molecular dynamics (MD) simulations.

The molecular composite system considered in the theoretic study is 9,9'-bifluorene (BF), where the identical fluorene moieties are held together by a

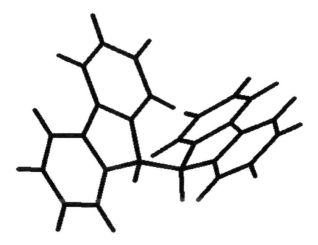

Figure 8.10 The molecular diagram of 9,9'-bifluorene. X-ray data show that the molecule has a *gauche* configuration. The dihedral angle referred in the text corresponds to rotation around the single bond that connects the two moieties

flexible saturated carbon–carbon bond (Fig. 8.10). Frequency-domain gas-phase [139] and condensed-phase [81] experiments have indicated that the two fluorene moieties are weakly coupled in BF. When excited, the excitation is delocalized due to the electronic excitation exchange coupling, and an exciton is formed. At short times, the excitation oscillates between the two moieties at twice the electronic excitation exchange coupling frequency. However, at long times, due to the contact with the thermal bath, the coherent excition dynamics becomes a hopping process, which leads to stochastic energy transfer. The timescale on which the coherent motion becomes incoherent is determined by the fluctuations induced by the solvent on the electronic excitation exchange coupling, and on the energy difference between the moieties. The fluctuations in electronic excitation exchange coupling are caused by the solvent-induced changes in the relative orientations and distances between moieties. At each instant, the environment of each moiety is distinguishable, producing fluctuations in their relative energies. As we have shown before, such dynamic effects have been observed experimentally in the anisotropy [181]. However, these observations only provide the timescale of the macroscopic excitation transfer dynamics. The microscopic features of exciton dynamics can be visualized from the classical molecular dynamics simulation, providing answers to questions concerning the extent of localization, the nature and the timescale of the site energy, and electronic excitation exchange coupling fluctuations.

The electronic excitation exchange coupling of the isolated BF was calculated from the transition charge distribution of the composite molecule obtained from semi-empirical electronic structure calculations. This is a case in which the dipole–dipole approximation is not reliable. At the X-ray structure geometry, the total electronic coupling was 99 cm^{-1}, compared to 97 cm^{-1} estimated from the gas-phase experiments [139]. As expected, the calculated coupling depends strongly on the dihedral angle, which corresponds to the rotation around the single bond that connects the two moieties (Fig. 8.11). Interaction with solvent molecules induces changes in the relative orientation of the two moieties, leading to a fluctuating electronic excitation exchange coupling. The solvent effects were investigated by carrying out a classical molecular dynamics simulation of bifluorene immersed in an equilibrated cubic box of 252 hexane molecules with periodic boundary conditions. The interaction between the solute and solvent was modeled with coulombic and Lennard–Jones potentials, and the inter-moiety torsional potential was parameterized to produce the known potential barriers. The simulations were carried out using CHARMM [19], and additional details are given elsewhere [173].

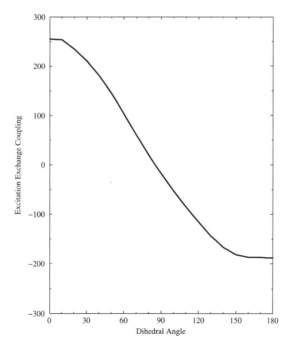

Figure 8.11 Electronic excitation exchange coupling as a function of inter-moiety dihedral angle: the X-ray geometry corresponds to a dihedral angle of ~ 60°

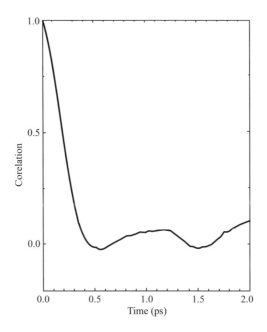

Figure 8.12   The normalized correlation function of the electronic coupling fluctuations for bifluorene in hexane from molecular dynamics simulation

As mentioned above, the influence of the time-dependent coupling on the excitation energy transfer can be understood by considering the time correlation function. Figure 8.12 shows the early part of the normalized correlation function of the calculated excitation exchange coupling (the *dressed coupling* is 20 cm^{-1}). The decay of the correlation function can be approximated by a single exponential with a lifetime of ~ 250 fs; the variance of the dressed coupling fluctuations is 94 cm^{-2}. These two quantities indicate that the process is in the fast modulation limit and will give rise to a T_2^* time of ~ 12 ps [173]. If the T_1^{-1} of Eqn 8.26 was zero, the anisotropy would then decay on the same time scale as T_2^*. This predicted decay time is much longer than the experimentally observed decay time of 0.3 ps in hexane [181]. The result demonstrates that the energy fluctuations of the individual moieties that determine T_1 has a much more important role in the excitation transfer process than coupling fluctuations. Since bifluorene moieties are nonpolar, energy fluctuations may be required to include polarizability, not considered in these simulations. Solvation studies of nonpolar solutes in nonpolar solvents [91, 144] show that rapid energy fluctuations occur, similar in nature to the polar case.

8.3   ENERGY TRANSFER AND LOCALIZATION IN ANTENNA COMPLEXES AND REACTION CENTERS

8.3.1   Time-resolved spectroscopic methods for characterization of excitonic systems

Energy transfer processes in molecular aggregates can be studied using ultrafast time-domain laser techniques to follow the evolution of optical excitations. Elucidation of the mechanism of energy transfer requires a detailed understanding of the energy and phase relaxation processes following the initial excitation, that can be studied using methods such as pump–probe and photon-echo spectroscopy [112]. Pump–probe spectroscopy involves preparation of a transient state of the system by excitation with a "pump" laser pulse and subsequently following evolution of this initial state by monitoring the absorbance of a temporally delayed "probe" pulse. The various frequency components of the probe can be analysed separately by placing a dispersing device before the detector, or they can all be measured simultaneously on the detector. Photon-echo techniques utilize a laser pulse to create a coherent superposition of the ground state with resonant electronic states: this initial state begins to decay due to optical dephasing, and a second, delayed, laser pulse is then applied to "rephase" the system. A coherent photon-echo signal is generated due to this rephasing, and the intensity of this signal reflects the ability to recover the initial state after the delay period.

For a multilevel system, the nature of the initial state prepared in a pump–probe experiment depends upon the duration of the excitation pulses compared with the timescales of the dynamical processes responsible for spectral broadening. Under conditions of incoherent excitation (i.e. when electronic dephasing is fast in comparison with the temporal duration of the pump pulses), a distribution of populations is established in states that are accessed from the ground state by the excitation pulse spectrum. Inter-level and intra-level energy relaxation processes can be studied from the evolution of magic-angle transient spectra, while orientational dynamics are conveniently followed from measurement of the anisotropy (linear dichroism) of the pump–probe signals.

In cases in which the duration of the laser pulses employed is considerably shorter than the timescale of electronic dephasing, a superposition of optically accessible eigenstates is initially prepared. Usually, there is a significant separation in the timescales between electronic and vibrational dephasing. This permits observation of the vibrational structures of the electronic transitions (in the time domain) even though the electronic spectrum is featureless. Coherent excitation of this type can have important consequences in pump–probe spectroscopy when the levels of the superposition are probed to a common final state. Interference between the individual transitions can lead to high initial values (greater than 0.4) and oscillations in the pump–probe anisotropy: the decay of these features directly reflects the timescale of dephasing of the initial superposition state (see in particular Eqn 8.26). Depending upon the number of transitions excited and the mutual orientations of the transition dipole moments, initial anisotropies between 0.4 and 1 (see Fig. 8.9) can be observed. Wynne and Hochstrasser [171, 172] and Knox and Gulen [84] have discussed the effects of coherence upon the anisotropy of optical transitions for the case of molecular dimers. Manifestations of coherent excitation have been observed in femtosecond pump–probe anisotropy studies conducted on porphyrins [47] and on dimers and aggregates comprised of linear tetrapyrroles [36] and bacteriochlorophylls [7, 89, 124].

Pump–probe methods provide a useful means for studying energy-transfer processes within pairs and larger assemblies of chromophores. Magic angle pump–probe signals can be examined to follow the migration of optical excitations between different types of chromophores, as well as amongst distinct spectral forms of a single pigment type. For example, pump–probe studies have been employed to determine the timescale of energy transfer between carotenoid and bacteriochlorophyll pigments within photosynthetic light-harvesting complexes of purple bacteria [131]. Energy transfer between pigments that are identical on average, such as the two halves of a molecular homodimer, will not affect the isotropic spectra, but can cause depolarization of pump–probe or emission signals if the transition moments of the chromophores are nonparallel. Kim *et al.* [81] have applied sub-picosecond emission anisotropy measurements to investigate the timescale of excitation transfer between the monomer units of the bifluorene dimer (see Section 8.2.10). Coherent effects in the anisotropy of pump–probe and emission signals of molecular aggregates provide important information on their geometry and electronic structure, such as the energy separation between optically allowed exciton levels. Photon-echo methods are useful for determining the origins of spectral broadening in molecular aggregates and revealing the influence of static and dynamic site energy disorder upon the exciton wavefunctions [112]. Two-pulse and three-pulse photon-echo measurements have been applied to the investigation of J-aggregates [39, 42, 156] and bacterial light-harvesting complexes [69, 73, 178]. In photon-echo experiments on J-aggregates, quantum beating between one-exciton and two-exciton transitions are apparently influenced by the coherence between one-exciton states [42, 156].

Probing the spectra and dynamics of transients in the near-infrared and infrared regions significantly expands the scope of pump–probe studies of molecular aggregates. Detailed investigation of the exciton level structure is made possible from examination of vibrational transitions or low-energy interband and charge-transfer transitions to states that are otherwise inaccessible directly from the ground state. Pump–probe measurements conducted on the "special pair" bacteriochlorophyll dimer of reaction centers (RCs) from purple bacteria in the 1–5 μm range have revealed transitions into levels not seen in the linear spectrum; important insights into the nature of electronic states of the RCs have been obtained from analysis of these transitions and their associated anisotropies [52, 174]. Investigation of transitions between bands provides a means to follow equilibration within exciton levels that cannot be directly observed from ground state, and the anisotropy of such band-to-band transitions can contain information on the extent to which excitations are delocalized [88]. Femtosecond band-to-band spectroscopy has proven useful in the study of excitons in bacterial light-harvesting systems, described in further detail in Section 8.3.2.

8.3.2 Energy transfer and localization in antenna complexes

8.3.2.1 *Exciton dynamics in the photosynthetic antenna complexes of purple bacteria*

The mechanism of energy migration within the LH-1 and LH-2 complexes of photosynthetic purple bacteria is currently a subject of considerable experimental and theoretical interest. These systems are comprised of a circular arrangement of minimum repeating sub-units that consist of a pair of helical polypeptides (α and β) that each bind bacteriochlorophyll (BChl) pigments [154]. Recent crystal structures for LH-2 complexes from *Rps. acidophila* [103] and *Rhodospirillum molischianum* [85] have shown the systems to be built from nine and eight sub-units, respectively, and reveal the close proximity between BChl pigments identified with the long-wavelength absorption band at 850 nm (these chromophores are referred to as B850). The structure of LH-1 complexes is similar but larger (12–16 sub-units) [75, 154] and contains only a single spectral form of BChl chromophores that absorb in the 870–890 nm range. A schematic depiction of the arrangement of LH-1 and LH-2 complexes is shown in Fig. 8.13: the core complexes enclose the reaction centers (RCs), while the smaller peripheral complexes are interspersed between them.

While the striking circular symmetry and close spacing between BChl pigments suggests that optical excitations will propagate within the complexes as coherently delocalized excitons, it should be kept in mind that the crystal structures simply represent an equilibrium configuration. At ambient temperatures, nuclear motions of the pigments and their protein environment about this geometry can cause variations of the individual site energies and restrict the formation of extended states. These energy variations can be classified as *static* or *dynamic* depending upon their characteristic timescale in relation to the inter-site interaction energy β. The primary objective of experimental studies on the LH-1 and LH-2 complexes has been to ascertain the extent to which their electronic states are delocalized, and the degree to which static and dynamic

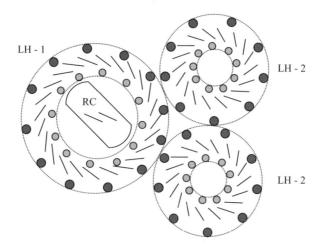

Figure 8.13   A schematic arrangement of peripheral (LH-2) and core (LH-1) complexes of purple bacteria shown from a top view of the membrane. Solid circles represent the *a*- and *b*-polypeptides, and the lines represent a side view of the bacteriochlorophyll pigments. RC = reaction center

inhomogeneity influence the mechanism of exciton migration. Although the underlying physical principles are identical to those considered in the case of molecular crystals, little is known regarding the motion of excitons within a protein environment. Delocalization strongly affects the optical and photophysical properties of each aggregate, thereby determining not only the mode of energy transfer *within* individual rings but also energy migration dynamics *between* complexes and to the final destination, the photosynthetic reaction center.

In recent years, a number of experimental techniques have been applied to elucidate the optical properties and energy-transfer mechanism in LH-1 and LH-2 systems. These studies have involved direct time-domain investigation of the dynamics of optical excitations within the antenna systems using ultrafast techniques, as well as examination of their spectral properties, applying methods such as hole-burning and polarized fluorescence excitation spectroscopy at low temperatures and photon-echo measurements at room temperature. In spite of the large number of complementary approaches which have been brought to bear in the study of LH-1 and LH-2 complexes, a consensus has not been reached in the literature regarding the extent to which energy is delocalized and the mechanism by which excitation is transferred within the antenna systems. Most studies to date have been conducted in the region of the Q_y optical transitions, where the static spectral features and dynamics observed in isotropic and polarized measurements are equally consistent with the presence of localized and delocalized electronic states. Conclusive evidence for the collective nature of excitations within the LH-1 and LH-2 antenna systems at room temperature is now beginning to emerge, although quantitative estimates of the mean and moments of the distribution of exciton delocalization lengths, and the importance of static and dynamic inhomogeneity, have not been firmly established. In this section, we present a brief discussion of the optical properties of the antenna structures and summarize the current status of spectroscopic studies on LH-1 and LH-2 complexes.

8.3.2.2   *Exciton states and optical properties of circular aggregates*

The eigenstates of a circular aggregate comprised of N identical pigments can be written in the familiar form of the Frenkel excitons $|k\rangle$, as given in Eqn 8.5, where the states $|n\rangle$ represent the excited states of the individual bacteriochlorophyll molecules. If only interactions between nearest neighbors are taken into account, the energies are given by Eqns 8.4 and 8.5. An illustration of the band-level structure of a circular aggregate comprised of 18 pigments is shown in Fig. 8.14 for a head-to-tail configuration of transition moments, in which case the inter-site interaction energy β is negative. Here $(N/2) - 1$ doubly degenerate levels span the band with a nondegenerate ($k = 0$) level present at lowest energy. The transition dipole moments for optical transitions $0 \rightarrow k$ from the ground state into the manifold of exciton levels can be written in this case as

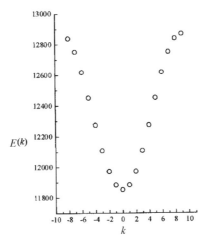

Figure 8.14   Dispersion of exciton energy $E(k)$ with k. The exciton levels span a band of width 4β, and their energies are given by Eqn 8.5. The inter-site interaction energy β was taken as $-254\,\text{cm}^{-1}$ for this calculation

$$\mu_{0 \to k} = \frac{\sqrt{N}}{2} [\delta_{k,-1}(a + ib)(\mathbf{x} - i\mathbf{y}) + \delta_{k,+1}(a - ib)(\mathbf{x} + i\mathbf{y}) + 2c(\delta_{k,0})\mathbf{z}], \quad (8.29)$$

where a, b, and c are the projections of the transition dipole moment for $0 \to n$ ground-state absorption along the x-, y-, and z-axes for the first pigment ($n = 0$) within the aggregate; $\mathbf{x}$, $\mathbf{y}$, and $\mathbf{z}$ are unit vectors along the respective Cartesian directions in the aggregate frame, and δ is the usual Kronecker delta function. Note that $a^2 + b^2 + c^2 = 1$. The salient optical properties of a circular aggregate of pigments are immediately apparent from Eqn 8.29. Transitions from the ground state are allowed only into the $k = 0$ and $k = \pm 1$ levels, which are polarized normal to the aggregate plane (along the z-axis) and within the aggregate (x–y) plane, respectively. It is evident from Eqn 8.29 that the only allowed levels for a coplanar circular configuration of transition moments (from the ground state) correspond to the $k = \pm 1$ states.

As described in Section 8.2, inhomogeneity within the molecular aggregates establishes eigenstates that are admixtures of the "ideal" excitons (of Eqn 8.5 in this case) and which become progressively localized on spatially distinct regions with increasing levels of disorder. The dipole strength associated with the $k = \pm 1$ states of a homogeneous circular aggregate is thus redistributed amongst levels within the exciton manifold, activating transitions into levels that are otherwise forbidden in the absence of disorder. For an assumed functional form of the distribution of site energies, the spectral [4, 39, 152] and dynamic [88] consequences of disorder can be modeled using random sampling techniques, as discussed in Section 8.2.4. The degree of localization can be studied from parameters such as the participation ratio or quantum connectedness length, which were discussed in Section 8.2.8. The dependence of these parameters upon the extent of disorder (relative to the inter-site interaction energy) was presented in Fig. 8.8 for a circular aggregate built from 18 pigments. The localization criteria provide insight into the degree to which the extended states of the homogeneous aggregate become confined by inhomogeneity. The difficulties in interpreting these criteria in terms of a physical length of delocalization were alluded to in Section 8.2.8.

8.3.2.3 *Dynamics of optical excitations within core and peripheral antenna complexes and sub-units*

In recent years, the availability of high-resolution information on the three-dimensional structure of light-harvesting complexes, together with continuing advances in laser technology, has motivated extensive experimental and computational studies of the LH-1 and LH-2 complexes. An issue of central interest has been the extent and characteristic timescales of inhomogeneity within the core and peripheral complexes, and the degree to which disorder acts to restrict coherent delocalization of excitation energy. Time-domain and frequency-domain laser methods have been applied to investigate the excitation dynamics and spectra of the intact antenna systems, as well as to characterize the properties of their dimeric sub-units, which constitute the individual building blocks of the circular aggregates. A number of studies have conclusively demonstrated the collective nature of excitations within LH-1 and LH-2 systems. The *extent* of energy delocalization still remains unresolved, however, and estimates appearing in the literature range from proposals of localization on single dimer sub-units to delocalization over the entire ring of pigments. We present here a summary of the important findings and discuss briefly the conclusions that have emerged from recent investigations.

The ability to reversibly dissociate LH-1 complexes into their constituent dimeric sub-units by detergent titration has provided the opportunity to study the intrinsic properties of individual building blocks of the antenna complexes [154, 161]. The sub-units (B820) display the optical properties of a strongly coupled dimer, with a strongly allowed transition to the lower dimer level at 820 nm and a weak transition to the upper state at 795 nm. From modeling of the B820 CD spectrum it has been concluded that the configuration of BChl pigments within B820 is similar to their relative positioning within the intact LH-1 complexes, i.e. the Q_y transitions of the monomers are oriented at an angle of $\sim 20°$ [161]. An important objective of investigations of the B820 dimer has been to determine the extent and timescales of sub-unit energy disorder, a key determinant of exciton dynamics within the intact antenna aggregates.

Transient spectra of the B820 sub-units from core complexes of *Rs. rubrum* were measured by Visser *et al.*, who reported an absence of discernable sub-picosecond dynamics within the ~ 250 fs resolution of their experiments [162]. Pump–probe measurements with 35 fs resolution by Kumble *et al.* [89] and 15 fs resolution by Arnett *et al.* [8] have subsequently revealed that (i) rapid stabilization of the excitation (dynamic Stokes shift) takes place on a sub-50 fs timescale, and (ii) excitations are coupled to nuclear motions with frequencies ranging from 20 to 1000 cm^{-1}. The observation of the ultrafast Stokes shift (of *ca.* 80 cm^{-1}) in the dimer implies that the influence of self-trapping upon the exciton wavefunctions (as discussed in Section 8.2.7) should be carefully considered.

Important insights into the dimer exciton level structure and the microscopic origins of inhomogeneity have recently been obtained from pump–probe anisotropy studies. Wavelength-integrated measurements have revealed a heavily damped oscillation in the anisotropy at a frequency corresponding to the separation between symmetric and antisymmetric dimer levels, arising from an initially prepared coherent superposition of these states [89]. Subsequent wavelength-selective studies with improved time resolution have clearly revealed underdamped oscillations in the anisotropy and *a pronounced dispersion of the oscillation frequency with detection wavelength* [8] (see Fig. 8.15). This has enabled the anisotropy measurements to be used as a room-temperature

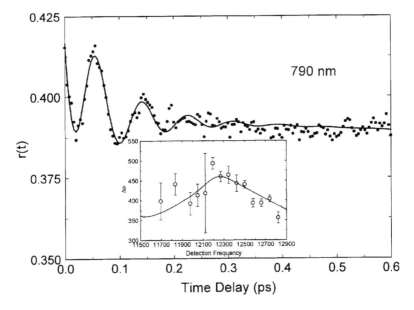

Figure 8.15 The time-dependent anisotropy $r(t)$ for the B820 bacteriochlorophyll dimer and fit (solid line), measured at 790 nm. Inset: variation of oscillation frequency ($\Delta\omega$) with detection frequency (hollow circles) Taken from reference 8.

line-narrowing technique, providing information on the joint distribution of dimer exciton levels.

Modeling of the observed dispersion profiles has led to two significant conclusions concerning the constituent bacteriochlorophylls: (i) distributions of the *asymmetry* (i.e. inequivalence of monomer transition energies) and *average transition frequency* of the two monomers are *simultaneously* present; and (ii) these distributions are *strongly correlated* in a fashion such that the degree of asymmetry becomes more pronounced as the average transition energy of the two pigments increases. These findings suggest that the stabilizing cross-interactions between the two pigment–polypeptide units of the dimer (such as hydrogen bonding) also act to *equalize* the transition energies of the two bacteriochlorophyll chromophores within the B820 sub-unit. Since similar interactions between adjacent sub-units are instrumental in the formation of the circular antenna complexes, a significant implication of these findings is that aggregation might serve to decrease the range of bacteriochlorophyll site energies present within the antenna systems.

The nature of electronic states and the mechanism of energy transfer within core and peripheral antenna complexes have been studied in a number of frequency-domain and time-domain measurements. Evidence for the existence of delocalized states at cryogenic temperatures has been obtained from hole-burning spectroscopy, which has revealed the absorption bands to be built from a small number (about five) of optical transitions [170]. This observation is consistent with the presence of delocalized states and a band-level structure that is similar to that of a homogeneous ring, but partly perturbed by the presence of weak or modest disorder. Taking motional narrowing effects into account, the inhomogeneous line widths of these bands (~ 120 cm^{-1}) are consistent with a modest degree of site energy disorder. Results from polarized fluorescence excitation spectroscopy at 4 K have, however, been interpreted differently by van Grondelle and coworkers [157]. A sharp rise in the emission anisotropy for long-wavelength excitation was taken to indicate incoherent energy transfer between states localized on dimer sub-units.

Conclusive support for the existence of delocalized states at room temperature has emerged from the observation of collective effects in absorption and emission spectra. The observation of anomalously large photobleaching intensities in the long-wavelength absorption bands of LH-1 and LH-2 complexes implies the presence of extended states with dipole strengths that are considerably larger than those of the constituent bacteriochlorophyll monomers [79, 175]. From analysis of the photoinduced absorbance changes in LH-2 complexes from *Chr. tepidum*, Kennis *et al.* have concluded the presence of strongly delocalized states which possibly extend over the entire B850 ring [79]. Corresponding superradiant emission at room temperature has been recently observed by Monshouwer *et al.* for LH-1 and LH-2 complexes from *Rb. sphaeroides*, for which the radiative rates were found to be about three times larger than for monomeric bacteriochlorophyll [111].

Direct investigation of the dynamics of optical excitations within core and peripheral antenna systems have been performed using polarized and magic angle pump–probe and fluorescence upconversion techniques with femtosecond time resolution [12, 23, 56, 57, 68, 89, 114, 118]. The majority of these measurements have been conducted in the 700–950 nm spectral region where the pump–probe signals contain contributions from transient absorption, emission and photobleaching. Ultrafast spectral evolution and biphasic depolarization occurring on a sub-500 fs timescale have been observed in all studies. While these results have provided important insights into the dynamics of relaxation processes occurring within the antenna systems, *assignment of the processes themselves has not proved to be straightforward*. Some investigators have interpreted their observations as reflecting the timescales of incoherent hopping of localized energy between sub-units, while others have proposed that the spectral evolution and depolarization represent inter- and intra-level relaxation within a manifold of delocalized exciton states. Extensive modeling of experimental pump–probe spectra [23, 107, 114, 118] and depolarization dynamics [12, 23, 68] has been performed and the effects of energy disorder have been considered; yet the estimates of the delocalization lengths vary from 1–2 subunits [118] to a substantial portion of the ring [114].

A powerful technique for probing exciton band structure and distinguishing localized and delocalized states is femtosecond band → band spectroscopy, which is currently being applied in our laboratory to study LH-1 and LH-2 complexes. Owing to the rigorous selection rules (such as $\Delta k = 0$ for a linear chain, or $\Delta k = 0, \pm 1$ for a circular aggregate) which are typical for optical transitions in molecular aggregates at sufficiently low temperatures, transitions from the nondegenerate ground state can usually access only a small subset of the N levels within the exciton manifold, where N is the total number of pigments. An elegant experimental technique for observing *all* states within the manifold involves the measurement of transitions between the optical exciton band and a second band also comprised of N states: the selection rules can be satisfied for each and every level in this fashion. In the case in which the width of the second exciton band is negligible in comparison with that of the first, the spectra of band → band transitions directly yield the density-of-states function for the first band, after appropriate correction for the Boltzmann distribution. Early examples of experimental determination of band-level structure by this method are the vibrational exciton (vibron) → electronic exciton absorption measurements [26] and the exciton luminescence studies [63] performed on crystalline forms of aromatic compounds.

The ability to follow the evolution of band → band spectra on ultrafast timescales provides a means of studying population dynamics within portions of the exciton band that cannot otherwise be observed directly from ground state. Further, consideration of the properties of band → band transitions reveals *a significant sensitivity of the transition anisotropy to exciton delocalization*. Coherent contributions to the anisotropy of the type described in Section 8.2 can arise *only* from a superposition of delocalized eigenstates, in contrast to the case of emission signals [88]. Femtosecond band → band measurements conducted on B820 dimer sub-units and re-assembled LH-1 complexes from *Rs. rubrum* have revealed a significantly higher initial anisotropy for LH-1 (0.4–0.5) than is seen for B820 (0.21) [90]. These results confirm the presence of exciton states within the LH-1 system which are delocalized over several bacterio-chlorophyll pigments.

8.3.3 Energy transfer in the photosynthetic reaction center

8.3.3.1 Introduction

Bacterial photosynthetic reaction centers use light energy to create charge separation across the bacterial membrane, which is the primary step in the photosynthetic reaction sequence [15, 16, 32, 43, 100]. The excitation of a reaction center (RC), either directly by light or by energy transfer from LH complexes (which surround RCs in the bacterial membrane), initiates an energy migration process between the cofactors of the RC. This process terminates when the energy gets into the lowest energy excited state of the RC, which is localized on a *pair of cofactors*, the "special pair", and electron transfer is generally initiated from this excited state.[†] The goal of the present section is to describe coherent and incoherent energy transfer phenomena in the RC, and put the relevant experimental and theoretical information into the framework of the general discussion of Section 8.3.

While RCs from many sources have been purified and studied, the three-dimensional structures of only two bacterial RCs have been obtained by X-ray crystallography techniques [30, 31, 37, 145, 176], and we will confine our discussion mostly to work performed on these. Both RCs contain eight cofactors, which are organized in two branches, L and M, within a protein matrix, and are related to each other by an approximate C_2 symmetry. The RC from *Rhodobacter (Rb.) sphaeroides* (Fig. 8.16), contains four bacteriochlorophyll (BChl)-*a* molecules, two bacteriopheophytin (BPh)-*a* molecules, and two quinones [37, 145, 176]. The RC from *Rhodopseudomonas (Rps.) viridis* contains bacteriochlorophyll-*b* and bacteriopheophytin-*b* cofactors arranged in a similar fashion [30, 32]. The "special pair" (P) mentioned above comprises two of the bacteriochlorophylls, which reside with their pyrrole rings parallel to each other and only ~ 3.5 Å apart (note, however, that the center-to-center distance

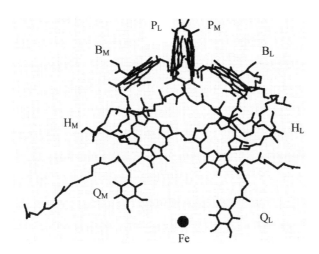

Figure 8.16 The 3-D arrangement of the cofactors in the *Rb. sphaeroides* reaction center [145]. P, special pair; B_L, B_M, accessory BChls; H_L, H_M, bacteriopheophytins; Q_L, Q_M, quinones; Fe, iron atom

[†] In certain mutant RCs, charge separation can be initiated from excited states of other cofactors [155].

of the two BChls is actually 8 Å). Excitation of the special pair by light, either directly or by energy transfer from LH complexes (which surround RCs in the bacterial membrane), leads to a fast electron transfer reaction to the BPh (H) on the L side only. The accessory BChl (B), located between the special pair and H_L, plays a crucial role in mediating this electron transfer step, either as a superexchange medium or as a real intermediate [10].

Elucidation of the molecular factors responsible for the efficiency and uni-directionality of the electron transfer reaction requires a thorough knowledge of the electronic structure of the special pair and its interactions with other cofactors and with protein side chains in the RC. Section 8.3.3.2 will be devoted to the special pair, describing in detail the excitonic and charge resonance interactions which lead to its unique characteristics. Section 8.3.3.3 will discuss the interaction between the accessory BChls and the special pair and the validity of a "supermolecule" picture, in which interchromophoric excitonic interactions dictate the excited state electronic structure of both B and P. Section 8.3.3.4 discusses similar issues related to energy transfer from the BPhs.

8.3.3.2 The special pair: electronic structure of a strongly coupled dimer

The two lowest energy bands in the absorption spectra of bacteriochlorophylls are assigned to the Q_Y and Q_X transitions [166]. The proximity of the two BChls of the special pair leads to an excitonic splitting of these transitions. A large amount of information exists concerning the excitonic states related to the Q_Y transition, since these states are involved in the electron-transfer reaction. The approximate C_2 symmetry of the special pair allows us to designate the two states as P_{Y+} and P_{Y-}, the latter being the lower-energy state [13, 116, 125, 127, 159, 160]. The direction of the transition dipoles of the two BChl mono-mers comprising the special pair, $\vec{\mu}_{P_L}$ and $\vec{\mu}_{P_M}$, inside the protein can be obtained from electronic structure calculations [166] in combination with the geometry of the cofactors as obtained from the crystal structure of the reaction center [37, 145, 176]. These dipoles can be combined to form the special pair excitonic transition dipoles, $\vec{\mu}_{P_{Y+}}$ and $\vec{\mu}_{P_{Y-}}$. The transition dipoles of the two excitonic levels, expressed in terms of the two monomeric transition dipoles, are then readily calculated from Eqn 8.3. The ratio of absorption cross-sections from ground state to P_{Y+} and P_{Y-} can then be written as

$$\frac{\sigma_{P_{Y+}}}{\sigma_{P_{Y-}}} = \frac{(1 + \sin 2\alpha \cos \theta)\omega_{P_{Y+}}}{(1 - \sin 2\alpha \cos \theta)\omega_{P_{Y-}}}. \tag{8.30}$$

Here $\omega_{P_{Y\pm}}$ is the frequency for the transition into $P_{Y\pm}$, θ is the angle between the monomer transition dipoles, which are assumed to have the same magnitude, and α is defined in Eqn. 8.3. If the two excited states $P_L^* P_M$ and $P_L P_M^*$ (the asterisk denotes an excited state) are degenerate, then $\vec{\mu}_{P_{Y+}}$ is parallel to the C_2

symmetry axis, while $\vec{\mu}_{P_{Y_-}}$ is perpendicular to it, and the absorption ratio turns out to be 1:7.5. However, the excitonic splitting can involve an asymmetry. This will arise if $P_L^* P_M$ and $P_L P_M^*$ are not degenerate, due to different interactions of the two cofactors with the surrounding protein matrix. It can also arise from mixing of the excitonic states with other nearby, nondegenerate states, such as charge transfer states of the type $P_L^+ P_M^-$ and $P_L^- P_M^+$, or accessory BChl states (see Hamiltonian matrix, Section 8.2.5). Both of these will be discussed in detail below. Any asymmetry will lead to a change in the directions of the excitonic transition dipoles, as well as their lengths, thereby affecting the ratio of Eqn 8.30.

Transition from the ground state into the P_{Y_-} state forms the lowest energy band in the absorption spectrum of reaction centers. This strong band is significantly shifted to the red compared to the Q_Y band of BChls in solution, and at room temperature it appears at 950 nm for *Rps. viridis* [159] and 860 nm for *Rb. sphaeroides* [160] (compare with the B absorption at 840 nm and 800 nm, respectively). The band due to the transition into the upper excitonic state is much weaker, and is superimposed on the Q_Y transitions of the B cofactor. A low-energy shoulder appearing on the B absorption band of *Rps. viridis* (at 850 nm) at cryogenic temperatures was assigned to the P_{Y_+} [159]. This assignment was based on the almost perpendicular polarization found for this band in photoselection experiments. A similar shoulder on the 800 nm band of the *Rb. sphaeroides* absorption spectrum was assigned in a similar way to P_{Y_+} [13, 14]. Hole-burning experiments [146] have confirmed the location of the upper exciton band; a photochemical hole burned in the P_{Y_-} band at 4.2 K led to the appearance of an additional hole, blue-shifted by 1900 cm^{-1} and 1300 cm^{-1} in *Rps. viridis* and Rb. sphaeroides, respectively. By changing the burn energy it was verified that the position of the secondary hole was positively correlated with that of the burnt hole, and it was therefore assigned to P_{Y_+}. In both species, it actually coincided with the spectral position of the shoulder discussed above. The above excitonic splitting values imply coupling energies of 850 cm^{-1} and 650 cm^{-1} respectively, which are similar to the those obtained from quantum-mechanical calculations of the special pair [116, 125, 147, 148]. How do nuclear degrees of freedom modulate the effect of this coupling?

The only vibrational modes coupled to special pair optical transitions with significant dimensionless shift seem to be low-frequency modes, and these were observed using several techniques. Transient photochemical hole-burning spectra of the special pair of *Rps. viridis* [55, 70, 71, 120, 138, 146] show a short Franck–Condon progression in a ~ 130 cm^{-1} mode, which was termed "the marker mode." This mode is missing in BChl hole-burning spectra, and was therefore assigned to an intermolecular vibration. Coupling to protein phonons having a mean frequency of 30–40 cm^{-1} (with a dimensionless shift of ~ 2) was also inferred from the hole-burning data. Near-infrared resonance Raman measurements also indicate that only low-frequency modes show a nonnegligible dimensionless shift [25, 115]. This implies that the relation $\beta \gg \hbar\omega\Delta$ of Section 8.2.6 holds for all the modes coupled to optical excitation in the special pair, and so it can be clearly defined as a *strongly coupled dimer*.

The exciton bandwidth of the special pair shows an unusual dependence on temperature. The positions of the two excitonic bands of the special pair at different temperatures are provided in Table 8.1. The P_{Y-} band of *Rps. viridis* shifts from 990 nm at 6 K to 950 nm at room temperature. Similar shifts are seen in the spectrum of *Rb. sphaeroides*. While the P_{Y+} band cannot be resolved in the ground-state absorption spectrum at temperatures higher than ~ 10 K, its position can be inferred from other experiments, such as circular dichroism [117, 132] and broad band femtosecond spectra [7, 163]. The location of the P_{Y+} band of the *Rb. sphaeroides* reaction center at 85 K was deduced from measurements on RC samples containing nickel-substituted accessory BChls [54]. In *Rps. viridis*, the P_{Y+} and location does not show any dependence on temperature, while in *Rb. sphaeroides* there is a disagreement with regard to the amount of band shift at room temperature, but in any case it is much smaller than the corresponding shift in the lower excitonic band.

Several different phenomena can account for the temperature dependence of the excitonic bands. First, the temperature effect might be due to modulation of the interaction energy by bath fluctuations, as discussed in Section 8.2.6. Intra-dimer vibrational modes, involving relative motion of the monomers, can also be involved in interaction energy modulation [179]. It is also possible that temperature-dependent mixing with CT states of the special pair, described in detail below, leads to shifts in band positions. The temperature dependence here involves the effect of local protein structural changes on the energetics and bandwidth of the CT states. A unified theoretic description of the electronic structure of the special pair as a function of temperature, based on all of the available information, is still absent.

The importance of CT states for the spectroscopic features of the special pair was underscored by quantum-mechanical calculations [116, 125, 126, 147, 148, 165, 169]. Within strict C_2 symmetry, the P_{Y-} band can mix only with the corresponding out-of-phase mixture of the charge-transfer states, i.e.

Table 8.1   Location of the excitonic bands of RCs at different temperature

Temperature (K)	Rb. sphaeroides		Rps. viridis	
	P_{Y+}	P_{Y-}	P_{Y+}	P_{Y-}
6–10 [13]	810	890	850	990
85 [54]	814	886		
100 [132]			850	980
297	813 [163], 825 [7]	860	850 [117]	950 [117]

$CT_- = P_L^+P_M^- - P_L^-P_M^+$ (see Section 8.2.5). Since the two-fold symmetry is only approximate in the reaction center, mainly due to different protein side chains on the two sides of the dimer, it is possible to have some component of CT_+ mixed into the P_{Y-} band, or in the extreme case to localize the charge on one of the two BChls.

Experimental measurements of the charge transfer character of the P_{Y-} band, using Stark effect spectroscopy [11, 96–98], show that upon excitation into the P_{Y-} state the dipole moment changes by 10 Da at room temperature, and the direction of the change is consistent with charge motion between the two BChls that form the special pair [97]. A measurement of low-lying transitions from P_{Y-} to CT-dominated states, predicted by the theoretic calculations, can provide a direct probe of the energetics and degree of mixing of the CT states. Such a measurement was recently performed [174] using femtosecond infrared spectroscopy, and the spectrum obtained is reproduced in Fig. 8.17. It includes two main bands, one peaking at 5300 cm^{-1} and the other at 2700 cm^{-1}. An additional low-intensity band which has been reported [164] is seen to peak at 1800 cm^{-1}. By comparing the pump–probe anisotropy of the two main bands with the anisotropy value predicted from the crystal structure of the *Rb. sphaeroides* RC [37, 145, 176] it was concluded that the transition dipoles of both bands are dominated by CT transitions. The observation of the excessive width of the 5300 cm^{-1} band (4500 cm^{-1}) led to the suggestion [174] that it actually consists of two or more bands, possibly due to mixing with Q_X states.

Scherer and Fischer [126] have recently modeled the main features of this infrared spectrum [174] and suggested that the 2700 cm^{-1} band is an excitation

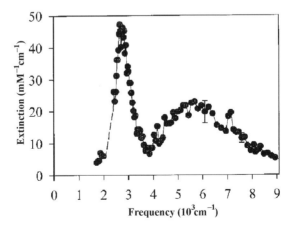

Figure 8.17   The infrared excited-state absorption spectrum of *Rb. sphaeroides* reaction centers [172]. The two bands are not related to any ground-state absorption feature, and are thought to arise from charge-transfer transitions of the special pair

to a singlet state formed by coupling the lowest triplet states of each of the BChls of the special pair. The transition derives its intensity from additional coupling to CT states. These authors have further elaborated the suggestion that the broadband peaking at 5300 cm^{-1} contains two transitions, by proposing that the excessive width is due to assymmetric splitting of the CT transitions. Their model also shows that the P_{Y-} state contains $\sim 15\%$ CT character, while the P_{Y+} state barely mixes with the CT states at all, due to conservation of the C_2 symmetry by these states. This result fits nicely with the experimental observation that the electron–phonon coupling of the P_{Y+} state is much weaker than that of the P_{Y-} state [138]. Magnetic resonance measurements on the radical cation of the special pair techniques [93, 94] imply a large asymmetry in charge distribution over the two monomers. However, the relation of these measurements to the neutral complex remains to be established.

Methods to modify the symmetry of the special pair have been discovered in recent years. The mutation of either of the histidines that coordinate the central Mg atoms of the special pair BChls to leucine leads to a replacement of the corresponding BChl with a BPh. Mutants obtained in this way are called the heterodimer RC (with the BPh on the M side) and the reverse heterodimer RC (with the BPh on the L side). Such mutants were prepared from both *Rb. capsulatus* [21, 22] and *Rb. sphaeroides* [51] reaction centers. Since BPh is easier to reduce than BChl by ~ 300 mV, it is reasonable to assume that the CT character of the excited states of the heterodimer will be enhanced. Indeed, measurement of Stark shifts [51] showed an increase of 30–70% in the excitation-related dipole moment change. The shape of the lowest-energy absorption band in heterodimer RCs is different from that of wild-type RCs. In the case of the *Rb. sphaeroides* heterodimer, room-temperature spectra show a broader and featureless band [180], while low-temperature spectra consist of two broad bands, both of which show the same electric field induced spectral changes [51]. Experimental data from mutants with disrupted protein–cofactor hydrogen bonding [180] was used in combination with spectral simulations to suggest that the lower excitonic state is coupled more strongly to BChl$^+$ BPh$^-$ than to the higher-energy BChl$^-$ BPh$^+$. Thus, measurements on the heterodimer directly show the importance of CT states on the spectroscopy of the special pair.

8.3.3.3 Interaction between the special pair and the accessory bacteriochlorophylls

The proximity of the cofactors in RCs has led to "supermolecule" descriptions of their excited states, in which the interchromophoric coupling induces the formation of mixed (excitonic) states involving not only the special pair dimer, but other cofactors as well [116, 125, 147, 149, 165]. Theoretical supermolecule descriptions typically provide a "zero-temperature" picture, and it remains to be examined whether finite-temperature effects, mainly intramolecular nuclear motion and interactions of the cofactors with their surroundings affect the mixing and localize the states. Recent experimental results touch primarily on the issue of mixing of P and B states, and these will be discussed in some detail in this section. Energy transfer from H will be reviewed in the following subsection.

A proper analysis of the ultrafast energy transfer amongst the cofactors of the reaction center requires the density matrix equations of motion, analogous to those in Eqn 8.23. The dynamics after excitation of B* involves at the very least the nine density matrix elements from the basis B_L^*, B_M^*, and P_Y^+. The dynamical matrix is readily obtainable using Redfield or stochastic theories. As shown in Sections 8.2.9 and 8.2.10, the relaxation parameters are anticipated to depend mainly on the fluctuations of the energy gaps between the eigenstates of this three-level system. The role of the coherences of this system of states in the energy transfer process has not yet been evaluated. However, given the rapid timescales of energy transfer (c. 100–200 fs), it is not obvious that kinetic (i.e. population only) modelling of these dynamics will adequately describe the behavior. The short optical pulses in such energy-transfer experiments will generate the initial condition given in Eqn 8.12, which will decay with dynamics such as that illustrated for two levels by Eqns 8.23 and 8.24.

The electronic coupling between P and B states can be calculated using the dipole approximation [53, 125, 165, 169], since the distances between the cofactors are rather large. In fact, calculations based on monopole transition charges [113] lead to similar values for the coupling [53], i.e. 70–100 cm^{-1}, depending on the type of RC. What kind of energy redistribution processes can be induced by interchromophoric coupling of this strength? Does it lead to Förster-type, incoherent hopping of the energy, or to a process of coherent migration within delocalized states? A series of experiments have addressed this issue over the past decade. Using multicolor pump–probe [17, 18, 53, 67, 95] or fluorescence upconversion spectroscopy [72, 142, 143], these experiments invariably measured energy transfer rates from B to P of ~ 100 fs. Recent experiments with very short laser pulses and multiwavelength detection have been interpreted with a two-step model, in which the first step involves a coherent energy transfer process, to be discussed in detail below, and the second step is due to internal conversion between P_{Y+} and P_{Y-} [7, 72, 163].

The dependence of the special-pair excitonic bandwidth and line shape on temperature, which was discussed above, leads to a large variation of the spectral overlap of B and P. Indeed, calculations based on spectral properties of the cofactors at 10 K [66] predict that Förster energy transfer between B and P cannot explain the sub-100 fs energy transfer seen in the experiments, while calculations based on room-temperature spectra provide more "generous" estimates for Förster energy-transfer rates [53]. However, the experimental data have established an independence of the rate on temperature

[17, 44, 66, 143], indicating that conventional Förster energy transfer cannot be the sole mechanism responsible for the decay of B* excitation.

Several authors [66, 72, 143] have suggested that short-range interactions such as exchange [33] and orbital penetration [128] should be considered as a possible cause for the energy redistribution. The exchange mechanism for excitation energy transfer between two molecules is based on close proximity of atoms participating in the excited states of the two. Further, the magnitude of the interaction depends critically on the relative orientation of the involved p_z orbitals (in the case of $\pi \rightarrow \pi^*$ transitions, as in BChls). From the RC crystal structure [30, 31, 37, 145, 176] it can be calculated that the smallest distance between pairs of special pair atoms that participate significantly in the electronic excitation [113] is ~ 5 Å, while most atom pairs are separated by much larger distances.

Calculations of the exchange interaction for a D_{2h} symmetry naphthalene dimer [128] have recently confirmed that in many cases the exchange interaction is small. While the geometry of this dimer is optimal for exchange (with p_z orbitals of the two monomers directed along the line connecting the neighboring carbon atoms), at a separation of 5 Å the total "short-range" interaction between the monomers is as low as 1 cm^{-1} [128]. In the RC, where the geometry is not optimal, the exchange interaction could be even weaker. In fact, using published values for exchange integrals [74] and BChl Q_Y transition density matrix elements [113], we have calculated that the total exchange interaction energy between B_L and P_L would be 0.1 cm^{-1}, and therefore completely negligible compared with the multipole interaction.

Recent experimental studies have provided evidence for the importance of mixed states for energy-transfer processes in the RC. Measured polarization anisotropy values of transient signals, obtained by pumping *Rb. sphaeroides* RCs at 810 nm with 50 fs pulses and probing at several near-IR and IR wavelengths [53], were consistently lower than the values calculated from the crystal structure of the *Rb. sphaeroides* RC [37, 145, 176], assuming localized pump-induced excitations. The probe wavelengths corresponded to stimulated emission from P_{Y-} (at 950 nm), triplet/doublet absorption of the charge-separated special pair [122, 123] (at 1200 nm) and CT-related transient absorption (at 3840 nm) [174].[†] It was suggested that the results could be explained if delocalized states, involving B* and P_{Y+}, are excited in the experiment. Energy transfer then involves localization of the excited state on P_{Y+} followed by internal conversion to P_{Y-}. Vos *et al.* [163] excited *Rb. sphaeroides* and *Rps. viridis* RCs at a series of wavelengths, using 30 fs pump pulses. Exciting *Rb.*

[†]The low anisotropy at probe wavelength of 950 nm, from a gain measurement, should be compared to CW (34) and time-resolved [143] fluorescence measurements, which show the expected, higher anisotropy value (close to 0.4). Thus it is likely that there are contributions from overlapping excited state transitions at this wavelength. There do not appear to be such contributions at the other two wavelengths mentioned in the text.

sphaeroides RCs at 770 nm, where both H and B are excited (at a 3:1 ratio), they measured a ~200 fs decay of the 800 nm bleach signal and a similar time for the appearance of the bleaching of the P_{Y-} band, attributing that time to "downhill" energy transfer from B^*. In contrast, excitation at 820 nm, which was estimated to involve both B^* (40%), P_{Y+} (40%), and P_{Y-} (20%), led to an essentially instantaneous (< 20 fs) appearance of 90% of the P_{Y-} bleaching signal. This was attributed to coherent excitation of B^* and P_{Y+}. Similar conclusions were obtained from recent multiwavelength *polarized* pump–probe experiments on *Rb. sphaeroides* RCs [7], in which it was observed that excitation of B^* leads to significant excited state spectral evolution preceding ground state recovery (with ~ 20 fs time constant). Again, this was attributed to coherent excited state dynamics involving both B^* and P_{Y+}.

Thus it is seen that an emergent view of excitation energy transfer from B to P stresses the supermolecule aspects of the dynamics, with the energy redistribution process regarded as internal conversion between strongly mixed states. The first step in the process is localization of the excitation (initially formed in a mixed $B^* + P_{Y+}$ state) on P_{Y+}. The localization step is induced by fluctuations of the bath (protein side chains) surrounding the cofactors, and involves a coherence decay and not a population decay. A model for this step, which follows the ideas of Section 8.2.7, is presented in Fig. 8.18. The B–P "dimer" forms an intermediate case between the two cases of Fig. 8.4. Since there is a driving force of ~ 100 cm^{-1} for energy transfer from B^* to P_{Y+},

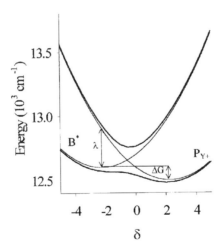

Figure 8.18 Model potential surfaces for the interaction of B^* and P_{Y+}, each having a potential energy surface of $\omega_0 = 40$cm^{-1}. Additional parameters are as follows: $\beta = 100$cm^{-1}, $\lambda = 300$cm^{-1}, and $\Delta G = 100$cm^{-1} (ΔG is defined in the figure); δ is a dimensionless solvent coordinate defined so as to produce the required reorganization energy between the two surfaces

solvent-induced relaxation on the lower eigenstate surface will lead to localization in P_{Y+}. The larger transition dipole of B^* entails preferential population of the upper surface, and thus internal conversion precedes localization. The rate for internal conversion between the excitonic states will be dictated by the time-correlation function of the fluctuations of the $B^* - P_{Y+}$ energy gap, as discussed above. This time-correlation function can in principle be obtained from molecular dynamics simulations, as was recently done in the case of the charge-separated states [99]. Also complicating the interpretation of experimental results is the possibility that a direct pathway for photo induced change-separation exists between the B and H pigments [M.E. Van Brederode, M.R. Jones and R. Van Grondelle 1997 *Chem. Phys. Lett.* 268; 143–149].

8.3.3.4 *Interaction between the accessory bacteriochlorophylls and the bacteriopheophytins*

The electronic coupling between H and B in the reaction center, calculated with the dipole approximation [125, 169], is 60–100 cm^{-1}, depending on the species. As in Section 8.3.3.3, we would like to be able to understand the importance of this coupling in determining the spectroscopic properties of the reaction center. In particular, the possibility of mixed states will be again discussed here, in view of recent experimental observations [143, 163].

Early measurements [18] suggested that, similar to the case of B^*, the transfer of the energy from H^* to the special pair occurs on a sub-100 fs time scale. Stanley *et al.* [143] excited *Rb. sphaeroides* RCs at 85 K with 760 nm pulses (in the BPh absorption band) and detected a 260 fs rise in the special pair fluorescence signal detected at 940 nm. They noted that this rate is much too fast to be attributed to direct, Förster-type energy transfer from H to P, and suggested a sequential process involving B. Within this scheme the H $\rightarrow$ B transfer rate is $\sim$ 160 fs, similar to the B $\rightarrow$ P rate. Multiwavelength pump–probe measurements [163] corroborated the proposed sequential scheme of energy transfer from H to P.

As in the case of B $\rightarrow$ P energy transfer, elucidation of the source of ultrafast H $\rightarrow$ B energy transfer requires careful consideration. While in *Rps. viridis* RCs the H^* and B^* absorption bands overlap to a great extent even at low temperatures [120], the overlap of these bands in *Rb. sphaeroides* RCs is rather small [143]. Calculations based on Förster theory, assuming that no Stokes shift of the emission spectrum of H^* occurs before energy transfer, show that the spectral overlap is too small to allow for the ultrafast transfer rates experimentally measured. If an instantaneous Stokes shift is introduced into the calculation (with a similar magnitude to the shift measured [9] in solution, $\sim$ 300 cm^{-1}), the spectral overlap is increased, and an energy transfer rate of the correct order of magnitude is obtained. An "instantaneous" Stokes shift can occur, for example, if an extremely fast intramolecular vibrational energy redistribution process leads to the formation of hot excited molecules, with the emission spectrum significantly shifted compared to the absorption spectrum. Since most experimental work was done with excitation energy close to the 0–0 transition of H, this is unlikely. Thus, the question of mixed states and coherent energy flow arises here, as in the case of B → P energy transfer. The results of Vos *et al.* [163], in which a large bleach component appears in the B band immediately after the pump pulse, seem to point in this direction. Contrary to the B → P case, where strong mixing of states was predicted by theory, calculations [125, 169] actually show quite a small mixing of H states with B states, mainly due to the larger energy gap. In any case, the persistence of delocalized eigenstates depends on the rigidity of the surrounding protein structure, which determines the rate of relaxation into localized states. It is clear that more experimental and theoretical work is required to resolve such issues.

Acknowledgments

We would like to thank Elisabeth Morlino for reading the manuscript. We also thank Peter Hamm and Dongbo Hu for helpful suggestions. This research was supported by grants from NSF and NIH.

References

1. Abrahams, E., P. W. Anderson, D. C. Licciardello, and T. V. Ramakrishnan 1979. Scaling theory of localization: absence of quantum diffusion in two dimensions. *Phys. Rev. Lett.* 42: 673–676.
2. Abram, I. I., and R. M. Hochstrasser 1980. Theory of the time evolution of exciton coherence in weakly disordered crystals. *J. Chem. Phys.* 72: 3617–3625.
3. Agranovich, V. M., and M. D. Galanin 1982. Electronic excitation energy transfer in condensed matter. In *Modern Problems in Condensed Matter Sciences*. North-Holland, Amsterdam.
4. Alden, R. G., E. Johnson, V. Nagarajan, W. W. Parson, C. J. Law, and R. G. Cogdell 1997. Calculations of spectroscopic properties of the LH$_2$ bacteriochlorophyll–protein antenna complex form *Rhodopseudomonas acidophila*. *J. Phys. Chem. B* 101: 4667–4680.
5. Allen, M. P., and D. J. Tildesley 1987. *Computer Simulation of Liquids*. Clarendon Press, Oxford.
6. Anderson, P. W. 1958. Absence of diffusion in certain random lattices. *Phys. Rev.* 109: 1492–1505.
7. Arnett, D. C., C. C. Moser, P. L. Dutton, and N. F. Scherer 1997. The first events in photosynthesis: electronic coupling and energy transfer dynamics in the photosynthetic reaction center from *Rhodobacter sphaeroides*. *J. Phys. Chem. B*, submitted.
8. Arnett, D. C., R. Kumble, R. W. Visschers, P. L. Dutton, R. M. Hochstrasser, and N. F. Scherer 1998 SPIE. vol. 3273, 244–254.

9. Becker, M., V. Nagarajan, and W. W. Parson 1991. Properties of the excited-singlet states of bacteriochlorophyll *a* and bacteriopheophytin *a* in polar solvents. *J. Am. Chem. Soc.* 113: 6840–6848.

10. Bixon, M., J. Jortner, and M. E. Michel-Beyerle 1995. A kinetic analysis of the primary charge separation in bacterial photosynthesis. Energy gaps and static heterogeneity. *Chem. Phys.* 197: 389–404.

11. Boxer, S. G., R. A. Goldstein, D. J. Lockhart, T. R. Middendorf, and L. Takiff 1989. Excited states, electron-transfer reactions, and intermediates in bacterial photosynthetic reaction centers. *J. Phys. Chem.* 93: 8280–8294.

12. Bradforth, S. E., R. Jimenez, F. V. Mourik, R. V. Grondelle, and G. R. Fleming 1995. Excitation transfer in the core light-harvesting complex (LH-1) of *Rhodobacter sphaeroides* – an ultrafast fluorescence depolarization and annihilation study. *J. Phys. Chem.* 99: 16 179–16 191.

13. Breton, J. 1985. Orientation of the chromophores in the reaction center of *Rhodopseudomonas viridis*. Comparison of low-temperature linear dichroism spectra with a model derived from X-ray crystallography. *Biochim. Biophys. Acta* 810: 235–245.

14. Breton, J. 1988. Low temperature linear dichroism study of the orientation of the pigments in reduced and oxidized reaction centers of *Rps. viridis* and *Rb. sphaeroides*. In *The Photosynthetic Bacterial Reaction Center. Structure and Dynamics*. Plenum Press, New York, pp. 59–69.

15. Breton, J., and A. Vermeglio 1988. The photosynthetic bacterial reaction center. Structure, spectroscopy and dynamics. *In* Nato ASI Series. Series A: Life Sciences. Plenum Press, New York.

16. Breton, J., and A. Vermeglio 1992. The photosynthetic bacterial reaction center II. Structure, spectroscopy and dynamics. *In* Nato ASI Series. Series A: Life Sciences. Plenum Press, New York.

17. Breton, J., J.-L. Martin, G. R. Fleming, and J.-C. Lambry 1988. Low-temperature femtosecond spectroscopy of the initial step of electron transfer in reaction centers from photosynthetic purple bacteria. *Biochemistry*, 27: 8276–8284.

18. Breton, J., J.-L. Martin, A. Migus, A. Antonetti, and A. Orszag 1986. Femtosecond spectroscopy of excitation energy transfer and initial charge separation in the reaction center of the photosynthetic bacterium *Rhodopseudomonas viridis*. *Proc. Natl Acad. Sci., USA* 83: 5121–5125.

19. Brooks, B. R., R. E. Bruccoleri, B. D. Olafson, D. J. States, S. Swaminathan, and M. Karplus 1983. CHARMM: a program for macromolecular energy, minimization, and dynamics calculations. *J. Comp. Chem.* 4: 187–217.

20. Bublitz, G. U., and S. G. Boxer 1997. Stark spectroscopy: applications in chemistry, biology and materials science. *Ann. Rev. Phys. Chem.* 48: 213–242.

21. Bylina, E. J., and D. C. Youvan 1988. Directed mutation affecting spectroscopic and electron transfer properties of the primary donor in the photosynthetic reaction center. *Proc. Natl Acad. Sci., USA* 85: 7226–7330.

22. Bylina, E. J., and D. C. Youvan 1990. Mutagenesis of reaction center histidine 1173 yields an 1-side heterodimer. In *Current Research in Photosynthesis*. Kluwer Academic Publishers, Dordrecht, pp. 53–59.

23. Chachisvilis, M., O. Kuhn, T. Pullerits, and V. Sundstrom 1997. Excitons in photosynthetic purple bacteria: wavelike motion or incoherent hopping? *J. Phys. Chem.* 101: 7275–7283.

24. Chang, J. C. 1977. Monopole effects on electronic excitation interactions between large molecules I. Application to energy transfer in chlorophylls. *J. Chem. Phys.* 67: 3901–3909.

25. Cherepy, N. J., A. P. Shreve, L. J. Moore, S. Franzen, S. G. Boxer, and R. A. Mathies 1994. Near-infrared resonance Raman spectroscopy of the special pair and the accessory bacteriochlorophylls in photosynthetic reaction centers. *J. Phys. Chem.* 98: 6023–2029.
26. Colson, S. D., D. M. Hanson, R. Kopelman, and G. W. Robinson 1968. Direct observation of the entire exciton band of the first excited singlet states of crystalline benzene and naphthalene. *J. Chem. Phys.* 48: 2215–2231.
27. Davydov, A. S. 1962. *Theory of Molecular Excitons.* McGraw-Hill, New York.
28. De Boer, S., and D. A. Wiersma 1989. Optical dynamics of exciton and polaron formation in molecular aggregates. *Chem. Phys.* 131: 135–144.
29. De Boer, S., K. J. Vink, and D. A. Wiersma 1987. Optical dynamics of condensed molecular aggregates: an accumulated photon-echo and hole-burning study of the J-aggregate. *Chem. Phys. Lett.* 137: 99–106.
30. Deisenhofer, J., and H. Michel 1989. The photosynthetic reaction center from the purple bacterium *Rhodopseudomonas viridis. Science* 245: 1463–1473.
31. Deisenhofer, J., and J. Norris 1993. *The Photosynthetic Reaction Center.* Academic Press, New York.
32. Deisenhofer, J., O. Epp, K. Miki, H. R., and H. Michel 1985. Structure of the protein subunits in the photosynthetic reaction centre of the *Rhodopseudomonas viridis*at 3Å resolution. *Nature* 318: 618–624.
33. Dexter, D. L. 1953. A theory of sensitized luminescence in solids. *J. Chem. Phys.* 21: 836–850.
34. Ebrey, T. G., and R. K. Clayton 1969. Polarization of fluorescence from bacterio-chlorophyll in castor oil, in chromatophores and as p870 in photosynthetic reaction centers. *Photochem. Photobiol.* 10: 109–117.
35. Economou, E. N. 1983. *Green's Functions in Quantum Physics.* Springer, New York.
36. Edington, M. D., R. E. Riter, and W. F. Beck 1995. Evidence for coherent energy transfer in allophycocyanin trimers. *J. Phys. Chem.* 99: 15 699–15 704.
37. Ermler, U., G. Fritzsch, S. K. Buchanan, and H. Michel 1994. Structure of the photosynthetic reaction center from *Rhodobacter sphaeroides* at 2.65 Å resolution: cofactors and protein–cofactor interactions. *Structure* 2: 925–936.
38. Fidder, H., and D. A. Wiersma 1991. Resonance-light-scattering study and line-shape simulation of the J band. *Phys. Rev. Lett.* 66: 1501–1504.
39. Fidder, H., and D. A. Wiersma 1993. Exciton dynamics in disordered molecular aggregates – dispersive dephasing probed by photon echo and Rayleigh scattering. *J. Phys. Chem.* 97: 11 603–11 610.
40. Fidder, H., J. Knoester, and D. A. Wiersma 1990. Superradiant emission and optical dephasing in J-aggregates. *Chem. Phys. Lett.* 171: 529–536.
41. Fidder, H., J. Knoester, and D. A. Wiersma 1991. Optical properties of disordered molecular aggregates: a numerical study. *J. Chem. Phys.* 95: 7880–7890.
42. Fidder, H., J. Terpstra, and D. A. Wiersma 1991. Dynamics of Frenkel excitons in disordered molecular aggregates. *J. Chem. Phys.* 94: 6895–6907.
43. Fleming, G. R., and R. van Grondelle 1994. The primary steps of photosynthesis. *Phys. Today* 48–55.
44. Fleming, G. R., J.-L. Martin, and J. Breton 1988. Rates of primary electron transfer in photosynthetic reaction centres and their mechanistic implications. *Nature* 333: 190–192.
45. Förster, T. 1967. Delocalized excitation and excitation transfer. In *Modern Quantum Chemistry. Istanbul Lectures. Part III: Action of Light and Organic Crystals.* Academic Press, New York, pp. 93–137.

46. Fulton, R. L., and M. Gouterman 1964. Vibronic coupling. II. Spectra of dimers. *J. Chem. Phys.* 41: 2280–2286.

47. Galli, C., K. Wynne, S. LeCours, M. J. Therien, and R. M. Hochstrasser 1993. Direct measurement of electronic dephasing using anisotropy. *Chem. Phys. Lett.* 206: 493–499.

48. Greene, B. I., J. Orenstein, and S. Schmitt-Rink 1990. All-optical nonlinearities in organics. *Science* 247: 679–687.

49. Grover, M., and R. Silbey 1971. Exciton migration in molecular crystals. *J. Chem. Phys.* 54: 4843–4851.

50. Haken, H., and G. Strobl 1968. Exact treatment of coherent and incoherent triplet exciton migration. In *The Triplet State*. Cambridge University Press, London, pp. 311–314.

51. Hammes, S. L., L. Mazzola, S. G. Boxer, D. F. Gaul, and C. C. Schenck 1990. Stark spectroscopy of the *Rhodobacter sphaeroides* reaction center heterodimer mutant. *Proc. Natl Acad. Sci., USA.* 87: 5682–5686.

52. Haran, G., K. Wynne, C. C. Moser, P. L. Dutton, and R. M. Hochstrasser 1996. Femtosecond infrared studies of photosynthetic reaction centers: new charge transfer bands and ultrafast energy redistribution. In *Ultrafast Phenomena X*. Springer-Verlag, Berlin, pp. 326–327.

53. Haran, G., K. Wynne, C. C. Moser, P. L. Dutton, and R. M. Hochstrasser 1996. Level mixing and energy redistribution in bacterial photosynthetic reaction centers. *J. Phys. Chem.* 100: 5562–5569.

54. Hartwich, G., M. Friese, H. Scheer, A. Ogrodnik, and M. E. Michel-Beyerle 1995. Ultrafast internal conversion in 13^2-OH-Ni-bacteriochlorophyll in reaction centers of *Rhodobacter sphaeroides* R26. *Chem. Phys.* 197: 423–434.

55. Hayes, J. M., P. A. Lyle, and G. J. Small 1994. A theory of the temperature dependence of hole-burned spectra. *J. Phys. Chem.* 98: 7337–7341.

56. Hess, S., E. Akesson, R. J. Cogdell, T. Pullerits, and V. Sundstrom 1995. Energy transfer in spectrally inhomogeneous light-harvesting pigment–protein complexes of purple bacteria. *Biophys. J.* 69: 2211–2225.

57. Hess, S., F. Feldchtein, A. Babin, I. Nurgaleev, T. Pullerits, A. Sergeev, and V. Sundstrom 1993. Femtosecond energy transfer within the LH2 peripheral antenna of the photosynthetic purple bacteria *Rhodobacter sphaeroides* and *Rhodopseudomonas palustris* LL. *Chem. Phys Lett.* 216: 247–257.

58. Hochstrasser, R. M. 1976. Triplet exciton states of molecular crystals. In *Int. Rev. Sci.: Phys. Chem.*, Ser. Two, Vol.3. Butterworth, London, pp. 1–36.

59. Hochstrasser, R. M., and I. I. Abram 1979. Time-resolved coherent anti-Stokes Raman scattering in weakly disordered molecular crystals. In *Light Scattering in Solids*. Plenum Press, New York.

60. Hochstrasser, R. M., and M. Kasha 1964. Application of exciton model to monomolecular lamellar systems. *Photochem. Photobiol.* 4: 317.

61. Hochstrasser, R. M., and C. A. Nyi 1980. Dynamical effects from resonance Raman and fluorescence studies of the molecular exciton system perylene. *J. Chem. Phys.* 72: 2591–2600.

62. Hochstrasser, R. M., and P. N. Prasad 1972. Phonon sidebands of electronic transitions in molecular crystals and mixed crystals. *J. Chem. Phys.* 56: 2814–2823.

63. Hochstrasser, R. M., T.-Y. Li, H.-N. Sung, J. Wessel, and A. H. Zewail 1974. Experimental studies of triplet exciton bands of molecular crystals. *Pure Appl. Chem.* 37: 85–96.

64. Holstein, T. 1959. Polaron motion. I. Molecular crystal model. *Ann. Phys.* 8: 325–342.

65. Holstein, T. 1959. Polaron motion. II. Small polaron. *Ann. Phys.* 8: 343–389.

66. Jean, J. M., C.-K. Chan, and G. R. Fleming 1988. Electronic energy transfer in photosynthetic bacterial reaction centers. *Isr. J. Chem.* 28: 169–175.

67. Jia, Y., D. M. Jonas, T. Joo, Y. Nagasawa, M. J. Lang, and G. R. Fleming 1995. Observation of ultrafast energy transfer from the accessory bacteriochlorophylls to the special pair in photosynthetic reaction centers. *J. Phys. Chem.* 99: 6263–6266.

68. Jimenez, R., S. N. Dikshit, S. E. Bradforth, and G. R. Fleming 1996. Electronic excitation transfer in the LH_2 complex of *Rhodobacter sphaeroides*. *J. Phys. Chem.* 100: 6825–6834.

69. Jimenez, R., F. v. Mourik, J. Y. Yu, and G. R. Fleming 1997. Three-pulse photon echo measurements on LH1 and LH2 complexes of *Rhodobacter sphaeroides*: a nonlinear probe of energy transfer. *J. Phys. Chem.* 101: 7350–7359.

70. Johnson, S. G., D. Tang, R. Jankowiak, J. M. Hayes, G. J. Small, and D. M. Tiede 1989. Structure and marker mode of the primary electron donor state absorption of photosynthetic bacteria: hole-burned spectra. *J. Phys. Chem.* 93: 5953–5957.

71. Johnson, S. G., D. Tang, T. Jankowiak, J. M. Hayes, G. J. Small, and D. M. Tiede 1990. Primary donor state mode structure and energy transfer in bacterial reaction centers. *J. Phys. Chem.* 94: 5849–5855.

72. Jonas, D. M., M. J. lang, Y. Nagasawa, T. Joo, and G. R. Fleming 1996. Pump–probe polarization anisotropy study of femtosecond energy transfer within the photosynthetic reaction center of *Rhodobacter sphaeroides* R26. *J. Phys. Chem.* 100: 12 660–12 673.

73. Joo, T., Y. W. Jia, J. Y. Yu, D. M. Jonas, and G. R. Fleming 1996. Dynamics in isolated bacterial light harvesting antenna (LH2) of *Rhodobacter sphaeroides* at room temperature. *J. Phys. Chem.* 100: 2399–2409.

74. Jortner, J., S. A. Rice, J. L. Katz, and S.-I. Choi 1965. Triplet excitons in crystals of aromatic molecules. *J. Chem. Phys.* 42: 309–323.

75. Karrasch, S., Bullough, P. A., and R. Ghosh 1995. The 8.5 Å projection map of the light-harvesting complex I from *Rhodospirillum rubrum* reveals a ring composed of 16 subunits. *EMBO J.* 14: 631–638.

76. Kasha, M. 1963. Energy transfer mechanisms and the molecular exciton model for molecular aggregates. *Rad. Res.* 20: 55–71.

77. Kenkre, V. 1982. *Exciton Dynamics in Molecular Crystals and Aggregates*. Springer-Verlag, Berlin.

78. Kenkre, V. M., and R. S. Knox 1974. Generalized master equation theory of excitation transfer. *Phys. Rev. B* 9: 5279–5290.

79. Kennis, J. T. M., A. M. Streltsov, T. J. Aartsma, T. Nozawa, and J. Amesz 1996. Energy transfer and exciton coupling in isolated B800–850 complexes of the photosynthetic purple sulfur bacterium *Chromatium tepidum*. The effect of structural symmetry on bacteriochlorophyll excited states. *J. Phys. Chem.* 100: 2438–2442.

80. Kim, Y. R., M. Lee, J. R. G. Thorne, R. M. Hochstrasser, and J. M. Zeigler 1988. Picosecond reorientations of the transition dipoles in polysilanes using fluorescence anisotropy. *Chem. Phys. Lett.* 145: 75–80.

81. Kim, Y. R., P. Share, M. Pereira, M. Sarisky, and R. M. Hochstrasser 1989. Direct measurements of energy transfer between identical chromophores in solution. *J. Chem. Phys.* 91: 7557–7562.

82. Klafter, J., and J. Jortner 1978. Effects of structural disorder on the optical properties of molecular crystals. *J. Chem. Phys.* 68: 1513–1522.

83. Knox, R. S. 1971. *Theory of Molecular Excitons*. Plenum Press, New York.

84. Knox, R. S., and D. Gulen 1993. Theory of polarized fluorescence from molecular pairs: Förster transfer at large electronic coupling. *Photochem. Photobiol.* 57: 40–43.

85. Koepke, J., X. Hu, C. Muenke, K. Schulten, and H. Michel 1996. The crystal structure of the light-harvesting complex II (B800–850) from *Rhodospirillum molischianum. Structure* 4: 581–597.
86. Kopelman, R. 1983. *Spectroscopy and Excitation Dynamics of Condensed Molecular Systems.* North-Holland, Amsterdam.
87. Kubo, R. 1969. A stochastic theory of the line shape. *Adv. Chem. Phys.* 15: 101–127.
88. Kumble, R., and R. M. Hochstrasser 1998. Disorder-induced exciton scattering in the light-harvesting systems of purple bacteria: influence on the anisotropy of emission and band → band transitions. *J. Chem. Phys.*, submitted.
89. Kumble, R., S. Palese, R. W. Visschers, P. L. Dutton, and R. M. Hochstrasser 1996. Ultrafast dynamics within the B820 subunit from the core (LH-1) antenna complex of *Rs. rubrum. Chem. Phys. Lett.* 262: 396–404.
90. Kumble, R., Y. Kholodenko, R. W. Visschers, E. Gooding, P. L. Dutton, and R. M. Hochstrasser 1998. in preparation.
91. Ladanyi, B. M., and R. M. Stratt 1996. Short-time dynamics of solvation: relationship between polar and nonpolar solvation. *J. Phys. Chem.* 100: 1266–1282.
92. Leggett, A. J., S. Chakravarty, A. T. Dorsey, M. P. A. Fisher, A. Garg, and W. Zwerger 1987. Dynamics of the dissipative two-state system. *Rev. Mod. Phys.* 59: 1–85.
93. Lendzian, F., B. Bonigk, M. Plato, K. Mobius, and W. Lubitz 1992. Nitrogen-15 ENDOR experiments on the primary donor cation radical D + . bul. in bacterial reaction center single crystals of *Rb. shaeroides* R-26. In *The Photosynthetic Bacterial Reaction Center II.* J. Breton and A. Vermeglio, editors., Plenum Press, New York. pp. 89–97.
94. Lendzian, F., M. Humber, R. A. Isaacson, B. Endeward, M. Plato, B. Bonigk, K. Mobius, W. Lubitz, and G. Feher 1993. The electronic structure of the primary donor cation-radical in *Rhodobacter sphaeroides* R-26 – ENDOR and triple-resonance studies in single crystals of reaction centers. *Biochim. Biophys. Acta* 1183: 139–160.
95. Lin, S., A. K. W. Taguchi, and N. W. Woodbury 1996. Excitation wavelength dependence of energy transfer and charge separation in reaction centers from *Rhodobacter sphaeroides*: evidence for adiabatic electron transfer. *J. Phys. Chem.* 100: 17 067–17 078.
96. Lockhart, D. J., and S. G. Boxer 1987. Magnitude and direction of the change in dipole moment associated with excitation of the primary electron donor in *Rhodopseudomonas sphaeroides* reaction centers. *Biochemistry* 26: 664–668.
97. Lockhart, D. J., and S. G. Boxer 1988. Stark effect spectroscopy of *Rhodobacter sphaeroides* and *Rhosopseudomonas viridis* reaction centers. *Proc. Natl Acad. Sci., USA* 85: 107–111.
98. Loesche, M., G. Feher, and M. Y. Okamura 1987. The stark effect in reaction centers from *Rhodobacter sphaeroides* R-26 and *Rhodopseudomonas viridis. Proc. Natl Acad. Sci., USA* 84: 7537–7541.
99. Marchi, M., J. N. Gehlen, D. Chandler, and M. Newton 1993. Diabatic surfaces and the pathway for primary electron-transfer in a photosynthetic reaction center. *J. Am. Chem. Soc.* 115: 4178–4190.
100. Martin, J.-L., and M. H. Vos 1992. Femtosecond biology. *Ann. Rev. Biophys. Biomol. Struct.* 21: 199–222.
101. Matro, A., and J. A. Cina 1995. Theoretical study of time-resolved fluorescence anisotropy from coupled chromophore pairs. *J. Phys. Chem.* 99: 2568–2582.
102. McClure, D. S. 1959. *Electronic Spectra of Molecules and Ions in Crystals. Part 1. Molecular Crystals.* Academic Press, New York.

103. Mcdermott, G., S. M. Prince, A. A. Freer, A. M. Hawthornthwaite-Lawless, M. Z. Papiz, R. J. Cogdell, and N. W. Isaacs 1995. Crystal structure of an integral membrane light-harvesting complex from photosynthetic bacteria. *Nature* 374: 517–521.

104. McRae, E. G. 1961. Molecular vibrations in the exciton theory for molecular aggregates. *Austral. J. Chem.* 14: 329–343, 344–353.

105. McRae, E. G. 1963. Molecular vibrations in the exciton theory for molecular aggregates. *Austral. J. Chem.* 16: 295–314, 315–333.

106. McRae, E. G., and M. Kasha 1964. The molecular exciton model. In *Physical Processes in Radiation Biology*. Academic Press, New York, 23–42.

107. Meier, T., V. Chernyak, and S. Mukamel 1997. Multiple exciton coherence sizes in photosynthetic antenna complexes viewed by pump–probe spectroscopy. *J. Phys. Chem. B* 101: 7332–7342.

108. Meier, T., Y. Zhao, V. Chernyak, and S. Mukamel 1997. Polarons, localization, and excitonic coherence in superradiance of biological antenna complexes. *J. Chem. Phys.* 107: 3876–3893.

109. Merrifield, R. E. 1963. Vibronic states of dimers. *Rad. Res.* 20: 154–158.

110. Michl, J., and V. Bonacık-Kouteckı 1990. *Electronic Aspects of Organic Photochemistry*. John Wiley, New York.

111. Monshouwer, R., M. Abrahamsson, F. van Mourik, and R. van Grondelle 1997. Superradiance and exciton delocalization in bacterial photosynthetic light-harvesting systems. *J. Phys. Chem. B* 101: 7241–7248.

112. Mukamel, S. 1995. *Nonlinear Optical Spectroscopy*. Oxford University Press, New York.

113. Nagae, H., T. Kakitani, T. Katoh, and M. Mimuro 1993. Calculation of the excitation transfer matrix elements between the S_2 or S_1 state of carotenoid and the S_2 or S_1 state of bacteriochlorophyll. *J. Chem. Phys.* 98: 8012–8023.

114. Nagarajan, V., R. G. Alden, J. C. Williams, and W. W. Parson 1996. Ultrafast exciton relaxation in the B850 antenna complex of *Rhodobacter sphaeroides*. *Proc. Natl Acad. Sci., USA* 93: 13 774–13 779.

115. Palaniappan, V., M. A. Aldema, H. A. Frank, and D. F. Bocian 1992. Q_y-excitation resonance Raman scattering from the special pair in *Rhodobacter sphaeroides* reaction centers. Implications for primary charge separation. *Biochemistry* 31: 11 050–11 058.

116. Parson, W. W., and A. Warshel 1987. Spectroscopic properties of photosynthetic reaction centers. 2. Application of the theory to *Rhodopseudomonas viridis*. *J. Am. Chem. Soc.* 109: 6152–6163.

117. Philipson, K. D., and K. Sauer 1973. Comparative study of the circular dichroism spectra of reaction centers from several photosynthetic bacteria. *Biochemistry* 12: 535–539.

118. Pullerits, T., M. Chachisvilis, and V. Sundstrom 1996. Exciton delocalization length in the B850 antenna of *Rhodobacter sphaeroides*. *J. Phys. Chem.* 100: 10 787–10 792.

119. Rackovsky, S., and R. Silbey 1973. Electronic energy transfer in impure solids I. Two molecules in embedded in a lattice. *Mol. Phys.* 25: 61–72.

120. Reddy, J. R. S., S. V. Kolaczkowski, and G. J. Small 1993. Nonphotochemical hole burning of the reaction center of *Rhosdopseudomonas viridis*. *J. Phys. Chem.* 97: 6934–6940.

121. Redfield, A. G. 1965. The theory of relaxation processes. *Adv. Mag. Res.* 1: 1–32.

122. Reimers, J. R., and N. S. Hush 1994. The influence of spin-forbidden monomer excitations on spin-allowed electron transfer and electron-localized states of

mixed-valence and single-valence dimeric systems. *Inorg. Chim. Acta* 226: 33–42.

123. Reimers, J. R., and N. S. Hush 1995. The nature of the near-infrared electronic absorption at 1250 nm in the spectra of the radical cations of the special pair of the photosynthetic reaction centers of *Rhodobacter sphaeroides* and *Rhodopseudomonas viridis*. *J. Am. Chem. Soc.* 117: 1302–1308.

124. Savikhin, S., D. R. Buck, and W. S. Struve 1997. Oscillating anisotropies in a bacteriochlorophyll protein: evidence for quantum beating between exciton levels. *Chem. Phys.* 223: 303–312.

125. Scherer, P. O. J., and S. F. Fischer 1991. Interpretation of optical reaction center spectra. In *Chlorophylls*. CRC Press, Boca Raton, pp. 1079–1093.

126. Scherer, P. O. J., and S. F. Fischer 1997. Interpretation of a low-lying excited state of the reaction center of *Rb. sphaeroides* as a double triplet. *Chem. Phys. Lett.* 268: 133–142.

127. Scherz, A., and W. W. Parson 1984. Exciton interactions in dimers of bacterio-chlorophyll and related molecules. *Biochim. Biophys. Acta* 766: 653–665.

128. Scholes, G. D., and K. P. Ghiggino 1994. Electronic interactions and interchro-mophore excitation transfer. *J. Phys. Chem.* 98: 4580–4590.

129. Schreiber, M., and Y. Toyozawa 1982. Numerical experiments on the absorption lineshape of the exciton under lattice vibrations. II. The average oscillator strength per state. *J. Phys. Soc. Japan* 51: 1537–1543.

130. Sension, R. J., S. T. Repinec, A. Z. Szarka, and R. M. Hochstrasser 1993. Femto-second laser studies of the *cis*-stilbene photoisomerization reactions. *J. Chem. Phys.* 98: 6291–6315.

131. Shreve, A. P., J. K. Trautman, H. A. Frank, T. G. Owens, and A. C. Albrecht 1991. Femtosecond energy-transfer processes in the B800–850 light-harvesting complex of *Rhodobacter-sphaeroides*-2.4.1. *Biochim. Biophys. Acta* 1058: 280–288.

132. Shuvalov, V. A., and A. A. Asadov 1979. Arrangement and interaction of pigment molecules in reaction centers of *Rhodopseudomonas viridis*. Photodichroism and circular cichroism of reaction centers at 100 K. *Biochim. Biophys. Acta* 545: 296–308.

133. Siebrand, W. 1964. Vibrational structure of electronic states of molecular aggre-gates. I. A variation approach to dimeric systems. *J. Chem. Phys.* 40: 2223–2235.

134. Silbey, R. 1976. Electronic energy transfer in molecular crystals. *Ann. Rev. Phys. Chem.* 27: 203–223.

135. Silbey, R. 1983. *Spectroscopy and Excitation Dynamics of Condensed Molecular Systems*. North-Holland, Amsterdam.

136. Silinsh, E. A., and V. Capek 1994. *Organic Molecular Crystals. Interaction, Local-ization and Transport Phenomena*. AIP Press, New York.

137. Skinner, J. L. 1994. Models of Anderson localization. *J. Phys. Chem.* 98: 2503–2507.

138. Small, G. J. 1995. On the validity of the standard model for primary charge separation in the bacterial reaction center. *Chem. Phys.* 197: 239–257.

139. Smith, P. G., S. M. Hong, A. J. Kaziska, A. L. Motyka, S. Gnanakaran, R. M. Hochstrasser, and M. R. Topp 1994. Electronic coupling and conformational barrier crossing of 9,9'-Bifluorenyl studied in a supersonic jet. *J. Chem. Phys.* 100: 3384–3393.

140. Soules, T. F., and C. B. Duke 1971. Resonant energy transfer between localized electronic states in a crystal. *Phys. Rev. B* 3: 262–274.

141. Spano, F. C., J. R. Kuklinski, and S. Mukamel 1991. Cooperative radiative dynamics in molecular aggregates. *J. Chem. Phys.* 94: 7534–7544.

142. Stanley, R. J., and S. G. Boxer 1995. Oscillations in the spontaneous fluorescence from photosynthetic reaction centers. *J. Phys. Chem.* 99: 859–863.

143. Stanley, R. J., B. King, and S. G. Boxer 1996. Excited state energy transfer pathways in photosynthetic reaction centers. I. Structural symmetry effects. *J. Phys. Chem.* 100: 12 052–12 059.

144. Stephens, M. D., J. G. Saven, and J. L. Skinner 1997. Molecular theory of electronic spectroscopy in nonpolar fluids: ultrafast solvation dynamics and absorption and emission line shapes. *J. Chem. Phys.* 106: 2129–2144.

145. Stowell, M. H., T. M. McPhillips, D. C. Rees, S. M. Soltis, E. Abresch, and G. Feher 1997. Light induced structural changes in the photosynthetic reaction center: implications for the mechanism of electron–proton transfer. *Science* 276: 812–816.

146. Tang, D., S. G. Johnson, R. Jankowiak, J. M. Hayes, G. J. Small, and D. M. Tiede 1990. Structure and marker mode of the primary electron donor state absorption of photosynthetic bacteria: hole burned spectra. In *Perspectives in Photosynthesis*. Kluwer Academic Publishers, Amsterdam, pp. 99–120.

147. Thompson, M. A., and G. K. Schenter 1995. Excited states of the bacteriochlorophyll *b* dimer of *Rhodopseuodomonas viridis*: a QM/MM study of the photosynthetic reaction center that includes MM Polarization. *J. Phys. Chem.* 99: 6374–6386.

148. Thompson, M. A., and M. C. Zerner 1991. A theoretical examination of the electronic structure and spectroscopy of the photosynthetic reaction center from *Rhodopseudomonas viridis*. *J. Am. Chem. Soc.* 113: 8210–8215.

149. Thompson, M. A., M. C. Zerner, and J. Fajer 1991. A theoretical examination of the electronic structure and excited states of the bacteriochlorophyll *b* dimer from *Rhodopseudomonas viridis*. *J. Phys. Chem.* 95: 5693–5700.

150. Thorne, J. R. G., S. T. Repinec, S. A. Abrash, R. M. Hochstrasser, and J. M. Zeigler 1990. Polysilane excited states and excited state dynamics. *Chem. Phys.* 146: 315–325.

151. Thouless, D. J. 1974. Electrons in disordered systems and the theory of localization. *Phys. Rep.* 13: 93–142.

152. Tilgner, A., H. P. Trommsdorff, J. M. Zeigler, and R. M. Hochstrasser 1992. Poly(di-*n*-hexyl-silane) in solid solutions: experimental and theoretical studies of electronic excitations of a disordered linear chain. *J. Chem. Phys.* 96: 781–796.

153. Toyozawa, Y. 1958. Theory of lineshapes of the exciton absorption bands. *Progr. Theor et. Phys.* 20: 53–81.

154. van Grondelle, R., J. P. Dekker, T. Gillbro, and V. Sundstrom 1994. Energy transfer and trapping in photosynthesis. *Biochim. Biophys. Acta* 1187: 1–65.

155. VanBrederode, M. E., M. R. Jones, F. VanMourik, I. H. M. VanStokkum, and R. VanGrondelle 1997. A new pathway for transmembrane electron transfer in photosynthetic reaction centers of *Rhodobacter sphaeroides* not involving the excited special pair. *Biochemistry* 36: 6855–6861.

156. Vanburgel, M., D. A. Wiersma, and K. Duppen 1995. The dynamics of one-dimensional excitons in liquids. *J. Chem. Phys.* 102: 20–33.

157. VanMourik, F., R. W. Visschers, and R. van Grondelle 1992. Energy transfer and aggregate size effects in the inhomogeneously broadened core light-harvesting complex of *Rhodobacter sphaeroides*. *Chem. Phys. Lett.* 193: 1–7.

158. Velsko, S., and R. M. Hochstrasser 1985. Theory of vibrational coherence and population decay in isotopically mixed crystal. *J. Chem. Phys.* 82: 2180–2190.

159. Vermeglio, A., and G. Paillotin 1982. Structure of *Rhodopseudomonas viridis* reaction centers absorption and photoselection at low temperature. *Biochim. Biophys. Acta* 681: 32–40.

160. Vermeglio, A., J. Breton, G. Paillotin, and R. Cogdell 1978. Orientation of chromophores in reaction centers of *Rhodopseudomonas sphaeroides*: a photoselection study. *Biochim. Biophys. Acta* 501: 514–530.

161. Visschers, R. W., M. C. Change, F. van Mourik, P. S. Parkes-Loach, B. A. Heller, P. A. Loach, and R. van Grondelle 1991. Fluorescence polarization and low-temperature absorption spectroscopy of a subunit form of light-harvesting complex from purple photosynthetic bacteria. *Biochemistry* 30: 5734–5742.

162. Visser, H. M., D. J. G. Somsen, F. van Mourik, S. Lin, I. H. M. van Stokkum, and R. van Grondelle 1995. Direct observation of sub-picosecond equilibration of excitation energy in the light-harvesting antenna of *Rhodospirillum rubrum. Biophys. J.* 69: 1083–1099.

163. Vos, M. H., J. Breton, and J.-L. Martin 1997. Electronic energy transfer within the hexamer cofactor system of bacterial reaction centers. *J. Phys. Chem. B* 101: 9820–9832.

164. Walker, G. C., S. Maiti, B. R. Cowen, C. C. Moser, P. L. Dutton, and R. M. Hochstrasser 1994. Time resolution of electronic transitions in the photosynthetic reaction center. *J. Phys. Chem.* 98: 5778–5783.

165. Warshel, A., and W. W. Parson 1987. Spectroscopic properties of photosynthetic reaction centers. 1. Theory. *J. Am. Chem. Soc.* 109: 6143–6152.

166. Weiss, C. 1972. The pi electron structure and absorption spectra of chlorophylls in solution. *J. Mol. Spectrose.* 44: 37–80.

167. Wertheimer, R., and R. Silbey 1980. On excitation transfer and relaxation models in low-temperature systems. *Chem. Phys. Lett.* 75: 243–248.

168. Witkowski, A., and W. Moffitt 1960. Electronic spectra of dimers: derivation of the fundamental vibronic equation. *J. Chem. Phys.* 33: 872–875.

169. Won, Y., and R. A. Friesner 1988. Simulation of optical spectra form the reaction center of *Rhodopseudomonas viridis. J. Phys. Chem.* 92: 2208–2214.

170. Wu, H.-M., M. Ratsep, I.-J. Lee, R. J. Cogdell, and G. J. Small 1997. Exciton level structure and energy disorder of the B850 ring of the LH2 antenna complex. *J. Phys. Bhem. B* 101: 7654–7663.

171. Wynne, K., and R. M. Hochstrasser 1993. Coherence effects in the anisotropy of optical experiments. *Chem. Phys.* 171: 179–188.

172. Wynne, K., and R. M. Hochstrasser 1995. Anisotropy as an ultrafast probe of coherence in degenerate systems exhibiting Raman, fluorescence, transient absorption and chemical reactions. *J. Raman Spectrose.* 26: 561–569.

173. Wynne, K., S. Gnanakaran, C. Galli, F. Zhu, and R. M. Hochstrasser 1994. Luminescence studies of ultrafast energy transfer oscillations in dimers. *J. Luminesence* 60&61: 735–738.

174. Wynne, K., G. Haran, G. D. Reid, C. C. Moser, P. L. Dutton, and R. M. Hochstrasser 1996. Femtosecond infrared spectroscopy of low lying excited states in reaction centers of *Rhodobacter sphaeroides. J. Phys. Chem.* 100: 5140–5148.

175. Xiao, W. H., S. Lin, A. K. W. Taguchi, and N. W. Woodbury 1994. Femtosecond pump-probe analysis of energy and electron transfer in photosynthetic membranes of *Rhodobacter capsulatus. Biochemistry* 33: 8313–8222.

176. Yeates, T. O., H. Komiya, A. Chirino, R. D. C., J. P. Allen, and G. Feher 1988. Structure of the reaction center from *Rhodobacter sphaeroides* R-26 and 2.4.1: Part 4. Protein–cofactor (bacteriochlorophyll, bacteriopheophytin, and carotenoid) interactions. *Proc. Natl Acad. Sci., USA* 85: 7993–7997.

177. Yoon, B., J. M. Deutch, and J. H. Freed 1975. Comparison of generalized cumulant and projection operator methods in spin-relaxation theory. *J. Chem. Phys.* 62: 4687–4696.

178. Yu, J. Y., Y. Nagasawa, R. v. Grondelle, and G. R. Fleming 1997. Three pulse echo peak shift measurements on the B820 subunit of LH1 of *Rhodospirillum rubrum. Chem. Phys. Lett.* 280: 404–410.
179. Zgierski, M. Z. 1988. On the temperature dependence of the P(−) absorption peak position in the optical spectra of the reaction center of *Rhodopseudomona viridis. Chem. Phys. Lett.* 153: 195–199.
180. Zhou, H., and S. G. Boxer 1997. Charge resonance effects on electronic absorption line shapes: application to the heterodimer absorption of bacterial photosynthetic reaction centers. *J. Phys. Chem.* 101: 5759–5766.
181. Zhu, F., C. Galli, and R. M. Hochstrasser 1993. The real-time intramolecular electronic excitation transfer dynamics of 9′,9-bifluorene and 2′,2-binaphthyl in solution. *J. Chem. Phys.* 98: 1042–1057.

9

Excitation energy transfer in photosynthesis

Rienk van Grondelle and Oscar J. G. Somsen

Vrije Universiteit, The Netherlands

9.1 INTRODUCTION

To harvest solar light, photosynthetic organisms are equipped with a light-harvesting antenna system. Photons absorbed by the antenna pigments are transferred to the photosynthetic reaction center with great speed. Once trapped by the reaction center, the excitation energy is efficiently converted into a stable charge separation. This complex machinery of pigmented proteins, in most cases intrinsic membrane proteins, thus converts solar energy with high efficiency into a transmembrane electrochemical potential that is used for the formation of ATP, the reduction of NAD(P)H, and the oxidation of water to form oxygen (for recent reviews, see [25, 77, 99]).

Evidence for the presence of photochemical reaction centers (RCs) in an amount much smaller than that of the other chlorophyll molecules was first obtained in experiments by Emerson and Arnold [3, 22, 23], who demonstrated that a maximum in the rate of photosynthesis is obtained at an excitation level when about 1 out of 200 chlorophylls is excited in a single turnover flash. The simplest interpretation is that these hundreds of chlorophyll molecules all transfer their excitation energy to the same reaction center. Since the basic description of the excitation energy transfer and trapping processes by Duysens in the early 1950s [19, 20], the research into this fascinating process has progressed along various lines. Using biochemical and genetic methods, many light-harvesting proteins have been purified to a great extent, and often we are able to express or modify these systems as precisely as we wish [40]. Using

Resonance Energy Transfer. Edited by David L. Andrews and Andrey A. Demidov.
© 1999 John Wiley & Sons Ltd

advanced femtosecond laser techniques, developed during the past few decades, the dynamics of energy transfer could be investigated over a timescale that extends from several tens of femtoseconds to many nanoseconds. However, the key development has probably been the resolution of the structure of several important light-harvesting complexes during the past few years; the peripheral light-harvesting complex of green plants [51], the peripheral light-harvesting complex (LH2) of *Rhodopseudomonas (Rps.) acidophila* [59], the LH2 of *Rhodospirillum (Rs.) molischianum* [47], and the core of Photosystem 1 of *cyanobacteria* [50]. All of these are intrinsic membrane proteins, and together with the known structures – the BChl *a* protein of green sulphur bacteria [24], the phycobiliproteins (e.g. [10]), and the recently resolved structure of the peridinin–carotenoid protein of dinoflagellates [38], all membrane-attached light-harvesting systems – we now have a multitude of structures available that exhibit an amazing variation, and will allow us to greatly extend our knowledge about the process of excitation energy transfer and the underlying physics.

9.1.1 Disordered versus ordered light-harvesting systems

Although the various structures exhibit a wide spread in organizational motifs, one striking aspect stands out. Comparing bacterial with plant light-harvesting systems, the bacterial LH2 peripheral and LH1 core antenna are structures with a very high degree of symmetry, while LHCII and even more so Photosystem 1 (PS 1) appear largely disordered – with an almost random pigment organization in the case of the latter. One of the major reasons for this variation is, of course, the size of the elementary building block. In LH1 and LH2 this is a pair of small transmembrane polypeptides, α and β, that carry two and three BChls, respectively. Assembly into a larger system will always lead to a structure with a high degree of symmetry. In contrast, the PS 1 core consists of a single pair of large polypeptides, the PsaA and PsaB gene products that, together with a large number of smaller sub-units, forms the PS1 core that binds about 100 chlorophylls. LHCII seems to be an intermediate case. Although monomeric LHCII still appears to be chaotic in comparison with the LH1 and LH2 rings, the basic unit of LHCII is a trimer of a ~ 25 kDa protein that exhibits perfect C_3 symmetry. The other apparent difference between plant and bacterial light-harvesting systems is the pigment density. In LH1 of purple bacteria, the density is 2 BChl *a*'s per 12 kDa, while in LHCII 12–14 Chl's occur per 25 kDa, a factor of three more. A similar number applies to PS1. Thus, in plant light-harvesting systems the various opportunities to bind chlorophyll molecules are optimally exploited.

Finally, although PS1 differs greatly from the LHI–RC core of purple bacteria, there is also a fundamental similarity. In the LH1–RC core, the rate-limiting step for trapping is energy transfer from any of the light-harvesting pigments to the special pair in the RC. In this scenario, the trapping efficiency is highest when as many antenna sites as possible are able to transfer to the RC. This is clearly optimized in a ring. In PS1, crudely speaking, the pigments are organized in a band around the electron transfer chain, and on average they are all at a distance of more than 2 nm. However, also in this structure the number of contact sites has been optimized, leading to efficient trapping. It is not the symmetry that is important – as long as no unfavorable assemblies are generated (e.g. stacked dimers) – but the proximity of as many as possible pigments to the site at which the electron transfer occurs.

9.2 THE STRUCTURE OF LIGHT-HARVESTING COMPLEXES

In the following, we will briefly discuss the structural models that have been obtained during the past few years for some of the major light-harvesting systems of photosynthesis. Understanding how their structure relates to their spectroscopic properties and their function, the transfer of excitation energy, is the major long-term goal of the research activity.

9.2.1 The peripheral light-harvesting complex (LH2) of photosynthetic purple bacteria

Figure 9.1 shows the structure of the LH2 or B800–850 complex of *Rps. acidophila* resolved by Cogdell and Isaacs and coworkers [29, 59], using X-ray diffraction on 3-D crystals. On the basis of this result, Michel and Schulten and coworkers [47] obtained the structure of the LH2 complex of *Rs. molischianum*. The LH2s of *Rps. acidophila* and *Rs. molischianum* are both highly symmetric rings, displaying C_9 symmetry in the case of *Rps. acidophila* and C_8 symmetry for *Rs. molischianum*. Although only a low-resolution structure is available for LH1 [45], it is evident that it too is organized as a ring, most probably with a 16-fold symmetry. A three-dimensional model of the reaction center snugly fits into the proposed LH1 ring [45, 71].

For both LH2 and LH1, the basic building block of the structure is a heterodimer of two small (5–6 kDa) polypeptides, α and β [111]. Both the α and the β polypeptides consist of a single transmembrane helix, and at about one-third of the way along the α-helical stretch, a highly conserved histidine is found that ligates the BChl. Thus, in the LH1 and LH2 pigment rings the basic element is a BChl-dimer. For LH1, the $\alpha\beta$-BChl$_2$ sub-unit can be purified (it is called B820 after its absorption maximum) and it has retained many of the essential spectral properties of LH1 [61, 102]. For LH2, such a sub-unit cannot be obtained. In LH2, and most probably also in LH1, the heterodimeric

Figure 9.1   A top view of the structure of the peripheral light-harvesting complex (LH2) of *Rps. acidophila*. The nine α-polypeptides constitute the inner ring of helices, and the nine β-polypeptides the outer ring. The 18 B850 bacteriochlorophylls (black) are sandwiched between the two concentric polypeptide rings and have their chlorin planes perpendicular to the plane of the LH2 ring. The nine B800 bacteriochlorophylls (gray) are positioned between the helices and have their chlorin planes almost parallel to the membrane plane. (Adapted with permission from McDermott *et al.* [59])

sub-units associate into a ring with the α-polypeptides on the inside, the β-polypeptides at the outside and the pigments sandwiched between the two concentric rings of polypeptides. As a result of the formation of the ring, the absorption shifts to 870 nm for LH1 and to 850 nm for LH2, although in the latter case different absorption maxima are found, depending on the presence or absence of H bonds [27, 28]. The β-polypeptide of LH2 binds a second pigment, nearer to the cytosolic side of the complex, and in LH2 of *Rps. acidophila* these form a nine-membered ring, absorbing at around 800 nm and positioned at a distance of about 1.7 nm from the B850 ring.

Within the ring of B850 dimers, all distances between the pigments are very similar, somewhat less than 10 Å. Nevertheless, within the αβ-sub-unit the electron density seems continuous, whereas electron density due to the two neighboring BChls on adjacent sub-units is discontinuous. The reason for this is that the β-BChl is clearly bent. One further important point is that within the $\alpha\beta$-dimer, overlap between the two BChls occurs between chlorin rings I, as in the special pair of the reaction center, while between BChls on adjacent sub-units the overlap is between chlorin rings III. As a consequence, one may view LH1 and most probably also LH1 as "rings of interacting dimers." This concept is in fact supported by many experimental observations ([6, 100, 101], and see below). The BChls in the B850 ring of LH2 and at B870 ring of LH1 all have their Q_y transition dipole almost parallel to the membrane plane and their Q_x perpendicular [49]. The estimated excitonic coupling between BChls in a sub-unit is about 250 cm^{-1}; between BChls on adjacent sub-units the coupling is somewhat less [81]. In contrast, the pigments in the B800 ring are "monomeric"; the distance between two neighbors is about 21 Å and the corresponding dipole–dipole coupling about 20 cm^{-1}. The interaction between a B800 pigment and the pigments in the B850 ring is of a similar magnitude. The B800s in LH2 are almost flat in the plane of the membrane, those in the LH2 of *Rs. molischianum* are tilted away from the membrane plane by about 30°. Finally, the LH2 rings of *Rps. acidophila* and *Rs. molischianum* contain two carotenoids per $\alpha\beta$-sub-unit.

In LH2 of *Rps. acidophila*, the carotenoid is rhodopin, with 12 double bonds, of which 11 are conjugated. In the structure of LH2, one rhodopin molecule is well-resolved, and there is evidence in the structure for the second carotenoid [29]. The former is in an all-*trans* conformation, as predicted from Raman experiments. However, this rhodopin molecule is strongly twisted and has an orientation that is more or less perpendicular to the plane of the membrane. The molecule starts from a hydrophilic pocket in the cytosol region of the complex, with contacts between the polar headgroup and some charged amino acids. Then there are a number of contacts between the carbon atoms C(7) to C(11) and the B800 ring. However, at the position that would be most favorable for a direct interaction between the two conjugated systems, the distance is too large for a significant amount of ultrafast energy transfer via electron exchange to take place. The orientation of the carotenoid S_2 transition dipole, and consequently also the S_1 transition dipole, is normal to the direction of the B800 Q_y dipole. For these reasons, energy transfer from carotenoid to B800 should be small. On its path through the membrane, the carotenoid makes several contacts with the phytyl chain of the α-B850 of the next sub-unit, and then makes close contacts with the α-B850, which would allow for a contribution of electron exchange to the ultrafast energy transfer. The orientation of the carotenoid S_2 transition dipoles is parallel to the Q_x transition dipole of both α and β-B850, allowing for fast $S_2 \rightarrow Q_x$ energy transfer to either. The extrapolated position of the second carotenoid in the structure of LH2 of *Rps. acidophila* mimics the B800 and B850 interactions of the first carotenoid, although the molecule starts at the periplasmic side of the membrane.

9.2.2 LHCII, the major chlorophyll binding light-harvesting complex of plants

LHCII is the major light-harvesting pigment protein of higher plants and algae, and is responsible for the binding of about 50% of all chlorophyll on land. It serves to feed excitation energy into the minor light-harvesting complexes (CP29, CP26, and CP24) and into the core of Photosystem 2, which is eventually used for charge separation. LHCII is a member of a family of light-harvesting complexes that includes the various forms of LHCII and the minor light-harvesting complexes. The basic unit of all these complexes is a membrane protein, with a molecular weight of about 25 kDa, that is known to fold into a structure with three transmembrane helices, A, B, and C. In LHCII the monomeric sub-unit binds 7–8 Chl *a*, 5–6 Chl *b*, two luteins, one neoxanthin, and trace amounts of violaxanthin [76]. In its native form LHCII is a trimer, and in 1994 the structure of the trimer of LHCIIb was resolved to a resolution of 3–4 Å by Kühlbrandt and coworkers, using cryo-electron microscopy and two-dimensional crystals of LHCII [51]. A schematic model of the structure and arrangement of the pigments is shown in Fig. 9.2. In the structure, the A and B helices are tilted at an angle of about 30° away from the membrane normal, while the C-helix is perpendicular. Further, two carotenoids and 12 chlorins can be identified in the electron density, which are assigned to the two luteins, seven Chl *a*'s and five Chl *b*'s. The luteins are arranged in a crossed structure, and most likely have a role in stabilizing the complex, although details of the interaction between the luteins and the protein are not resolved. At the current resolution, Chl *a* and Chl *b* are indistinguishable. Also, the phytyl tails of the chlorophylls cannot be observed and, as a consequence, the orientation of the Q_x and Q_y transition dipoles within each of the chlorin planes is not known.

In the proposed model, the assignment of the Chl *a*'s and Chl *b*'s is based on the following argument. After excitation, there is a small but finite chance that a triplet is formed selectively on one of the Chl *a*'s due to the assumed fast Chl *b* to Chl *a* energy transfer. Since one role of the carotenoids is to quench these Chl *a* triplets with high efficiency, to prevent the formation of harmful oxygen radicals, the Chl *a*'s must be positioned in Van der Waals contact with the luteins. Therefore, the seven chlorophylls in the core of LHCII, all of which make close contact with the luteins, have been assigned to the seven Chl *a*'s; and the remaining ones to Chl *b*. In view of recent reconstitution experiments with LHCII, it may be possible that some of the binding sites are promiscuous and can be occupied by either a Chl *a* or a Chl *b*.

In the pigment arrangement, there is a pseudo-twofold symmetry axis that relates the two luteins and the Chl's *a*1, *a*2, and *a*3 with *a*4, *a*5, and *a*6. The distances between the various chlorophylls vary between 0.8 and 1.1 nm. The pair *b*3, *a*3 seems to be "isolated" from the other pigments – the same is true for the pigments *a*6, *a*7, and *b*6. Although the room-temperature absorption

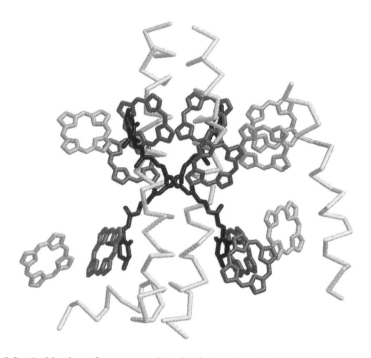

Figure 9.2 A side view of a monomeric unit of the major plant light-harvesting complex LHCII. The protein contains three transmembrane helices labeled A, B, and C. The cross-brace of xantophylls consists of two luteins (black), which are in Van der Waals contact with seven chlorophyll *a* molecules (dark gray porphyrins). The five remaining chlorophyll *b* molecules (light gray porphyrins) are all at a greater distance to the luteins. Note that in this model the directions of the Q_y and Q_x transitions of each chlorophyll are not specified. In nature, this complex occurs in a trimeric form, which was also the form that was crystallized. (Adapted with permission from Kühlbrandt *et al.* [51])

spectrum only shows two dominant bands at about 650 nm (Chl *b*) and 675 nm (Chl *a*), by a variety of polarized spectroscopic techniques 11 bands can be identified in the 640–690 nm spectral region. LHCII exhibits intense circular dichroism (CD), indicative for Chl *a*–Chl *a* and Chl *b*–Chl *b* excitonic interaction; the interaction between Chl *a*'s and *b*'s is most likely weak [70, 95].

9.2.3 The core antenna and reaction center of Photosystem 1

The core of Photosystem 1 is the most complex photosynthetic light-harvesting plus electron transfer system for which a 3-D structure is now available [50]. The functional unit of the PS1 core of *Synechococcus elongatus* consists of 11 sub-units, including the two major sub-units, PSaA and PSaB, each of which has a molecular weight of about 80 kDa, with known sequence and binding about 100 Chl a's, 10–25 carotenoids and three FeS clusters. The structure has been resolved up to 4 Å and the positions of about 90 Chls have been determined. A top view of the pigment arrangement in the PS 1 core is shown in Fig. 9.3. As in the case of LHCII, no information about the direction of the Q_y and Q_x dipoles is available. The core of PS1 is characterized by 22 transmembrane helices, 11 for each large sub-unit, that exhibit a C_2 symmetry around an axis that passes through the centrally located special pair of the electron transfer chain, P700, and the iron sulphur cluster F_x. The electron transfer chain is embedded in a structure of ten transmembrane helices, five from each sub-unit, the arrangement of which is strongly reminiscent of that of the L and M sub-units in the purple bacterial reaction center. All the other 90 antenna Chl a's are dispersed in a band around this core, and for a large part associated with the remaining six transmembrane helices on each of the large sub-units. For all

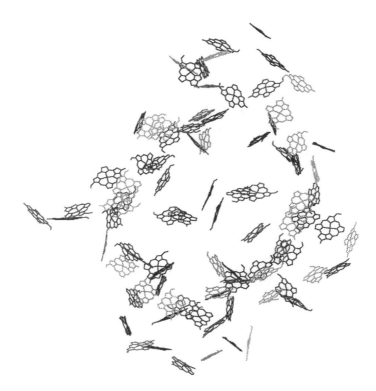

Figure 9.3 A top view of the chlorophyll pigment arrangement in photosystem I. The centrally located special pair (P700) and the electron transfer chain, is surrounded by 90 other Chl a molecules. The gray level indicates the vertical positions of the chlorophyll molecules. (Based on the X-ray structure of Krauss *et al.* [50])

of the Chls except two, the distance to any of the pigments in the electron transfer chain exceeds 2 nm, making energy transfer slow (10–20 ps). Two chlorophylls are found that seem to connect the antenna with the second and third pair of Chls of the electron transfer system, and it has been suggested that these form a special entry for excitation energy. There is a remarkable sequence analogy between the antenna part of the large PS1 sub-units and the core proteins of PS2, CP47, and CP43, and for that reason it has been suggested that the pigment–protein arrangement of the six outer transmembrane helices and their associated Chls may be a good model for the PS2 core.

9.3 THE MECHANISM OF ENERGY TRANSFER AND TRAPPING IN PHOTOSYNTHESIS

9.3.1 The Förster equation

One of the major conclusions drawn from the early work of Duysens was that the resonance energy transfer mechanism as proposed by Förster [19, 26] was very well able to explain all of the relevant results. He calculated that, assuming an effective chlorophyll concentration of 0.2 M and a ratio of RCs to antenna chlorophylls of 1 : 200, the trapping efficiency would be larger than 0.95, i.e. 3000 "hops" would be made during the fluorescence lifetime, with a single hop time in the picosecond range.

The physical mechanism underlying the Förster expression is Fermi's Golden Rule, which is valid when the coupling between the initial and final states is sufficiently weak, and phase correlation is lost upon transfer. Other essential assumptions are the absence of overlapping electron density between the donor and acceptor molecule, and fast thermal relaxation in the donor (see Chapter 2 for details). This last assumption assures mono-exponential decay, and allows one to use the stationary fluorescence emission spectrum in calculating the Franck–Condon factors for the transition. The expression for the energy transfer rate k_{DA} as given by Förster [26] is based on a dipole–dipole approximation. This can be avoided by expressing the transfer rate directly in terms of the excitonic interaction strength (V_{DA}) between the donor and the acceptor [83]:

$$k_{DA} = 4\pi^2 c V_{DA}^2 \frac{\int F_D(\nu) A_A(\nu) \nu^{-4} \, d\nu}{\int F_D(\nu) \nu^{-3} \, d\nu \int A_A(\nu) \nu^{-1} \, d\nu} \tag{9.1}$$

Here, F_D is the emission spectrum of the donor, A_A is the absorption spectrum of the acceptor, V_{DA} is the excitonic interaction strength (in cm^{-1}) between donor and acceptor, ν is the wavenumber, and c is the speed of light (in $cm s^{-1}$).

For a pair of bacteriochlorophyll molecules in the B800 ring of LH2, reasonable assumptions are as follows: $V_{DA} = 20\text{cm}^{-1}$, (approximately) Gaussian emission and absorption spectra with a FWHM bandwidth of 200 cm^{-1} and a Stokes shift of 50 cm^{-1}. This yields a Förster hopping rate $k_{DA} = 1.4\text{ps}^{-1}$, in good agreement with experiment.

With the high pigment concentrations found in photosynthetic proteins, it is no longer certain that the conditions for which the Förster equation was originally derived still apply. For example, for center-to-center distances less than 1.5 nm, the dipole model for calculating the interaction strength may have to be replaced by a point charge model [55], although recent calculations illustrate that the deviations are small [81]. This problem is avoided in Eqn 9.1, which is valid for any type of electronic interaction, and has the additional advantage of being expressed in more accessible parameters than the original equation by Förster [26]. More seriously, the short distances lead to overlapping electron densities, which may induce other transfer mechanisms such as exchange transfer (see Chapter 6). In addition, strong interaction leads to delocalized states – and while these states may be regarded as donors and acceptors instead, this complicates the calculation of the interaction between them. Finally, ultrafast spectroscopic experiments indicate that vibrational coherence decay [9] may take place on the same timescale as the energy transfer itself. In that light, it is remarkable that the transfer rates observed so far can still be calculated from the Förster equation, with reasonable values for the structural and spectral parameters. We propose that the Förster equation can be used to estimate the magnitudes of excitation energy transfer rates.

9.3.2   Trapping by the reaction center

In the incoherent limit, the excited-state dynamics of a photosynthetic system follows the master equation:

$$\mathrm{d}p_i/\mathrm{d}t = -k_l p_i + \sum_j (p_j W_{ij} - p_i W_{ij}) - \delta_{iRC} k_{RC} p_{RC}. \qquad (9.2)$$

The p_i are time-dependent probabilities (fractional populations) of finding the excitation at site i in the lattice. In particular, p_{RC} is the population of the site at which charge separation occurs; for instance, the special pair of the RC. Further, k_l is the rate of loss from a site due to fluorescence, internal conversion and inter-system crossing, k_{RC} is the rate of charge separation, and W_{ij} is the Förster rate of energy transfer from site i to site j.

The time-dependent solution of Eqn 9.2 is a sum of N exponential decays, which merge into a continuous distribution of decay times in an ensemble of nonidentical systems, e.g. as a result of inhomogeneous broadening. The faster decay times are observed as equilibration phenomena, while trapping experiments, i.e. with largely a-select detection and excitation, detect only the mono-exponential decay from thermal equilibrium. Both types of dynamics can be studied by solving the master equation numerically (see, e.g., [103]). The most relevant parameters are the structure of the lattice and the spectroscopic properties of the pigments, such as the shape of the homogeneous spectrum, the Stokes shift, and the degree of inhomogeneous broadening. Below, we describe analytic approximations that provide insight as to how the exciton dynamics depends on various structural and spectral parameters.

The lifetime of an exciton in a regular lattice has been expressed with random walk [74] and perturbation [85, 93] approaches, and these have been shown to be compatible [84]. A remarkable result is that the exciton lifetime can be expressed as the sum of the time needed for migration through the antenna, delivery to and trapping in the RC. We shall prove the generality of this principle, by deriving it from a much simpler steady state flux calculation. This may also be more applicable to less regular systems such as PS1. The only required assumption is that the trapping is indeed mono-exponential. As indicated in Fig. 9.4, we use a simplified sequential model for the RC–LHA complex, and distinguish three different sites: the RC with population p_{RC}, the z delivery sites which are connected to the RC with total population p_D, and the $N - z - 1$ other antenna sites with total population p_A. The parameter z is also referred to as the coordination number of the RC. Delivery to, and escape from, the RC occurs at rates W_D and W_E respectively. The transfer between the delivery and the other antenna pools can be described by two as yet unknown transfer rates a and b, which depend not only on the transfer between the pools, but also on the migration within them.

First, we consider a situation in which a constant flux J is absorbed by the antenna sites, as indicated in Fig. 9.4. A steady state develops in which the fluxes from the antenna sites through the delivery sites to the RC, as well as that of charge separation, all become equal to J:

$$J = k_{RC} p_{RC} = W_D p_D - z W_E p_{RC} = a p_A - b p_D. \qquad (9.3)$$

The steady state populations are also related to the trapping kinetics. In the case of mono-exponential trapping, the total population becomes $J\tau$, where τ is the exciton lifetime. The steady state populations are obtained by solving Eqn 9.3, and thus

$$\tau = \frac{p_{RC} + p_D + p_A}{J} = \frac{1}{k_{RC}} \left[1 + \frac{z W_E}{W_D} \left(1 + \frac{b}{a} \right) \right] + \frac{1}{W_D} \left(1 + \frac{b}{a} \right) + \frac{1}{a}. \qquad (9.4)$$

This is a complicated expression which, moreover, has two unknown parameters, a and b. However, these can be eliminated by considering the limiting

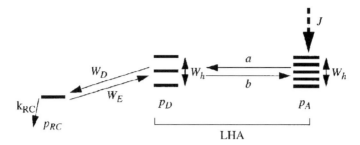

Figure 9.4 A sequential model of the energy trapping in the RC–LHA complex. The model distinguishes the RC, delivery sites, and other sites in the antenna. A constant flux J is absorbed by the other antenna sites

cases of fast and slow charge separation. The resulting expression of the exciton lifetime is much simpler, and separates naturally into a sum of three contributions, each of which has a clearly defined function:

$$\tau = \tau_{\text{trap}} + \tau_{\text{del}} + \tau_{\text{mig}}. \tag{9.5}$$

First, τ_{trap} dominates the exciton lifetime in the case of very slow charge separation. It depends only on k_{RC} and the equilibrium population ρ_{EQ} of the RC:

$$\tau_{\text{trap}}^{-1} = k_{RC}\rho_{EQ}, \quad \text{with } \rho_{EQ} = \left[\sum_i \exp\frac{E_{RC} - E_i}{k_B T}\right]^{-1}. \tag{9.6}$$

Equation 9.6 has been successfully used to explain the trapping time in Photosystem 1, where the trap, a chlorophyll a dimer, P700, is red-shifted relative to the bulk antenna, while at the same time there exist pigments in PS1 that are at lower energy than P_{700} [15]. However, it is debatable to what extent the red spectral forms in PS1 are equilibrated with the main antenna [90].

Secondly, the migration time, τ_{mig}, is the average "first passage time" [74, 85, 93] for an excitation to the RC. It dominates the exciton lifetime in the second limiting case of very fast charge separation, where $W_D = W_E = W_h$, where W_h is the average transfer (hopping) rate within the antenna. For various regular light-harvesting antenna structures with different dimensions d and sizes N, it can be expressed as

$$\tau_{\text{mig}} = 0.5Nf_d(N)W_h^{-1}. \tag{9.7}$$

The structure function $f_d(N)$ can be obtained analytically [85], and is of the order of magnitude of unity, as demonstrated in Fig. 9.5. Finally the delivery

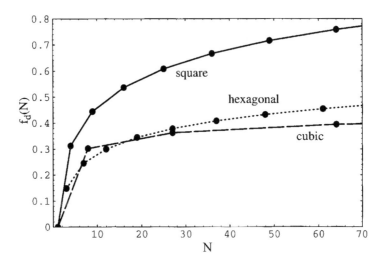

Figure 9.5 The structure function $f_d(N)$ for an LHA organized as a two-dimensional square lattice (solid), a three-dimensional cube (dashed), or a two-dimensional hexagonal lattice (dotted): N is the total number of pigments per RC. The structure function $f_d(N)$ gives the relation between τ_{hop}, N and the migration time as in Eqn 9.7. (Reproduced from Somsen *et al.* [85]).

term, τ_{del}, describes the additional time needed for transfer to the RC, when W_D and W_E differ from W_h:

$$\tau_{\text{del}} = \left(\frac{1}{\rho_{EQ}} - 1\right)\frac{W_D}{z W_E}\left(\frac{1}{W_D} - \frac{1}{W_h}\right), \tag{9.8}$$

Indeed, τ_{del} disappears for $W_D = W_h$. The ratio W_E/W_D in Eqn 9.8 reflects detailed balance, which depends only on the excited state energies of the RC, E_{RC} and the delivery sites, E_D (assuming that it can be approximated with a single value):

$$W_D/W_E = e^{(E_D - E_{RC})/k_B T}. \tag{9.9}$$

In the typical case in which the delivery and other antenna sites have approximately equal energies, Eqn 9.8 simplifies to

$$\tau_{\text{del}} \approx \frac{N-1}{z}\left(\frac{1}{W_D} - \frac{1}{W_h}\right), \tag{9.10}$$

i.e. the average additional delivery time for an equilibrated antenna, when W_D differs from W_h. In that case (see, e.g., Fig. 9.6), the expression for the excition

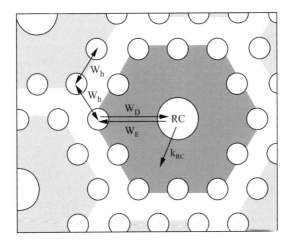

Figure 9.6   A hexagonal LHA–RC model for excitation migration at RT. Each photo-synthetic unit contains one RC and 12 LHA sites and is surrounded by six other units. Transfer between neighboring sites, either within one unit or between neighboring units, takes place with the hopping rate W_h, while W_D and W_E are the rates of transfer to and from the RC respectively. The charge separation rate is k_{RC}. (Reproduced from Somsen *et al.* [85])

lifetime is equivalent to those obtained with random walk [74] and perturbation [85, 93] approaches. However, the expression presented here remains valid when the delivery sites differ from those of the rest of the antenna, or when the antenna does not form a regular lattice. In the latter case, Eqn 9.7 gives only a rough estimate of the migration time.

In our present derivation, the flux is entered only through the nondelivery antenna sites (see Fig. 9.4). The extension to the more general case of a distribution over all three pools is trivial. When the flux is distributed over the delivery and other antenna sites, the final result (Eqns 9.5–9.8) is not affected. A different result is obtained when the flux is entered through the RC. This should not be used for the present calculation, however, because the exciton decay is not mono-exponential when the RC is excited (but see Eqn 9.11).

9.3.3   Escape from the reaction center

It is well documented that, following selective excitation of the RC pigments of photosynthetic bacteria, there is a finite chance for the excitation to return to the LHA and be emitted as fluorescence [30, 88]. One can define the escape ratio as

$$\alpha = \frac{QY_{RC}}{QY_{LHA}} = \frac{\tau_{\text{trap}}}{\tau_{\text{trap}} + \tau_{FP}}. \tag{9.11}$$

In Eqn 9.11, QY_{RC} and QY_{LHA} are the fluorescence quantum yields following RC and LHA excitation, respectively. The $\tau_{FP} = \tau_{\text{del}} + \tau_{\text{mig}}$ is the average time for the excitations within the LHA to reach the RC for the first time. The second equality in Eqn 9.11 is exact for the sequential model defined in the previous subsection, and can be used to estimate the parameters W_D, W_E, k_{RC}, W_h, z, and N.

As an example, let us now apply Eqns 9.5 and 9.11 to analyse how various combinations of the parameters lead to reasonable estimates for the excited state lifetime and the escape efficiency. For the photosynthetic bacterium *Rhodospirillum rubrum*, the PSU consists only of a RC surrounded by a single ring of $\alpha\beta$-pairs that form the core antenna or LH1. In recent models, LH1 contains 24–32 BChl *a*'s, and their distance to the special pair of the RC is about 4 nm. The rate of charge separation in the membrane bound RC is taken as $(4.5 \text{ ps})^{-1}$ [7] and the excited-state lifetime in membranes of *Rs. rubrum* is 50–60 ps [87]. From singlet–singlet annihilation experiments, the hopping time between adjacent neighbors was estimated to be 0.65 ps or faster ([4, 90], and see below). The escape ratio was determined from the size of the RC absorption bands in the fluorescence excitation spectrum, and estimated to be between 0.2 and 0.3. From these parameters we estimate that $\tau_{\text{trap}} = 10$–18 ps and $\tau_{FP} = 35$–48 ps. From Eqn 9.6, $\rho_{EQ} = 0.25$–0.45, somewhat larger than $1/N$, which indicates that the RC is red-shifted with respect to the main LHA absorption (note that this refers to the position of the zero-phonon lines).

The first-passage time is the sum of the time needed for migration in the LHA and transfer to the RC: $\tau_{FP} = \tau_{\text{mig}} + \tau_{\text{del}}$. The former may be evaluated; e.g. in a hexagonal lattice and with $\tau_h \leqslant 0.65$ ps and $N = 12$, $\tau_{\text{mig}} \leqslant 1.1$ ps. Other lattices yield comparable numbers. This indicates that the first-passage time is largely determined by τ_{del}.

Estimates for W_D and W_E require knowledge of the coordination number z. However, we can rephrase the whole problem in terms of the parameters $\bar{W}_D$ and $\bar{W}_E$, the effective trapping and detrapping rates [84]:

$$\bar{W}_D = \frac{z W_D}{N - 1}, \qquad \bar{W}_E = z W_E, \tag{9.12}$$

with the aforementioned conditions, and those of Eqns 9.10, 9.11, and 9.5 become

$$\alpha = \frac{\bar{W}_D + \bar{W}_E}{k_{RC} + \bar{W}_D + \bar{W}_E}, \qquad \tau = \frac{\bar{W}_D + \bar{W}_E}{\bar{W}_D} \frac{1}{k_{RC}} + \frac{1}{\bar{W}_D}. \tag{9.13}$$

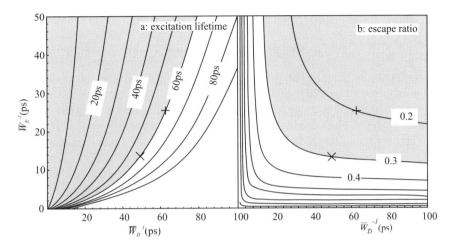

Figure 9.7 A contour plot of the excitation lifetime (a) and the escape ratio (b) in LH1 ($N = 13, Z = 6, k_{RC}^{-1} = 4.5 \, \text{ps}, \tau_h = 0.65 \, \text{ps}$) as a function of trapping and detrapping time, according to the local trap approximation. Marked areas are $\tau < 60$ ps and $\alpha < 0.3$. The outcome of our estimate (see text) is marked by "×" ($\alpha = 0.33$) and "+" ($\alpha = 0.2$). (Reproduced from Somsen *et al.* [84])

Figure 9.7 shows an analysis of exciton trapping kinetics of *Rs. rubrum*, which leads to the estimates $\bar{W}_D^{-1} = 63$–50 ps and $\bar{W}_E^{-1} = 26$–14 ps. For $z = N - 1$, as appears to be the case in LH1, also $W_D^{-1} = 63$–50 ps. The estimates for $\bar{W}_D$, and to a lesser extent the equilibrium RC excitation density, are insensitive to most of the experimental parameters mentioned above. In particular, they are more or less independent of the hopping rate, unless it slows down dramatically [85]. The estimate of the effective detrapping rate, $\bar{W}_E = z W_E$, depends strongly on α, but is almost independent of the actual value of τ. In contrast, $\bar{W}_D$ scales approximately proportional to τ and depends only weakly on α. Note that these estimates break down for very slow detrapping rates.

This analysis can be extended to include explicitly the finite rate of excitation transfer in the LHA. Within the values of the experimental parameters, one arrives at very similar estimates for the effective trapping and detrapping rates [84]. Consequently, we conclude that the excited-state decay in the LH1 core antenna of photosynthetic bacteria is limited by the average rate at which excitations are transferred to the RC from its neighboring pigments in the LHA. We propose that this is a general property of all LHA–RC systems, including those of Photosystem 1 [94] and Photosystem 2.

9.4 DYNAMICS OF EXCITATION ENERGY TRANSFER

The timescale of energy transfer is easily established. Given that the intrinsic lifetime of the excited state of chlorophyll and bacteriochlorophyll is only a few nanoseconds, and since the quantum yield of stable charge separation may easily exceed 95%, the energy transfer and charge separation process must be fully completed in 100 ps or less. It is remarkable that this is precisely realized by many of the photosynthetic systems. Because the energy-transfer process may involve many steps, each of the individual steps must be much faster than 100 ps. Below, we will see that, indeed, individual energy-transfer events can be much faster, even faster than 100 fs, and we will discuss some of the experimental evidence that often the rate-limiting step is to get from the antenna into the reaction center.

9.4.1 Energy transfer in the peripheral and core antennae of photosynthetic purple bacteria

LH1, LH2, and the intact photosynthetic unit of purple photosynthetic bacteria have been subjected to a large number of spectroscopic studies, notably polarized light spectroscopy (circular dichroism (CD), linear dichroism (LD), polarized fluorescence (FP), etc.) [49], a variety of line-narrowing techniques (hole-burning [16, 79, 107, 108], site-selective fluorescence [63, 100]), IR and Raman spectroscopy, etc. Also, using the structural information, several of the spectroscopic features have been modeled [2, 48, 81]. From these studies, two opposing views have emerged that we will briefly discuss.

In the first view, LH1 and LH2 are considered to be rings of interacting dimers, in which many of the essential spectroscopic features, including the dramatic red shift, largely originate within the dimer. The excitonic interaction between neighboring sub-units within the ring is considered to be a relatively weak perturbation, i.e. relative to the intradimer excitonic interaction, the mixing of charge transfer states, the possible (and so far unknown) contribution from electron exchange arising from ring I overlap, the intrinsic energetic disorder and the electron–phonon or electron–vibration coupling. It is now recognized that the spectra of all photosynthetic pigment–proteins are strongly inhomogeneously broadened, and estimates of the amount of inhomogeneous broadening range between 200 and 500 cm^{-1}. Also, the electron–phonon coupling is estimated to be of that order of magnitude. The general idea behind the "ring of dimers" model is that following excitation, any phase relation between excitations on different dimers is rapidly destroyed – either dynamically, due to the coupling to vibrations or phonons, or as a consequence of the interference of the pure eigenstates due to energetic disorder.

In the alternative view, the spectroscopic features are totally determined by the set of excitonic eigenstates of the full ring. In this model, the excitonic interaction between adjacent BChls is the dominant term that completely determines the red shift observed upon formation of the ring. Also, in this model, the lowest state of the exciton manifold is almost optically forbidden, due to the in-plane orientation of the Q_y-transition dipoles, and all the oscillator strength is equally divided between two orthogonal transitions slightly above the lowest one. Experimental evidence to support this model includes hole-burning experiments [79, 107, 108], the interpretation of the low-temperature absorption spectrum [81], and estimates of the absorption cross-section of the major transition at 850 nm or 870 nm [57, 69]. We are of the opinion that the latter view is less accurate, mainly because it ignores all the nonexcitonic contributions that all have the effect of destroying the fully delocalized coherent states. Also, as will be shown below, the "ring of dimers" model provides a simple and elegant explanation for many of the dynamic results.

From a variety of spectroscopic studies (reviewed in Van Grondelle *et al.* [99]) it was concluded that the energy transfer dynamics within LH1 and LH2 had to be ultrafast. With the advent of femtosecond laser spectroscopy, in particular using Ti : sapphire lasers, many of the elementary energy-transfer steps have been time-resolved. The B800 → B850 energy transfer at room temperature takes about 700–800 fs for LH2 from *Rb. sphaeroides*, and this time constant is not very dependent on the species [36, 44, 58, 65, 89, 93, 97]. The energy-transfer time is only weakly dependent on temperature, being about 1 ps at 77 K and about 2 ps at 4 K [96]. A typical experimental result is shown in Fig. 9.8.

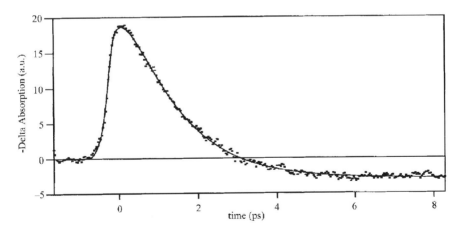

Figure 9.8 Two-color pump–probe measurement in the B800 band of *Rb. sphaeroides* at 77 K. Excitation was at 791 nm and detection at 810 nm. The solid line is a fit with a rise of 440 fs, reflecting downhill energy transfer among B800s, and a 1.26 ps decay due to B800 → B850 transfer

The B800 $\rightarrow$ B850 energy transfer has been modeled in terms of a Förster process [64, 93, 97]. The weak temperature-dependence of this energy-transfer step suggests the involvement of some vibronic level of B850, or possibly the higher excitonic states of the B850 ring [58, 108]. Previously, from fluorescence polarization experiments, it was concluded that efficient energy transfer occurs within the B800 ring [49]. More recently, from polarized pump–probe spectroscopy, a time constant of about 0.5–1.0 ps was estimated for energy transfer between neighboring B800s [36, 63, 65].

For energy transfer within the B850 and B875 rings, single-site lifetimes of the order of a few hundred femtoseconds were estimated, for instance, from an analysis of the efficiency of singlet–singlet annihilation [4]. Also, the observation that within a few picoseconds transient absorption changes were almost fully depolarized were interpreted as sub-picosecond energy transfer among B850s in LH2 and B970s in LH1 [87, 98]. The energy migration in B850 and B870 has been recorded directly by fluorescence depolarization [9, 43], and by equilibration of the transient absorption spectrum [105]. In both cases, a similar interpretation was used, based on hopping between dimers in a ring. The site energies of the dimers were taken at random from an inhomogeneous distribution of $\sim 400\,\mathrm{cm}^{-1}$ width, and average hopping times of about 100 fs were obtained. The fluorescence anisotropy decays faster in LH2 than in LH1, and in this model this arises simply from the smaller number of units in the ring of LH2 (larger angle change per hop). The model could be extended to low temperatures where the site energy variation impedes the energy transfer over more than a few sites [103]. Remarkably, Chachisvilis *et al.* [12] and Bradforth *et al.* [9] found that oscillations at 105 cm^{-1}, assigned to vibrational wavepacket motion, dephased significantly more slowly than the observed depolarization timescale, suggesting vibrational coherence transfer [41] in the energy transfer process.

As discussed above, the extent of exciton delocalization in LH1 and LH2 has been extensively debated. Key quantities are the electronic coupling between the BChls, the electron–phonon coupling (reorganization energy and timescale), the temperature and the disorder. From the difference in position between the pump-induced bleaching (ground state to one-exciton state) and pump-induced absorption (one-exciton state to two-exciton state), Sundström and coworkers [11, 78] estimate a delocalization length of 4 ± 2 units in LH1 and LH2, more or less independent of temperature. A measurement of the superradiance in LH1 and LH2 gave an even smaller number [64]. On the other hand, ultrafast decay in the transient absorption and emission of LH2 was taken as an indication for relaxation between fully delocalized states [68].

An incisive discussion of how delocalization influences different observables has been given by Leegwater [56], and more recently by Meier *et al.* [60]. In an attempt to provide experimental characterization of the electron–phonon coupling, Jimenez *et al.* carried out three-pulse photon-echo peak shift (3PEPS) measurements on LH1 and LH2 [42]. They concluded that on a 50 fs timescale, fluctuations in the environment and vibrations lead to the dynamic localization on a dimeric sub-unit of LH1 and LH2. A similar conclusion was drawn from the ultrafast reorganization as observed by the formation of the Stokes shift in a few tens of femtoseconds [53]. In the peak shift decay this initial phase was followed by an exponential phase that was interpreted as a loss in the rephasing capability of the system due to energy transfer. During the energy transfer the system samples all of the various environments that contribute to the inhomogeneous broadening, and as a consequence the information about the original environment is lost. Again, the model that assumes hopping on an inhomogeneously broadened ring of dimers gave a fit to the results. The homogeneous broadening was estimated about 200 cm^{-1}, the inhomogeneous broadening about 500 cm^{-1} and the hopping time about 100 fs. This interpretation is very much supported by a 3PEPS experiment on the LH1 sub-unit, B820, in which the 100 fs phase in the peak shift decay ascribed to energy transfer was absent, and was replaced by a nondecaying component arising from inhomogeneous broadening [109]. The striking similarity between the 3PEPS of LH1 and B820 further supports the idea that excitations in these antenna complexes are delocalized over only a dimer unit.

Energy transfer from carotenoid to BChl in LH2 of *Rb. sphaeroides* can occur on a timescale of a few hundred fs [82]. Recent fluorescence upconversion experiments have demonstrated that in LH1 and LH2 of *Rb. sphaeroides* the S_2 lifetime of sphaeroidene is shortened to about 55 fs for the former and 80 fs for the latter. This should be compared to a 150–250 fs internal conversion time from $S_2 \rightarrow S_1$, dependent on the solvent [80]. For a B800–830 complex of *Chromatium purpuratum*, Gillbro and coworkers [1] report an $S_2 \rightarrow$ BChl (Q_x) transfer time of 100 fs and $S_1 \rightarrow Q_y$ transfer times of 3.8 ps and 0.5 ps for the carotenoid that transfers to B830 and the carotenoid that transfers to B800, respectively.

In the intact bacterial photosynthetic unit, the energy transfer from one LH2 to another, and from LH2 to LH1, takes place on a timescale of a few picoseconds [31, 37, 67, 110]. Assuming Förster energy transfer between BChl's on neighboring rings, this would imply a closest distance of about 3 nm between the two pigment rings. Long before the crystal structure of LH2 became available, it was realized that the rate-limiting step in excitation trapping was the step from the LH1 pigments to the special pair in the RC. A transfer time of about 35 ps was obtained for the LH1 to P energy transfer [5], and this was interpreted as a distance of 4.5 nm between the special pair and the LH1 ring – in good agreement with models suggested for the RC–LH1 core, assuming that LH1 is organized as a ring of 16 $\alpha\beta$-BChl$_2$ sub-units with a structure as in LH2 of *Rps. acidophila* [71, 77]. A summary of the timescales is given in Fig. 9.9.

LH2   LH1 - RC

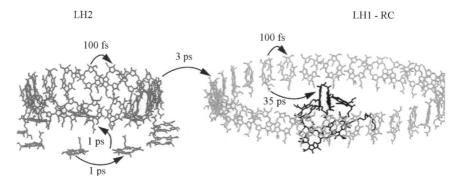

Figure 9.9 A schematic overview of structural organization and energy transfer in the bacterial light-harvesting antenna. Shown are the light-harvesting 1 ring with a reaction center in its center, and one of the surrounding light-harvesting 2 rings. In LH2, transfer occurs within the B800 ring (1 ps) and to the B850 ring (1 ps), within the B850 ring (100 fs) and to LH1 (3 ps). In LH1, transfer occurs within the B880 ring (100 fs) and to the RC (35 ps)

9.4.2 The major peripheral plant light-harvesting complex LHCII

A variety of pico- and femtosecond studies have been performed to explore the dynamics of energy transfer within LHCII. In a pioneering fluorescence upconversion study by Eads *et al.*, the dominant time constant for energy was estimated to be 0.5 ± 0.2 ps [21]. In a low-intensity pump–probe study, in addition to the ultrafast process a slower Chl $b \rightarrow$ Chl a energy-transfer time in the range of a few ps was obtained [54]. Transient absorption measurements with shorter pulses [8] revealed Chl $b \rightarrow$ Chl a energy transfer times of 160 fs and also the slow process of 5 ± 2 ps, similar to [54]. In a fluorescence upconversion study by Du *et al.* [17], two lifetimes in the rise, presumably in Chl a fluorescence, were detected upon excitation at 650 nm: $\pm$ 250 fs and 5 ps, in rather good agreement with the pump–probe results. On the other hand, Pålsson *et al.* [72] using a one-color pump–probe technique, detected a major 500 fs and a minor 2–3 ps Chl $b \rightarrow$ Chl a transfer time, and inspite of the superior time resolution they could not distinguish any transfer component faster than 500 fs.

More recently, Visser *et al.* [104] and later Connelly *et al.* [14] resolved all three phases in the Chl $b \rightarrow$ Chl a energy transfer: 180 fs, 600 fs, and about 5 ps with relative amplitude ratios of 40%, 40%, and 20%, respectively. A study on LHCII monomers demonstrated that all three decay times are associated with Chl $b \rightarrow$ Chl a energy transfer within a monomeric unit of LHCII [46]. According to Visser and coworkers [104], in the trimer all of the energy transfer occurred to the major red-absorbing species at 676 nm. Connelly *et al.* [14] concluded from their data, obtained with an excellent signal-to-noise ratio, that probably the 175 fs component partly reflected energy transfer between "blue" and "red" Chl *b*'s [92]. Measurements of singlet–singlet and singlet–triplet annihilation suggest that inter-monomer energy transfer occurs on a timescale of 10–20 ps [8, 104].

Very recently, Gradinaru *et al.* [34] have studied the Chl *b* → *a* transfer in one of the minor LHCs, CP29. In CP29, six of the Chl *a* and two of the Chl *b* binding sites are conserved [32], suggesting a pigment stoichiometry of six Chl *a* : two Chl *b* : one lutein : one neoxanthin : one violaxanthin. Using femtosecond pump–probe spectroscopy at 77 K, Gradinaru *et al.* obtained a detailed picture of the energy transfer dynamics in CP29. Some of their results are shown in Fig. 9.10. The kinetics of Chl *b* to Chl *a* transfer in CP29 contain many components similar to those in LHCII and most likely reflect the same energy transfer processes. Specifically, Gradinaru *et al.* [34] could assign a slow, 2–3 ps phase to energy transfer from a Chl *b* absorbing at 650 nm, most likely Chl *b5* or Chl *b6* in the LHCII assignment, and a fast, 0.2–0.3 ps phase to energy transfer from a Chl *b* absorbing around 640 nm, most likely Chl *b3*. They were also able to distinguish fast (few hundred fs) and slow (few ps) energy transfer steps within the pool of Chl *a* pigments.

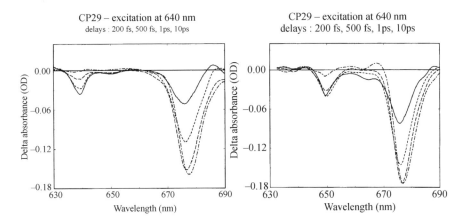

Figure 9.10 Energy transfer from each of the two Chl's *b* in the CP29 complex was probed in a multicolor transient absorption study at 77 K. In panel A some typical difference absorption spectra upon excitation at 640 nm are shown, for delays between pump and probe pulses of 200 fs (——), 500 fs (– – –), 1 ps (––), and 10 ps (-.-.-). Similarly, in the case of 650 nm excitation, spectra for delays of 250 fs (——), 500 fs (– – –), 1.5 ps (––), and 10 ps (-.-.-.-) are presented in panel B. It is evident that the "blue" Chl *b* transfers excitation with a sub-picosecond time constant (about 300 fs), while for the "red" Chl *b* this is much slower (about 2.2 ps). This information was correlated with the structural model for LHCII, with the protein sequence homology between LHCII and CP29, and with linear dichroism features, to yield a spatial and spectral assignment of some Chl molecules in both antenna complexes. (Reproduced from Gradinaru *et al.* [34])

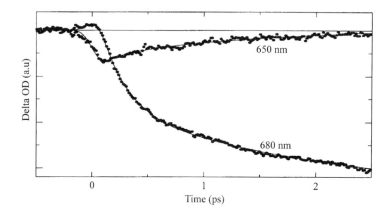

Figure 9.11 Ultrafast ingrowth of chlorophyll bleaching upon excitation at 500 nm in LHCII of spinach at 77 K. Pump–probe traces detected at 650 and 680 nm. The instrument response was 160 fs. Shown are the data (points) together with a double-exponential fit (0.22 and 2.1 ps). The 650 nm trace shows instantaneous bleaching due to direct chlorophyll *b* excitation. This bleaching decays with a time constant of mainly 2.1 ps, due to excitation transfer from chlorophyll *b* to *a*. The 680 nm trace shows a bi-exponential ingrowth of the bleaching, with time constants of 220 fs and 2.1 ps. The 220 fs component is mainly due to carotenoid to chlorophyll *a* excitation transfer: the 2.1 ps component to chlorophyll *b* to *a* and *a* to *a* excitation transfer. (Reproduced from Peterman *et al.* [75])

Carotenoid to Chl energy transfer in LHCII is highly efficient. Recently, two conflicting reports appeared on the dynamics and pathway of carotenoid to Chl *a* transfer. Peterman *et al.* [75] argued that no direct carotenoid to Chl *b* transfer occurred, while carotenoid to Chl *a* energy transfer took place in about 220 fs. The experiment is shown in Fig. 9.11. In contrast, Connelly *et al.* [13] claimed that the carotenoids exclusively transferred energy to Chl b, followed by Chl *b* → *a* energy transfer. The latter would be inconsistent with the assignment by Kühlbrandt, where only close contacts between Chl *a*'s and carotenoids exist.

9.4.3   Photosystem 1

From ultrafast fluorescence depolarization, Du *et al.* [18] estimated that the major hopping process within the bulk antenna of PS1 occurs on a timescale of 100 fs. On a timescale of a few picoseconds, the excitation energy is seen to equilibrate between a pool of very "red" pigments, absorbing around 720–730 nm, and the major PS1 core pigments. Figure 9.12 shows the fluorescence decay associated spectra observed in PS1 isolated from *Synechococcus elongatus*, and clearly two equilibration processes taking place on a 1–10 ps timescale can be

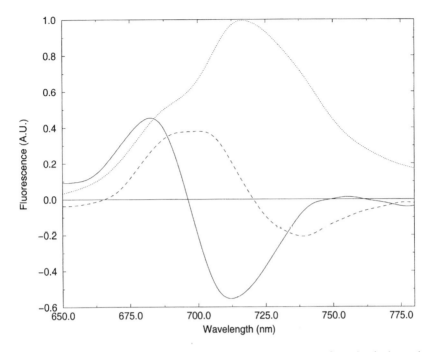

Figure 9.12 Decay associated spectra (DAS) of fluorescence of a-selectively excited trimeric PSI particles isolated from the cyanobacterium *Synechococcus elongatus*. The solid and dashed DAS exhibit decays with time constants of 3.2 ps and 10.3 ps, respectively, and represent spectral equilibrium of the major pool of antenna chlorophylls with two distinct pools of red-absorbing chlorophyll *a* molecules, that are present in these PSI particles. The dotted spectrum shows the (spectrally) equilibrated state which decays with a time constant of 36 ps

discerned. The process of energy transfer to P700 must occur at the same rate as this equilibration between core and red pigments. This is because even at very low temperatures, where escape from the red states is impossible, a reasonably high quantum yield for charge separation is still observed upon excitation of the core pigments ([33, 73] and Pålsson *et al.*, unpublished data). Trapping in Photosystem 1 is fast (20–25 ps in PS1 of plants and in *Synechocystis*, 35 ps in *Synechococcus elongatus* that has a relatively large amount of red chlorophylls) and charge separation is irreversible [35, 39, 99]. This has led to a model in which essentially all sites within the PS1 core are more or less equally efficient in transferring their energy to P700 (or any other pigment of the electron transfer chain) and which may be viewed as the 3-D version of the "2-D-ring-to-P" energy-transfer model that seems to operate for purple bacteria [94]. In our view, it is highly unlikely that the two chlorophylls that were proposed to act as a special entry for excitation energy [50] indeed have that role. They have more rapid energy transfer to the pigments in the electron transfer chain but are simply outnumbered by the other Chls. A simple simulation of the trapping kinetics in PS1 shows that leaving out the pair of connecting Chls hardly changes the trapping time [G. S. Beddard, personal communication].

The process of energy migration and charge separation cannot be experimentally separated in PS1. Kumazaki *et al.* [52], Trinkunas and Holzwarth [90], and White *et al.* [106] have used modeling to extract the intrinsic electron transfer rate. In a very recent study using a Photosystem 1 mutant that seemed to affect the special pair P700, but not the antenna spectra or dynamics, it was observed that the excited state lifetime approximately doubled [62]. This was taken by the authors as evidence for a model in which the charge separation rate by P700 is the rate-limiting step, in contrast to the "transfer-to-the-trap" limited model discussed above.

9.5   CONCLUSIONS

The combination of high-resolution structural data and ultrafast spectroscopy has enabled the development of a fairly complete picture of the light-harvesting process in purple bacteria. The efficiency of the overall process is based on individual energy transfer steps of 80–100 fs. In LH1, the core antenna surrounding the reaction center, several hundred energy transfer steps occur before the final transfer to the special pair (35 ps) and the initiation of charge separation. The observation of < 100 fs energy transfer, along with the retention of coherence and the enhanced radiative rates in LH1 and LH2, raises many challenging theoretic issues that will provide stimulus for theory and experiment for years to come. Inspite of the high symmetry and potential for strong intermolecular coupling, it does not appear that extensive electronic delocalization is necessary for achieving the near-unit efficiency of the light-harvesting process.

In green plant and cyanobacterial antennas, structural information is not yet sufficient for the most detailed molecular modeling of energy migration. Enough is known, however, to reveal both striking similarities with and differences from the purple bacteria. In particular, antenna molecules are held away from close contact with the primary electron donor and efficiency is achieved by using large numbers of antenna molecules with roughly similar transfer rates to perform the final transfer step to the primary donor.

Acknowledgments

The authors thank R. J. Cogdell for the atomic coordinates of the bacterial light-harvesting antenna (LH2), W. Kühlbrandt for the atomic coordinates of the plant light-harvesting complex (LHCII), and G. S. Beddard for the adaptation of the structure of Photosystem I (PSI), by N. Krauss and P. Fromme.

References

1. Andersson, P. O., R. J. Cogdell, and T. Gillbro 1996. Femtosecond dynamics of carotenoid-to-bacteriochlorophyll *a* energy transfer in the light-harvesting antenna complexes from the purple bacterium *Chromatium purpuratum. Chem. Phys.* 210: 195–217.
2. Alden, R. G., E. Johnson, V. Nagarajan, W. W. Parson, C. J. Law, and R. J. Cogdell 1997. Galculations of spectroscopic properties of the LH2 bacteriochlorophyll–protein antenna complex from *Rhodopseudomonas acidophila. J. Phys. Chem. B* 101: 4667–4680.
3. Arnold, W., and H. I. Kohn 1934. The chlorophyll unit in photosynthesis. *J. Gen. Physiol.* 18: 109–112.
4. Bakker, J. G. C., R. van Grondelle, and W. T. F. den Hollander 1983. Trapping, loss and annihilation of excitations in a photosynthetic system. II. Experiments with the purple bacteria *Rhodospirillum rubrum* and *Rhoclopseuclomonas capsulatus. Biochim. Biophys. Acta* 725: 508–518.
5. Beekman, L. M. P., F. van Mourik, M. R. Jones, H. M. Visser, C. N. Hunter, and R. van Grondelle 1994. Trapping kinetics in mutants of the photosynthetic purple bacterium *Rhodobacter sphaeroides*: Influence of the charge separation rate and consequences for the rate-limiting step in the light-harvesting process. *Biochemistry* 33: 3143–3147.
6. Beekman, L. M. P., M. Steffen, I. H. M. van Stokkum, J. D. Olsen, C. N. Hunter, S. G. Boxer, and R. van Grondelle 1997. Characterization of the light harvesting antennas of photosynthetic purple bacteria by Stark spectroscopy: 1. LH1 antenna complex and the B820 subunit from *Rhodospirillum rubrum. J. Phys. Chem. B* 101: 7284–7292.
7. Beekman, L. M. P., R. W. Visschers, R. Monshouwer, M. Heer-Dawson, T. A. Mattioli, P. McGlynn, C. N. Hunter, B. Robert, I. H. M. van Stokkum, R. van Grondelle, and M. R. Jones 1995. Time-resolved and steady-state spectroscopic analysis of membrane-bound reaction centers from *Rhodobacter sphaeroides*: comparisons with detergent-solubilised complexes. *Biochemistry* 34: 14 712–14 721.
8. Bittner, T., K. D. Irrgang, G. Renger, and M. R. Wasielewski 1994. Ultrafast excitation energy transfer and exciton–exciton annihilation processes in isolated light harvesting complexes of photosystem II (LHCII) from spinach. *J. Phys. Chem.* 98: 11 821–11 826.
9. Bradforth, S. E., R. Jiminez, F. van Mourik, R. van Grondelle, and G. R. Fleming 1995. Excitation transfer in the core light-harvesting complex (LH-1) of *Rhodobacter sphaeroides*: an ultrafast fluorescence depolarization and annihilation study. *J. Phys. Chem.* 99: 16 179–16 191.
10. Brejc, K., R. Ficner, R. Huber, and S. Steinbacher 1995. Isolation, crystallization, crystal structure analysis and refinement of allophycocyanin from the cyanobacterium *Spirulina platensis* at 2.3 Å resolution. *J. Mol. Biol.* 249: 424–440.
11. Chachisvilis, M., O. Kühn, T. Pullerits, and V. Sundström 1997. Excitons in photosynthetic purple bacteria. Wavelike motion or incoherent hopping? *J. Phys. Chem. B* 101: 7275–7283.

12. Chachisvilis, M., T. Pullerits, M. R. Jones, C. N. Hunter, and V. Sundsbröm 1994. Vibrational dynamics in the light harvesting complexes of the photosynthetic bacterium *Rhodobacter sphaeroides*. *Chem. Phys. Lett.* 224: 345–351.

13. Connelly, J. P., M. G. Müller, R. Bassi, R. Croce, and A. R. Holzwarth 1997. Femtosecond transient absorption study of carotenoid to chlorophyll energy transfer in the light-harvesting complex II of photosystem II. *Biochemistry* 36: 281–287.

14. Connelly, J. P., M. G. Müller, M. Hucke, G. Gatzen, C. W. Mullineaux, A. V. Ruban, P. Horton, and A. R. Holzwarth 1997. Ultrafast spectroscopy of trimeric light-harvesting complex II from higher plants. *J. Phys. Chem. B* 101: 1902–1909.

15. Croce, R., G. Zucchelli, F. M. Garlaschi, R. Bassi, and R. Jennings 1996. Excited state equilibration in the photosystem I light-harvesting complex: P700 is almost isoenergetic with its antenna. *Biochemistry* 35: 8572–8579.

16. De Caro, C., R. W. Visschers, R. van Grondelle, and S. Volker 1994. Inter- and intraband energy transfer in LH2-antenna complexes of purple bacteria. A fluorescence line-narrowing and hole-burning study. *J. Phys. Chem.* 98: 10 584–10 590.

17. Du, M., X. Xie, L. Mets, and G. R. Fleming 1994. Direct observation of ultrafast energy-transfer processes in light-harvesting complex II. *J. Phys. Chem.* 98: 4736–4741.

18. Du, M., X. Xie, Y. Jia, L. Mets, and G. R. Fleming 1993. Direct observation of ultrafast energy transfer in PSI core antenna. *Chem. Phys. Lett.* 201: 535–542.

19. Duysens, L. N. M. 1952. Transfer of excitation energy in photosynthesis. Doctoral thesis. Utrecht University, The Netherlands.

20. Duysens, L. N. M. 1964. Photosynthesis. *Progr. Biophys.* 14: 1–104.

21. Eads, D. D., Jr. E. W. Gastner, R. S. Alberte, L. Mets, and G. R. Fleming 1990. Direct observation of energy transfer in a photosynthetic membrane: chlorophyll *b* to chlorophyll *a* transfer in LHCII. *J. Phys. Chem.* 93: 8271–8275.

22. Emerson, R., and W. Arnold 1932. A separation of the reactions in photosynthesis by means of intermittent light. *J. Gen. Physiol.* 15: 391–420.

23. Emerson, R., and W. Arnold 1932. The photochemical reaction in photosynthesis. *J. Gen. Physiol.* 16: 191–205.

24. Fenna, R. E., and B. W. Matthews 1975. Chlorophyll arrangement in a bacteriochlorophyll protein from *Chlorobium limicola*. *Nature* 258: 573–577.

25. Fleming, G. R., and R. van Grondelle 1994. The primary steps of photosynthesis. *Phys. Today* (February) 48–55.

26. Förster, Th. 1965. Delocalized excitation and excitation transfer. In *Modern Quantum Chemistry*. O. Sinanoglu, editor. Academic Press, New York, pp. 93–137.

27. Fowler, G. J. S., Sockalingum G. D., B. Robert, and C. N. Hunter 1994. Blue shifts in bacteriochlorophyll absorbance correlate with changed hydrogen bonding patterns in light-harvesting LH2 mutants of *Rhodobacter sphaeroides* with alterations at αTyr44 and 45. *Biochem. J.* 299: 695–700.

28. Fowler, G. J. S., R. W. Visschers, G. G. Grief, R. van Grondelle, and C. N. Hunter 1992. Genetically modified photosynthetic antenna complexes with blue-shifted absorbance bands. *Nature* 355: 848–850.

29. Freer, A. A., S. Prince, K. Sauer, M. Z. Papiz, A. M. Hawthornthwaite-Lawless, G. McDennott, R. J. Cogdell, and N. W. Isaacs 1996. Pigment–pigment interaction and energy transfer in the antenna complex of the photosynthetic bacterium *Rps. acidophila*. *Structure* 4: 449–462.

30. Freiberg, A., J. P. Allen, J. C. Williams, and N. W. Woodbury 1996. Energy trapping and detrapping by wild type and mutant reaction centers of purple nonsulfur bacteria. *Photosynth. Res.* 48: 309–319.

31. Freiberg, A., V. I. Godik, T. Pullerits, and K. Timpmann 1988. Directed picosecond excitation transport in purple photosynthetic bacteria. *Chem. Phys.* 128: 227–235.

32. Giuffra, E., D. Cugini, R. Croce, and R. Bassi 1996. Reconstitution and pigment-binding properties of recombinant CP29. *Eur. J. Biochem.* 238: 112–120.

33. Gobets, B., H. van Amerongen, R. Monshouwer, J. Kruip, M. Rogner, R. van Grondelle, and J. P. Dekker 1994. Polarized site-selection spectroscopy of isolated photosystem 1 particles. *Biochim. Biophys. Acta* 1188: 75–85.

34. Gradinaru, C. C., A. A. Pascal, F. van Mourik, B. Robert, P. Horton, R. van Grondelle, and H. van Amerongen 1998. Ultrafast evolution of the excited states in the minor chlorophyll *a/b* complex CP29 from green plants studied by energy selective pump–probe spectroscopy. *Biochemistry*, 37: 1143–1149.

35. Hastings, G., S. Hoshina, A. N. Webber, and R. E. Blankenship. 1995. Universality of energy and electron transfer processes in photosystem I. *Biochemistry* 34: 15 512–15 522.

36. Hess, S., E. Akesson, R. J. Cogdell, T. Pullerits, and V. Sundström 1995. Energy transfer in spectrally inhomogeneous light-harvesting pigment–protein complexes of purple bacteria. *Biophys. J.* 69: 2211–2225.

37. Hess, S., M. Chachisvilis, K. Timpmann, M. R. Jones, G. J. C. Fowler, C. N. Hunter, and V. Sundström 1995. Temporally and spectrally resolved subpicosecond energy transfer within LH2 and from LH2 to LH1 in photosynthetic purple bacteria. *Proc. Natl. Acad. Sci. USA* 92: 12 333–12 337.

38. Hoffmann, E., P. M. Wrench, F. P. Sharples, R. G. Hiller, W. Welte, and K. Diederichs 1996. Structural basis of light-harvesting by carotenoids: peridinin–chlorophyll–protein from *Amphinidium carterae*. *Science* 272: 1788–1791.

39. Holzwarth, A. R., G. Schatz, H. Brock, and E. Bittersmann 1993. Energy transfer and charge separation kinetics in photosystem 1. Part 1: Picosecond transient absorption and fluorescence study of cyanobacterial photosystem I particles. *Biophys. J.* 64: 1813–1822.

40. Hunter, C. N. 1995. Genetic manipulation of the antenna complexes of purple bacteria. In *Anoxygenic Photosynthetic Bacteria*. R. E. Blankenship, M. T. Madigan, and C. E. Bauer, editors. Kluwer Academic Publishers, Dordrecht, pp. 473–501.

41. Jean, J. M., and G. R. Fleming 1995. Competition between energy and phase relaxation in electronic curve crossing processes. *J. Chem. Phys.* 103: 2092–2101.

42. Jimenez, R., F. van Mourik, and G. R. Fleming 1997. Three pulse echo peak shift measurements on LH1 and LH2 complexes of *Rhodobacter sphaeroides*: a nonlinear spectroscopic probe of energy transfer. *J. Phys. Chem. B* 101: 7350–7359.

43. Jiminez, R., S. N. Dikshit, S. E. Bradforth, and G. R. Fleming 1996. Electronic excitation transfer in the LH2 complex of *Rhodobacter sphaeroides*. *J. Phys. Chem.* 100: 6825–6834.

44. Joo, T., Y. Jia, J.-Y. Yu, D. M. Jonas, and G. R. Fleming 1996. Dynamics in isolated bacterial light harvesting antenna (LH2) of *Rhodobacter sphaeroides* at room temperature. *J. Phys. Chem.* 100: 2399–2409.

45. Karrasch, S., P. A. Bullough, and R. Ghosh 1995. The 8.5 Å projection map of the light-harvesting complex I from *Rhodospirillum rubrum* reveals a ring composed of 16 subunits. *EMBO J.* 14: 631–638.

46. Kleima, F. J., C. C. Gradinaru, F. Calkoen, I. H. M. van Stokkum, R. van Grondelle, and H. van Amerongen 1997. Energy transfer in LHCII monomers at 77 K studied by sub-picosecond transient absorption spectroscopy. *Biochemistry*, 36: 15262–15268.

47. Koepke, J., X. Hu, C. Muenke, K. Schulten, and H. Michel 1996. The crystal structure of the light-harvesting complex II (B800–850) from *Rhodospirillum molischianum*. *Structure* 4: 581–597.

48. Koolhaas, M. H. C., G. van der Zwan, R. N. Frese, and R. van Grondelle 1997. The red shift of the zero-crossing in the CD spectra of the LH2 antenna complex of *Rhodopseudomonas acidophila*; a structure based study. *J. Phys. Chem. B* 101: 7262–7270.

49. Kramer, H. J. M., R. van Grondelle, C. N. Hunter, W. H. J. Westerhuis, and J. Amesz 1984. Pigment organization of the B800–850 antenna complex of *Rhodopseudomonas sphaeroides*. *Biochim. Biophys. Acta* 765: 156–165.

50. Krauss, N., W.-D. Schubert, O. Klukas, P. Fromme, H. T. Witt, and W. Saenger 1996. Photosystem I at 4 Å resolution represents the first structural model of a joint photosynthetic reaction centre and core antenna system. *Nat. Struct. Biol.* 3: 965–973.

51. Kühlbrandt, W., D. N. Wang, and Y. Fujiyoshi 1994. Atomic model of plant light-harvesting complex by electron crystallography. *Nature* 367: 614–621.

52. Kumazaki, S., I. Ikegami, and K. Yoshihara 1997. Excitation and electron transfer from selectively excited primary donor chlorophyll (P700) in a Photosystem I reaction center. *J. Phys. Chem. A* 101: 597–604.

53. Kumble, R., S. Palese, R. W. Visschers, P. L. Dutton, and R. M. Hochstrasser 1996. Ultrafast dynamics within the B820 subunit from the core (LH-1) antenna complex of *Rs. rubrum. Chem. Phys. Lett.* 261: 396–401.

54. Kwa, S. L. S., H. van Amerongen, S. Lin, J. P. Dekker, R. van Grondelle, and W. S. Struve 1992. Ultrafast energy transfer in LHC-II trimers from the Chl *a/b* light-harvesting antenna of photosystem II. *Biochim. Biophys. Acta* 1102: 202–212.

55. Lalonde, D. E., J. D. Petke, and G. M. Maggiora 1989. Evaluation of approximations in molecular exciton theory. 2. Applications to oligomeric systems of interest in photosynthesis. *J. Phys. Chem.* 93: 608–614.

56. Leegwater, J. A. 1996. Coherent versus incoherent energy transfer and trapping in photosynthetic antenna complexes. *J. Phys. Chem.* 100: 14 403–14 409.

57. Leupold, D., H. Stiel, K. Teuchner, F. Nowak, W. Sandner, B. Ucker, and H. Scheer 1996. Size enhancement of transition dipoles to one and two-exciton bands in a photosynthetic antenna. *Phys. Rev. Lett.* 77: 4675–4678.

58. Ma, Y.-Z., R. J. Cogdell, and T. Gillbro 1997. Energy transfer and exciton annihilation in the B800–850 antenna complex of the photosynthetic bacterium *Rhodopseudomonas acidophila* (strain 10050). A transient femtosecond absorption study. *J. Phys. Chem. B* 101: 1087–1095.

59. McDermott, G., S. M. Prince, A. A. Freer, A. M. Hawthornthwaite-Lawless, M. Z. Papiz, R. J. Cogdell, and N. W. Isaacs 1995. Crystal structure of an integral membrane light-harvesting complex from photosynthetic bacteria. *Nature* 374: 517–521.

60. Meier, T., V. Chernyak, and S. Mukamel 1997. Multiple exciton coherence sizes in photosynthetic antenna complexes viewed by pump–probe spectroscopy. *J. Phys. Chem. B* 101: 7332–7342.

61. Miller, J. F., S. B. Hinchigeri, P. S. Parkes-Loach, P. M. Callahan, J. R. Sprinkle, J. R. Riccobono, and P. A. Loach 1987. Isolation and characterization of a subunit form of the light-harvesting complex of *Rhodospirillum rubrum*. *Biochemistry* 26: 5055–5062.

62. Melkozernov, A. N., H. Su, S. Lin, S. Bingham, A. N. Webber, and R. E. Blankenship 1997. Specific mutation near the primary donor in photosystem 1 from *Chlamydamonas reinhardtii* alters the trapping time and spectroscopic properties of P700. *Biochemistry* 36: 2898–2907.

63. Monshouwer, R., and R. van Grondelle 1996. Excitations and excitons in bacterial light-harvesting complexes. *Biochim. Biophys. Acta* 1275: 70–75.

64. Monshouwer, R., M. Abrahamsson, F. van Mourik, and R. van Grondelle 1997. Superradiance and exciton delocalisation in bacterial photosynthetic light-harvesting systems. *J. Phys. Chem. B* 101: 7241–7248.

65. Monshouwer, R., I. Ortiz de Zarate, F. van Mourik, and R. van Grondelle 1995. Low-intensity pump–probe spectroscopy on the B800 to B850 transfer in the light harvesting 2 complex of *Rhodobacter sphaeroides*. *Chem. Phys. Lett.* 246: 341–346.

66. Monshouwer, R., R. W. Visschers, F. van Mourik, A. Freiberg, and R. van Grondelle 1995. Low-temperature absorption and site-selected fluorescence of the light-harvesting antenna of *Rhodopseudomonas viridis*. Evidence for heterogeneity. *Biochim. Biophys. Acta* 1229: 373–380.

67. Nagarajan, V., and W. W. Parson 1997. Excitation energy transfer between the B850 and B875 antenna complexes of *Rhodobacter sphaeroides*. *Biochemistry* 36: 2300–2306.

68. Nagarajan, V., R. G. Alden, J. C. Williams, and W. W. Parson 1996. Ultrafast exciton relaxation in the B850 antenna complex of *Rhodobacter sphaeroides*. *Proc. Natl Acad. Sci., USA* 93: 13 774–13 779.

69. Novoderezhkin, V. I., and A. P. Razjivin 1995. Exciton dynamics in circular aggregates: application to antenna of photosynthetic purple bacteria. *Biophys. J.* 68: 1089–1100.

70. Nussberger, S., J. P. Dekker, W. Kühlbrandt, B. M. van Bolbuis, R. van Grondelle, and H. van Amerongen 1994. Spectroscopic characterization of three different monomeric forms of the main chlorophyll *a/b* binding protein from chloroplast membranes. *Biochemistry* 33: 14 775–14 783.

71. Papiz, M. Z., S. M. Prince, A. M. Hawthornthwaite-Lawless, G. McDermott, A. A. Freer, N. W. Isaacs, and R. J. Cogdell 1996. A model for the photosynthetic apparatus of purple bacteria. *Trends Plant Sci.* 1: 198–206.

72. Pålsson, L. O., M. D. Spangfort, V. Gulbinas, and T. Gillbro 1994. Ultrafast chlorophyll *b* – chlorophyll *a* excitation energy transfer in the isolated light harvesting complex, LHCII, of green plants. Implications for the organization of chlorophylls. *FEBS Letts* 339: 134–138.

73. Pålsson, L. O., J. P. Dekker, E. Schlodder, R. Monshouwer, and R. van Grondelle 1996. Polarized site-selective fluorescence spectroscopy of the long-wavelength emitting chlorophylls in isolated Photosystem I particles of *Synechococcus elongatus*. *Photosynth. Res.* 48: 239–246.

74. Pearlstein, R. M. 1982. Exciton migration and trapping in photosynthesis. *Photochem. Photobiol.* 35: 835–844.

75. Peterman, E. J. G., R. Monshouwer, I. H. M. van Stokkum, R. van Grondelle, and H. van Amerongen 1997. Ultrafast single excitation transfer from carotenoids to chlorophylls via different pathways in light-harvesting complex II of higher plants. *Chem. Phys. Lett.* 264: 279–284.

76. Peterman, E. J. G., C. C. Gradinaru, F. Calkoen, J. C. Borst, R. van Grondelle, and H. van Amerongen 1997. The xanthophylls in light-harvesting complex II of higher plants: light harvesting and triplet quenching. *Biochemistry* 36: 12 208–12 215.

77. Pullerits, T., and V. Sundström 1996. Photosynthetic light-harvesting pigment-proteins: toward understanding how and why. *Acc. Chem. Res.* 29: 381–389.

78. Pullerits, T., M. Chachisvilis, and V. Sundström 1996. Exciton delocalization length in the B850 antenna of *Rhodobacter sphaeroides*. *J. Phys. Chem.* 100: 10 787–10 792.

79. Reddy, N. R. S., R. Picorel, and G. J. Small 1992. B896 and B870 components of the *Rhodobacter sphaeroides* antenna: a hole-burning study. *J. Phys. Chem.* 96: 6458–6464.

80. Ricci, M., S. E. Bradforth, R. Jiminez, and G. R. Fleming 1996. Internal conversion and energy transfer dynamics of spheroidene in solution and in the LH-1 and LH-2 light-harvesting complexes. *Chem. Phys. Lett.* 259: 381–390.

81. Sauer, K., R. J. Cogdell, S. M. Prince, A. Freer, N. W. Isaacs, and H. Scheer 1996. Structure-based calculations of the optical spectra of the LH2 bacteriochlorophyll–protein complex from *Rhodopseudomonas acidophila*. *Photochem. Photobiol.* 64: 564–576.

82. Shreve, A. P., J. K. Trautman, H. A. Frank, T. G. Owens, and A. C. Albrecht 1991. Femtosecond energy-transfer processes in the B800–850 light-harvesting complex of *Rhodobacter sphaeroides* 2.4.1. *Biochim. Biophys. Acta* 1058: 280–288.

83. Somsen, O. J. G. 1995. Excitonic interaction in photosynthesis. Migration and spectroscopy. Ph.D. thesis. Free University, Amsterdam.

84. Somsen, O. J. G., L. Valkunas, and R. van Grondelle 1996. A perturbed two-level model for exciton trapping in small photosynthetic systems. *Biophys. J.* 70: 669–683.

85. Somsen, O. J. G., F. van Mourik, R. van Grondelle, and L. Valkunas 1994. Energy migration and trapping in a spectrally and spatially inhomogeneous light-harvesting antenna. *Biophys. J.* 66: 1580–1596.

86. Sturgis, J. N., and B. Robert 1996. The role of chromophore coupling in tuning the spectral properties of the peripheral light-harvesting protein of purple bacteria. *Photosynth. Res.* 50: 5–10.

87. Sundström, V., R. van Grondelle, H. Bergström, E. Åkesson, and T. Gillbro 1986. Excitation-energy transport in the bacteriochlorophyll antenna systems of *Rhodospirillum rubrum* and *Rhodobacter sphaeroides* studied by low-intensity picosecond absorption spectroscopy. *Biochim. Biophys. Acta* 851: 431–446.

88. Timpmann, K., A. Freiberg, and V. Sundström 1995. Energy trapping and detrapping in photosynthetic bacteria *Rhodopseudomonas viridis*. *Chem. Phys.* 194: 275–283.

89. Trautman, J. K., A. P. Shreve, C. A. Violette, H. A. Frank, T. G. Owens, and A. C. Albrecht 1990. Femtosecond dynamics of energy transfer in B800–850 light-harvesting complexes of *Rhodobacter sphaeroides*. *Proc. Natl Acad. Sci., USA* 87: 215–219.

90. Trinkunas, G., and A. R. Holzwarth 1996. Kinetic modeling of exciton migration in photosynthetic systems: 3. Application of genetic algorithms to simulations of excitation dynamics in three-dimensional Photosystem I core antenna/reaction center complexes. *Biophys. J.* 71: 351–364.

91. Trinkunas, G., and L. Valkunas 1989. Exciton–exciton annihilation in picosecond spectroscopy of molecular systems. *Exp. Tech. Physik.* 37: 455–458.

92. Trinkunas, G., J. P. Connelly, M. G. Muller, L. Valkunas, and A. R. Holzwarth. 1997. A model for the excitation dynamics in the light-harvesting complex II from higher plants. *J. Phys. Chem. B* 101: 7313–7320.

93. Valkunas, L. 1986. Influence of structural heterogeneity on energy migration in photosynthesis. *Laser Chem.* 6: 253–267.

94. Valkunas, L., V. Liuolia, J. P. Dekker, and R. van Grondelle 1995. Description of energy migration and trapping in photosystem I by a model with two distance scaling parameters. *Photosynth. Res.* 43: 149–154.

95. Van Amerongen, H., S. L. S. Kwa, B. M. van Bolhuis, and R. van Grondelle 1994. Polarized fluorescence and absorption of macroscopically aligned light-harvesting complex II. *Biophys. J.* 67: 837–847.

96. Van der Laan, H., Th. Schmidt, R. W. Visschers, K. J. Visscher, R. van Grondelle, and S. Völker 1990. Energy transfer in the B800–850 antenna complex of the purple bacterium *Rhodobacter sphaeroides:* a study by spectral hole-burning. *Chem. Phys. Lett.* 170: 231–238.

97. Van Grondelle, R., H. Y. M. Kramer, and C. P. Rijgersberg 1982. Energy transfer in the B800–850-carotenoid light-harvesting complex of various mutants of *Rhodopseudomonas sphaeroides* and of *Rhodopseudomonas capsulatus. Biochim. Biophys. Acta* 682: 208–215.

98. Van Grondelle, R., H. Bergström, V. Sundström, and T. Gillbro 1987. Energy transfer within the bacteriochlorophyll antenna of purple bacteria at 77 K studied by picosecond absorption recovery. *Biochim. Biophys. Acta* 894: 313–326.

99. Van Grondelle, R., J. P. Dekker, T. Gillbro, and V. Sundström 1994. Energy transfer and trapping in photosynthesis. *Biochim. Biophys Acta* 1187: 1–65.

100. Van Mourik, F., R. W. Visschers, and R. van Grondelle 1992. Energy transfer and aggregate size effects in the inhomogeneously broadened core light-harvesting complex of *Rhodobacter sphaeroides. Chem. Phys. Lett.* 193: 1–7.

101. Visschers, R. W., F. van Mourik, R. Monshouwer, and R. van Grondelle 1993. Inhomogeneous spectral broadening of the B820 subunit form of LH1. *Biochim. Biophys. Acta* 1141: 238–244.

102. Visschers, R. W., M. C. Chang, F. van Mourik, P. S. Parkes-Loach, B. A. Heller, P. A. Loach, and R. van Grondelle 1991. Fluorescence polarization and low-temperature absorption spectroscopy of a subunit form of light-harvesting complex I from purple photosynthetic bacteria. *Biochemistry* 30: 5734–5742.

103. Visser, H. M., O. J. G. Somsen, F. van Mourik, and R. van Grondelle 1996. Excited-state energy equilibration via sub-picosecond energy transfer within the inhomogeneously broadened light-harvesting antenna of the LH-1 only *Rhodobacter sphaeroides* mutant M2192 at room temperature and 4.2 K. *J. Phys. Chem.* 100: 18 859–18 867.

104. Visser, H. M., F. J. Kleima, I. H. M. van Stokkum, R. van Grondelle, and H. van Amerongen. 1996. Probing the many energy-transfer processes in the photosynthetic light-harvesting complex II at 77K using energy-selective sub-picosecond transient absorption spectroscopy. *Chem. Phys.* 210: 297–312.

105. Visser, H. M., O. J. G. Somsen, F. van Mourik, S. Lin, I. H. M. van Stokkum, and R. van Grondelle 1995. Direct observation of sub-picosecond equilibration of excitation energy in the light-harvesting antenna of *Rhodospirillum rubrum. Biophys. J.* 69: 1083–1099.

106. White, N. T. H., G. S. Beddard, J. R. G. Thorne, T. M. Feehan, T. E. Keyes, and P. Heathcote 1996. Primary charge separation and energy transfer in the photosystem I reaction center of higher plants. *J. Phys. Chem.* 100: 12 086–12 099.

107. Wu, H. M., N. R. S. Reddy, R. J. Cogdell, C. Muenke, H. Michel, and G. J. Small 1996. A comparison of the LH2 antenna complex of three purple bacteria by hole-burning and absorption spectroscopies. *Mol. Cryst. Liq. Cryst.* 291: 163–173.

108. Wu, H. M., Savikhin S., N. R. S. Reddy, R. Jankowiak, R. J. Cogdell, W. S. Struve, and G. J. Small 1996. Femtosecond and hole-burning studies of B800's excitation energy relaxation dynamics in the LH2 antenna complex of *Rhodopseudomonas acidophila* (strain 10050). *J. Phys. Chem.* 100: 12 022–12 033.

109. Yu, J.-Y., Y. Nagasawa, R. van Grondelle, and G. R. Fleming 1997. Three pulse echo peak shift measurements on B820 subunit of LH1 of *Rhodospirillum rubrum. Chem. Phys. Lett.*, 280: 404–409.

110. Zhang, F. G., R. van Grondelle, and V. Sundström 1992. Pathways of energy flow through the light-harvesting antenna of the photosynthetic purple bacterium *Rhodobacter sphaeroides. Biophys. J.* 61: 911–920.
111. Zuber, H., and R. J. Cogdell 1995. Structure and organization of purple bacterial antenna complexes. In *Anoxygenic Photosynthetic Bacteria*. R. E. Blankenship, M. T. Madigan, and C. E. Bauer, editors. Kluwer Academic Publishers, Dordrecht, pp. 315–348.

10

The Fenna–Matthews–Olson protein: a strongly coupled photosynthetic antenna

Sergei Savikhin, Daniel R. Buck, and Walter S. Struve
Ames Laboratory–USDOE and Iowa State University, USA

10.1 INTRODUCTION

The Fenna–Matthews–Olson complex is a BChl (bacteriochlorophyll) *a –* protein antenna found in green sulfur bacteria such as *Prosthecochloris aestuarii, Chlorobium tepidum*, and *Chlorobium limicola* [4]. Its three-dimensional X-ray structure was the first determined for any photosynthetic pigment–protein complex [13, 30, 31]. The FMO (Fenna–Matthews–Olson) pigment organization resembles that of many other photosynthetic antennas (e.g. the LHC-II peripheral antenna of photosystem II of higher green plants [22], the LH2 antenna from purple bacteria [32], and the photosystem I core antenna in green plants and cyanobacteria [21], in that the nearest-neighbor pigment separations are on the order of 10 Å, leading to strong resonance couplings between pigments. The FMO protein is also typical in that its pigments are spatially grouped into identical clusters. It has long been postulated [45] that the pigments within such clusters are strongly coupled, and that Förster-type energy transfers occur between clusters (the "pebble mosaic" model). However, the FMO protein organization is unique among antennas with known structure: its secondary structure is dominated by β-sheets, whereas the pigments in other antennas are coordinated largely to α-helices. It is considerably less evolved than the LHC-II Chl a/b antenna of spinach, since it uses 50 protein residues to bind each pigment (versus 15 for LHC-II [22]).

Resonance Energy Transfer. Edited by David L. Andrews and Andrey A. Demidov
© 1999 John Wiley & Sons Ltd

The early revelation of the FMO protein structure led to expectations that this system would play a pivotal role in illuminating antenna structure–function relationships. This situation has not yet materialized; studies of the the cyano-bacterial C-phycocyanin and allophycocyanin antennas and the LH2 antennas from the purple bacteria *Rhodobacter sphaeroides* and *Rhodopseudomonas acid-ophila* have proven far more fruitful to date [60]. This is because primary energy-transfer processes in the FMO protein are considerably faster than in C-phycocyanin or allophycocyanin, and because the FMO subunit exhibits a larger number of inequivalent pigment sites (seven, versus two and three in the LH2 and C-phycocyanin antennas respectively). The accessibility since 1993 of pump–probe laser techniques with 40–80 fs resolution has alleviated the first problem; the second still poses a fundamental challenge to spectroscopists and to theorists attempting to simulate FMO electronic structure (see Section 10.3).

The principal light-harvesting antenna in green photosynthetic bacteria is the chlorosome, which is appressed to the inner cytoplasmic membrane [3, 34, 52]. Encased in a phospholipid/glycolipid envelope, the ellipsoidal chlorosome is largely built up of BChl $c/d/e$ aggregates, organized into rod-like elements containing ~ 1000 pigments (depending on growth conditions) and absorbing at ~ 740 nm (depending on species). Electronic excitations in the BChl $c/d/e$ antenna travel through a longer-wavelength, crystalline BChl a – protein base-plate antenna on the way to the reaction centers [34]. In green thermophilic bacteria (e.g. *Chloroflexus aurantiacus*), the baseplate antenna is believed to be one that has been spectroscopically characterized as B795 (so named because the band maximum of its Q_y absorption feature occurs at 795 nm); its spatial organization is unknown [12]. In green sulfur bacteria (e.g. *Pc. aestuarii, Cb. tepidum, Cb. limicola*), the baseplate antenna is believed to be the FMO protein [30], whose room-temperature Q_y absorption spectrum peaks at 809 nm for *Pc. aestuarii* (Section 10.2). Aside from structural similarities in their oligmeric BChl $c/d/e$ light-harvesting antennas (which are unique in photosynthesis because pigment–pigment interactions rather than proteins appear to control their local BChl organization), the green thermophilic and sulfur bacteria are not closely related [3].

Olson and Romano [36] demonstrated that some 5% of the bacteriochlor-ophyll in green bacteria is present as a water-soluble BChl a–protein complex. The fluorescence properties of this 809 nm-absorbing protein from *Pc. aestuarii* suggested that it conveys electronic energy from the ~ 740 nm BChl light-harvesting antenna to the P840 reaction centers [54–56]. Olson succeeded in crystallizing this protein as hexagonal rods, yielding the first chlorophyll pro-tein available for X-ray studies. The water-soluble FMO protein from *Pc. aestuarii* (Fig. 10.1) is a trimer of identical 50 kDa subunits [30, 58]. The trimers are packed in the $P6_3$ space group, with cell dimensions $a = b = 111.9$ Å, $c = 98.6$ Å (each unit cell contains two trimers). Each subunit within a trimer contains a β-sheet with 16 strands, folded like a taco shell and enclosing

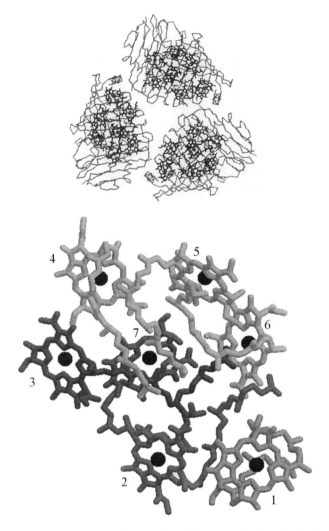

Figure 10.1   The structure of FMO trimer from *Pc. aestuarii*, showing folded protein β-sheets enclosing 21 BChl *a* pigments (top), and a spatial arrangement of seven BChl *a* pigments in one subunit (bottom). Ribbons (top) are protein β-sheet strands; dark spheres are Mg atoms. Structures are plotted from Brookhaven PDB coordinates of Tronrud *et al.* [58]

seven pentacoordinate BChl *a* pigments with well-defined positions and orientations. An individual FMO trimer in the crystal structure exhibits perfect C$_3$ symmetry. Six of the seven inequivalent BChls are liganded by protein residues in the interior: BChls 1, 3, 4, 6 and 7 are coordinated to His residues, while

BChl 5 is coordinated to a Leu side chain. (We follow the BChl numbering scheme of Matthews and Fenna [30], as shown in Fig. 10.1. The BChls rotationally equivalent to BChl 1 – i.e. belonging to neighboring protein subunits of the trimer – are termed BChls 8 and 15, etc.). BChl 2 is liganded to a solvent molecule. Nearest-neighbor Mg–Mg distances are 11–15 Å; there are edge-to-edge contacts between BChls, but no close sandwich–dimer interactions analogous to those in purple bacterial reaction centers. The phytyl chains are packed in the central space between BChls to form a hydrophobic core.

The sequences of the FMO proteins from *Pc. aestuarii* and the closely related species *Cb. tepidum* [63] are 78% homologous, and all pigment-binding residues are conserved between these species [10, 11]. The X-ray structure of the FMO protein from *Cb. tepidum* has very recently been determined [23]. The 80 unconserved residues in *Cb. tepidum* are either on the protein periphery or in the hydrophobic interior, but do not affect interactions with the pigments. The positions and orientations of the macrocycles in *Pc. aestuarii* and *Cb. tepidum* are nearly super-posable; the BChl separations are essentially the same. Deviations from congruence between BChl rings in the respective species are dominated by differences in nonplanarity (e.g. out-of-plane Mg positions). These deviations apparently suffice to cause noticable changes in the distribution of exciton band intensities in the Q_y absorption spectrum.

10.2   STEADY STATE SPECTROSCOPY

Prior to the X-ray structure determination, Philipson and Sauer [41] (hereafter PS) had already inferred that the BChl pigment separations in the FMO protein from *Pc. aestuarii* were 12–15 Å, on the basis of exciton splittings (the separations between the lowest- and highest-energy components) in its low-temperature Q_y absorption spectra. (At the time, the protein was believed to originate from the green bacterium *Chloropseudomonas ethylica*; its parentage is now known to be *Pc. aestuarii*.) While essentially featureless at 300 K, the absorption spectrum resolved into at least five bands at 77 K, with maxima located at 792, 804, 813, 817, and 824 nm. The corresponding circular dichroism (CD) bands were located at 787, 800, 812, 814 and 823 nm. The multiple components in the CD spectrum were attributed to exciton interactions; it was believed at the time that the BChl *a* protein was tetrameric, with five pigments per subunit. The absorption bands exhibited full widths at half maximum (fwhm) ranging from 210 cm^{-1} (for the 792 nm band) to 120 cm^{-1} (824 nm). Olson, Ke and Thompson [37] (hereafter OKT) later analysed the Q_y absorption and CD spectra at 2 K. While the OKT absorption spectrum somewhat resembles the PS spectrum (with similar band positions and fwhms), the CD spectra are significantly different. The PS and OKT absorption and spectra have since been used as criteria for exciton simulations of FMO electronic spectra. The longest-wavelength absorption bands appear at similar positions for FMO trimers from *Pc. aestuarii* and *Cb. tepidum* (near 825, 815, and 805 nm), but with different relative intensities; the 814 nm band exhibits the largest oscillator strength in *Pc. aestuarii*, while the 805 nm band is the most intense one in *Cb. tepidum*.

Small and co-workers investigated the spectral hole-burning of FMO trimers from *Pc. aestuarii* [18] and *Cb. tepidum* [44] at 4.2 K. The earlier hole-burning study revealed the presence of at least eight exciton components in the Q_y spectrum (including two at 824 and 827 nm, separated by 40 cm^{-1}, that are responsible for the long-wavelength band observed in steady state absorption spectra.) Since each subunit contains only seven pigments, this number implied that resonance couplings between pigments belonging to different subunits are an important determinant of FMO exciton structure (Section 10.3). The hole widths of higher exciton components were ~ 50 cm^{-1}, corresponding to level relaxation times of ~ 100 fs. (This lifetime broadening masked the possible presence of splittings in higher exciton components arising from interactions between subunit.) The inhomogeneous broadening in the lowest-energy exciton components (824 and 827 nm) was ~ 20 cm^{-1}. While this diagonal energy disorder is small (less than observed in any other pigment–protein complex to date), it has major implications for the symmetries of exciton states prepared in real FMO trimers (see below). The inhomogeneous broadening observed in the 827 nm band of FMO trimers from *Cb. tepidum* was ~ 70 cm^{-1} [44].

The triplet–singlet (T–S) absorption difference spectrum of FMO trimers from *Pc. aestuarii* also gives strong evidence for the excitonic nature of their excited states, because all of the absorption bands are found to be sensitive to triplet state formation [61]. The BChl *a* triplet state forms on one of the pigments with the lowest site energy (currently believed to be pigments 7, 14 and 21 in *Pc. aestuarii*). Hence, the effective T–S spectrum is that for the ground-state trimer, minus the absorption spectrum arising from the exciton states that stem from resonance couplings among the remaining 20 pigments. Consequently, the difference spectrum is not localized to an absorption band for an uncoupled BChl with the lowest site energy, but exhibits positive and negative peaks spanning from ~ 795 to 925 nm.

10.3 FMO EXCITON SIMULATIONS

The first reports of FMO X-ray structures were soon followed by attempts to simulate the observed absorption and CD spectra, using resonance interactions computed from the known BChl positions and orientations. The total electronic Hamiltonian for N coupled chromophores in a pigment–protein complex is

$$\hat{H} = \sum_p^N \hat{H}_p + \sum_{p<q}^N V_{pq}, \tag{10.1}$$

where $\hat{H}_p$ is the electronic Hamiltonian for pigment p and V_{pq} represents the interaction between pigments p and q. As a first approximation, the N one-exciton states may be expanded in the basis of the states $|1^*23\ldots\rangle, |12^*3\ldots\rangle,$ $|123^*\ldots\rangle$, which localize Q_y excitation on pigments 1, 2, 3, etc. We use $|\chi_i^{(1)}\rangle$ to denote the one-exciton basis function that localizes the excitation on pigment i, i.e. $|\chi_1^{(1)}\rangle = |1^*23\ldots\rangle, |\chi_2^{(1)}\rangle = |12^*3\ldots\rangle$, etc. The one-exciton states $|\psi_i^{(1)}\rangle$ and their energies $E_i^{(1)}$ are obtained by diagonalizing the $N \times N$ Hamiltonian matrix with elements

$$H_{ij} = \langle\chi_i^{(1)}| \sum_{p=1}^N \hat{H}_p + \sum_{p<q}^N V_{pq}|\chi_j^{(1)}\rangle. \tag{10.2}$$

The diagonal elements of this matrix are $H_{ii} = E_i$, the Q_y excited-state energies of pigments i in the absence of resonance interactions. The off-diagonal elements $(i \neq j)$ are

$$H_{ij} = \langle\chi_i^{(1)}|V_{ij}|\chi_j^{(1)}\rangle \equiv V_{ij}, \tag{10.3}$$

because all other contributions in Eqn 10.2 vanish through orthogonality. For pigments i, j with separation large compared to the molecular size, and with excited states connected to the ground states by strongly allowed electric dipole transitions, V_{ij} is dominated by the dipole–dipole interaction. For pigments separated by ~ 10 Å (e.g. nearest neighbors in a subunit of an FMO trimer), contributions from higher-multipole interactions gain importance. The expansion coefficients c_j^i in the one-exciton states,

$$|\psi_i^{(1)}\rangle = \sum_{j=1}^N c_j^i|\chi_j^{(1)}\rangle \tag{10.4}$$

(which appear in the Hamiltonian matrix of eigenvectors), may be used to calculate the absorption coefficients $B_i^{(1)}$ for the absorption lines corresponding to transitions from the N-pigment ground state $|0\rangle = |123\ldots\rangle$ to the one-exciton states $\psi_i^{(1)}$,

$$B_i^{(1)} = \left|\sum_{j=1}^N c_j^i\langle 0|\mu|\chi_j^{(1)}\rangle\right|^2$$

$$= |c_1^i\hat{\mu}_1 + c_2^i\hat{\mu}_2 + \ldots + c_N^i\hat{\mu}_N|^2. \tag{10.5}$$

Here the unit vectors $\hat{\mu}_j \equiv \langle 0|\mu_j|\chi_j^{(1)}\rangle$ are the transition moment directions for the lowest electronic transitions on pigments j. The CD signal corresponding to the same one-exciton transition in this approximation is proportional to

$$CD_i = \sum_{j,j'=1}^{N} c_j^i c_{j'}^i (\hat{\mu}_j \times \hat{\mu}_{j'}) \bullet (\vec{R}_j - \vec{R}_{j'}) \qquad (10.6)$$

where $\vec{R}_j$ is the position of pigment j. Computation of the absorption and CD spectra thus requires knowledge of the N pigment diagonal energies E_i, the $N(N-1)/2$ interactions V_{ij}, the pigment positions $\vec{R}_j$ (Mg atom coordinates), and transition moment directions $\hat{\mu}_j$. The interactions V_{ij} have been computed to varying degrees of approximation from BChl transition monopoles, derived from Q_y wavefunctions obtained in semi-empirical SCF–MO–PPP calculations [9, 64]. These monopoles q_k are "transition charges" localized at the chromophore nuclei k; the resonance interaction between pigments i, j may be summed over all pairwise Coulomb energies between monopoles belonging to apposite pigments. Alternatively, the total transition dipole for pigment j may be summed over the point monopoles as $\mu_i = \sum q_k \vec{r}_k$, where $\vec{r}_k$ is the position of atom k; the resonance energy between pigments i, j may then be expressed as that for the interaction of two point dipoles located at the respective Mg atom sites. In the "extended dipole" method, each of the transition dipoles is represented as a pair of charges separated by a finite distance, and the interaction between two pigments is computed as the sum of four resulting Coulomb energies. The sets of BChl a transition monopoles evaluated by Weiss [64] and Chang [9] predict that the lowest electronic transition is directed largely along the Q_y-axis in the macrocycle plane (the ratio of its x- to y-components is less than 0.03 and 0.002 in the respective calculations). Hence, the transition moment directions may be taken as essentially passing through BChl nitrogen atoms N1 and N3 (or through the midpoints of carbons C2, C3 and C12, C13) in the atomic numbering system of Fenna, Ten Eyck, and Matthews [31]. The same system is used in the Brookhaven protein data base coordinates from the structure of the Tronrud *et al.* [58].

The seven independent diagonal energies E_i are not spectroscopically measurable, and there is no *ab initio* theory for predicting how the known ligands and protein environment determine them. Gudowska-Nowak *et al.* [16] have also shown that the Q_y energies are sensitive to BChl ring conformation (which varies significantly within this protein, according to [58]). The seven parameters E_i have therefore been freely varied to fit the PS and OKT absorption and CD spectra using Eqns 10.5 and 10.6. The earliest simulations assumed that interactions between BChls in different FMO subunits were insignificant, and thus limited the exciton calculations to the seven pigments in one subunit [40]. The BChl–BChl interactions for y-polarized transitions (evaluated with the point monopole method) were as large as 252 cm^{-1}. Pearlstein and Hemenger were unable to fit the PS absorption and CD spectra simultaneously under these constraints for any combination of diagonal energies, and proposed that the lowest BChl a transition may be Q_x-polarized instead. Using x-polarized transition moments with interactions evaluated by the extended dipole method, simultaneous fits could be obtained. This proposal did not gain wide acceptance (and the authors themselves cautioned that there was no precedent for such drastic protein effects on transition moment orientations). After the FMO protein X-ray structure was refined to 1.9 Å resolution [58], Pearlstein [38] recognized that several of the BChls are extremely close to the outer wall of the polypeptide. Hence, if aggregation occurs between FMO trimers in solution, causing van der Waals contact between the β-sheets of subunits belonging to different trimers, strong interactions can occur between BChls belonging to different subunits. By assuming a plausible mutual geometry between contacting subunits, he was able to obtain improved simulations of absorption and CD spectra [41, 66] using y-polarized transitions. However, while Whitten *et al.* [66] reported that FMO trimers exist as aggregates containing tens of trimers in solution, it now appears likely that they do not aggregate under these conditions [61].

The Johnson and Small hole-burning study [18] detected more than seven Q_y exciton components in trimers from *Pc. aestuarii*, indicating that the exciton states are delocalized over more than seven pigments. Following this, Pearlstein [39] and Lu and Pearlstein [27] were able to obtain good quality simulations of the OKT absorption and CD spectra by extending their exciton calculations to all 21 BChls in the trimer. The largest coupling between subunits (which occurs among the rotationally equivalent pigments 7, 14, and 21) is ~ 20 cm^{-1}. While this is dwarfed by the largest intra-subunit interaction (~ 190 cm^{-1} as evaluated by the point monopole method, [39]), it considerably affects the CD spectrum. Each of the seven haphazardly polarized exciton levels in one subunit becomes split into three closely spaced states in the trimer [39, 44]. In a trimer with perfect C_3 symmetry, one of these is nondegenerate and polarized along the trimer C_3 axis; the other two are degenerate and polarized in the normal plane. Hence, while the couplings between subunits cause relatively small perturbations to the absorption spectrum, their influence is magnified in anisotropic optical properties such as CD and pump–probe anisotropies (Section 10.4).

The 21-pigment exciton simulations of the OKT and PS spectra yield quite different views of FMO electronic structure [27]. In the OKT fits (which are considerably better than the PS fits, as shown in Fig. 10.2), BChl 7 has the lowest optimized diagonal energy, followed by BChl 6. The lowest exciton states, responsible for the bands at 824 and 827 nm, are thus dominated by

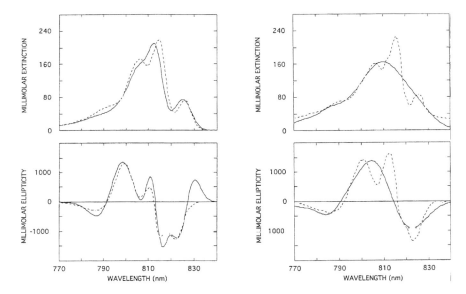

Figure 10.2 Lu and Pearlstein simulations [27] of absorption and CD spectra of Olson *et al.* [37] and Philipson and Sauer [41], left and right respectively. Solid curves are simulations; dashed curves are experimental spectra

excitations on BChls 6, 7 (and on the symmetry-equivalent BChls 13, 14 and 20, 21). In the PS fits, BChl 3 has the lowest site energy, and the lowest three levels are dominated by excitations on BChls 3 and 4. The optimized diagonal energies and trimer eigenvalues are shown in perspective for both the OKT and PS fits in Fig. 10.3. These examples illustrate the sensitivity of the optimized diagonal energies to the experimental spectra (the OKT and PS CD spectra are fairly dissimilar; cf. Fig. 10.2). They also remind us that our knowledge of BChl diagonal energies relies entirely on these simulations; we have no independent knowledge of their ordering or spacing. While BChl 7 has been cited as the lowest-energy pigment in several reviews [3, 60], there is still no empirical basis for assigning spectroscopic phenotypes (such as absorption or CD features) to specific (groups of) pigments identified in the electron density map.

More recently, Gülen [17] pointed out that CD rotational strengths are unusually sensitive to small changes in BChl organization, such as going from crystal to solution environments (cf. Eqn 10.6). She suggested T–S absorption difference spectra and LD (linear dichroism) as alternative spectral criteria; both of these are directly influenced by exciton structure. By simulating the latter spectra [61] along with the absorption spectrum for FMO trimers from *Pc. aestuarii*, Gülen concluded that BChl 6 has the lowest diagonal energy, at 816 nm.

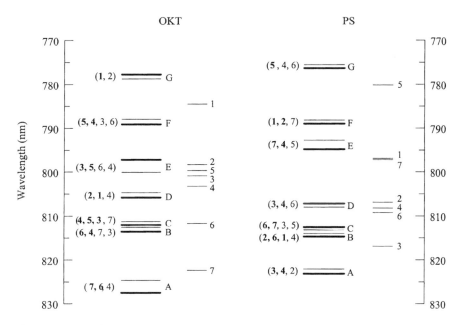

Figure 10.3 Exciton levels and diagonal energies (long and short bars, respectively) from the OKT and PS Hamiltonians for *Pc. aestuarii* trimers. Exciton level groups are labeled A, B, ..., G in order of energy; thick bars denote degenerate levels. Sets of BChl pigments containing > 75% of the excitation density in each level group are given by numbers in parentheses; pigments containing > 50% of the density given by boldface numbers. Diagonal energies are labeled with pigment numbers

In a new anisotropy study of FMO trimers from *Cb. tepidum*, Savikhin *et al.* [49] found evidence that the laser-excited states do not reflect the full C_3 symmetry of the trimer X-ray structure; they are better described as states localized to the seven BChls within a subunit. This is not surprising, since the ~ 70 cm^{-1} inhomogeneous broadening (arising from random diagonal energy disorder) of the lowest exciton levels in *Cb. tepidum* [44] is several times larger than the maximum couplings between subunits (~ 20cm^{-1}). Hence, the electronic states of typical FMO trimers are considerably distorted from C_3 symmetry. Excitation in the Q_y band system prepares a statistical distribution of states: a small minority of trimers are prepared in states with near-C_3 symmetry, but many more will have excitations that are essentially localized within a subunit. The effects of diagonal disorder on the absorption spectrum are minor, but they must be considered in the CD spectrum. Figure 10.4 shows simulations of FMO absorption and CD spectra, using the Pearlstein interactions [39] and the Lu and Pearlstein diagonal energies [27] from OKT fits (D. R. Buck, S. Savikhin, and W. S. Struve, unpublished work). The

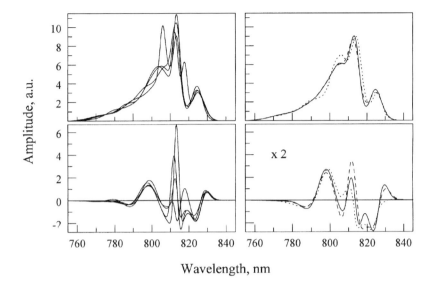

Figure 10.4 Representative absorption and CD spectral simulations for individual FMO trimers from *P. aestuarii* (left), and spectra averaged over 1000 trimers (right, solid curves). Each of the 21 BChl *a* input diagonal energies has an independent 80 cm^{-1} Gaussian noise component. Dashed spectra at right are computed without diagonal disorder; dotted curves are experimental spectra from Olson *et al.* [37]. The resonance couplings are from Pearlstein [39]; the nonrandom diagonal energy components are from from the Lu and Pearlstein fits [27] to OKT absorption and CD spectra

diagonal energies were augmented with 80 cm^{-1} fwhm Gaussian noise. Individual trimers show contrasting CD spectra, and the ensemble-averaged CD spectrum differs considerably from the spectrum computed without diagonal disorder. Diagonal disorder must clearly be considered in FMO trimers from *Cb. tepidum*. Its importance is less certain in *Pc. aestuarii*, where the inhomogeneous broadening is reportedly only ~ 20 cm^{-1} [18, 61].

In our view, exciton simulations for FMO trimers are in need of independent data on the BChl diagonal energies. Mutant FMO proteins lacking specific BChls would be informative; R. Blankenship and co-workers have recently attempted to mutate proteins from *Cb. tepidum* (private communication). The evaluation of transition charges for the calculation of resonance couplings needs to be updated, because (for example) the BChl *a* atomic coordinates for the semiempirical SCF–MO–PPP calculations (see [9]) were taken from crystal data on tetraphenylporphyrin. This molecule, which lacks the two dihydro features that distinguish the BChl *a* macrocycle from porphyrin, is not very similar to BChl *a*.

10.4   FMO PRIMARY PROCESSES

10.4.1   Energy transfer

Most of the spectral equilibration among FMO exciton levels is complete within < 1 ps at room temperature, and thus remained inscrutable until the arrival of Ti : sapphire lasers in photosynthesis laboratories during the past 3–4 years. Two of the earliest characterizations of energy transfers in FMO trimers from *Pc. aestuarii* were achieved by Causgrove *et al.* [6] and Lyle and Struve [28]. In the latter work, samples in Tris buffer were excited in the Q_y band system between 790 and 825 nm with tunable pulses from a Styryl 8 dye laser (2–3 ps autocorrelation), and probed at the same wavelength. The one-color anisotropy $r(t)$ was computed from the absorption difference signals $\Delta A_\parallel(t), \Delta A_\perp(t)$ measured with probe polarizations parallel and perpendicular to the pump polarization,

$$r(t) = \frac{\Delta A_\parallel(t) - \Delta A_\perp(t)}{\Delta A_\parallel(t) + 2\Delta A_\perp(t)}. \tag{10.7}$$

This decay function reflects the kinetics of energy transfers between excited states with contrasting polarization. The nominal initial anisotropy $r(0)$, measured at 814 nm, was considerably less than 0.4, suggesting that the anisotropy had an substantial sub-picosecond component buried under the instrument function. The subsequent decay in $r(t)$ was fitted with a single-exponential function with a lifetime of 2.3 ps. This was attributed to energy transfers between lowest-energy excitons belonging to adjacent subunits within a trimer, because (in view of Förster energy transfer rates expected for nearest-neighbor BChls separated by 11–15 Å within a subunit) intra-subunit energy transfers were expected to be considerably faster than ~ 2 ps [15, 19]. In this scenario, the 2.3 ps anisotropy lifetime corresponds to energy transfer steps between subunits with 6.9 ps kinetics, because equilibration among subunits of a trimer with rate constant k yields an anisotropy decay with a lifetime of $1/3k$ [29].

The first femtosecond studies were done on FMO trimers from *Cb. tepidum* at room temperature, yielding a wealth of information on the timescales of spectral equilibration and energy transfers between subunits in that protein [46, 50]. The instrument functions for one- and two-color experiments were ~ 60 and 250 fs respectively. A two-color isotropic ΔA profile for trimers pumped at 800 nm and probed at 820 nm – dominated at all times by photobleaching (PB) and stimulated emission (SE) – showed a SE rise time of 370 fs, yielding a first look at the kinetics of downhill energy transfers between states with SE maxima near 800 and 820 nm. An 821 nm one-color anisotropy could be fitted with a biexponential decay with lifetimes of 130 fs and 1.7 ps; the latter lifetime was similar to the 2.3 ps anisotropy obtained by Lyle and Struve [28] for *Pc. aestuarii* trimers. The anisotropy dropped below 0.4 within less than 100 fs, and the residual anisotropy at long times was 0.04. A more extensive study yielded one-color anisotropies at several wavelengths from 798 to 825 nm. Biexponential fits in 8 ps windows yielded long components with lifetimes from 1.4 to 2.0 ps; the latter value occurred at 825 nm, which overlaps the lowest three exciton components.

Two-color isotropic profiles were accumulated for several combinations of pump and probe wavelengths. Several of the downhill pump–probe experiments yielded nonexponential PB/SE rise behaviors; examples of lifetimes from biexponential fits to the rising edges of these profiles were 159, 412 fs (800 → 830 nm); and 67, 347 fs (790 → 820 nm). (It should be cautioned that the very short components may arise in part from pseudo-two-color optical coherences near the temporal region of pump–probe overlap [7]). However, a 770 → 800 nm profile showed no PB/SE rise behavior, but exhibited a fast PB/SE decay component with a lifetime of 405 fs. This example suggests that the time evolution of absorption difference signals in these two-color experiments does not arise simply from red-shifting of a monomeric BChl *a* ΔA spectrum [2] that accompanies downhill energy transfers between individual pigments. Otherwise, all three of the above profiles (in which the probe wavelength was shifted 30 nm to the red of the pump wavelength) should exhibit qualitatively similar early behavior, albeit with different kinetics. Analysis of the femtosecond components from two-color profiles at other wavelengths yielded components spanning from 55 to 990 fs, indicating that the spectral equilibration among FMO levels is kinetically heterogeneous. This work suggests a dynamic scenario in which spectral equilibration within a subunit is essentially complete by 1 ps; subsequent energy transfers between subunits (signaled by 1.7–2.0 ps anisotropy decays) occur with 5–7 ps kinetics. (Recall from the previous section that the laser-prepared states in FMO trimers from *Cb. tepidum* are typically localized largely within a subunit, rather than reflecting the crystallographic C_3 symmetry.) The PB/SE at long times is typically dominated by a 40–60 ps lifetime component, consistent with reports that FMO trimers at in the absence of sodium dithionite at 300 K show excitation trapping at redox sites with ~ 60 ps kinetics [4]. Similar long-time PB/SE decay was observed by Causgrove *et al.* [6] in trimers from *Pc. aestuarii*.

While informative, these room-temperature studies cannot be interpreted in terms of specific level-to-level relaxation scenarios, because the absorption spectrum at this temperature is structureless. However, individual absorption bands arising from transitions to well-defined groups of levels (Fig. 10.3) become clearly resolved in the absorption spectrum at 19 K. Buck *et al.* [5] accumulated two-color profiles for *Cb. tepidum* trimers excited at 789 nm (overlapping one of the highest-lying groups of exciton levels; cf. Fig. 10.3),

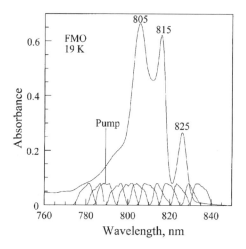

Figure 10.5   Laser spectra (for 789 nm excitation pulses and for probe pulses spaced by ~ 5 nm from 780 to 835 nm), superimposed on steady state absorption spectrum of *Cb. tepidum* trimers at 19 K [5]

and probed at ~ 5 nm intervals from 780 to 835 nm (Fig. 10.5). These profiles were assembled into a three-dimensional ΔA versus time and probe wavelength surface. Slices through a cubic spline fit to this surface are shown for several time delays from 40 fs to 80 ps in Fig. 10.6; the absorption difference spectrum shows complex spectral evolution at this temperature. At 40 fs, most of the PB/SE is concentrated near the 789 nm excitation wavelength; even at this time, the spectrum already exhibits a marked PB/SE peak at 805 nm. The spectrum is essentially equilibrated by 10 ps, yielding a PB/SE maximum at ~ 825 nm. The intervening spectra (like the steady state absorption spectrum) show considerable structure throughout, with PB/SE features at 790, 805, 815, and 825 nm. These wavelengths resemble the positions of bands in the steady state spectrum (Fig. 10.5). A global analysis of two-color profiles accumulated in 8, 80, and 566 ps time windows (to facilitate identification of short- as well as long-lifetime components) yielded six lifetimes: 170 fs, 630 fs, 2.5 ps, 11 ps, 74 ps, and 840 ps. (The longest component is not well defined in this analysis, owing to the finite window.) These lifetimes underscore the fact that spectral equilibration (which requires less than 1 ps at 300 K) is considerably decelerated at 19 K [47]. The decay-associated spectra (DAS) for these lifetimes are shown in Fig. 10.7. Here a 170 fs PB/SE decay component for probe wavelengths near 795 nm is coupled with a 170 fs PB/SE rise component near 815 nm. Similarly, 630 fs PB/SE decay and rise components (corresponding to negative and positive amplitudes) are found at 805 nm and 820–825 nm, respectively.

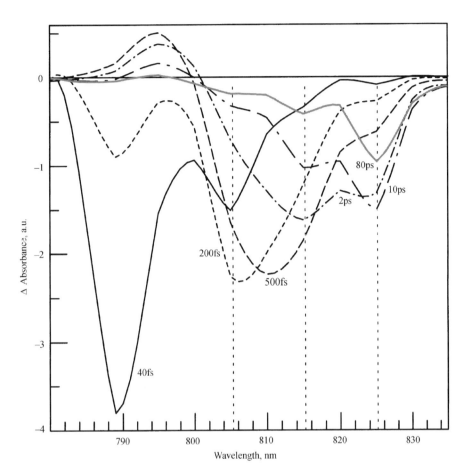

Figure 10.6   Sections of cubic spline fit to ΔA versus time and wavelength surface, at fixed time delays from 40 fs to 80 ps. Positive and negative signals correspond to ESA and PB/SE respectively; surface assembled from 12 two-color profiles of *Cb. tepidum* trimers excited at 789 nm at 19 K [5]

10.4.2   Exciton simulations of absorption difference spectra

These absorption difference spectra can be compared with ΔA spectra simulated in exciton calculations to yield a kinetic model for primary processes in FMO trimers [5]. The photobleaching component in the absorption difference spectrum is evaluated directly from Eqn 10.5; any excitation wavelength will uniformly bleach the entire steady state spectrum arising from the set of N ground $\rightarrow$ one-exciton transitions ($N = 7$ and 14 for exciton states restricted to

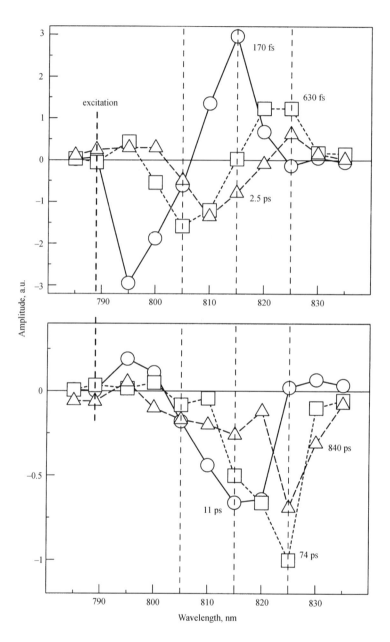

Figure 10.7 Decay-associated spectra (DAS) from global fits to FMO two-color absorption difference profiles under 789 nm excitation. The abscissa is the probe wavelength. Positive and negative DAS amplitudes correspond to PB/SE decay and rise components, respectively [5]

a single subunit and delocalized over the trimer, respectively). Lu and Pearlstein [27] simulated their OKT absorption spectra with Gaussian components exhibiting widths from 95 to 342 cm^{-1}; we arbitrarily assigned a width of 150 cm^{-1} to all levels. The stimulated emission profile for each exciton level is represented by a symmetrical Gaussian profile that is congruent with the corresponding absorption profile: the Einstein coefficients for absorption and SE are equal if the equilibrium geometries are similar in the ground and Q_y states, and no major Stokes shift is expected between absorption and SE for pigments in the hydrophobic protein interior.

Determining the contribution from excited-state absorption arising from one-exciton $\rightarrow$ two-exciton transitions requires evaluation of the $N(N-1)/2$ two-exciton states $\psi_i^{(2)}$. These are expanded in terms of the $N(N-1)/2$ doubly excited basis functions $|\chi_{12}^{(2)}\rangle = |1^*2^*3\cdots\rangle$, $|\chi_{13}^{(2)}\rangle = |1^*23^*\cdots\rangle$, $|\chi_{23}^{(2)}\rangle = |12^*3^*\cdots\rangle$, etc., which describe excitations localized on pairs of pigments. The energies $E_i^{(2)}$ of states $\psi_i^{(2)}$ are obtained by diagonalizing the matrix of the Hamiltionian (Eqn 10.1) in this basis. The matrix elements are

$$\langle\chi_{ij}^{(2)}|\hat{H}|\chi_{ij}^{(2)}\rangle = E_i + E_j,$$
$$\langle\chi_{ij}^{(2)}|\hat{H}|\chi_{ik}^{(2)}\rangle = V_{jk}, \qquad j \neq k,$$
$$\langle\chi_{ij}^{(2)}|\hat{H}|\chi_{kl}^{(2)}\rangle = 0, \qquad i \neq k,l, \quad j \neq k,l. \tag{10.8}$$

In this basis, the diagonal matrix elements are pairwise sums of single-pigment excitation energies E_i, E_j; the nonvanishing off-diagonal matrix elements V_{jk} in Eqn 10.8 are identical to certain off-diagonal matrix elements in the one-exciton basis (Eqn 10.3). This general two-exciton matrix becomes further simplified in specialized cases (e.g. one-dimensional J-aggregates in which only nearest-neighbor resonant interactions are considered [24]). In our FMO simulations, all $N(N-1)/2$ interactions were considered. The two-exciton states are linear combinations of the doubly excited basis functions $|\chi_{ij}^{(2)}\rangle$, i.e.

$$|\psi_k^{(2)}\rangle = \sum_{ij} d_{ij}^k |\chi_{ij}^{(2)}\rangle, \tag{10.9}$$

where the real-valued coefficients d_{ij}^k are taken from the matrix of eigenvectors of the real symmetric two-exciton Hamiltonian. The absorption coefficient for any one- to two-exciton ESA (excited-state absorption) transition $|\psi_i^{(1)}\rangle \rightarrow |\psi_k^{(2)}\rangle$ is then

$$B_{l\rightarrow k}^{(2)} = |\langle\psi_l^{(1)}|\mu|\psi_k^{(2)}\rangle|^2 = \left|\sum_{ijm} d_{ij}^k c_m^l \langle\chi_m^{(1)}|\mu|\chi_{ij}^{(2)}\rangle\right|^2, \tag{10.10}$$

where

$$\langle \chi_m^{(1)} \mid \mu \mid \chi_{ij}^{(2)} \rangle = \hat{\mu}_j \delta_{im} + \hat{\mu}_i \delta_{jm}. \tag{10.11}$$

At least one of the terms always vanishes on the right-hand side of Eqn 10.11, because the doubly excited basis functions $\mid \chi_{ij}^{(2)} \rangle$ are undefined for $i = j$.

The normalized absorption coefficients for transitions from the ground state to the one-exciton states $\psi_i^{(1)}$ follow the sum rule

$$\sum_{i=1}^{N} B_i^{(1)} = N, \tag{10.12}$$

i.e. the total oscillator strength remains conserved in the presence of exciton couplings. The ESA absorption coefficients for the one-exciton $\rightarrow$ two-exciton transitions $\psi_i^{(1)} \rightarrow \psi_j^{(2)}$ obey the analogous sum rule

$$\sum_{i=1}^{N} \sum_{j=1}^{N} B_{i \rightarrow j}^{(2)} = N(N-1). \tag{10.13}$$

This states that the sum of absorption coefficients for all one-exciton $\rightarrow$ two-exciton transitions originating from a particular one-exciton state $\psi_i^{(1)}$, averaged over the N one-exciton states, is

$$\left\langle \sum_{j=1}^{N} B_{i \rightarrow j}^{(2)} \right\rangle = N - 1. \tag{10.14}$$

In Fig. 10.8, we show prompt absorption difference spectra simulated using the 21-pigment exciton model for *Pc. aestuarii* trimers excited at 780, 790, 815, and 825, nm. The spectra in the right-hand column were computed using the Pearlstein [39] off-diagonal Hamiltonian matrix elements and OKT diagonal energies [27]; the left-hand spectra were computed with zero off-diagonal matrix elements. The latter spectra represent an idealized FMO complex with noninteracting pigments; they are *not* the spectra that would be observed if coherence decays rapidly before observation. The spectral evolution produced by tuning the laser from 780 to 825 nm in the absence of resonance couplings exhibits red-shifting of a monomeric BChl $a\Delta A$ spectrum. Upon turning the couplings on, the spectral evolution becomes more complicated; secondary PB/SE maxima now appear at wavelengths other than the laser wavelength, and arise from holes in the one-exciton $\rightarrow$ two-exciton ESA spectrum that otherwise tends to cancel much of the PB/SE spectrum. The ΔA spectrum excited at 825 nm (which essentially pumps only the lowest three levels, because the next

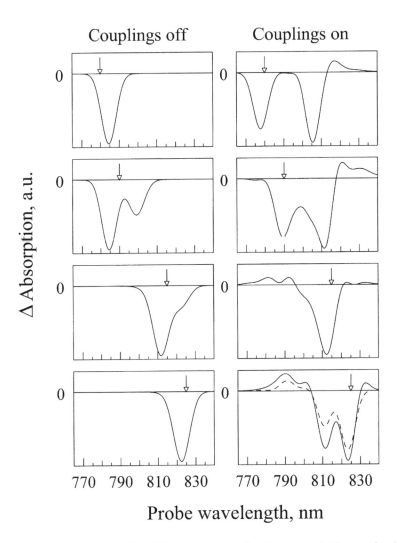

Figure 10.8   Prompt absorption difference spectra for *Pc. aestuarii* trimers, simulated using the OKT diagonal energies. Excitation wavelengths (shown by arrows) are 780, 790, 815, and 825 nm. Spectra at left obtained using zero off-diagonal matrix elements; spectra at right computed using off-diagonal elements from [39]. Positive and negative signals are ESA and PB/SE, respectively [5]

group of levels absorbs at 812–813 nm) shows a major secondary PB/SE peak at ~ 812 nm in addition to the principal maximum at ~ 825 nm; this contrasts with the corresponding ΔA spectrum obtained with zero off-diagonal elements (left side of Fig. 10.8).

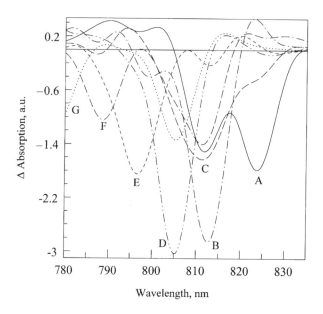

Figure 10.9   OKT absorption difference spectra for FMO trimers excited in the level groups A, B, ..., G. Each spectrum is averaged over the three states belonging to the pertinent group (cf. Figure 3, [5])

Figures 10.9 and 10.10 show absorption difference spectra evaluated using the OKT and PS diagonal energies, respectively. Each of the curves is averaged over a group of three exciton levels (the level groups are labeled A through G in order of ascending energy, and are identified in the energy level diagram of Fig. 10.3). The OKT ΔA spectrum for group A (the lowest three exciton levels near 825 nm) shows dual PB/SE peaks near 825 and 812 nm, like the prompt OKT spectrum excited at 825 nm in Fig. 10.8. The PS spectrum for group A is dominated by a broad PB/SE peak near 810 nm. Hence, the physical differences between the OKT and PS ΔA spectra in Figs 10.9 and 10.10, like those between the corresponding CD spectra (Fig. 10.2), are considerable. The spectra in Figs 10.9 and 10.10 (simulated using 21-pigment exciton models) closely resemble the ΔA spectra computed with seven-pigment models that limit exciton states to a single subunit (not shown); these isotropic spectra are not strongly perturbed by the couplings between subunits.

10.4.3   Kinetic models for exciton relaxation in FMO trimers

The simulated ΔA spectra have been combined with kinetic models from relaxation between exciton level groups to generate three-dimensional surfaces

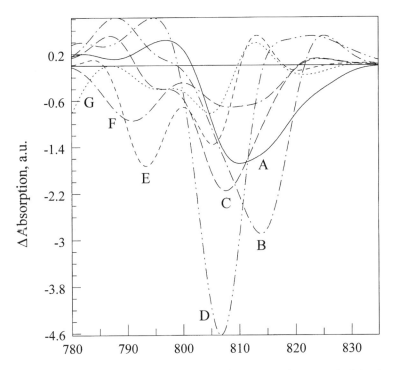

Figure 10.10   PS absorption difference spectra for FMO trimers excited in the level groups A, B, ..., G. Each spectrum is averaged over the three states belonging to the pertinent group [5]

of absorption difference versus time and probe wavelength under 789 nm excitation. Figure 10.11 compares two such surfaces, generated from the OKT simulations in the presence and absence of resonance couplings, with the experimental ΔA surfaces. The relevant kinetic models are given in Fig. 10.12. In the presence of resonance couplings, group F levels excited near 789 nm relax to group D levels within 100 fs; the latter are responsible for the steady state absorption band near 805 nm. The group D levels then branch equally into the group B/C levels clustered near 815 nm (with 170 fs kinetics) and the group A levels (with 630 fs kinetics). The group B/C levels then relax to group A with 2.5 ps kinetics. This model yields a ΔA surface that reproduces several of the major features of the experimental surface (bottom of Fig. 10.11). The simulations without resonance couplings (middle of Fig. 10.11) cannot generate realistic ΔA surfaces, primarily because they cannot replicate the bimodal long-time spectrum with PB/SE maxima near 815 and 825 nm (bottom of Fig. 10.8). PS simulations likewise failed to yield realistic ΔA surfaces for any kinetic model. None of our simulations predicts the intense experimental PB/SE maximum near 790 nm at zero time, but this feature is most likely due to strong coherent coupling artifacts, which are not incorporated in our simulations [8, 46].

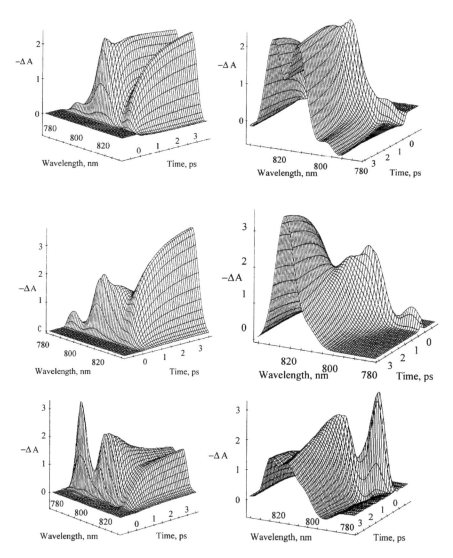

Figure 10.11 ΔA versus time and wavelength surfaces for *Pc. aestuarii* trimers excited at 789 nm: OKT simulation with resonance couplings (top), OKT simulation without resonance couplings (middle), and experimental (bottom). Left and right columns give different perspectives for visualization. Positive vertical axis corresponds to PB/SE [5]

Modeling of the kinetics at times prior to 100 fs was problematic with either the OKT or PS models, because of the secondary PB/SE peak that appears at ∼ 805 nm by 40 fs (Fig. 10.6). The simulated ΔA spectra for group F levels (which are excited at 789 nm) show no such feature; the group E levels do show intense PB/SE near this wavelength. This suggested an early-time kinetic model like the one with the sequential 30 fs and 70 fs steps shown on the left-hand side of Fig. 10.12. The need to invoke these steps may be an artifact of using an exciton model for *Pc. aestuarii* to simulate experimental ΔA spectra for *Cb. tepidum*; the FMO electronic structure may differ in these two species, particularly in the higher exciton levels [44]. For this reason, comparable experimental data for *Pc. aestuarii* trimers would be valuable. Our global analyses of experimental profiles yielded no components with lifetimes shorter than 100 fs, since the laser cross-correlations in these two-color experiments were 200–250 fs wide.

Some variations of the kinetic model yield similar OKT surfaces. Varying the proportions of groups B and C excited by energy transfers from group D has relatively little effect, because the former two groups have qualitatively similar OKT spectra (Fig. 10.9). A scenario postulating that group F branches within 100 fs into groups D and E (which then relax to groups A and B with 630 and 170 fs kinetics, respectively) works equally well.

DAS components with lifetimes similar to some of ours have been observed by Freiberg *et al.* [14] in low-temperature absorption difference spectra of FMO trimers from *Cb. tepidum*. More recently, Vulto *et al.* [62] characterized the excited-state relaxation kinetics of FMO trimers from *Pc. aestuarii* at 10 K.

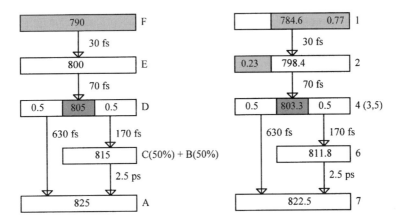

Figure 10.12 Kinetic models for simulations in Figure 10.11: in the presence of resonance couplings (left) and without resonance couplings (right). Levels are labeled with level groups A, B, ..., G on the left, and with pigment numbers on the right. Shading identifies levels excited near 789 nm [5]

Their conclusions resemble ours in that relaxation occurs by cascading down the manifold of Q_y states. The lifetimes of their DAS components were 0.5, 1.7, 5.5 and 30 ps, which differ considerably from the ones observed in *Cb. tepidum* (cf. Fig. 7; [5]). These discrepancies may stem from species differences between *Pc. aestuarii* and *Cb. tepidum*. Alternatively, they may be influenced by the different temperatures at which the studies were done. Like Buck *et al.* [5], Vulto *et al.* [62] assigned the shorter-lifetime components to relaxation steps involving higher energy levels (cf. Fig. 10.12).

10.4.4   Anisotropies and exciton localization

Our simulations of isotropic ΔA spectra do not differentiate sharply between 7- and 21-pigment exciton states [5]. Optical anisotropies are apt to be more sensitive to localization in FMO trimers, because the exciton transition moment directions depend markedly on whether the states are localized to one or three subunits (Section 10.2). By evaluating the PB, SE, and ESA contributions to the polarized absorption differences $\Delta A_\parallel$ and $\Delta A_\perp$ (cf. Eqns 10.5 and 10.10) and computing the resulting anisotropy $r(t)$ from Eqn 10.7, the theoretic anisotropy at early times can be simulated if the relevant exciton states are known [49]. Such simulations are shown in Fig. 10.13 for hypothetic two-color anisotropies, evaluated under the assumption that exciton states are delocalized over the whole trimer (21 pigments) or over one subunit of the trimer (seven pigments). The excitation wavelengths 825 and 815 nm overlap the lowest two features in the absorption spectrum of *Cb. tepidum* trimers. For one-color experiments (where the probe and pump wavelengths are the same), the predicted initial anisotropy is between 0 and 0.4; in these cases, the absorption difference signals are dominated by PB/SE. In two-color experiments (where one-exciton → two-exciton excited-state absorption can dominate), the anisotropy can reach negative values, as well as values higher than 0.4. Experimental anisotropies measured in 815 → 815 and 825 → 815 nm experiments [48] are +0.35 and −0.30, respectively. These values are inconsistent with the 21-pigment model (Fig. 10.13), which predicts $r(0) \sim 0.05$ and +0.25, respectively. They are more consistent with the seven-pigment model. This result suggests that the $\sim 70\,\mathrm{cm}^{-1}$ diagonal energy disorder in *Cb. tepidum* [44] causes the exciton levels in typical trimers to be essentially localized to one subunit (cf. Section 10.3).

10.4.5   Quantum beats and coherence decay

Femtosecond coherent oscillations sometimes appear in the low-temperature ΔA signals of FMO trimers. They are not conspicuous in the isotropic signals (although these are noticeable under high signal/noise); they are quite

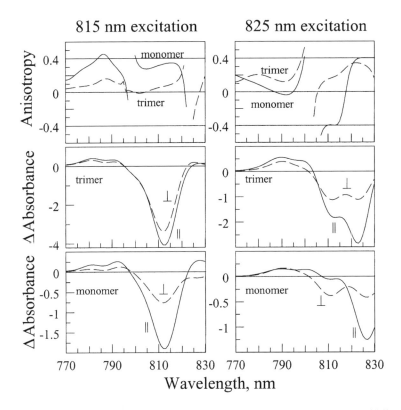

Figure 10.13 Two-color anisotropies (top) and polarized ΔA spectra (middle and bottom) simulated for *P. aestuarii* trimers under pump wavelengths of 815 and 825 nm (left and right columns respectively). Monomer and trimer curves denote 7- and 21-pigment models. The symbols $\parallel$ and $\perp$ indicate parallel and perpendicular signals. Laser pulse spectra are 7 nm fwhm [49]

pronounced in the anisotropy decays for some pump–probe wavelengths. Figure 10.14 shows one-color polarized absorption difference signals $\Delta A_{\parallel}(t)$ and $\Delta A_{\perp}(t)$ and the anisotropy decay $r(t)$ for FMO trimers from *Cb. tepidum* at 19 K, excited at 820 nm with 9.7 nm bandwidth [48]. This laser spectrum overlaps the two longest-wavelength absorption bands at 815 and 825 nm (inset of Fig. 10.14). The polarized signals oscillate during the first few hundred femtoseconds with $\sim 180°$ phase difference and relative amplitudes in ratio ~ 2; the oscillations in the isotropic decay are therefore weak. The 220 fs period of the oscillations corresponds to $\Delta E/hc \sim 150\,\mathrm{cm}^{-1}$, which is the energy separation between level groups A and (B, C) that are responsible for the 825 and 815 nm absorption bands. The experimental anisotropy reaches -0.11 at 100 fs, and swings to $+0.55$ at 220 fs; the latter time is well beyond the 87 fs autocorrelation function width in these one-color experiments.

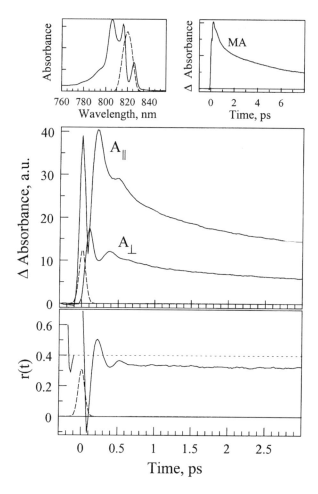

Figure 10.14 Polarized absorption difference signals and anisotropy for *Cb. tepidum* trimers excited at 820 nm at 19 K with 9.7 nm bandwidth [48]. Laser autocorrelation is given by the dashed curve. PB/SE dominates the ΔA signal at all times. Insets give a laser spectrum superimposed on an FMO steady state spectrum (left) and an isotropic ΔA signal (right)

Fits of the anisotropy in Fig. 10.14 with a triexponential monotonic decay plus a sum of four damped sinusoidal terms yielded only one oscillating term with damping time comparable to or longer than the period; the oscillations show no evidence for Fourier components other than 220 fs. The damping time is ~ 160 fs. The sensitivity of the oscillations to the laser excitation spectrum is shown in Fig. 10.15, where the laser center wavelength was swept across the FMO absorption spectrum from 812 to 828 nm. Minimal oscillations are

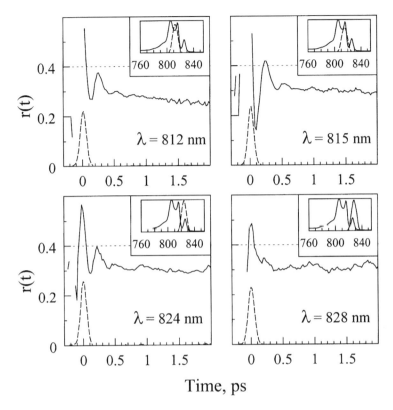

Figure 10.15   Variation of $r(t)$ with pump–probe wavelength from 812 to 828 nm. Insets show laser spectra superimposed on a steady state spectrum [48]

observed when the laser spectrum overlaps only one of the two bands at 825 and 815 nm. In view of the oscillations' sensitivity to the laser spectrum, the frequency match, and the fact that they occur predominantly in the $r(t)$ rather than in the isotropic decay, they are unlikely to arise from coherent nuclear motion [8]. Vulto *et al.* [62] did not observe coherent oscillations in their one-color anisotropies of FMO trimers from *Pc. aestuarii*, and suggested that their appearance in *Cb. tepidum* trimers was caused by accumulated photon echoes. However, their excitation wavelength were 815 nm or shorter. The evidence in Figs 10.14 and 10.15 strongly suggests that the oscillations are quantum beats between levels belonging to groups A and (B,C), respectively; the initial coherence between these levels is typically lost within 160 fs.

The oscillations can be modeled by assuming that the laser spectrum overlaps exciton levels $|1\rangle$ and $|2\rangle$ belonging to the level groups A and (B,C), respectively, as shown in Fig. 10.16. (The three level groups combined contain nine

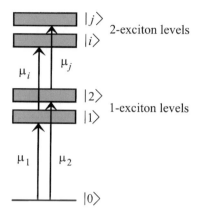

Figure 10.16 A conceptual level diagram for analysis of anisotropy beats in FMO trimers. Levels $|1\rangle$ and $|2\rangle$ are one-exciton levels optically connected to the ground state, while levels $|i\rangle$ and $|j\rangle$ are two-exciton levels optically connected to $|1\rangle$ and $|2\rangle$, respectively

levels – cf. Fig. 10.3 – but this treatment is straightforwardly generalized.) Our theory resembles several earlier discussions of optical anisotropies [20, 43, 67], aside from its incorporation of excited state absorption (see below). Prior to damping, the laser-prepared state propagates as the superposition [1]

$$|\psi(t)\rangle = a_1|1\rangle e^{-iE_1 t/\hbar} + a_2|2\rangle e^{-iE_2 t/\hbar}, \qquad (10.15)$$

since the 815 nm levels relax to lower-lying levels with no faster than ~ 2.5 ps kinetics [5]. Our analysis does not extend to the region of pump–probe pulse overlap; this would have to be treated using a time-ordered perturbation expansion of the third-order nonlinear polarization [7]. The expansion coefficients,

$$a_1 = \vec{E}_{u1} \cdot \langle 1|\vec{\mu}|0\rangle \equiv \vec{E}_{u1} \cdot \vec{\mu}_1, \qquad a_2 = \vec{E}_{u2} \cdot \langle 2|\vec{\mu}|0\rangle \equiv \vec{E}_{u2} \cdot \vec{\mu}_2, \qquad (10.16)$$

contain the pump pulse electric field amplitudes $\vec{E}_{u1}$ and $\vec{E}_{u2}$ for wavelengths λ_1 and λ_2 near the ground state $\rightarrow$ one-exciton transitions $|0\rangle \rightarrow |1\rangle$ and $|0\rangle \rightarrow |2\rangle$, respectively. The stimulated emission signal is

$$SE = -\left|\left(\vec{E}_{r1} \cdot \vec{\mu}_1\right)\left(\vec{E}_{u1} \cdot \vec{\mu}_1\right)e^{-iE_1 t/\hbar} + \left(\vec{E}_{r2} \cdot \vec{\mu}_2\right)\left(\vec{E}_{u2} \cdot \vec{\mu}_2\right)e^{-iE_2 t/\hbar}\right|^2, \qquad (10.17)$$

where $\vec{E}_{r1}$ and $\vec{E}_{r2}$ are the probe pulse electric field amplitudes at the respective wavelengths. The polarized stimulated emission signals will then be

$$SE_{\parallel} = -\left|E_{r1}E_{u1}\mu_{1x}^2 e^{-iE_1 t/\hbar} + E_{r2}E_{u2}\mu_{2x}^2 e^{-iE_2 t/\hbar}\right|^2,$$

$$SE_{\perp} = -\left|E_{r1}E_{u1}\mu_{1x}\mu_{1y}e^{-iE_1 t/\hbar} + E_{r2}E_{u2}\mu_{2x}\mu_{2y}e^{-iE_2 t/\hbar}\right|^2, \tag{10.18}$$

and the corresponding photobleaching signals are

$$PB_{\parallel} = -E_{r1}^2 E_{u1}^2 \mu_{1x}^4 - E_{r2}^2 E_{u2}^2 \mu_{2x}^4 - E_{r1}^2 E_{u2}^2 \mu_{1x}^2\mu_{2x}^2 - E_{r2}^2 E_{r1}^2 \mu_{1x}^2\mu_{2x}^2,$$

$$PB_{\perp} = -E_{r1}^2 E_{u1}^2 \mu_{1x}^2\mu_{1y}^2 - E_{r2}^2 E_{u2}^2 \mu_{2x}^2\mu_{2y}^2 - E_{r1}^2 E_{u2}^2 \mu_{1y}^2\mu_{2x}^2 - E_{r2}^2 E_{r1}^2 \mu_{1x}^2\mu_{2y}^2. \tag{10.19}$$

Unlike the stimulated emission signals, the photobleaching signals do not reflect coherence between states $|1\rangle$ and $|2\rangle$; the entire steady state absorption spectrum is uniformly bleached by pump pulses at either of the wavelengths λ_1 or λ_2. We assume that the trimer-fixed transition moment directions are

$$\vec{\mu}_1 = \mu_1 \begin{bmatrix} 0 \\ 0 \\ 1 \end{bmatrix}, \qquad \vec{\mu}_2 = \mu_2 \begin{bmatrix} \alpha \\ 0 \\ \gamma \end{bmatrix}, \tag{10.20}$$

with $\alpha^2 + \gamma^2 = 1$. Rotational averaging of the SE and PB signals over the random orientations of FMO trimers in the low-temperature glass yields the pump–probe anisotropy

$$r(t) = 0.4 \frac{2I_{r1}I_{u1} + 2I_{r2}I_{u2} + (I_{r1}I_{u2} + I_{r2}I_{u1})\frac{1}{2}(3\gamma^2 - 1)}{2I_{r1}I_{u1} + 2I_{r2}I_{u2} + (I_{r1}I_{u2} + I_{r2}I_{u1})}$$

$$\frac{+\frac{1}{2}(\gamma^2 + 3)\sqrt{I_{r1}I_{r2}I_{u1}I_{u2}}\cos(\Delta E t/\hbar)}{+2\gamma^2\sqrt{I_{r1}I_{r2}I_{u1}I_{u2}}\cos(\Delta E t/\hbar)}. \tag{10.21}$$

Here $I_{r1} = E_{r1}^2\mu_1^2$ is proportional to the probe pulse intensity at wavelengths near the $|0\rangle \to |1\rangle$ transition, multiplied by that transition's oscillator strength, etc. For parallel transition moments ($\gamma = 1$), the anisotropy remains at 0.4 for all times in this regime. While no oscillations are observed in $r(t)$ in this case, they do appear in the isotropic signal (which is the denominator in Eqn 10.21). For perpendicular transition moments ($\gamma = 0$), there are maximal oscillations in $r(t)$, but they are absent in the isotropic signal. Since the FMO oscillations are far stronger in $r(t)$ than in the isotropic signals (Fig. 10.14), our analysis suggests that $\gamma \sim 0$ for the present coherences. For uniform excitation of the two-exciton components excited in a one-color experiment ($I_{u1} = I_{u2} = I_{r1} = I_{r2}$), the anisotropy would oscillate between 0.30 and 0.10 about $r = 0.2$ prior to damping. This amplitude and nonoscillating component are both considerably lower than is observed in Fig. 10.14. They are also lower than would be predicted if stimulated emission were the only component contributing to ΔA, because the photobleaching component does not contribute to the coherent signal. In the latter case (which is analogous to the situations described by Wynne and Hochstrasser [67] and Knox and Gülen [20], $r(t)$ would oscillate between 0.7 and 0.1 for $\gamma \sim 0$.

This analysis ignores one exciton $\rightarrow$ two-exciton excited-state absorption, which is important in FMO trimers. If excited-state absorption from one-exciton levels $|1\rangle$ and $|2\rangle$ occurs to the two-exciton levels $|i\rangle$ and $|j\rangle$, respectively, for wavelengths in the laser probe spectrum (Fig. 10.16), the excited-state absorption signal will be

$$ESA = |\vec{E}_{r3} \cdot \langle 1|\vec{\mu}|i\rangle a_1 e^{-iE_1 t/\hbar} + \vec{E}_{r4} \cdot \langle 2|\vec{\mu}|j\rangle a_2 e^{-iE_2 t/\hbar}|^2. \tag{10.22}$$

Here $\vec{E}_{r3}$ and $\vec{E}_{r4}$ are the probe amplitudes at the wavelengths for the excited-state absorption transitions $|1\rangle \rightarrow |i\rangle$ and $|2\rangle \rightarrow |j\rangle$. The excited-state absorption signals are

$$\begin{aligned} ESA_\parallel &= |E_{r3} E_{u1} \mu_{ix} \mu_{1x} e^{-iE_1 t/\hbar} + E_{r4} E_{u2} \mu_{jx} \mu_{2x} e^{-iE_2 t/\hbar}|^2, \\ ESA_\perp &= |E_{r3} E_{u1} \mu_{iy} \mu_{1x} e^{-iE_1 t/\hbar} + E_{r4} E_{u2} \mu_{jy} \mu_{2x} e^{-iE_2 t/\hbar}|^2 \end{aligned} \tag{10.23}$$

with

$$\vec{\mu}_i = \langle 1|\mu|i\rangle = \mu_i \begin{bmatrix} \alpha_1 \\ \beta_1 \\ \gamma_1 \end{bmatrix}, \qquad \vec{\mu}_j = \langle 2|\mu|j\rangle = \mu_j \begin{bmatrix} \alpha_2 \\ \beta_2 \\ \gamma_2 \end{bmatrix} \tag{10.24}$$

and $\alpha_i^2 + \beta_i^2 + \gamma_i^2 = 1$. The anisotropy is then

$$r(t) = 0.4 \frac{\begin{aligned} &2I_{r1}I_{u1} + 2I_{r2}I_{u2} + (I_{r1}I_{u2} + I_{r2}I_{u1})\tfrac{1}{2}(3\gamma^2 - 1) + A \\ &+ \left[\tfrac{1}{2}(\gamma^2 + 3)\sqrt{I_{r1}I_{r2}I_{u1}I_{u2}} + B\sqrt{I_{r3}I_{r4}I_{u1}I_{u2}}\right]\cos(\Delta E t/\hbar) \end{aligned}}{\begin{aligned} &2I_{r1}I_{u1} + 2I_{r2}I_{u2} + (I_{r1}I_{u2} + I_{r2}I_{u1}) + C \\ &+ \left(2\gamma^2\sqrt{I_{r1}I_{r2}I_{u1}I_{u2}} + D\sqrt{I_{r3}I_{r4}I_{u1}I_{u2}}\right)\cos(\Delta E t/\hbar) \end{aligned}}, \tag{10.25}$$

where the terms arising from excited-state absorption are

$$A = -I_{r3}I_{u1}\tfrac{1}{2}(3\gamma_1^2 - 1) - I_{r4}I_{u2}\left[\frac{\gamma_2^2}{2}(3\gamma_2^2 - 1) + \frac{\alpha^2}{2}(3\alpha_2^2 - 1) + 3\alpha\gamma\alpha_2\gamma_2\right],$$

$$B = -\gamma_1\gamma_2\gamma + \frac{\gamma}{2}(\alpha_1\alpha_2 + \beta_1\beta_2) - \frac{3\alpha}{4}(\alpha_1\gamma_2 + \alpha_2\gamma_1),$$

$$C = -I_{r3}I_{u1} - I_{r4}I_{u2}, \qquad D = -\gamma(\gamma_1\gamma_2 + \alpha_1\alpha_2 + \beta_1\beta_2). \tag{10.26}$$

Hence, in the presence of one-exciton → two-exciton ESA, the isotropic signal still shows no oscillations when $\gamma = 0$, i.e. when $\vec{\mu}_1 \cdot \vec{\mu}_2 = 0$. For this case, the anisotropy simplifies to

$$r(t) = 0.4 \frac{\begin{array}{c} 2I_{r1}I_{u1} + 2I_{r2}I_{u2} + (I_{r1}I_{u2} + I_{r2}I_{u1})\frac{1}{2}(3\gamma^2 - 1) - I_{r3}I_{u1}P_2(\cos\gamma_1) \\ -I_{r4}I_{u2}P_2(\cos\alpha_2) + \left[2\sqrt{I_{r1}I_{r2}} - \frac{3\alpha}{4}(\alpha_1\gamma_2 + \alpha_2\gamma_1)\sqrt{I_{r3}I_{r4}}\right]\sqrt{I_{u1}I_{u2}}\cos(\Delta Et/\hbar) \end{array}}{2I_{r1}I_{u1} + 2I_{r2}I_{u2} + I_{r1}I_{u2} + I_{r2}I_{u1} - I_{r3}I_{u1} - I_{r4}I_{u2}},$$

$$(10.27)$$

where α (which depends on the direction of $\vec{\mu}_2$) is ± 1, and $P_2(x)$ is the second-order Legendre polynomial in x. Owing to the presence of the Legendre polynomials in the numerator, the nonoscillating part of $r(t)$ can now differ from 0.2. Depending on the laser spectra and moment orientations, r can literally range from $-\infty$ to $+\infty$. It is somewhat less than 0.4 for the anistropies in Figs 10.14 and 10.15. The amplitude of the oscillating part similarly depends on the transition moment directions and spectra. For suitable transition moments, the presence of strong ESA at the laser wavelengths easily produces oscillations as large as is observed in Fig. 10.14.

This analysis does not account for the observed 160 fs damping time. This damping is unlikely to be due to loss of exciton coherence; low-temperature accumulated photon-echo measurements on *Pc. aestuarii* trimers [25, 26] suggest that their exciton coherence lifetimes range from ∼300 fs (for levels near 795 nm) to several hundred picoseconds (near 825 nm). Instead, the damping stems from the superposition of coherences from group A and B/C levels in the presence of diagonal energy disorder. Separate calculations (not shown) predict that for 80 cm^{-1} Gaussian disorder, quantum beats between levels separated by 150 cm^{-1} will become damped with a time constant of the same order of magnitude as the 220 fs frequency, as is observed.

10.5   EPILOG AND FUTURE PROSPECTS

The FMO dynamic scenario that emerges from our ultrafast spectoscopy studies is somewhat reminiscent of the Sauer pebble mosaic model [45]. At room temperature, the sub-picosecond events are dominated by spectral equilibration among Q_y exciton states that are localized essentially within one sub-unit of the trimer. Subsequent transfers occur with 5–7 ps kinetics among the lowest-energy states (dominated by excitations on pigments 6 and 7 according to the Pearlstein fits to the OKT spectra), and are revealed by 1.7–2.3 ps anisotropy decay components. The spectral equilibration and anisotropy decay are considerably decelerated at low temperatures.

A current issue for the strongly coupled B850 antennas from purple photo-synthetic bacteria (such as *Rps. acidophila* and *Rb. sphaeroides*) is the extent of exciton localization during excited-state population decay due to static and dynamic disorder [53]. Recent superradiance experiments on B850 complexes from *Rps. acidophila* [33] suggest that while the effective exciton domain size can initially be as large as 18 BChl *a* pigments for aggregates with C_9 symmetry, the presence of disorder typically reduces this to $\sim$2.8 pigments at room temperature. A similar conclusion ($\sim$4 BChls) was drawn for *Rb. sphaeroides*, based on comparisons between measured ΔA spectra and simulations analo-gous to the ones described in Sections 10.3 and 10.4 [42]. In FMO trimers from *Cb. tepidum*, the range of typical domain sizes is much more limited, because the $\sim 70\text{cm}^{-1}$ diagonal energy disorder effectively localizes the exciton size to within one subunit for the majority of trimers (Section 10.3). Since each of the exciton levels in a seven-pigment model in turn concentrates $> 50\%$ of the excitation density in two to three BChls at most (cf. Fig. 10.3), the laser-prepared states (which are technically 21-pigment states) are in fact already highly localized in FMO trimers from *Cb. tepidum*. They may be less localized in the corresponding protein from *Pc. aestuarii*, where the inhomogeneous broadening is reported to be smaller [18, 61].

The principal unresolved questions revolve around the unknown BChl diag-onal energies. The assignments of empirical exciton levels to states with known excitation densities on specific pigments are therefore uncertain; this precludes comparisons between experimental and theoretical energy transfer rates in this intriguing pigment–protein complex. The recently determined crystal structure of the FMO protein from *Cb. tepidum* [23] should enable comparisons between exciton simulations of optical spectra between the two species; these may illuminate the effects of the now-known protein environments and Bchl con-formations on the diagonal energies and interactions. Site-directed mutations at the BChl binding and neighboring sites would provide a valuable independent method for inferring which pigment groups are primarily responsible for the features observed in optical spectra.

Acknowledgments

We are indebted to Timothy Causgrove, Shumei Yang, Paul Lyle, Herbert van Amerongen, Su Lin, and Robert Blankenship, who assisted us in the early experiments and exciton modeling. We thank Roger Fenna, Robert Blankenship, and Wenli Zhou for sharing their FMO protein preparations. The Ames Laboratory is operated for the U.S. Department of Energy by Iowa State University, under Contract No. W-7405-Eng-82. This work was supported by the Division of Chemical Sciences, Office of Basic Energy Sciences.

Endnote

Since this manuscript was submitted in August 1997, H. van Amerongen and T. Aartsma have compared absorption difference spectra, triplet-singlet absorption difference spectra, and steady-state spectra for FMO trimers from *Pc. aestuarii* with exciton simulations. They have concluded that the lowest diagonal energy belongs to BChl$_3$.

References

1. Avouris, P., W. M. Gelbart, and M. A. El-Sayed 1977. Nonradiative electronic relaxation under collision-free conditions. *Chem. Rev.* 77: 793–833.
2. Becker, M., V. Nagarajan, and W. W. Parson 1991. Properties of the excited-singlet states of bacteriochlorophyll *a* and bacteriopheophytin *a* in polar solvents. *J. Am. Chem. Soc.* 113: 6840–6848.
3. Blankenship, R. E., J. M. Olson, and M. Miller 1995. Antenna complexes from green photosynthetic bacteria. In *Anoxygenic Photosynthetic Bacteria*. R. E. Blankenship, M. T. Madigan, and C. E. Bauer, editors. Kluwer, Dordrecht, pp. 399–435.
4. Blankenship, R.E., P. Cheng, T. P. Causgrove, D. C. Brune, S. H.-H. Wang, J.-U. Choh, and J. Wang 1993. Redox regulation of energy transfer efficiency in antennas of green photosynthetic bacteria. *Photochem. Photobiol.* 57: 103–107.
5. Buck, D. R., S. Savikhin, and W. S. Struve 1997. Ultrafast absorption difference spectra of the Fenna–Matthews–Olson protein at 19 K: experiment and simulations. *Biophys. J.* 72: 24–36.
6. Causgrove, T. P., S. Yang, and W. S. Struve 1988. Polarized pump–probe spectroscopy of exciton transport in bacteriochlorophyll *a*-protein from *Prosthecochloris aestuarii. J. Phys. Chem.* 92: 6790–6795.
7. Chachisvilis, M., and V. Sundström 1996. The tunnelling contributions to optical coherence in femtosecond pump–probe spectroscopy of the three level system. *J. Chem. Phys.* 104: 5734–5744.
8. Chachisvilis, M., T. M. Pullerits, R. Jones, C. N. Hunter, and V. Sundström 1994. Coherent nuclear motions and exciton-state dynamics in photosynthetic light-harvesting pigments. In *Ultrafast Phenomena IX: Proceedings of the 9th International Conference*. P. F. Barbara, W. H. Knox, G. A. Mourou, and A. H. Zewail, editors. Dana Point, CA, May 2–6, 1994. Springer-Verlag, Berlin, pp. 435–436.
9. Chang, J. C. 1977. Monopole effects on electronic excitation interactions between large molecules. I. Application to energy transfer between chlorophylls. *J. Chem. Phys.* 67: 3901–3909.
10. Daurat-Larroque, S. T., K. Brew, and R. E. Fenna 1986. The complete amino acid sequence of a bacteriochlorophyll *a*-protein from *Prosthecochloris aestuarii. J. Biol. Chem.* 261: 3607–3615.
11. Dracheva, S., J. C. Williams, and R. E. Blankenship 1992. Sequencing of the FMO-protein from *Chlorobium tepidum*. In *Research in Photosynthesis*. N. Murata, editor. Kluwer, Dordrecht, Vol. 1, pp. 53–56.
12. Feick, R. G., and R. C. Fuller 1984. Topography of the photosynthetic apparatus of *Chloroflexus aurantiacus. Biochemistry* 23: 3693–3700.
13. Fenna, R. E., B. W. Matthews, J. M. Olson, and E. K. Shaw 1974. Structure of a bacteriochlorophyll–protein from the green photosynthetic bacterium *Chlorobium limicola*: crystallographic evidence for a trimer. *J. Mol. Biol.* 84: 231–240.

14. Freiberg, A., S. Lin, W. Zhou, and R. E. Blankenship 1996. Ultrafast relaxation of excitons in the bacteriochlorophyll antenna proteins from green photosynthetic bacteria. In *Ultrafast Processes in Spectroscopy*. O. Svelto, S. De Silvestri, and G. Denardo, editors. Plenum Press, New York.

15. Förster, T. 1948. Intermolecular energy transfer and fluorescence. *Ann. Phys.* 2: 55–75.

16. Gudowska-Nowak, E., M. D. Newton, and J. Fajer 1990. Conformational and environmental effects on bacteriochlorophyll optical spectra: correlations of calculated spectra with structural results. *J. Phys. Chem.* 94: 5795–5801.

17. Gülen, D. 1996. Interpretation of the excited-state structure of the Fenna–Matthews–Olson pigment protein complex of *Prosthecochloris aestuarii* based on the simultaneous simulation of the 4 K absorption, linear dichroism, and singlet–triplet absorption difference spectra: A possible excitonic explanation? *J. Phys. Chem.* 100: 17 683–17 689.

18. Johnson, S. G., and G. J. Small 1991. Excited state structure and energy transfer dynamics of the bacteriochlorophyll *a* antenna complex from *Prosthecochloris aestuarii*. *J. Phys. Chem.* 95: 471–479.

19. Knox, R. S. 1975. Excitation energy transfer and migration: theoretical considerations. in *Bioenergetics of Photosynthesis*. Govindjee, editor. Academic Press, New York. 183–221

20. Knox, R. S., and D. Gülen 1993. Theory of polarized fluorescence from molecular pairs: Förster transfer at large electronic coupling. *Photochem. Photobiol.* 57: 40–43.

21. Krauss, N., W. Hinrichs, I. Witt, P. Fromme, W. Pritzkow, Z. Dauter, C. Betzel, K. S. Wilson, H. T. Witt, and W. Saenger 1993. Three-dimensional structure of system I of photosynthesis at 6 Å resolution. *Nature* 361: 326–331.

22. Kühlbrandt, W., and D. N. Wang 1991. Three dimensional structure of plant light-harvesting complex determined by electron crystallography. *Nature* 350: 130–134.

23. Li, Y.-F., W. Zhou, R. E. Blankenship, and J. P. Allen 1997. Crystal structure of the bacteriochlorophyll *a* protein from *Chlorobium tepidum*. *J. Mol. Biol.* 271: 456–471.

24. Lin, S., H. van Amerongen, and W. S. Struve 1991. Ultrafast pump–probe spectroscopy of bacteriochlorophyll *c* antennae in bacteriochlorophyll *a*-containing chlorosomes from the green photosynthetic bacterium *Chloroflexus aurantiacus*. *Biochim. Biophys. Acta* 1060: 13–24.

25. Loewe, R. J. W., and T. J. Aartsma 1994. Optical dephasing and excited state dynamics in photosynthetic pigment–protein complexes. *J. Luminescence* 58: 154–157.

26. Loewe, R. J. W., and T. J. Aartsma 1995. In *Photosynthesis: from Light to Biosphere*. P. Mathis, editor. Kluwer Academic Publisher, Dordrecht, p. 363.

27. Lu, X., and R. M. Pearlstein 1993. Simulations of *Prosthecochloris* bacteriochlorophyll *a*-protein optical spectra improved by parametric computer search. *Photochem. Photobiol.* 57: 86–91.

28. Lyle, P. A., and W. S. Struve 1990. Evidence for ultrafast exciton localization in the Q_y band of bacteriochlorophyll *a*-protein from *Prosthecochloris aestuarii*. *J. Phys. Chem.* 94: 7338–7339.

29. Lyle, P. A., and W. S. Struve 1991. Dynamic linear dichroism in chromoproteins. *Photochem. Photobiol.* 53: 359–365.

30. Matthews, B. W., and R. E. Fenna 1980. Structure of a green bacteriochlorophyll protein. *Acc. Chem. Res.* 13: 309–317.

31. Matthews, B. W., R. E. Fenna, M. C. Bolognesi, M. F. Schmid, and J. M. Olson 1979. Structure of a bacteriochlorophyll *a*-protein from the green photosynthetic bacterium *Prosthecochloris aestuarii*. *J. Mol. Biol.* 131: 259–285.

32. McDermott, G., S. M. Prince, A. A. Freer, A. M. Hawthornthwaite-Lawless, M. Z. Papiz, R. J. Cogdell, and N. W. Isaacs 1995. Crystal structure of an integral membrane light-harvesting complex from photosynthetic bacteria. *Nature* 374: 517–521.

33. Monshouwer, R., M. Abrahamsson, F. van Mourik, and R. van Grondelle 1997. Superradiance and exciton delocalization in bacterial photosynthetic light-harvesting systems *J. Phys. Chem.*, 37: 7241–7248.

34. Olson, J. M. 1980. Chlorophyll organization in green photosynthetic bacteria. *Biochim. Biophys. Acta* 594: 33–51.

35. Olson, J. M. 1980. Bacteriochlorophyll *a*-proteins of two green photosynthetic bacteria. *Meth. Enzymol.* 69: 336–344.

36. Olson, J. M., and C. A. Romano 1962. A new chlorophyll protein from green bacteria. *Biochim. Biophys. Acta* 59: 726–728.

37. Olson, J. M., B. Ke, and K. H. Thompson 1976. Exciton interactions among chlorophyll molecules in bacteriochlorophyll *a* proteins and bacteriochlorophyll *a* reaction center complexes from green bacteria. *Biochim. Biophys. Acta* 430: 524–537.

38. Pearlstein, R. M. 1988. Interpretation of optical spectra of bacteriochlorophyll antenna complexes. In *Photosynthetic Light-harvesting Systems*. H. Scheer and S. Schneider, editors. Walter de Gruyter, Berlin, pp. 555–566.

39. Pearlstein, R. M. 1992. Theory of the optical spectra of the bacteriochlorophyll *a* antenna protein trimer from *Prosthecochloris aestuarii*. *Photosynth. Res.* 31: 213–226.

40. Pearlstein, R. M., and R. P. Hemenger 1978. Bacteriochlorophyll electronic transition moment directions in bacteriochlorophyll *a*-protein. *Proc. Natl Acad. Sci., USA* 75: 4920–4924.

41. Philipson, K. D., and K. Sauer 1972. Exciton interaction in a bacteriochlorophyll protein from *Chloropseudomonas ethylica*. Absorption and circular dichroism at 77 K. *Biochemistry* 11: 1880–1885.

42. Pullerits, T., and V. Sundström 1996. Photosynthetic light-harvesting pigment–protein complexes: toward understanding how and why. *Acc. Chem. Res.* 29: 381–389.

43. Rahman, T. S., R. S. Knox, and V. M. Kenkre 1979. Theory of depolarization of fluorescence in molecular pairs. *Chem. Phys.* 44: 197–211.

44. Reddy, N. R. S., R. Jankowiak, and G. J. Small 1995. High-pressure hole-burning studies of the bacteriochlorophyll *a* antenna complex from *Chlorobium tepidum. J. Phys. Chem.* 99: 16 168–16 178.

45. Sauer, K. 1975. Primary events in the trapping of energy. In *Bioenergetics of Photosynthesis*. Govindjee, editor. Academic Press, New York, pp. 115–181.

46. Savikhin, S., and W. S. Struve 1994. Ultrafast energy transfer in FMO trimers from the green bacterium *Chlorobium tepidum. Biochemistry* 33: 11 200–11 208.

47. Savikhin, S., and W. S. Struve 1996. Low-temperature energy transfer in FMO trimers from the green photosynthetic bacterium *Chlorobium tepidum. Photosynth. Res.* 48: 271–276.

48. Savikhin, S., D. R. Buck, and W. S. Struve 1997. Oscillating anisotropies in bacteriochlorophyll protein: Evidence for quantum beating between exciton levels. *Chem. Phys.*, 223: 303–312.

49. Savikhin, S., D. R. Buck, and W. S. Struve 1997. Pump–probe anisotropies of FMO trimers from *Chlorobium tepidum*: A diagnostic for exciton localization? *Biophys. J.* 73: 2090–2096.

50. Savikhin, S., W. Zhou, R. E. Blankenship, and W. S. Struve 1994. Femtosecond energy transfer and spectral equilibration in bacteriochlorophyll *a*-protein trimers from the green bacterium *Chlorobium tepidum. Biophys. J.* 66: 110–114.

51. Savikhin, S., Y. Zhu, S. Lin, R. E. Blankenship, and W. S. Struve 1994. Femtosecond spectroscopy of chlorosome antennas from the green bacterium *Chloroflexus aurantiacus. J. Phys. Chem.* 98: 10 322–10 334.

52. Staehelin, L. A., J. R. Golecki, R. C. Fuller, and G. Drews 1978. Visualization of the supramolecular architecture of chlorosomes (*Chlorobium* type vesicles) in freeze-fractured cells of *Chloroflexus aurantiacus. Arch. Microbiol.* 119: 269–277.

53. Sundström, V., and R. van Grondelle 1995. Kinetics of excitation transfer and trapping in purple bacteria. In *Anoxygenic Photosynthetic Bacteria.* R. E. Blankenship, M. T. Madigan, and C. E. Bauer, editors. Kluwer Academic Publishers, Dordrecht, pp. 349–372.

54. Sybesma, C., and J. M. Olson 1963. Transfer of chlorophyll excitation energy in green photosynthetic bacteria. *Proc. Natl Acad. Sci., USA* 49: 248–253.

55. Sybesma, C., and W. J. Vrendenberg 1963. Evidence for a reaction center P840 in the green photosynthetic bacterium. *Biochim. Biophys. Acta* 75: 439–441.

56. Sybesma, C., and W. J. Vrendenberg 1964. Kinetics of light-induced cytochrome oxidation and P840 bleaching in green photosynthetic bacteria under various conditions. *Biochim. Biophys. Acta* 88: 205–207.

57. Tronrud, D. E., and B. W. Matthews 1993. Refinement of the structure of a water-soluble antenna complex from green photosynthetic bacteria with incorporation of the chemically determined amino acid sequence. In *The Photosynthetic Reaction Center.* J. Norris and H. Deisenhofer, editors. Academic Press, San Diego, Vol. 1, pp. 13–21.

58. Tronrud, D. E., M. F. Schmid, and B. W. Matthews 1986. Structure and X-ray amino acid sequence of a bacteriochlorophyll *a* protein from *Prosthecochloris aestuarii* refined at 1.9 Å resolution. *J. Mol. Biol.* 188: 443–454.

59. van Amerongen, H., and W. S. Struve 1991. Excited state absorption in bacteriochlorophyll *a*-protein from the green photosynthetic bacterium *Prosthecochloris aestuarii*: reinterpretation of the absorption difference spectrum. *J. Phys. Chem.* 95: 9020–9023.

60. van Grondelle, R., J. P. Dekker, T. Gillbro, and V. Sundström 1994. Energy transfer and trapping in photosynthesis. *Biochim. Biophys. Acta* 1187: 1–65.

61. van Mourik, F., R. R. Verwijst, J. M. Mulder, and R. van Grondelle 1995. Singlet–triplet spectroscopy of the light-harvesting BChl *a* complex of *Prosthecochloris aestuarii*. The nature of the low-energy 825 nm transition. *J. Phys. Chem.* 98: 10307–10312.

62. Vulto, S., A. M. Streltsov, and T. J. Aartsma 1997. Excited state energy relaxation in the FMO complexes of the green bacterium *Prosthecochloris aestuarii* at low temperatures. *J. Phys. Chem. B* 101, 4845–4850.

63. Wahlund, T. M., C. R. Woese, R. W. Castenholz, and M. T. Madigan 1991. A thermophilic green sulfur bacterium from New Zealand hot springs, *Chlorobium tepidum* sp. nov. *Arch. Microbiol.* 156: 81–90.

64. Weiss, C. 1972. The pi electron structure and absorption spectra of chlorophylls in solution. *J. Mol. Spectrosc.* 44: 37–80.

65. Whitten, W. B., J. M. Olson, and R. M. Pearlstein 1980. Seven-fold exciton splitting of the 810-nm band in bacteriochlorophyll *a*-proteins from green photosynthetic bacteria. *Biochim. Biophys. Acta* 591: 203–207.

66. Whitten, W. B., R. M. Pearlstein, E. F. Phares, and N. E. Geacintov 1978. Linear dichroism of electric field oriented bacteriochlorophyll *a*-protein from green photosynthetic bacteria. *Biochim. Biophys. Acta* 503: 491–498.

67. Wynne, K., and R. M. Hochstrasser 1993. Coherence effects in the anisotropy of optical experiments. *Chem. Phys.* 171: 179–188.

Use of a Monte Carlo method in the problem of energy migration in molecular complexes

Andrey A. Demidov
Northeastern University, USA

11.1 INTRODUCTION

Incoherent energy migration in a complex molecular ensemble (CME) is a process wherein individual excitations jump from one molecule or chromophore to another in a random manner (see Fig. 11.1), and the resulting (macroscopic) flow of excitation is governed by the statistics of the individual (microscopic) jumps. The Monte Carlo method is one that can directly simulate this process and generate data that are most closely related to real physical experiments. In this chapter we will discuss in detail the Monte Carlo method in its application to rigid CMEs. We will use both the terms *molecule* and *chromophore*, when referring to the donor or acceptor of the excitation, but with the same underlying physical description – the choice will be based on the current context and the type of CME under consideration. Each CME is represented as a rigid complex, i.e. a complex containing molecules (chromophores) that have fixed positions and orientations in a molecular frame. In general, this limitation can be omitted, but the ensuing case falls outside of our consideration.

At the heart of the Monte Carlo method is a "gambling" with the movement of excitations at the microscopic level. We will represent such excitations as

Resonance Energy Transfer. Edited by David L. Andrews and Andrey A. Demidov.

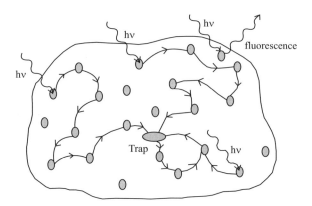

Figure 11.1 A scheme of a complex molecular ensemble (CME) with a trap. The scheme shows capture of excitation photons $h\nu$, a random walk of excitations between molecules in the CME, trapping, escape from a trap, and fluorescence

"localized excitons" – that is, quasi-particles that carry excitation energy from one chromophore to another. In contrast to the concept of delocalized excitons, which is especially used in the physics of crystalline matter, a localized exciton can be found (localized) on some particular chromophore at any given time. For brevity, later we will omit the descriptor "localized" and just use the term *exciton*. In our calculations, the outcome of exciton jumping from donor to acceptor chromophores is defined by "rolling dice" – i.e. through operation of a random number generator. To obtain the macroscopic response, one has to perform an average over millions of microscopic events – the larger the number of averages, the better (in terms of signal-to-noise ratio) the resulting data (see below). There are two principal averaging procedures: (i) averaging over a *large number* of CME or by generating a *large number* of excitons in a single CME (*number averaging*); and (ii) averaging over a *long time (time averaging)* for a limited number of CME – or even one. In the latter case it could be just one CME that is populated by a limited number of excitons, once per calculation cycle, and the cycles are "played" over and over until the required level of signal-to-noise (S/N) ratio is reached. From the ergodic theorem, it transpires that both kinds of averaging procedures produce the same result.

In our research we use the approach of time averaging for the following reasons. Our CMEs have a limited number of chromophores (from a few to up to hundreds) and this is not enough to obtain reasonable statistics under the number averaging model. We cannot generate a large enough number of excitons in our CMEs to give satisfactory statistics: moreover, exciton–exciton interactions would also be a problem. The generation of a large number of CMEs is not a solution either. Indeed, for better statistics one needs a large number of CMEs that must be computed at the same time, including the data arrays associated with them. However, one cannot simulate and compute more CMEs than are allowed by computer resources. Thus the most reasonable way is to use the time averaging model. In the latter case, far fewer limitations are imposed by computer memory, and the calculation performance is of key importance: the faster the computer, the better are the statistics that can be obtained within a reasonable computing time.

The first version of our Monte Carlo computer program was developed in 1986 [2–4] and ran on the BESM-6 computer (Moscow University), then on a personal AT-286 computer, and the latest version on a DEC-3000 Athena Unix station. The program is written in Fortran-77, with a text size of about 55 kByte. Fortran had been chosen for historical reasons, and proved to be very fast, flexible, and computer platform independent.

11.2   AN ILLUSTRATION OF MONTE CARLO CALCULATIONS IN THE PROBLEM OF FLUORESCENCE DECAY

To get a feeling for the Monte Carlo method, let us first consider the simple case of noninteracting molecules; for example, a low-concentration solution of a common dye. In this case there is no energy transfer between the dye molecules;

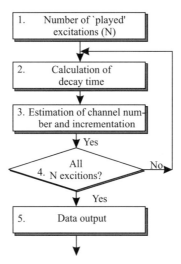

Figure 11.2   A scheme of a Monte Carlo algorithm for calculation of the fluorescence response in the simple case of a dye solution, when molecules do not interact with each other

fluorescent energy transfer is negligible. We will calculate the decay of fluorescence emitted by the dye molecules. The calculations follow the algorithm presented in Fig. 11.2: (*block 1*) setting the number N of absorbed photons to be "played"; (*block 2*) calculation of an excitation decay time for one individual molecule; (*block 3*) estimation of the channel number for statistics accumulation (see details below); (*block 4*) condition check: "Are all N excitations played?" – answer "No," go back to *block 2* to play next excitation; answer "Yes," go to *block 5* (data output) and then exit.

Let the fluorescence lifetime τ of our dye be 0.5 ns. The distribution function for the decay time of individual excitations is defined by the formula

$$t_{\text{decay}} = -\tau \ln(Ran(1)), \tag{11.1}$$

where $Ran(1)$ is a function that generates random numbers in the range $[0, 1]$. The value t_{decay} is the output of *block 2*. Let Δt be the value of the time step that we will call the time step discriminator. The number $K = integer(t_{\text{decay}}/\Delta t)$ is used to determine the "channel" that will serve as the current accumulator of the considered event of excitation decay. In other words, each decay is

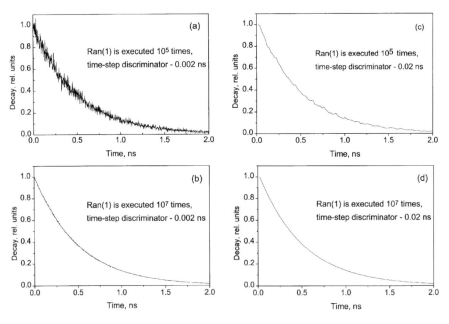

Figure 11.3 Kinetics of fluorescence decay calculated by the Monte Carlo method based on the algorithm shown in Figure 11.2. *Ran*(1) is the function that generates random numbers in the range [0, 1]. Mean value of excitation decay $\tau = 0.5$ ns in all cases. Graphs are calculated using LabView 4.0

considered as a single event that happens at the time t_{decay}; for example, 1 ns. If $\Delta t = 0.01$ ns, then $K = 100$ and the content of an array element (channel) $\text{SUM}_{K=100}$ will be incremented, etc. This method works is a similar way to photon counting techniques. Upon accumulation of statistics, the content of the array SUM_K will reveal the kinetics of fluorescence decay.

The quality of calculation (S/N ratio) significantly depends on the number N of "played" excitations and the size of Δt, as demonstrated in Fig. 11.3. The best S/N ratio is obtained in Fig. 11.3d, but the number of channels represented in this figure is only 100 (2 ns/0.02 ns), whereas Figs 11.3a, b have 1000 channels. The improvement of the S/N ratio with the decrease of Δt happens because of the tenfold increase of the averaging statistics per channel. The cost of this is a decrease in the time resolution (ten times worse in the case considered). The Monte Carlo calculations yield the mono-exponential decay of dye fluorescence with the same fluorescence decay time 0.5 ns, but with a differing S/N ratio and time resolution depending on the number of excitations N and the time step discriminator Δt. The calculations decribed above were performed using Lab View 4.0 programming tools.

11.3 ENERGY TRANSFER IN CME: MAJOR ALGORITHM

In the previous subsection we demonstrated the basic idea of Monte Carlo calculations. In any real CME, the physical processes are more complicated. They involve the statistics of excitation energy absorption, the dynamics of exciton migration within the CME and the different mechanisms of excitation decay. In Fig. 11.4 we present the general algorithm of a Monte Carlo program designed for simulation of the above processes. The calculation begins (Fig. 11.4, *block 1*) with a declaration of a total number of excitations N to be simulated by the program. The next step (*block 2*) is to generate a structure for the CME by assigning position coordinates to all molecules (chromophores) of this complex, and orientations to their transition dipole moments. The latter assignments are necessary if we intend to account for the influence of the orientation factor χ on the efficiency of energy transfer and/or the anisotropy of polarization (of either fluorescence or absorption) in the complex; otherwise, the mean value of $\langle \chi^2 \rangle = 3/2$ is used. The coordinate assignment takes into account the spectral type of each individual molecule. The definition and physical meaning of the orientation factor are described elsewhere in this book.

Let us say that we have M molecules in the CME and that they consist of three spectral types ($s = 1, 2, 3$). In *block 2* each molecule i will be labeled with a "tag" T_i that contains its molecular coordinates $\{\mathbf{r}_i\} = \{x_i, y_i, z_i\}$, the orientation of its transition dipole moment $\{\mathbf{n}_i\} = \{n_{xi}, n_{yi}, n_{zi}\}$, and its spectral assignment s_i. This labeling could be done by using, for example, data from X-ray crystallography, or from some model structure, or just using a random

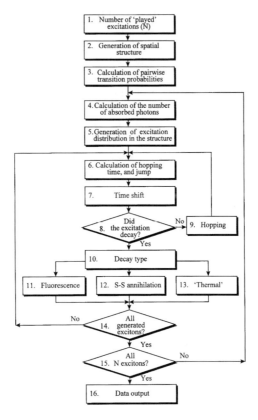

Figure 11.4   The basic algorithm of a Monte Carlo program designed to analyse energy migration in a CME. Detailed description in text

distribution. Each tag T_i is completed by adding spectral properties of spectral type s_i, such as the absorption and fluorescence spectra, the quantum yield of fluorescence and the fluorescence lifetime (or the lifetime of the chromophore/ molecule in the excited state). Thus, from the chromophore number i we obtain its explicit structural–spectral information from the tag T_i. The program can easily be modified for analysis of different CME structures by a simple change of just one subroutine responsible for the generation of $\{\mathbf{r}_i\}$ and $\{\mathbf{n}_i\}$.

Block 3 performs calculation of the probabilities of energy hopping between all pairs of molecules $i, j(i = 1, 2, \cdots, M; j = 1, 2, \cdots, M; i \neq j)$ by calling on their tags T_i and T_j. These tags provide all of the information required for the calculations. For example, let us consider that the energy transfer happens via the Förster mechanism [11] (see also the extensive coverage of other possible transfer mechanisms elsewhere in this book):

$$w_{ij} = \frac{2\chi_{ij}^2}{3} \frac{1}{\tau_{0i}} \left(\frac{R_{0(ij)}}{R_{ij}} \right)^6 = \frac{2\chi_{ij}^2}{3} \frac{\eta_i}{\tau_i} \left(\frac{R_{0(ij)}}{R_{ij}} \right)^6, \tag{11.2}$$

where τ_{0i} is the radiative energy decay time of donor molecule i. From the tag T_i we know that this molecule belongs to the spectral group s_i. The decay time τ_{0i} corresponds to the fluorescence decay time τ_i as $\tau_{0i} = \tau_i / \eta_i$, where η_i is the fluorescence quantum yield. The parameter χ_{ij} is the orientation factor and $R_{0(ij)}$ the Förster radius defined for energy transfer from the donor molecule i to the acceptor molecule j through the overlapping integral of the donor fluorescence and acceptor absorption spectra and the random orientation of their transition dipole moments. The parameter $R_{ij} = |\mathbf{r}_i - \mathbf{r}_j|$ is the distance between donor and acceptor. The tags defined in the previous *block 2* contain all of the information required for the calculations, thus the program can determine all pairwise rates of energy transfer. At this point it should be emphasized that the developed program is *not limited to the Förster mechanism of energy transfer*! Any alternative or combined mechanisms of energy transfer can be accommodated in *block 3* by a simple switch in the subroutine that calculates the rates w_{ij}, the rest of the Monte Carlo program stays intact.

The matrix w_{ij} of pairwise energy-transfer rates defines all possible routes for energy escape from molecule i through hopping to another molecule. This matrix is further extended by including the alternative possibilities of de-excitation via various mechanisms such as fluorescence, thermal decay, capturing by traps, singlet–singlet annihilation, etc.; we can do it through the inclusion of extra elements such as $w_{i,M+1}$ 'thermal decay', $w_{i,M+2} = 1/\tau_{0i}$ (radiative decay), etc. The extended matrix w_{ij} is the master matrix to be used in the body of the Monte Carlo procedure (see below). The best way to carry on further calculations is to use the full matrix w_{ij}, but this may be unreasonable in the cases when the analysed CME is too large and computer resources too limited. Some of our most complicated calculations have run for several weeks without termination.

To overcome this problem, the following method of closest neighbors is used. We can declare the number of nearest neighbors L and make a submatrix $\tilde{w}_{ik}$, where $k = 1, 2, \cdots, L, L+1, L+2, \cdots$, from the master matrix w_{ij}. Here the index i has the same definition as above, whereas the index k defines the number of the current neighbor molecule ($k = 1, 2, \cdots, L$), $\tilde{w}_{iL+1} = w_{iM+1}$, $\tilde{w}_{iL+2} = w_{iM+2}$, etc. A special subroutine was written to make an estimation of $\tilde{w}_{ik}$. This subroutine makes a search around the donor molecule i, and goes through all $j \neq i$, to find L closest acceptors. The number L chosen by the programmer can vary from 1 to $M - 1$. Each of the selected molecules is then given a secondary tag $T2_{ik}$ that allows us to define the global number of the kth neighbor of molecule i. In other words, the global number of the kth acceptor molecule in the closest neighborhood of the donor molecule i is $j = T2_{ik}$, and the latter index defines the principal tag T_j. The method of closest (even relatively remote) neighbors allows a significant reduction of the calculation time, especially for a large CME that has tens or hundreds of chromophores. Instead of making calculations in a space $M \times M$ we can work in a space $\sim L \times L$ (of course, to be more precise we should add decay channels $L + 1$, etc.), where $L \ll M$. If the number L is chosen reasonably, the results of the calculations are practically the same, because remote molecules are very weak acceptors and their contribution to the total dynamics of energy migration can be neglected.

The following stage is to "populate" the CME with excitations. This is performed in *blocks 4 and 5*. The first step is to calculate the probability of each chromophore i capturing the incident photon. This probability is defined by the equation

$$A_i = \sigma_{S_i} (\mathbf{e} \cdot \mathbf{n})^2 N_{\text{flux}} = \sigma_{S_i} \cos^2 \theta \, N_{\text{flux}}, \tag{11.3}$$

where N_{flux} is the flux of incident photons, σ_{S_i} is the absorption cross-section (where s_i defines the spectral type of the chromophore), $\mathbf{e}$ is the polarization of the excitation beam, and θ is the angle between this polarization and the orientation of the transition dipole moment.

The next step depends on the actual intensity of excitation N_{flux}: if $N_{\text{ex}} = \sum_{i=1}^{M} A_i > 0.1$ we go through *branch 1*; otherwise, we go through *branch 2* (both are inside *block 4* and are not shown). *Branch 1* is activated in the case of intensive excitation, $N_{\text{ex}} > 0.1$, where there is a real possibility of having two or more excitons created during the excitation pulse. One might argue that there is a "contradiction" between the latter statement and a seemingly "weak" condition $N_{\text{ex}} > 0.1$, rather than the stronger $N_{\text{ex}} > 1$. We have performed a number of calculations that reveal the appearance of nonlinear phenomena such as singlet–singlet annihilation even at 7% excitation [5]. *Branch 2* is activated for weak excitation, $N_{\text{ex}} < 0.1$, and the molecular complex exhibits a linear response to the excitation intensity. Below, we will consider mainly the latter case, let us say $N_{\text{ex}} = 10^{-4}$. Under the latter condition only one photon is absorbed per 10 000 molecular complexes. Since we have only one molecular complex in the simulation, at first glance it seems that we have to go through 10 000 calculations and that only one will be productive – but in fact this is not so. Indeed, if we admit that the CME linearly responds to the excitation, than we can design *block 4* (explicitly), the "weak" *branch 2*) in such a way that *exactly one* exciton is created per excitation light pulse. That can significantly decrease the computation time without any impact on the result of the calculations. The user can choose any level of threshold of a switch between *branch 1* and *branch 2*. In the above example, we used $N_{\text{ex}} = 0.1$.

Block 5 is the first structure in which we use the Monte Carlo procedure. In this block the program defines the particular molecule that captures the excitation. This is a purely statistical process. Let us consider it in our simple case of low-energy excitation, when we have only one captured exciton per complex. The probability P_i for an individual chromophore to capture excitation at the wavelength λ_{ex} is defined above as $\sigma_{S_i} \cos^2 \theta$. We could use the simple method of rolling a dice to define the "lucky" chromophore. Figure 11.5 illustrates this process, first as the program calculates the sum $P = \sum_{i=1}^{M} P_i$, and then as it generates the random number Ran (1) in the range $[0, 1]$. If all probabilities P_i are lined up by their indices, then the end of the "line-pointer" $R = P * Ran$ (1) would point to the chromophore that captures the photon of excitation. Evidently, the larger P_j is, the better are the chances for a chromophore j to be a "winner". The same principle will be applied later when we face the problem of choosing where the excitation will go from the excited molecule: either to another chromophore, through decay channels or elsewhere.

When it is defined which chromophore has captured the excitation, this exciton is given its own individual number (excitation tag), containing a record of the chromophore occuped by the excitation. The next block, *block 6*, deals with excitation hopping from one chromophore to another (see below) and the jump is recorded as the change of chromophore number on the excitation tag. In *block 6* we have the Monte Carlo procedure again. It follows the same pattern as described above and illustrated in Fig. 11.5. Briefly, let us say that the exciton is located on chromophore i; the excitation has a number of probabilities w_{ij} (Eqn 11.2) of jumping to another molecule ($j = 1, 2, \ldots, M$; $j \neq i$) and a number of decay probabilities $w_{i,M+1}, w_{i,M+2}$, etc. Thus the total probability for excitation to leave chromophore i is

$$W_i = \sum_j w_{ij}, \qquad j = 1, 2, \ldots M, M + 1, M + 2, \ldots, \quad j \neq i. \qquad (11.4)$$

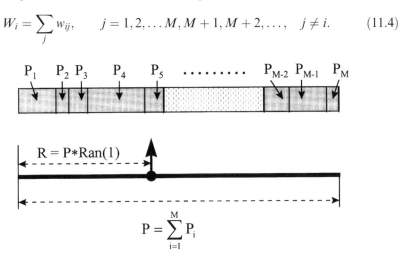

Figure 11.5 A scheme to demonstrate the Monte Carlo procedure of choice selection. The lengths of the pieces marked $P_1, P_2 \cdots$ are proportional to the subsequent probabilities (see text). $Ran(1)$ generates a random number in the range $[0, 1]$. The arrow points to the "winner"

The search for escape routes is conducted either in the domain of the nearest neighbors (see the above definition) or in the whole molecular complex. The parameter W_i is a rate constant for the departure of excitation from chromophore i, and thus the *mean* lifetime of *excitation localization* on this chromophore is defined as $1/W_i$. The next step is to generate the random number $Ran(1)$ and multiply it by W_i: if we had all "extended" probabilities w_{ij} lined up, then the product $W_i Ran(1)$ would point to the acceptor chromophore or decay channel for the excitation. The completion of exciton hopping will be registered by overwriting the record "molecule i is the host" on the exciton tag with, let us say, "molecule k is the host," or "fluorescence decay has happened," etc.

Thus we have a "winner" selected and we know how to update the records on the exciton tag (*block 9*), but we do not yet know the right "moment" to effect this update! This problem is considered in *block 7*. The mean lifetime of excitation localization defines the average time over which the exciton "sits" on the molecule, "looks around," and "decides" which chromophore to go to. The actual jump is considered as an instantaneous process, and the time of exciton localization, t_{loc}, in each particular case is statistically described by the distribution given above:

$$t_{\text{loc}} = -\frac{1}{W_i} \ln(Ran(1)). \tag{11.5}$$

At this stage, we again roll our dice ($Ran(1)$) to define the value of t_{loc}.

The concept of the exciton localization time is very important: it allows us to use the actual dynamics of excitation movement in the molecular complex and avoid the limitation that would be imposed by a fixed time step. In other words, we do not force the excitation to move at fixed (given *a priori*) intervals, but allow it to decide how large the time step is, the decision being acknowledged at each particular event of excitation hopping. The program has a special parameter T that tracks the absolute time counted from the moment ($T = 0$) when the excitation light pulse hits the CME. This parameter T increments in steps t_{loc}. The iteration process loops (*Loop #1*) through the blocks $6 \rightarrow 7 \rightarrow 8 \rightarrow 9 \rightarrow 6 \dots$ until *block 8* triggers a flag that exciton goes into the decay channel. At this point we have a great deal of information about the history of the "played" exciton; in particular, on which chromophore it was "born," how long it "lived" (T), which is the last chromophore on which it was localized, what caused its "death" (fluorescence, etc.), how many jumps it made during its lifetime, etc. We have tested this algorithm on a CME with all chromophores having the same fluorescence decay time τ, and with fluorescence being the only channel of excitation decay – no reabsorption was allowed. The kinetics of fluorescence emitted by such a CME is found to be mono-exponential with a mean time τ that is independent of the dynamics of excitation migration within the CME.

Now, consider what happens if the exciton goes into the decay channels (exit "Yes" in *block 10*). This is its last move. In *block 10* it is verified what the decay type is (fluorescent, thermal etc.) and the program will correspondingly proceed to one of the "decay" blocks (11, 12, 13, etc.). Let us consider first the case of "fluorescent" decay; the cases of "singlet–singlet annihilation" and "trapping" (the latter block not shown on the scheme) will be discussed later.

11.3.1   The case of fluorescent decay

When an exciton moves to the fluorescence channel (*block 11*) the following happens. The exciton tag provides information on from what particular molecule (and consequently its spectral type, position, and orientation) the fluorescence photon is emitted and at what particular time it happens. The counter of fluorescence events makes a record of this particular event. One part of this record is the incrementation of the appropriate accumulator in the specified channel SUM_k. This process is absolutely identical to one described above for a solution of fluorescent dye. The exciton tag is then disabled; this means that the exciton ceases to exist and the number of excitons generated by the preceding light pulse decrements. Obviously, in the case of weak excitation the result of decrementation is zero, but in the case of intense excitation it may not be. *Block 14* checks the routes of further calculations.

If *block 14* finds some excitons with enabled tags, it sends the program back into *block 6* to "play" these excitons. Let us call this *Loop # 2*. This situation may happen under powerful excitation ($N_{ex} > 0.1$), when the number of generated excitons per pulse (see *block 4*) could be more than one. *Loop # 2* proceeds until all excitons are taken out of the CME via any combination of decays. In the case of low-intensity excitation, we have only one exciton "born" in *block 4*; thus, evidently, in this case control is passed immediately from *block 14* to *block 15*. The latter block is responsible for execution of the global loop (*Loop # 3*) – back to *block 4*. This happens when the number of excitons that has been simulated is less than the number N declared in *block 1*. We have used a number N in the range 10^4–10^6; the larger the number, the better is the result of calculations (better S/N ratio).

Bearing in mind our example of fluorescent decay, we will find that after completion of the global loop we will have a lot of information about exciton statistics, dynamics, and kinetics of fluorescence after initial excitation. By including the various counters in the program, we can track all of the details of the excitation dynamics. In particular, in the case of fluorescence we can make a selection over the chromophore types and account for the orientation of their transition dipole moments. This allows us to define the decay of fluorescence emitted from the chromophore of specified spectral types, and the anisotropy of fluorescence. The program finishes the calculations with an output of the calculated data (*block 16*).

11.3.2 The case of singlet–singlet annihilation

This kind of excitation annihilation or decay (*block 12*) happens when we have more than one exciton existing in the CME simultaneously, and one of the excitons attempts to jump from one excited molecule to another, which is already occupied by another exciton (of course, the probability for such a transition is different from a transition from an excited to an unexcited molecule). Upon such an event, one of the excitons will be annihilated and its tag disabled.

11.3.3 The case of a trap

The CME may contain a trap (not shown) for the excitation. If the trap is "black," in the sense that there is no escape from it, decay of excitation happens following the algorithm similarly to the case of fluorescent decay. If the trap is "gray" there is a finite probability that excitation would escape back into the molecular complex and continue its "travel" between chromophores (see Fig. 11.1). The probability of escape is played according to the same Monte Carlo rules as discussed above.

11.4 APPLICATIONS OF MONTE CARLO SIMULATIONS

11.4.1 Kinetics of β155 fluorescence in C-phycocyanin, with no traps

One of the prime goals of our Monte Carlo program is to calculate the kinetics of excitation decay in a CME. This decay can be revealed through the decay of fluorescence. One of the advantages of computer simulation is its ability to target a specific site of a CME and filter out its contribution from the rest of the ensemble. Even when different types of chromophores give strongly overlapped spectra, the fluorescence decay from a selected group may be revealed by making appropriate index selection in the statistical averaging. In particular, it is very easy to filter out the fluorescence signal associated with, say, emission from the $\beta155$ chromophores in C-phycocyanin (C-PC). C-PC is a CME with a well-defined molecular structure and spectroscopic properties [1, 10, 16, 19, 20]; it is part of a light-harvesting complex that can be isolated from cyanobacteria. C-PC can be extracted in the form of a sub-unit, a monomer, a trimer, a hexamer, and a few hexamer rods. A detailed description of C-PC aggregates can be found in Chapter 5. Briefly, C-PC contains chromophores of three spectral types, $\alpha 84$, $\beta 84$, and $\beta 155$, all of which can absorb light and fluoresce, and there is efficient energy transfer between them. The C-PC monomer contains three chromophores, one of each type; the trimer, three of each type; and the hexamer, six of each type. The $\beta 155$ chromophores serve as sensitizing sites that make a minor contribution to the bulk fluorescence of C-PC, because excitation quickly migrates from these chromophores on to $\alpha 84$ and $\beta 84$ chromophores. In [7, 8] we have used Monte Carlo methods to calculate the kinetics of $\beta 155$ fluorescence in monomer, trimer and hexamer C-PC (see Fig. 11.6). In these numerical experiments we selectively "excited" only $\beta 155$ chromophores.

In Fig. 11.6, the solid curves are least-square fits that show the principle dependence of the fluorescence decay on time. The results of our simulation are in good agreement with experimental data [15, 18, 23]. In a number of works, it has been determined that the fast component of C-PC fluorescence in various aggregates is associated with the $\beta 155$ chromophores, falling into the following ranges for the monomer (τ_m), trimer (τ_t) and hexamer (τ_h): $\tau_m, \tau_t, \tau_h = 50$ ps, 36 ps, 10 ps [15]; $\tau_t = 36$ ps [23]; $\tau_m, \tau_t = 57$ ps, 27 ps [18]. Our calculation yields $\tau_m, \tau_t, \tau_h = 48$ ps, 33 ps, 11 ps. Here $\tau_m = 48$ ps is defined as the mean value that would be obtained from a mono-exponential fit in Fig. 11.6.

11.4.2 Calculation of averaged rates of energy transfer in molecular complexes

In real physical experiments, instruments can detect only macroscopic responses from an investigated object. The detected signal is an average, either over a statistically significant number of ensembles or over a period of time. In experiments with energy flow from chromophores of one spectral type to chromophores of another spectral type, the physical instrument generally registers an averaged (bulk) response from numerous chromophores in various molecular complexes – it cannot recognize the contribution of any individual (microscopic) event of energy transfer, but only the averaged result of such events. These are the macroscopic energy-transfer rates.

Let us consider Fig. 11.7. The complex shown consists of two molecules of the first spectral type, and six molecules of the second spectral type. Let the first type of molecules be donors and the second acceptors. As defined above, the rates of energy transfer from the donor molecules 1 and 2 to the acceptor molecules are given by

$$W_{1a} = \sum_{j=1}^{6} w_{1j}, \qquad W_{2a} = \sum_{j=1}^{6} w_{2j}. \qquad (11.6)$$

The overall rate of energy transfer is $k_{da} = \frac{1}{2}(W_{1a} + W_{2a})$. The hypothetic macroscopic detector cannot distinguish between the contributions of

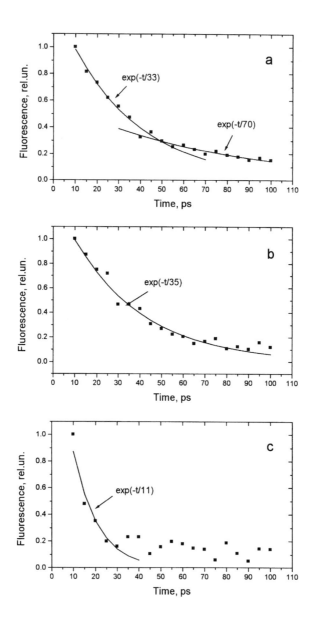

Figure 11.6 Kinetics of fluorescence emitted by $\beta 155$ chromophores in C-PC: (a) monomer; (b) trimer; and (c) hexamer. Points are data calculated by the Monte Carlo method and solid lines are mono- and bi-exponential fits. Excitation conditions described in text

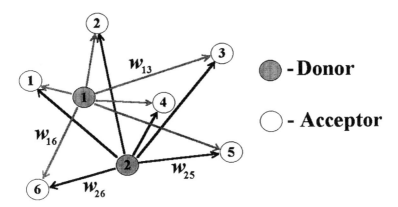

Figure 11.7 A model of CME consisting of molecules of two spectral pools. The first "donor" pool contains two molecules; the second "acceptor" pool, six molecules. Parameters w_{ij} give the probability of excitation transfer from molecule i to molecule j

molecules 1 and 2. Thus, the microscopic scheme of Fig. 11.7 can be simplified to give the macroscopic scheme shown in Fig. 11.8. The macroscopic rates of energy transfer k_{da} and k_{ad} shown in the latter figure fully describe the general features of energy transfer between two spectral pools of chromophores. In general, when the CME contains N donor and M acceptor molecules, the averaged rate k_{da} of energy transfer from the donor to acceptor molecules is [7, 8]

$$k_{da} = \frac{1}{N} \sum_{i=1}^{N} \sum_{j=1}^{M} w_{ij}. \tag{11.7}$$

In such complexes the energy equilibration is described by a well-known system of balance equations, which is based on the "bulk" model (see Fig. 11.8):

$$\frac{dn_d}{dt} = -\frac{1}{\tau_d} n_d - k_{da} n_d + k_{ad} n_a + F_d, \qquad \frac{dn_a}{dt} = -\frac{1}{\tau_a} n_a + k_{da} n_d - k_{ad} n_a + F_a. \tag{11.8}$$

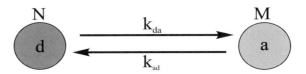

Figure 11.8 The "bulk" model of energy transfer between two pools of molecules D (donor) and A (acceptor)

Here $n_{d,a}$ are the excited states populations of donor (d) and acceptor (a) molecules, $\tau_{d,a}$ are the intrinsic decay times of excitation, and $F_{d,a}$ represents excitation of molecules by an external source. The rates $k_{da}(k_{ad})$ are the same as calculated in Eqn 11.7. Our Monte Carlo program can calculate these macroscopic rates. Moreover, it can define the *dynamics* of rate "constants." What does this mean, and why is the word "constant" in quotes? In fact, the complete formula has the form [7, 9]:

$$k(t)_{da} = \frac{1}{N} \sum_{i=1}^{N} \sum_{j=1}^{M} w_{ij} g(i, t), \qquad (11.9)$$

where $g(i, t)$ is a function proportional to the probability of finding an exciton on the donor molecule i. When this distribution is uniform and time-independent (which is commonly the case) $g(i, t)$ is a constant and the rate of energy transfer is constant too. Nevertheless, in some cases one can find that energy-transfer rates are time-dependent [11, 12, 14, 24]. Our program covers both cases, and we will demonstrate this kind of time-dependence below.

11.4.3   Averaged rates of energy transfer in C-PC aggregates, with no traps

In aggregates of C-PC without traps, the function $g(i, t)$ is a constant [7, 9] and the macroscopic rates of energy transfer can be calculated using Eqn 11.7. The result of these calculations is presented in Table 11.1. These rates can be used directly with the system of balance equations similar to Eqn 11.8 [7, 8].

The procedure of rate calculation in our Monte Carlo program has the flexibility to reveal the major channels of energy transfer in the CME. This is achieved by calculating the energy transfer between different subgroups of the chromophores, i.e. rate calculations can be conducted within and between various domains of chromophores (molecular indices). We have used this feature to find the major energy flow channels in the C-PC rod (see Fig. 11.9). Figure 11.9 shows energy-transfer rates greater than 35 ns^{-1}; weaker rates are not shown, because they do not make a significant contribution to the overall energy flow in the C-PC rod. One can see that in the C-PC trimer the fastest process is energy equilibration between $\alpha 84$ and $\beta 84$ chromophores. The major paths of energy flow between trimers are the $\alpha 84$–$\alpha 84$ and $\beta 155$–$\beta 155$ channels. We have mentioned above that the latter chromophores mainly serve as sensitizers – they do not play a significant role in energy flow along the C-PC rod. Thus, the prime channel of energy transfer inside the hexamer is the $\alpha 84$–$\alpha 84$ path. Energy flow *between* hexamers proceeds through the $\beta 84$–$\beta 84$ channel. The latter channel is a "bottle neck" for the overall flow of energy along the C-PC rod.

Table 11.1   Rates of energy transfer among C-PC chromophores (ns^{-1})

	Acceptor		
Donor	$\alpha 84$	$\beta 84$	$\beta 155$
Monomer			
$\alpha 84$	0	12.4	0.445
$\beta 84$	10.5	0	7.54
$\beta 155$	2.04	34.1	0
Trimer			
$\alpha 84$	2.28	1400	0.951
$\beta 84$	1180	18.8	7.87
$\beta 155$	4.36	35.6	0.584
Hexamer			
$\alpha 84$	361	1430	16.4
$\beta 84$	1210	43.8	10.9
$\beta 155$	75	49.5	195
Two-hexamer rod			
$\alpha 84$	370	1440	16.6
$\beta 84$	1220	129	11.5
$\beta 155$	77.2	53.5	198

11.4.4   Efficiency of energy delivery in C-PC rods

C-phycocyanin is a part of the more complex CME called phycobilisome (PBS). In PBS the C-PC assembly, in the form of rods, is attached to allophycocyanin (APC) by the terminal C-PC trimer. The opposite side of the phycocyanin rod is extended by phycoerythrin. Energy absorbed by C-PC chromophores migrates along C-PC rods toward APC, and then from APC to chlorophyll, in the light-harvesting antenna of photosystem 2 [13]. The efficiency of energy utilization captured by C-PC strongly depends on how fast this energy can be delivered to the terminal chromophores of the C-PC rod that is coupled to APC. The conducted analysis of energy flow in C-PC gives a clear idea about the major channels of energy transfer, but not about the overall efficiency. To target the latter problem, we have carried out a special investigation using the Monte Carlo program [7, 8]. For evaluation purposes, we modeled the presence of the APC as introduction of a perfect trap (100% trapping efficiency) on the either of the chromophores ($\alpha 84$, $\beta 84$, or $\beta 155$) in the terminal trimer in the four hexamer rods of C-PC.

Without the trap, the lifetime of excitation in C-PC is defined as $\tau = 1.5$ ns [19] and is independent of the C-PC aggregation and the dynamics of energy

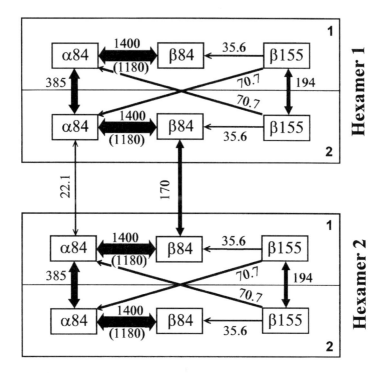

Figure 11.9 A scheme of major energy-transfer channels in C-phycocyanin, showing two adjoining hexamers. Each hexamer consists of two trimers, marked 1 and 2. Direct and backward (bracketed) rates of energy transfer are given in ns^{-1}

transfer. When the trap is present, the lifetime drops significantly because of the energy quenching that it introduces. The lifetime of excitation (τ or $\tau_{tr}^{(k)}$) is calculated as

$$\tau = \frac{1}{N} \sum_{m=1}^{N} \Delta t^{(m)} = \frac{1}{N} \sum_{m=1}^{N} \left[t_d^{(m)} - t_0^{(m)} \right], \qquad (11.10)$$

where $t_0^{(m)}$ is the time-point of the *m*-exciton "birth," and $t_d^{(m)}$ is the time of its "death," *N* being the number of simulated excitons.

In the case of the four-hexamer rod and a trap located on the $\beta 84$ chromophore, the lifetime of excitation $\tau_{tr}^{(4)}$ is about 126 ps, in good agreement with the experimental result of 120–130 ps [21, 22]. This timescale is mainly defined by the mean time of energy flow along the C-PC rod. The selection of $\beta 84$ or $\alpha 84$ as a trap host is not important [7, 8]; the efficiency is practically the same, due to a strong coupling between these chromophores in the same trimer (see Table 11.1 and Fig. 11.9). On the other hand, a model with the trap located on the

$\beta155$ chromophores failed to show a significant efficiency of energy flow – the efficiency is about three times less. Thus, we can exclude the $\beta155$ chromophores as candidates for an energy flow bridge between C-PC and APC. A similar calculation for shorter rods with "$\beta84$" traps [7] yields $\tau_{tr}^{(1)}$, $\tau_{tr}^{(2)}$, $\tau_{tr}^{(3)} \simeq 7.6\text{ps}, 29\text{ps}, 70\text{ps}$. In the case of three-hexamer rods, Suter *et al.* [21, 22] have measured $\tau_{tr}^{(3)} = 83$ ps. It worth mentioning that if the trap is "gray," i.e. not 100% efficient, the calculated τ_{tr} is larger (see below).

11.4.5   Kinetics of fluorescence in C-PC rods with a trap

Let us now consider the kinetics of C-PC fluorescence when there is a trap at the terminal trimer in the four-hexamer rod, at the position of the $\beta - 84$ chromophore. Our Monte Carlo program allows us to calculate the dynamics of energy transfer and its impact on the kinetics of fluorescence. We have found [9] that whereas the energy migration in C-PC between chromophores of various spectral types is practically uniform (at least in close proximity), the rate of energy transfer from chromophores directly to the trap is time-dependent (see Fig. 11.10). This time-dependence can be understood from the following considerations. Initially ($t = 0$), all chromophores are excited homogeneously (of course, the absorption is weighted properly according to the spectroscopic and structural characteristics). Then the excitation population on chromophores in close proximity to the trap is depleted because of fast energy transfer from these chromophores to the trap. Thus we have a build-up gradient of excitation distribution along the C-PC rod (see Fig. 11.11) – there is a lower population of excited molecules close to the traps and a higher population distant from them. This phenomenon causes a decrease in the energy-transfer rates. In other words, we have the dynamics of the $g(i, t)$ function introduced in Eqn 11.9. The described calculations for $k(t)$ were performed with different efficiencies of excitation trapping, varying from 0% to 100%. A value of 10%, for example, means that an exciton has a 90% probability of escaping from the trap after capture.

Now, we will verify our earlier statement that the averaged rates of energy transfer calculated following the recipe of Eqn 11.9 can be used for macroscopic calculations of energy migration. We will analyse the C-PC rod with a trap located on the peripheral trimer and associated with the $\beta84$ chromophores as described above. The C-PC is excited by a δ-pulse at $t = 0$. In this case, the master system contains four balance equations:

$$\frac{dn_1}{dt} = -\frac{1}{\tau_1}n_1 - k_{12}n_1 - k_{13}n_1 - k_1n_1 + k_{21}n_2 + k_{31}n_3 + k_1'n_4,$$

$$\frac{dn_2}{dt} = -\frac{1}{\tau_2}n_2 - k_{21}n_2 - k_{23}n_2 - k_2n_2 + k_{12}n_1 + k_{32}n_3 + k_2'n_4,$$

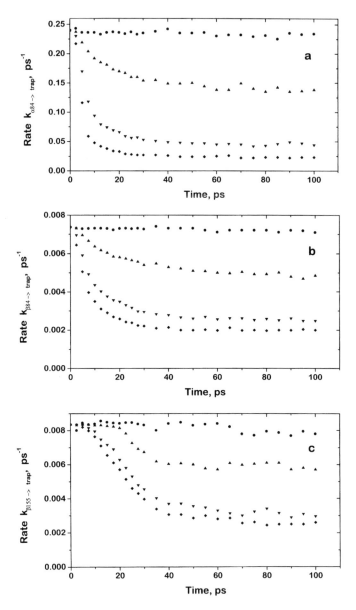

Figure 11.10   The calculated time-dependence of energy-transfer rates in a three hex-amer C-PC rod from $\alpha84$, $\beta84$, and $\beta155$ chromophores directly to the trap – located on the $\beta84$ chromophores of the first (terminal) trimer. C-PC is uniformly excited by a δ-pulse. Cases (a), (b), and (c) represent rates $k_1(t) = k_{\alpha84\to\text{trap}}$, $k_2(t) = k_{\beta84\to\text{trap}}$, and $k_3(t) = k_{\beta155\to\text{trap}}$ respectively. Data calculated for a quantum efficiency of trapping φ_0 equalling (●) 0%, (▲) 10%, (▼) 50%, and (◆) 100%

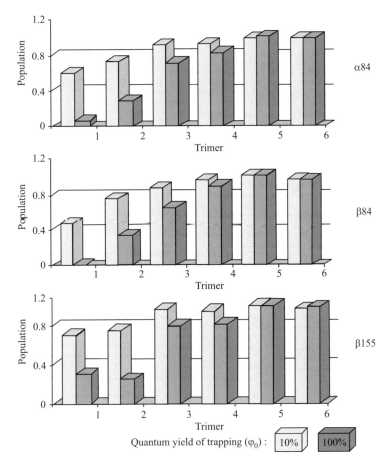

Figure 11.11 Distribution of excitation along a C-PC rod consisting of three hexamers, each hexamer containing two trimers, with a trap located in the first trimer on the $\beta 84$ chromophore. Time is 11 ps after δ-pulse excitation when all chromophores were excited homogeneously

$$\frac{dn_3}{dt} = -\frac{1}{\tau_3}n_3 - k_{31}n_3 - k_{32}n_3 - k_3 n_3 + k_{13}n_1 + k_{23}n_2 + k'_3 n_4,$$

$$\frac{dn_4}{dt} = k_1 n_1 + k_2 n_2 + k_3 n_3 - (k'_1 + k'_2 + k'_3 + q + \tau_4^{-1})n_4. \tag{11.11}$$

Here, n_1, n_2, and n_3 are the excited-state populations of the chromophores $\alpha 84$, $\beta 84$, and $\beta 155$; n_4 is the population of excitation in the trap; k_{ij} is the rate of energy transfer between a donor of spectral form i to an acceptor of spectral form j; k_i is the rate of energy transfer from type i chromophores directly to the trap, and k_i' is the rate of backward energy transfer; $\tau_i (i = 1, \ldots, 4)$ is the intrinsic lifetime of excitation, and q is the rate of excitation quenching in the trap.

As a condition of the simulations, conducted by the Monte Carlo method and using Eqn 11.11, it can be chosen that all chromophores of C-PC are excited with equal probabilities. The rates k_{ij} are taken from Table 11.1 $k_1' \simeq k_{\beta 84,\beta 84} = 140\mathrm{ns}^{-1}$, $k_2' \simeq k_{\beta 84,\alpha 84} = 1210\mathrm{ns}^{-1}$ and $k_3' \simeq k_{\beta 84,\beta 155} = 11.7\mathrm{ns}^{-1}$ (calculated from the Monte Carlo program); the rate parameters k_1, k_2, and k_3 are time-dependent [9], as shown in Fig. 11.10, and the intrinsic lifetimes $\tau_i = 1.5$ ns $(i = 1, \ldots, 4)$. It should be emphasized that no fitting parameters are involved in the calculations! Results from both types of calculations are presented in Fig. 11.12; both are in good agreement, and that indicates the equivalence of the Monte Carlo and master equation (Eqn 11.11)

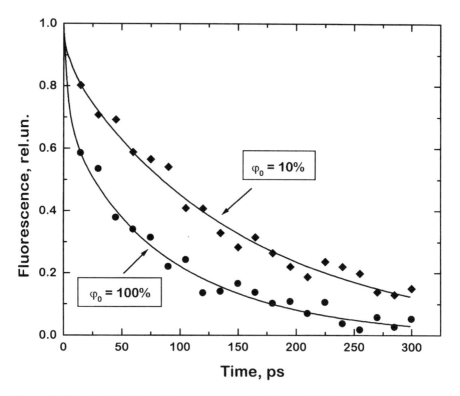

Figure 11.12 The kinetics of fluorescence emitted by a C-PC rod (three hexamers) having a trap in the terminal trimer at the $\beta 84$ chromophore. The trap has quantum efficiency $\varphi_0 = 10\%$ ($\blacklozenge$), 100% ($\bullet$). The solid lines are calculated by a Runge–Kutta method using Eqns 11.11 and 11.12; points are obtained from Monte Carlo calculations, with the number of "played" excitons $\sim 10^5$

calculations. Traps of two efficiencies, $\varphi_0 = 10\%$ and $\varphi_0 = 100\%$, were used. For reference, the value of φ_0 is related to other parameters by

$$\varphi_0 = \frac{\tau_4^{-1} + q}{k_1' + k_2' + k_3' + \tau_4^{-1} + q}. \tag{11.12}$$

We also independently calculate the total quantum yield of exciton trapping both by the Monte Carlo method and from the formula

$$\varphi(\varphi_0) = \frac{\int_0^\infty n_4(t)\mathrm{d}t}{n_1(0) + n_2(0) + n_3(0)}. \tag{11.13}$$

In Monte Carlo calculations the program can simply count how many excitons are "born" in C-PC, and how many of them "die" in the trap. In Eqn 11.13, $n_4(t)$ is defined through solution of Eqn 11.12 and $n_1(0) + n_2(0) + n_3(0)$ is the total number of excitations at $t = 0$. We find good agreement between $\varphi(\varphi_0)$ defined by both methods [9].

11.4.6   Total versus local efficiencies of trapping

Our simulations of energy migration in CMEs with a trap tackle another problem: the dependence of the overall (total) excitation trapping efficiency on quantum (local) trapping efficiency. Figure 11.1 demonstrates the process of an excitation random walk, where the exciton can visit a trap a number of times and escape from it if $\varphi_0 < 100\%$. The Monte Carlo program can give us an answer as to what extent the total efficiency φ of excitation capture depends on the local efficiency of trapping as a single event, φ_0. The latter is defined as

$$\varphi_0 = \frac{\tilde{q}}{\tilde{q} + q_{\mathrm{detrap}}}, \tag{11.14}$$

where $\tilde{q}$ is the rate of excitation quenching and q_{detrap} is the rate of excitation escape back to the molecular complex. With C-PC, Eqn 11.14 converts to Eqn 11.12 with $\tilde{q} = q + \tau_4^{-1}$ and $q_{\mathrm{detrap}} = k_1' + k_2' + k_3'$. After escape, the exciton can migrate for a while and visit the trap repeatedly. Its chances of survival are inversely proportional to the number of such visits and the value φ_0. Let us make a rough evaluation, assuming that $\varphi_0 = 0.8$ and that the exciton "visits" a trap twice. At the first visit, its chance of survival is $\varphi_{\mathrm{survive}} = 1 - \varphi_0$, and after the second visit only $(1 - \varphi_0)^2$. Thus the efficiency of excitation "utilization" after two visits is $1 - (1 - \varphi_0)^2$ that gives us $\varphi_2 \simeq 0.96$. If the number of visits is N, we should expect that φ is proportional to $1 - (1 - \varphi_0)^N$. The more accurate

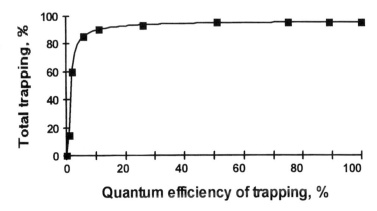

Figure 11.13   The dependence of the total efficiency of excitation trapping φ on the quantum efficiency of trapping φ_0. The quantum efficiency denotes the probability of excitation utilization during the single act of capturing by a trap, i.e. $\varphi_0 = 10\%$ means 90% probability for the exciton to escape back to the antenna. Calculations performed for C-PC rod; trap located in the terminal trimer on the $\beta84$ chromophore

calculations can be conducted by the Monte Carlo method because they involve a random walk of the excitation in the CME. We carried out this kind of analysis with C-PC rods [9] and found the dependence presented in Fig. 11.13. One can see that, beginning from $\varphi_0 = 10\%$, the total efficiency of excitation trapping φ exceeds 90%.

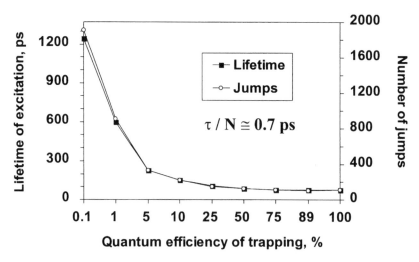

Figure 11.14   The dependence of the excitation lifetime (τ) and number of jumps (N_{jumps}) on trap quantum efficiency φ_0 in a three-hexamer rod. Mean time of one jump $t_{jump} = \tau/N_{jumps} \simeq 0.7$ ps

Now we have another question: What is the dependence of the excitation lifetime on the trap "grayness"? It is obvious that the smaller φ_0 is, the longer is the excitation lifetime. This can be manifest, for example, in a prolonged fluorescence decay time. The calculated dependence of the excitation lifetime in C-PC rods and the number of excitation jumps versus φ_0 is presented in Fig. 11.14 [9]. One can see that the lifetime of excitation is inversely proportional to φ_0, mainly because of the increased number of hops an exciton would make during its life, whereas the mean value of the hopping time is essentially constant. In Fig. 11.14, the curves showing the lifetime of excitation and the number of excitation jumps are strongly correlated, and their ratio, which represents the mean excitation jump time, is roughly constant and equal to 0.7 ps. In our terms, the hopping time is actually the time of excitation localization.

11.4.7   Statistics of the hopping time

In [4, 8, 9] we have found that the statistics of the exciton hopping time (in the investigated molecular complexes) is independent of CME structure and can be described by

$$P = \text{Const.} \left(\frac{t}{t_{\text{jump}}} \right) \exp^{-t/t_{\text{jump}}}, \tag{11.15}$$

where t_{jump} is the mean time of excitation localization on molecules of some particular spectral type. Figure 11.15 gives an example of the function P calculated for C-PC; the dots are data "directly" calculated by the Monte Carlo program, whereas the solid line is plotted using Eqn 11.15. The parameter t_{jump} used in this calculation is determined by the Monte Carlo program as a mean time of excitation localization on the chromophores of either of the spectral types: $\alpha 84$, $\beta 84$, and $\beta 155$ (see Table 11.2). Statistics calculated for, say, $\alpha 84$ and $\beta 84$ are the same, and are independent on aggregation. Again, no fitting parameters are involved in these plots. Both the Monte Carlo and Eqn 11.15 data are normalized to one at $t = t_{\text{jump}}$. In all cases, Eqn 11.15 holds, and we obtain precise agreement between the Monte Carlo data and Eqn 11.15 over a range of at least *three orders* of magnitude. Figure 11.15 tells us that, in fact, the hopping time is spread out over a wide range; many hops occur much faster than t_0 and also many much slower than that, but the mean value is t_{jump}.

11.4.8   Nonlinear phenomena

11.4.8.1   Saturation of absorption

Saturation is a nonlinear phenomenon that happens when the number of excited molecules becomes a significant fraction of the total (10% is commonly

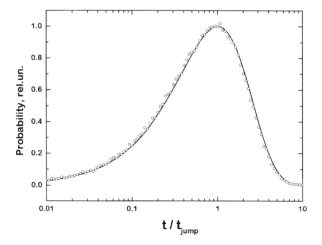

Figure 11.15   The distribution of the probability of exciton jump time (t) in a C-PC rod; t_{jump} is the mean value of excitation localization on chromophores of a selected (or all) spectral pool. Points are calculated by the Monte Carlo method, and the solid line by using Eqn 11.15

Table 16.2   Mean values of exciton hopping (localization) time, t_{jump}, from specific C-PC chromophores (ps)

C-PC aggregate	$\alpha 84$	$\beta 84$	$\beta 155$
Monomer	74	54	27
Trimer	0.71	0.83	24
Hexamer	0.56	0.81	3.3
Two-hexamer rod	0.55	0.76	3.2
Four-hexamer rod	0.55	0.74	3.2

adequate for a threshold). The physical reason for this phenomenon is the decreased number of molecules in the ground state. This phenomenon is also commonly called *bleaching*. The effect becomes apparent under intense excitation (Fig. 11.16). We have covered this possibility in *block 4* by use of the following simple procedure. For each pulse, the program defines how many photons can be absorbed if the absorption is linear. Then, we begin to play these excitations one after another. Let us say that in one considered pulse we obtain K photons that can potentially be absorbed by N molecules, and $K > 1$, let us say $K = 4$. After playing the first excitation absorption we have one molecule occupied and the number of potential excitations decremented by one; then the second excitation comes into play – again amongst *all N* molecules. If the program finds that this excitation hits the molecule that has already been

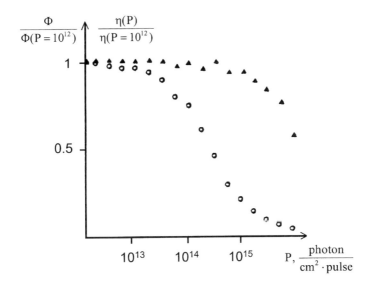

Figure 11.16 The calculated dependence of absorption efficiency – absorption saturation, η (◆), and fluorescence quantum yield Φ (○) on excitation intensity [5]

excited, then this excitation is discarded and dropped from consideration. The procedure is repeated until the last (fourth) excitation, and this guarantees that at any level of excitation the number of photons actually absorbed is no larger than N. The initial number K is defined statistically and may vary from pulse to pulse.

11.4.8.2 Singlet–singlet annihilation

The phenomenon of singlet–singlet (S–S) annihilation is extensively investigated in Chapter 7. Examples of Monte Carlo simulation of the process can be found in [5, 6]. Annihilation happens when two excitons are found in close proximity and one of them moves on to another excited molecule. This results in de-excitation of the donor molecule, with the acceptor rapidly losing the excess energy by nonradiative means, so that the acceptor is occupied by only *one* exciton; as a result, one exciton is annihilated. Annihilation may occur between excitons hosted by chromophores of either the same or of different spectral types. It does not matter which particular chromophore captured the excitation first – on migration, the exciton forgets which was its "parent." To some extent, this resembles the case of trapping, but where the traps are excitons themselves. These "traps" are free to move, however, i.e. they are dynamically distributed in the CME.

Let us first consider the simple case of a homogeneous system with all molecules of the same spectral type. We will focus on the rate of exciton transfer on to another excited molecule. Let the number of excitons in the CME be E. Then the total rate of annihilation of the excitation i is defined by

$$G_i^{SS} = \frac{1}{E-1} \sum_{\substack{j=1 \\ i \neq j}}^{E} w_{ij}^{SS}, \qquad (11.16)$$

where w_{ij}^{SS} is the pairwise rate of S–S annihilation for excitons i and j. At each particular moment, we know the exact position of both excitons because of the contents of their tags. The value of w_{ij}^{SS} can be calculated, for example, using the Förster mechanism, in which the fluorescence spectrum of the donor molecule is as already defined and the absorption spectrum of the excited acceptor relates to absorption from the excited state to a higher excited state. The averaged pairwise rate of S–S annihilation is defined as

$$\gamma_{SS} = \frac{1}{E(E-1)} \sum_{i=1}^{E} \sum_{\substack{j=1 \\ i \neq j}}^{E} w_{ij}^{SS}. \qquad (11.17)$$

In the case of spectrally heterogeneous structures, the rates of S–S annihilation must be calculated both for chromophores of the same spectral types and also for excitations located in different spectral pools. The latter case involves the selection of various sums $\sum \sum w_{ij}^{SS}$ over different pools. Our Monte Carlo program supports all types of these calculations [5, 6].

The common way to describe S–S annihilation is to use balance equations with an annihilation component, $-\gamma n^2$, where n is the density of excitation population and γ is the S–S constant – in our terms, $\gamma = \gamma_{SS}$. One must be careful with the units of γ_{SS}. Equation 11.17 defines calculation of γ_{SS} in terms of s^{-1}, whereas in the balance equation n is measured in cm^{-3} and γ in $s^{-1} cm^3$; conversion is by multiplying the value of γ_{SS} from (Eqn 11.17) by the volume V_{CME} of the CME in cm^3 [6]. Now, with reference to the earlier statement that S–S annihilation is a case of dynamically distributed traps, the question arises as to whether or not γ_{SS} is a constant. The answer is that it is time-dependent: this is revealed by the Monte Carlo program. Figure 11.17 shows this kind of dependence in a model system of phycoerythrin (a pigment of PBS) [5]. Similar behavior is observed experimentally in light-harvesting antennae [17]. The effect can be explained by the same effect as discussed earlier with traps – depletion of the exciton population in the vicinity of a trap [6, 17].

The efficiency of S–S annihilation depends on the structure of the molecular complex [6] and the energy of excitation [5]. The former dependence is due to restrictions that the CME structure may impose on the mobility of excitons,

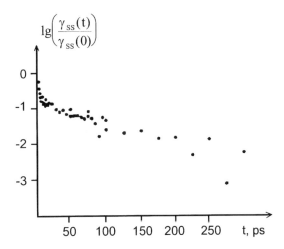

Figure 11.17 The calculated time-dependence of the macroscopic parameter of singlet–singlet annihilation (Eqn 11.17) in phycoerythrin (PE) pigments of phycobilisomes [5]. Excitation intensity 10^{15} photons cm^{-2} pulse^{-1}, $\gamma_{SS}^{PE-PE}(0) = 3.9 \times 10^{-2}$ ps^{-1}

whereas the latter defines the number of excitations and thus the number of dynamical traps created at $t = 0$. Being one of the channels of excitation quenching, S–S annihilation competes with radiative decay.

The nonlinear dependence of fluorescence output on the intensity of excitation pulses is shown in Fig. 11.16. This figure also shows the absorption saturation curve. One can see that the nonlinear effect of S–S annihilation reveals itself at a lower level of excitation than absorption saturation. Our Monte Carlo calculations [5] allow us to estimate that S–S anihilation may start at unexpectedly low levels of excitation. For example, we found in the model of PBS that nonlinear phenomena occur at the excitation intensity 10^{13} photon cm^{-2} pulse^{-1}. At this level, the probability of generation of one exciton per whole PBS is about 7% for a single pulse. In four cases, per 100 *absorbed* excitons there is a simultaneous generation of two or more excitons per one CME during excitation pulse.

11.5   CONCLUSION

In this chapter, the basic principles of the Monte Carlo method for describing energy migration have been demonstrated. We have shown the flexibility and power of this method for the analysis of energy migration in complex molecular systems with a rigid structure. Monte Carlo calculations can provide spectroscopic information similar to that determined by alternative methods such as "master equations" approaches, but the Monte Carlo method can give us more – statistical data on the excitation dynamics. In particular, one can calculate the statistics of excitation jumps, etc. In the current work, we have covered some linear and nonlinear phenomena in energy absorption and migration but, obviously, Monte Carlo calculations can go beyond the analysis of these processes. Our program is flexible and can be readily adapted to a variety of phenomena and to specific CMEs.

Acknowledgments

I would like to acknowledge a great deal of support and contribution for this work by my friends and co-workers, Dr. Ya. L. Kalaidzidis, Professor A. Yu. Borisov, and Professor D. L. Andrews.

References

1. Debreczeny, M. P., K. Sauer, J. H. Zhou, and D. A. Bryant 1993. Monomeric C-phycocyanin at room temperature and 77 K resolution of the absorption and fluorescence spectra of the individual chromophores and the energy transfer rate constants. *J. Phys. Chem.* 97: 9852–9862.
2. Demidov, A. A. 1986. Application of numerical methods for investigations of energy migration in organic complexes. In IV *Soviet Colloquium "Laser Application in Biology,"* Kishinev (USSR), October 2–6, pp. 59–63.
3. Demidov, A. A. 1987. Application of the Monte Carlo method for investigating of energy migration in complex organic compounds. *Moscow University Physics Bulletin* 42(3): 74–80 (English translation from *Vestnik Moskovskogo Universiteta. Fizika,* 1987, 28(3): 63–68). Allerton Press, Inc.
4. Demidov, A. A. 1988. A Monte Carlo study of energy transfer time dependence in organic complexes. *Moscow University Physics Bulletin* 43(2): 38–41 (English translation from *Vestnik Moskovskogo Universiteta. Fizika,* 1988, 29(2): 38–42). Allerton Press, Inc.
5. Demidov, A. A. 1988. Simulating nonlinear optical processes in organic complexes. *Moscow University Physics Bulletin* 43(5): 61–65. (English translation from *Vestnik Moskovskogo Universiteta. Fizika,* 1988, 29(5): 56–61). Allerton Press, Inc.
6. Demidov, A. A. 1989. Numerical analysis of the connection between the singlet–singlet annihilation pair coefficient and the spatial structure of a complex of a limited number of molecules. *Journal of Applied Spectroscopy* 50(1): 87–89. (English translation from *Zhurnal Prikladnoi Spektroskopii* (USSR) 1989, 50(1): 103–106). Plenum Press New York.
7. Demidov, A. A., and A. Yu. Borisov 1993. Computer simulation of energy migration in the C-phycocyanin of the blue–green algae *Agmenellum quadruplicatum. Biophys. J.* 64: 1375–1384.
8. Demidov, A. A., and A. Yu. Borisov 1993. Numerical modeling of energy migration in C-phycocyanin of the blue–green alga *Agmenellum quadruplicatum. Biofizika* (Russia) 38(1): 133–143.

9. Demidov, A. A., and A. Yu. Borisov 1994. Computer simulation of exciton jumping statistics and energy flow in C-phycocyanin of algae *Agmenellum quadruplicatum* in the presence of traps. *Photochem. Photobiol.* 60: 46–52.

10. Demidov, A. A., and M. Mimuro 1995. Deconvolution of C-phycocyanin β-84 and β-155 chromophore absorption and fluorescence spectra of cyanobacterium *Mastigocladus laminosus. Biophys. J.* 68: 1500–1506.

11. Förster, Th. 1948. Zwischenmolekulare Energiewanderung und Fluoreszenz. *Ann. Phys.* 6: 55–75.

12. Galanin, M. D. 1955. On the question about influence of concentration on the solution fluorescence. *Zh. Eksp. Teoret. Fiz.* (Russia) 28: 485–495.

13. Gantt, E. 1981. Phycobilisomes. *Ann. Rev. Plant Phys.* 32: 327–347.

14. Gösele, U., M. Hauser, U.K.A. Klein, and R. Frey 1975. Diffusion and long range energy transfer. *Chem. Phys. Lett.* 34: 519–522.

15. Holzwarth, A. R., J. Wendler, and G. W. Suter 1987. Studies on chromophore coupling in isolated phycobiliproteins. 2. Picosecond energy transfer kinetics and time resolved fluorescence spectra of C-phycocyanin from *Synechococcus-6301* as a function of the aggregation state. *Biophys. J.* 51. 1–12.

16. Mimuro, M., P. Füglistaller, R. Rumbeli, and H. Zuber 1986. Functional assignment of chromophores and energy transfer in C-phycocyanin isolated from the thermophilic cyanobacterium *Mastigocladus-laminosus. Biochim. Biophys. Acta* 848: 155–166.

17. Rubin, L. B., O. V. Braginskaya, M. L. Isakova, and N. A. Efremov 1985. Investigation of migration parameters of electron excitation energy in photosynthetic and artificial systems. *Photochem. Photobiol.* 42: 77–87.

18. Sandström, A., T. Gillbro, V. Sundström, R. Fischer, and H. Scheer 1988. Picosecond time-resolved energy-transfer within c-phycocyanin aggregates of *Mastigocladus-laminosus. Biochim. Biophys. Acta* 933: 42–53.

19. Sauer, K., and H. Scheer 1988. Excitation transfer in C-phycocyanin Förster transfer rate and exciton calculations based on new crystal structure data for C-phycocyanins from *Agmenellum-quadruplicatum* and *Mastigocladus-laminosus. Biochim. Biophys. Acta* 936: 157–170.

20. Schirmer, T., W. Bode, and R. Huber 1987. Refined three dimensional structures of two cyanobacterial C-phycocyanins at 2.1 and 2.5 Å resolution. A common principle and phycobilin–protein interaction. *J. Mol. Biol.* 196: 677–695.

21. Suter, G. W., and A. R. Holzwarth 1987. A kinetic-model for the energy transfer in phycobilisomes. *Biophys. J.* 52: 673–683.

22. Suter, G. W., P. Mazzola, J. Wendler, and A. R. Holzwarth 1984. Fluorescence decay kinetics in phycobilisomes isolated from the blue–green alga *Synechococcus 6301. Biochim. Biophys. Acta* 766: 269–276.

23. Wendler, J., W. John, H. Scheer, and A. R. Holzwarth 1986. Energy-transfer in trimeric C-phycocyanin studied by picosecond fluorescence kinetics. *Photochem. Photobiol.* 44: 79–85.

24. Yokota, M., and O. Tanimoto 1967. Effects of diffusion on energy transfer by resonance. *J. Phys. Soc. Japan* 22: 779–784.

Index